Vol. 8

The ARRL Antenna Compendium

Editor
Steven R. Ford, WB8IMY

Editorial Assistant
Maty Weinberg, KB1EIB

Cover Design
Sue Fagan, KB1OKW

Production
Michelle Bloom, WB1ENT
Jodi Morin, KA1JPA

Technical Illustrations
David Pingree, N1NAS

Published by:
ARRL

the national association for Amateur Radio
Newington, CT 06111 USA

Printed in Canada

ISBN: 978-0-87259-099-1

First Edition
First Printing

Contents

HF Portable

HF Directional

HF Omnidirectional

VHF/UHF Portable

VHF/UHF Omnidirectional

VHF/UHF Directional

Foreword

It has been several years since the ARRL published the last *Antenna Compendium*. Much has taken place since then. The new A*ntenna Compendium*, Volume 8, showcases some of the best, most innovative antenna projects published in *QST* magazine from 2005 through 2009.

You'll find the work of many well-known authors, some of whom have been active for quite a while and others who've stepped into the *QST* spotlight more recently. Examples include the late L.B. Cebik, W4RNL (several of his last articles appear in this volume); Phil Salas, AD5X; John Portune, W6NBC; Floyd Koontz, WA2WVL and Skip Teller, KH6TY.

We've grouped the 60 articles into several broad categories, depending on the design characteristics and applications:

- HF Portable
- HF Directional
- HF Omnidirectional
- VHF/UHF Portable
- VHF/UHF Omnidirectional
- VHF/UHF Directional

Every project was originally evaluated before being published in *QST*. In those few instances where corrections were necessary after publication, those corrections have been incorporated into these articles.

Regardless of your operating preferences, there is something in this compendium that is sure to appeal to you!

David Sumner, K1ZZ
Executive Vice President
Newington, Connecticut
October 2010

Guide to ARRL Member Services

ARRL, 225 Main Street Newington, Connecticut 06111-1494, USA

tel. 860-594-0200, Mon-Fri 8 AM to 5 PM ET (except holidays)

FAX: 860-594-0303 ♦ **e-mail:** hqinfo@arrl.org ♦ **ARRLWeb:** www.arrl.org

Public Service

Advocacy

Education

Technology

Membership

Interested in Becoming a Ham?

www.arrl.org/hamradio e-mail: newham@arrl.org
tel. 1-800-326-3942 (toll-free in the US)

News Center

ARRLWeb: www.arrl.org

ARRL Letter and Audio News:
www.arrl.org/arrlletter

Public Relations/Advocacy

Government Relations and Spectrum Protection:
www.arrl.org/govrelations
e-mail: govrelations@arrl.org

Public and Media Relations:
www.arrl.org/pio

Membership Benefits

Membership Benefits (all):
www.arrl.org/benefits

ARRL "All Risk" Ham Radio Equipment-Insurance:
www.arrl.org/insurance

ARRL Visa® Credit Card:
www.arrl.org/visa

ARRL.NET E-mail Forwarding:
www.arrl.org/arrlnet

Awards: www.arrl.org/awards

Contests: www.arrl.org/contests

FCC License Renewal / Modification:
www.arrl.org/arrlvec

Hamfests/Conventions:
www.arrl.org/hamfests

QSL Service: www.arrl.org/qsl

Regulatory Information
www.arrl.org/regulations

Technical Information Service
www.arrl.org/tis
e-mail: tis@arrl.org
tel. 860-594-0214

Contributions, Grants and Scholarships

ARRL Development Office:
www.arrl.org/development
e-mail: mhobart@arrl.org
tel. 860-594-0397

- ♦ARRL Diamond Club
- ♦Spectrum Defense Fund
- ♦Education & Technology Fund
- ♦Planned Giving/Legacy Circle
- ♦Maxim Society

ARRL Foundation Grants and Scholarships:
www.arrl.org/arrlf

Public Service

Public Service Programs:
www.arrl.org/publicservice

Amateur Radio Emergency Service® (ARES®):
www.arrl.org/ares

ARRL Field Organization:
www.arrl.org/volunteer

Clubs, Recruitment, Instructors and Teachers

Find an Affiliated Club:
www.arrl.org/clubsearch

Mentor Program:
www.arrl.org/mentor

Find a Licensing Class:
www.arrl.org/coursesearch

Support to Instructors:
www.arrl.org/instructor

Find an Exam Session:
www.arrl.org/examsearch

Volunteer Examiner Coordinator (VEC):
www.arrl.org/arrlvec

Publications & Education

QST — Official Journal of ARRL:
www.arrl.org/*QST*
e-mail: *QST*@arrl.org

QEX — Forum for Communications Experimenters:
www.arrl.org/qex
e-mail: qex@arrl.org

NCJ — *National Contest Journal*:
www.arrl.org/ncj
e-mail: ncj@arrl.org

Books, Software and Operating Resources:
tel. 1-888-277-5289
(toll-free in the US);
www.arrl.org/shop

Advertising:
www.arrl.org/ads
e-mail: ads@arrl.org

Certification and Continuing Education / Online Courses: www.arrl.org/cce

JOIN or RENEW
or ORDER Publications
Toll Free 1-888-277-5289 (US)
International callers
tel. +1 (860) 594-0355

About the ARRL

The seed for Amateur Radio was planted in the 1890s, when Guglielmo Marconi began his experiments in wireless telegraphy. Soon he was joined by dozens, then hundreds, of others who were enthusiastic about sending and receiving messages through the air—some with a commercial interest, but others solely out of a love for this new communications medium. The United States government began licensing Amateur Radio operators in 1912.

By 1914, there were thousands of Amateur Radio operators—hams—in the United States. Hiram Percy Maxim, a leading Hartford, Connecticut inventor and industrialist, saw the need for an organization to band together this fledgling group of radio experimenters. In May 1914 he founded the American Radio Relay League (ARRL) to meet that need.

Today ARRL, with approximately 155,000 members, is the largest organization of radio amateurs in the United States. The ARRL is a not-for-profit organization that:

- promotes interest in Amateur Radio communications and experimentation
- represents US radio amateurs in legislative matters, and
- maintains fraternalism and a high standard of conduct among Amateur Radio operators.

At ARRL headquarters in the Hartford suburb of Newington, the staff helps serve the needs of members. ARRL is also International Secretariat for the International Amateur Radio Union, which is made up of similar societies in 150 countries around the world.

ARRL publishes the monthly journal *QST*, as well as newsletters and many publications covering all aspects of Amateur Radio. Its headquarters station, W1AW, transmits bulletins of interest to radio amateurs and Morse code practice sessions. The ARRL also coordinates an extensive field organization, which includes volunteers who provide technical information and other support services for radio amateurs as well as communications for public-service activities. In addition, ARRL represents US amateurs with the Federal Communications Commission and other government agencies in the US and abroad.

Membership in ARRL means much more than receiving *QST* each month. In addition to the services already described, ARRL offers membership services on a personal level, such as the Technical Information Service—where members can get answers by phone, email or the ARRL website, to all their technical and operating questions.

Full ARRL membership (available only to licensed radio amateurs) gives you a voice in how the affairs of the organization are governed. ARRL policy is set by a Board of Directors (one from each of 15 Divisions). Each year, one-third of the ARRL Board of Directors stands for election by the full members they represent. The day-to-day operation of ARRL HQ is managed by an Executive Vice President and his staff.

No matter what aspect of Amateur Radio attracts you, ARRL membership is relevant and important. There would be no Amateur Radio as we know it today were it not for the ARRL. We would be happy to welcome you as a member! (An Amateur Radio license is not required for Associate Membership.) For more information about ARRL and answers to any questions you may have about Amateur Radio, write or call:

ARRL—The national association for Amateur Radio
225 Main Street
Newington CT 06111-1494
Voice: 860-594-0200
Fax: 860-594-0259
E-mail: **hq@arrl.org**
Internet: **www.arrl.org/**

Prospective new amateurs call (toll-free):
800-32-NEW HAM (800-326-3942)
You can also contact us via e-mail at **newham@arrl.org**
or check out *ARRLWeb* at **www.arrl.org/**

A Portable Inverted V Antenna

A portable antenna that is great for ARRL Field Day, offers directional control and may be useful where fixed antennas are not allowed.

By Joseph R. Littlepage, WE5Y

If necessity is the mother of invention then restrictive covenants must be the mother of desperate measures. I recently moved into a new neighborhood and found that the covenants made it difficult for me to erect or install any type of ham radio antenna structure. If you are faced with a similar situation or simply have need for a portable antenna that you can erect quickly and easily, this may be what you need.

The problem I faced is common to many living in newer subdivisions and is quite restrictive of Amateur Radio operations. I was determined, however, to find a way to get back on the air. My challenge was to develop a portable antenna system that would be easy and quick to erect and enable me to work a reasonable amount of DX without being obvious to the neighbors. In addition, I needed to determine a low enough power level to avoid TVI and other potential interference that could draw attention to my Amateur Radio activity. QRP and a good antenna proved to be the answer!

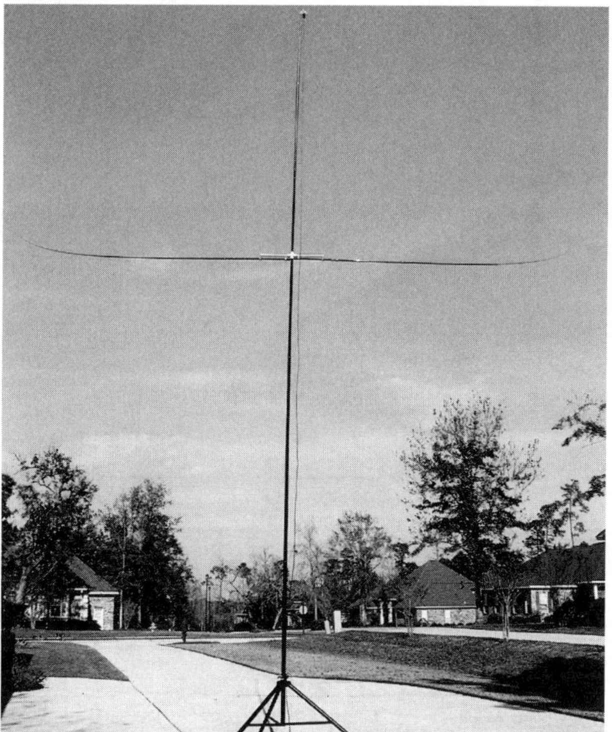

The Inverted V

I tried end-fed long wires, difficult-to-support dipoles and finally decided to try an inverted V cut for 17 meters. But how could I support it without a sky hook? That's when I hit upon the idea of using a lightweight telescopic pushup pole for a support mast. To serve as the base support for the pushup mast, I purchased a portable antenna tripod. I spread the tripod legs to about 40 inches apart and locked them in place. This combination is my sky hook!

Table 1

Wire Half-Element Lengths, Portable Inverted V Antenna

Band (Meters)	*Design Frequency (MHz)*	*Length*
20	14.175	16' 6$^1/_8$"
17	18.1	12' 11$^1/_8$"
15	21.175	11' $^5/_8$"
12	24.94	9' 4$^5/_8$"
10	28.4	8' 2$^7/_8$"

The top of the antenna should bring together the feed line and two wire elements angled at least 90° apart. Each of the wire elements is cut for a $^1/_4$ λ on the desired band. I chose to try the 17 meter band and cut each element for 18.1 MHz using the formula 234/f, with f the frequency in MHz for each quarter wave as shown in Table 1 for compromise CW and phone operation. You may wish to change the design frequency to suit your operational preferences. Final measurement and trimming was accomplished after the finished antenna was erected on site using an MFJ-259B antenna analyzer.

The next step was to devise a spreader assembly to hold the lower ends of the wire elements apart and keep the spread angle near 90° without resorting to ground anchor points. I wanted to be able to rotate the antenna using the "Armstrong" method to maximize its efficiency in selected directions. The solution was to use two 10 foot lightweight telescoping fiberglass fishing poles held end-to-end. A simple support arm assembly made of two 12 inch lengths of $^1/_2$ inch thin-wall PVC pipe joined to a $^3/_4$ inch PVC X connector. This holds the spreaders in a horizontal position and enables them to ride up and down

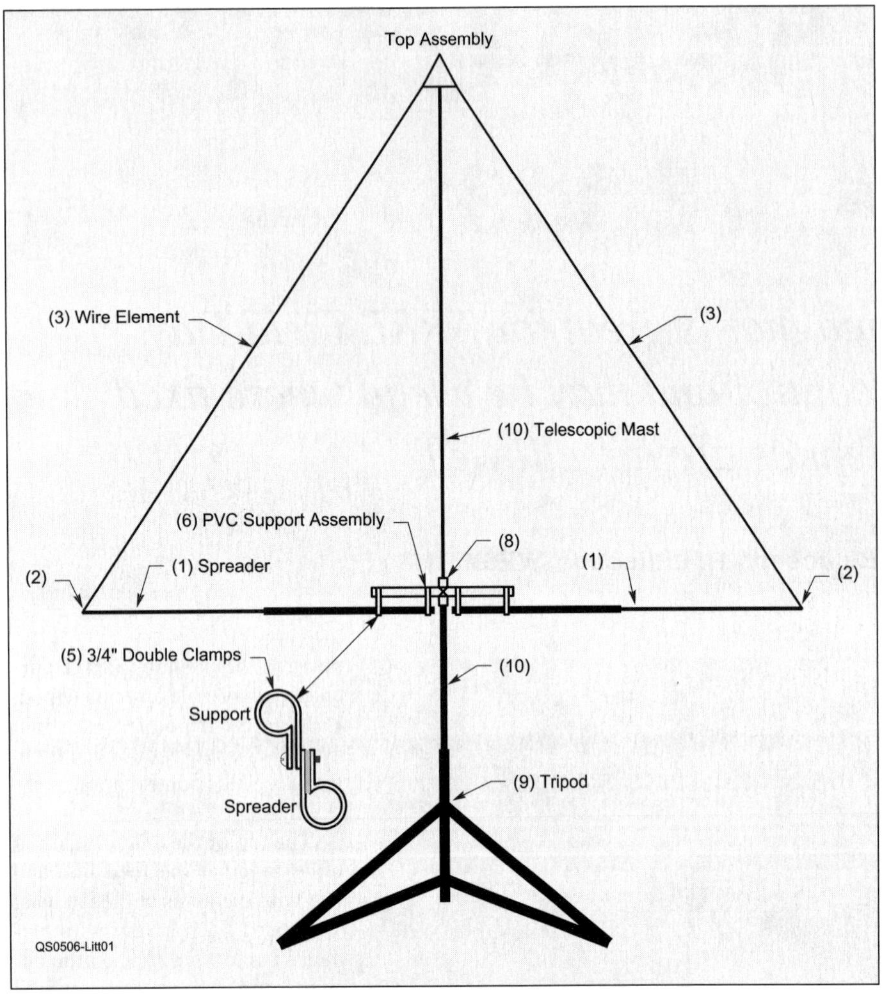

Figure 1—General arrangement of the completed inverted V antenna.

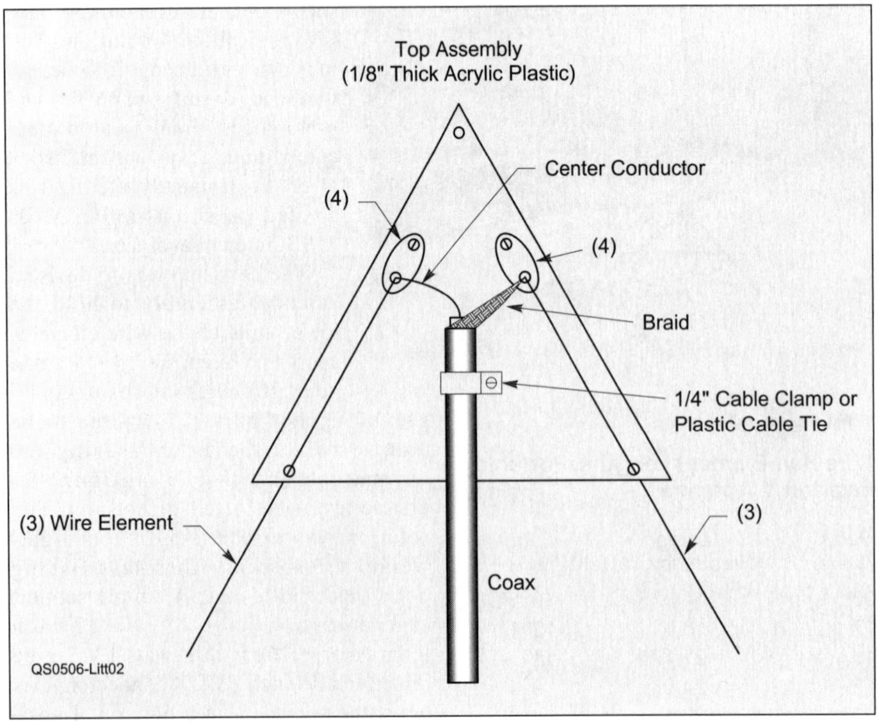

Figure 2—Details of the top assembly.

on the main support mast. The free ends of the wire elements are attached to the eyelets on the outer ends of the two spreader poles that are extended to full length as shown in Figure 1. Table 2 provides the bill of materials and sources for all the antenna's components.

Top Piece Assembly

A top fitting is required to bring the elements and transmission line together and to fix them to the top of the support. I constructed this assembly using a triangular piece of $^{1}/_{8}$ inch thick clear acrylic plastic or similar dielectric material. Its dimensions are not critical. Drill holes to accommodate a top hanger, two solder lugs, a nylon cable clamp for holding the coax and a hole in each lower corner to support the antenna wire. Neither the construction method nor the dimensions are critical.

Solder one end of an element wire and the center conductor of the coax to one solder lug and secure that to the top piece. Route the free end of that element half down through the lower hole on its side of the top piece. Solder the other element wire and the coax braid to the other solder lug and secure to the top piece in a like manner. Route that element half's free end through the other lower hole. Secure the coax to the top piece using a small nylon cable clamp as a strain relief. A small cable tie can be used instead if you carefully drill a hole on either side of the coax to route it through. The details are shown in Figures 2 and 3.

Spreader Support Assembly

A $^{3}/_{4}$ inch PVC X connector with $^{3}/_{4}$ to $^{1}/_{2}$ inch bushings pressed into opposite side

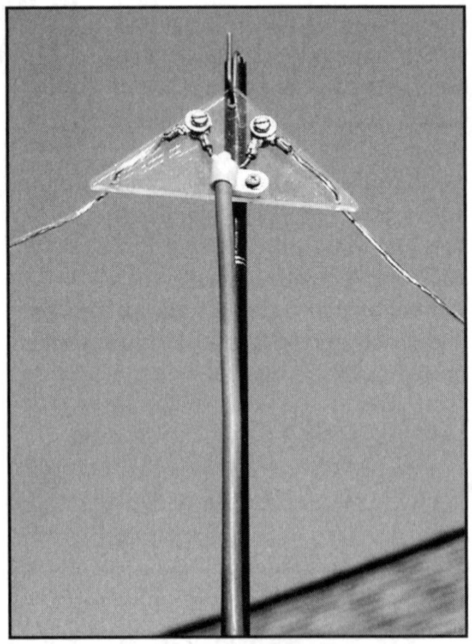

Figure 3—Photo of the top assembly.

Table 2
Materials List for Portable Inverted V Antenna

Item	*Quantity*	*Description*	*Source*	*Part Number*	*Unit Price*
1	2	B'n'M Black Widow crappie fishing pole, 3 sections, 10 feet.	Wal-Mart	BW3RR	$ 8.50
2	Package	Black interlock snap swivels, size 5.	Wal-Mart	BISS-5	$ 0.50
3	Roll	Insulated stranded #20 wire, 75 feet.	RadioShack	278-1219	$ 2.99
4	Pack	Crimp-on ring tongues, 24 assorted.	RadioShack	64-3030	$ 1.49
5	Pack	One-hole plastic clamp, 3/4" PP-1575UVB, six in pack.	Home Depot	E449-176	$ 1.49
6	10 feet	Silver line (thin wall) PVC pipe, 1/2"	Home Depot	717141340512	$ 1.80
7	2	PVC bushings, 3/4" × 1/2"	Home Depot	1287162647	$ 0.25
8	1	PVC "X" fitting, 3/4"	Home Depot	1287162483	$ 1.14
9	1	Portable antenna tripod.	MFJ	MFJ-1918	$39.95
10	1	Telescoping antenna mast, 33 feet.	MFJ	MFJ-1910	$79.95

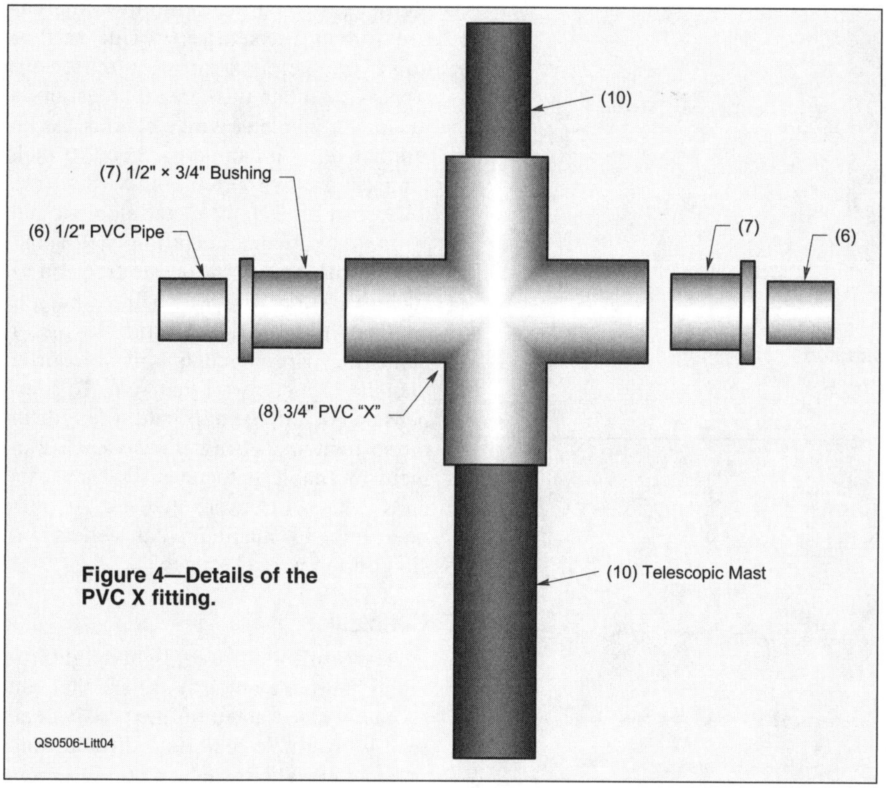

Figure 4—Details of the PVC X fitting.

Figure 5—Photo of the PVC X fitting.

holes is used to hold the two 12 inch long support arms as shown in Figures 4 and 5. The support arms are cut from thin wall 1/2 inch PVC pipe, used in preference to regular schedule 40 pipe to reduce weight. Press the bushings into opposite holes of the X piece until they are firmly seated. Then firmly press a 12 inch support arm into each bushing. Cement all joints using PVC cement. The two remaining holes will allow the spreader assembly to ride up and down the vertical mast. A slight raised lip molded inside these holes should be dressed down flush with the inside surface using a round file. To each support arm install two sets of back-to-back cable clamps. One end of each set is firmly attached around the support arm and the other end of each set around a fishing pole spreader. The outer clamp sets should be placed very near the end of each 12 inch PVC support arm, as detailed in Figure 6.

Stringing the Elements

When the top piece and the spreader support assemblies are completed it is time to string the elements. First, remove the innermost small top mast section with the eyelet from the support mast and set it aside. It will not be used. Place the large bottom end of the vertical mast, with the remaining nested sections inside, into the base supporting tripod. Next, raise the remaining centermost section of the telescopic support mast out about a foot and hold it in place with a spring-type clothespin.

Place the spreader support assembly down over the raised mast section with the spreader poles on the bottom side and against the clothespin. Next, attach the top assembly to the top opening of the small raised mast section with a bent hook fashioned from coat hanger wire run down inside the mast. The ends of the spreader poles can now be fully extended and twist locked in position. The free end of each wire element is then attached to the tip end eyelets on the spreader poles. Small fishing snap swivels can serve as attach hooks for the element ends.

Finally, remove the clothespin as you slowly start to raise the top mast section until it can be twist-locked into the next section below, which will follow it up. All the while the spreader support assembly will be moving down the vertical support mast to a point where the wire elements are straight and the spreader tips bend upward about 6 inches. This will maintain sufficient tension on the elements to prevent them from sagging. At this point, the spreader support assembly X connector should form a snug fit against the vertical mast. Failure to achieve a snug fit can result in the spreader assembly "weather vaneing" in a strong breeze.

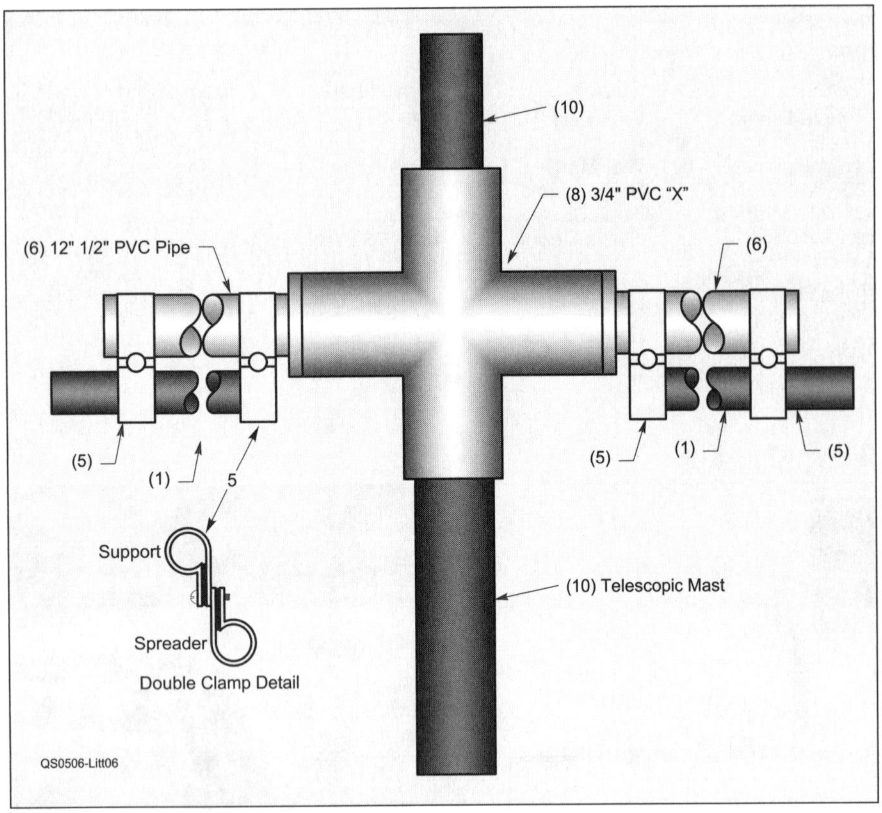

Figure 6—Details of the spreader attachment method.

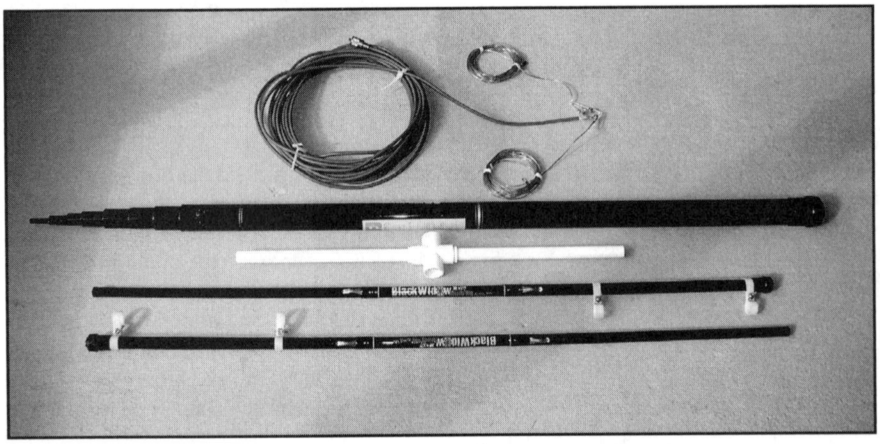

Figure 7—The completed antenna broken down for storage or travel.

Erecting the Antenna

With everything set up it is simply a matter of raising the remaining vertical support mast sections and twist-locking each one to the section below it. You should take every precaution to prevent the mast sections from loosening and falling back down. This can result in fracturing the lower ends of the sections as they hit the bottom of the support tripod tube. Although the fiberglass sections are strong, they are thin and very brittle and will chip or crack when overstressed. Simply placing a couple of inches of cushioning material inside the bottom end of the tripod tube will prevent damage. The coax cable can be attached to the mast as it goes up. The final height depends on how many sections your support mast has. The pushup pole that I use has 11 sections. I use all but the smallest top section (with the small eyelet tip) since it is too weak to support the antenna assembly without bending over.

Results

My first contacts were on 17 meter SSB running 5 W with an ICOM IC-703 powered by a 33 Ah gel-cell battery. A Connecticut station gave me a Q5 report during our 25 minute QSO. Two days later I worked Costa Rica, Guatemala, Colorado, New Mexico and Wisconsin with the same setup. Two of the QSOs were with pedestrian mobile stations.

The real test came a few weeks later when I worked ON6WA in Belgium and GI3DZE in Ireland using the same setup and power. I have received many good signal reports from the stations I have worked. When not in use, the antenna system can be broken down for storage or transport as shown in Figure 7.

For the 20 meter phone band, you can simply substitute a 1 inch PVC X connector with 1 inch to 1/2 inch bushings and assemble the support arms as described above. This modification allows the entire spreader assembly to ride farther down the vertical support mast and thereby maintain the spreader tension on the longer 20 meter wire elements. As an alternative, you can add a 3 foot 6 inch long drop-down extension wire to the lower end of each 17 meter element and retune the antenna accordingly. Placing a small fishing weight on the free end of each extension wire will help to keep it vertical. For the 10, 12 and 15 meter bands the wire elements will be shorter than the 17 meter wire that I cut. To compensate for this, simply add a length of monofilament fishing line to each element to enable it to reach the spreader ends. This will ensure that the 90° element spread is maintained as well as tension on the wire elements.

Conclusion

The result is a strong, lightweight, rotatable portable antenna system that can be easily constructed of inexpensive and readily available materials. In addition, you can elevate the apex of the V to any convenient height to clear an obstruction such as the edge of a roof or lower it to avoid neighborhood scrutiny. This antenna project has gotten me back on the air and could be the answer to some of your problems. The prospects for ARRL Field Day, emergency operations or backpacking to some remote site are limited only by your imagination.

Joe Littlepage, WE5Y, was first licensed in 1969 as WB5AAI. He upgraded to Advanced Class in 1975 and more recently to Amateur Extra in 2002. He holds a degree in physics and is retired from the USAF. Joe is an active member of the Mississippi Coast Amateur Radio Association (MCARA) and serves as editor of the club newsletter. He enjoys QRP, kit building, homebrewing, building wire antennas and solar power experimentation. Joe can be reached at **jlextra@netzero.net**.

The Inverted-U

L. B. Cebik, W4RNL

A simple rotatable field dipole for 20 through 10 meters.

The ARRL Field Day and other operations in the field tend to lean on three antenna principles—simplicity, small size and light weight. Complex assemblies increase the number of things that can go wrong. Large antennas are hard to transport and sometimes do not fit the space available. Heavy antennas require heavy support structures, so the overall weight seems to increase exponentially with every added pound of antenna.

In recent years, a number of light collapsible masts have hit the scene. When properly guyed with ropes, some will support antennas in the 5 to 10 pound range. Most are suitable for 10 meter tubular dipoles and as supports for wire antennas. The masts also allow the user to hand-rotate some antennas so that they are broadside to the desired target station. If we could only extend the range of the antenna to cover 20 through 10 meters, we might put these 20 to 30 foot masts to even better use. The inverted-U is an old idea that may provide one answer.

The Basic Idea

A dipole's highest current occurs within the first half of the distance from the feed point to the outer tips. We lose very little performance if we fold, bend or mutilate the outer end sections to fit an available space. If we start with a tubular 10 meter dipole—a little over 200 inches in overall length—we might add extensions for 12, 15, 17 or 20 meters, shaping them to suit the area in which we are operating.

If we can find space to erect a 10 meter rotatable dipole at least 20 feet above the ground, with a clear area to permit us to rotate the dipole, then we can simply let the extensions hang down. Figure 1 shows the relative proportions of the antenna on all bands from 10 to 20 meters. The 20 meter extensions are the length of half the 10 meter dipole. Safety dictates an antenna height of at least 20 feet to keep the tips above the 10 foot level. At any power level, the ends of a dipole are at a high RF voltage while transmitting and we must keep them out of contact with human body parts.

In principle, we do not lose very much signal strength by drooping up to half the overall element length straight down. What we lose in bidirectional gain shows up in decreased side-nulls as we increase the length of the drooping section. Figure 2 shows the free-space E-plane (azimuth) patterns of the inverted-U with a 10 meter horizontal section. There is an undetectable decrease in gain between the 10 meter and 15 meter versions. The 20 meter version shows a little over a half dB broadside gain decrease and a signal increase off the antenna ends. On 20 meters, the current in the vertical wires becomes significant, rounding the pattern.

The real limitation of an inverted-U is a function of the height of the antenna above ground. With the feed point at 20 feet above ground, we obtain the elevation patterns shown in Figure 3. The 10 meter pattern is typical for a dipole that is about $^{5}/_{8}$ λ above ground. On 15 meters, the antenna is only 0.45 λ high, with a resulting increase in the overall elevation angle of the signal and a reduction in gain. At 20 meters, the angle grows still higher, and the signal strength diminishes as the antenna height drops to under 0.3 λ. Nevertheless, the signal is certainly usable. A full-size dipole at 20 meters would show only a little more gain at the same height, and the elevation angle would be similar to that of the inverted-U,

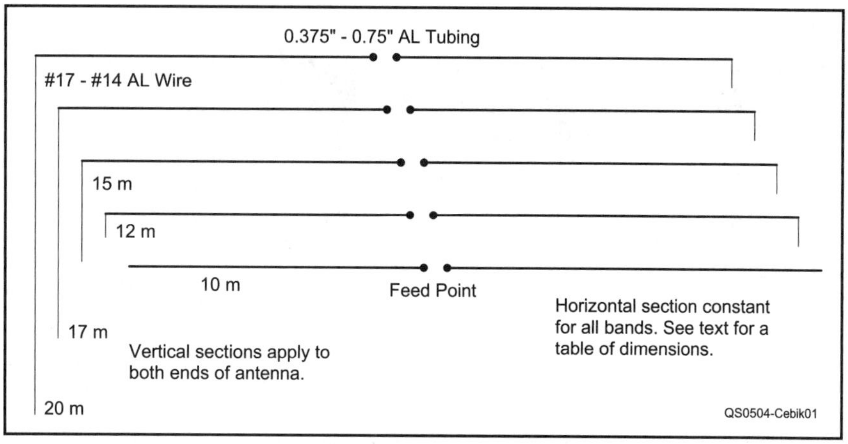

Figure 1—The general outline of the inverted-U field dipole for 20 through 10 meters. Note that the vertical end extension wires apply to both ends of the main 10 meter dipole.

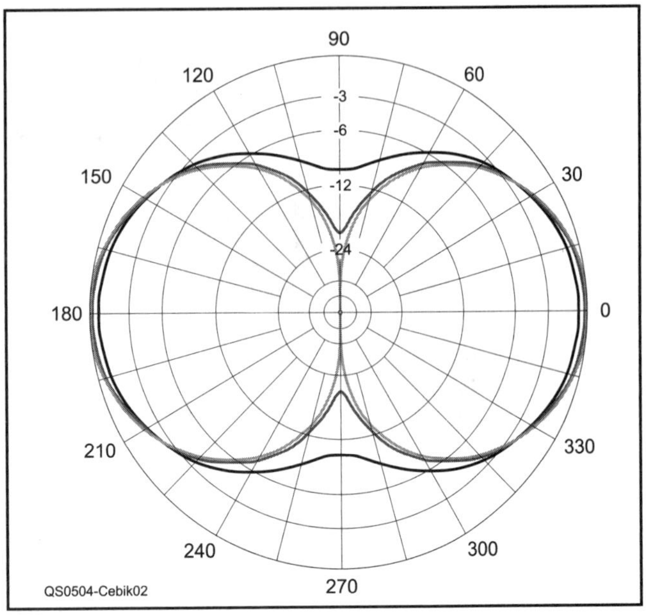

Figure 2—Free-space E-plane (azimuth) patterns of the inverted-U for 10 (green), 15 (red) and 20 (black) meters, showing the pattern changes with increasingly longer vertical end sections.

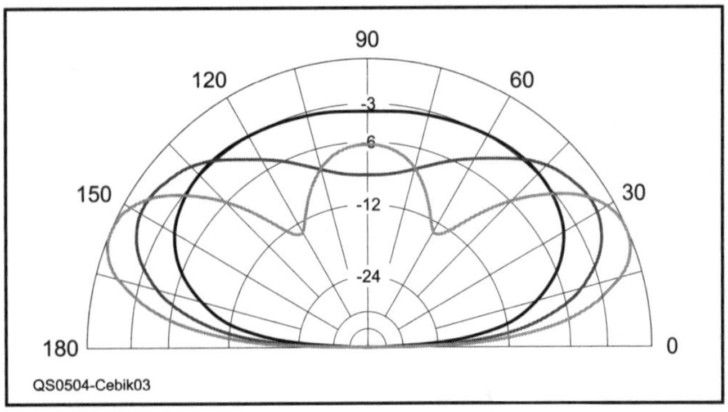

Figure 3—Elevation patterns of the inverted-U for 10 (green), 15 (red) and 20 (black) meters, with the antenna feed point 20 feet above average ground. Much of the decreased gain and higher elevation angle of the pattern at the lowest frequencies is due to its lower height as a fraction of a wavelength.

despite the difference in antenna shape.

As we decrease our frequency, there is no substitute for antenna height. Any horizontal antenna below about 3/8 λ in height will show a rapid decrease in low angle performance relative to heights above that level. If we raise the inverted-U to 40 feet, the 20 meter performance would be very similar to that shown by the 10 meter elevation plot in Figure 3. Table 1 summarizes the free-space and 20 foot performance characteristics of the inverted-U. Of special note is the fact that the feed-point impedance of the inverted-U remains well within acceptable limits for virtually all equipment, even at a height as low as 20 feet above ground. The SWR curves are also acceptably broad. Finding exact dimensions, even in special field conditions, becomes a non-critical task.

If we can accept the performance potential of a dipole for any band in the 20 meter to 10 meter range at the anticipated height of our mast possibilities, then the inverted-U provides a compact way of achieving that performance. The next step is fabricating one that we can use in the field.

Building a Field Version of the Inverted-U

Let's approach the construction of a field inverted-U in three steps: First, the tubing arrangement, second the center hub and feed-point assembly and third the drooping extensions. A parts list appears in Table 2.

1. The aluminum tubing dipole for 10 meters. Each half of the aluminum tubing dipole consists of three longer detachable sections of tubing and a short section mounted permanently to the feed-point plate, as shown in Figure 4. Let's consider each half of the element separately. Counting from the center of the plate—the feed point—the element extends 5 inches using 3/4 inch aluminum tubing. Then we have two 33 inch exposed tubing sections, with an additional 3 inches of tubing overlap per section. These sections are 5/8 and 1/2 inch in diameter, respectively. The exposed outer section is 30 inches long (with at least a 3 inch overlap) and consists of 3/8 inch diameter tubing.

Since the 5/8 and 1/2 inch sections are 36 inches long, you can make the outer 3/8 inch section the same overall length and use more overlap, or you can cut the tubing to 33 inches and use the 3 inch overlap. That much overlap is sufficient to ensure a strong junction while minimizing excess weight. When not in use, the three outer tubing sections will nest inside each other for storage. A 36 inch length for the outer section is a bit more convenient to un-nest for assembly. (I keep the end hitch pin on the 3/8 inch tubing as an easy way of pulling it into final position.) I prefer to use the readily available 6063-T832 aluminum tubing that nests well and has a long history of antenna service.

The only construction operation that you need to perform on the tubing is to drill a hole at about the center of each junction to pass a hitch pin clip. Obtain hitch pin clips (also called hairpin cotter pin clips in some literature) that fit snugly over the tubing.

Table 1
Anticipated Performance of the Inverted-U for 20 through 10 Meters

Using the tubular 10 meter dipole described in the text and #17 to #14 vertical wire element extensions.

	Free-Space			*20' Above Average Ground*		
Band	*Gain (dBi)*	*Resonant Impedance (Ohms)*	*Wire Length (inches) #17*	*Gain (dBi)*	*Elevation Angle (deg)*	*Impedance R ± jX (Ohms)*
10	2.1	73	—	7.6	24	65 – *j*2
12	2.0	71	16	7.2	27	67 – *j*8
15	1.9	64	38	6.4	32	69 – *j*8
17	1.7	55	62	5.7	38	65 – *j*4
20	1.4	41	108	4.8	50	52 + *j*4

Note: The wire length for the drooping ends is measured from the end of the 202 inch tubular dipole to the tip for #17 wire. Little change in length occurs as a function of a change in wire size. However, attachment of the extension to the element and special field conditions may require a few extra inches of wire.

Table 2
Parts List

Substitutions and alternative construction methods are possible and encouraged—as long as the overall antenna weight is not increased.

Qty	*Part*	*Comments*
6'	0.375" OD aluminum tubing	Two - 3' pieces
6'	0.5" OD aluminum tubing	Two - 3' pieces
6'	0.625" OD aluminum tubing	Two - 3' pieces
10"	0.75" OD aluminum tubing	Two - 5" pieces
4"	0.5" nominal (5/8" OD) CPVC	
50'	#17 wire	See Table 1 for the length of each piece. Aluminum preferred, but copper usable.
8	Hitch pin clips	Sized to fit tubing junctions.
1	4"×4"×1/4" Lexan plate	Other materials suitable.
2	SS U-bolts	Sized to fit support mast.
2	Sets SS #8/10 1.5" bolt, nut, washers	SS = stainless steel.
2	Sets SS #8 1" bolt, nut, washers	
2	Sets SS #8 0.5" bolt, nut, washers	
1	Coax connector bracket, 1/16" aluminum	See text for dimensions and shape.
1	Female coax connector	
2	Solder lugs, #8 holes	
2	Short pieces copper wire	From coax connector to solder lugs.

Note: 6063-T832 aluminum tubing is preferred and can be obtained from Texas Towers, **www.texastowers.com**, and other outlets. Lexan (polycarbonate) is available from McMasters-Carr, **www.mcmasters.com**, as are the hitch pin clips (if not locally available). Other items should be available from local home centers and radio parts stores.

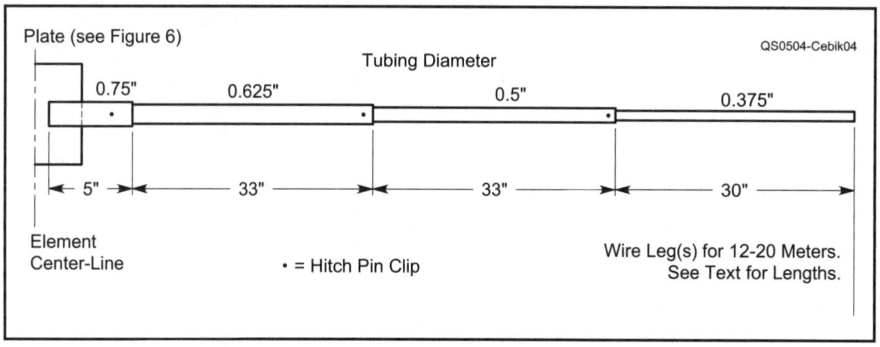

Figure 4—The general tubing layout for the inverted-U for each half element. The opposite side of the dipole is a mirror image of the one shown.

One size will generally handle about two or three tubing sizes. In this antenna, I used 3/32 inch (pin diameter) by 2 5/8 inch long clips for the 3/4 to 5/8 inch and the 5/8 to 1/2 inch junctions, with 3/32 by 1 5/8 inch pins for the 1/2 to 3/8 inch junction and for the final hitch pin clip at the outer end of the horizontal part of the antenna. Drill the holes (1/8 inch diameter) for the clips with the adjacent tubes in position relative to each other. I generally tape the junction temporarily for the drilling. Carefully deburr the holes so that the tubing slides easily when nested.

For this field antenna, the hitch pin clip junctions, shown in Figure 5, hold the element sections in position. The overlapping portions of the tube make the actual electrical contact between sections. Due to the effects of weather, junctions of this type are not suitable for a permanent installation, but are completely satisfactory for short-term field use. Good electrical contact requires clean, dry aluminum surfaces, so do not use any type of lubricant to assist the nesting and un-nesting of the tubes. Instead, clean both the inner and outer surfaces of the tubes before and after each field use.

Hitch pin clips are fairly large and harder to lose in the grass of a field site than most nuts and bolts. However, you may wish to attach a short colorful ribbon to the loop end of each clip. Spotting the ribbon on the ground is simpler than probing for the clip alone.

Each element half is 101 inches long, for a total 10 meter dipole element length of 202 inches. Length is not critical within about ±1 inch, so you may pre-assemble the dipole using the listed dimensions. However, if you wish a more precisely tuned element, tape the outer section in position and test the dipole on your mast at the height that you will use in the field. Adjust the length of the outer tubing segments equally at both ends for the best SWR curve on the lower 1 MHz of 10 meters. Even though the impedance will be above 50 Ω throughout the band, you should easily obtain an SWR curve under 2:1 that covers the entire band segment. However, you cannot perform this test until you construct a feed point and mounting plate.

2. *The center hub: mounting and feedpoint assembly.* I constructed the plate for mounting the element and the mast from a 4×4×1/4 inch thick scrap of polycarbonate (trade name Lexan), as shown in Figure 5. You may use other materials as long as they will handle the element weight and stand up to field conditions.

At the top and bottom of the plate are holes for the U-bolts that fit around my mast. Since field masts may vary in diameter at the top, size your U-bolts and their holes to suit the mast.

The element center, consisting of two 5 inch lengths of 3/4 inch aluminum tubing, is just above the centerline of the plate (to allow room for the coax fitting below). CPVC tubing of 1/2 inch nominal size has an outside diameter of about 5/8 inches and makes a snug fit inside the 3/4 inch tubing. The CPVC aligns the two aluminum tubes in a straight line and allows for a small (about 1/2 inch) gap between them. When centered between the two tubes, the CPVC is the same width as the plate. A pair of 1 1/2 inch #8 or #10 stainless steel bolts—each bolt with washers and a nut—secures the element to the plate.

Note in the sketch that you may insert the 5/8 inch tube as far into the 3/4 inch tube as it will go and be assured of a 3 inch overlap. I drilled all hitch pin clip holes perpendicular to the plate. Although this alignment is not critical to the junctions of the tubes, it is important to the outer ends of the tubes when we use the antenna below 10 meters.

I mounted a single-hole female UHF connector on a bracket made from a scrap of 1/16 inch thick L-stock that is 1 inch on a side. I drilled the UHF mounting hole first, before cutting the L-stock to length and trimming part of the mounting side. Then I drilled 2 holes for 1/2 inch long #8 stainless steel bolts about 1 inch apart, for a total length of L-stock of about 1 1/2 inches. The reason for the wide strip is to place the bolt heads for the bracket outside the area where the mast will meet the plate on the back side. Note in Figure 5 that the bracket nuts are on the bracket-side of the main plate, with the heads facing the mast. The bracket-to-plate mounting edge of the bracket needs to be only about 3/4 inch wide, so you may trim that side of the L-stock accordingly.

With the element center sections and the bracket in place, I drilled two holes for 1 inch long #8 stainless steel bolts at right angles to the mounting bolts and as close as feasible to the edges of the tubing at the gap. These bolts have solder lugs attached for short leads to the coax fitting. Solder lugs do not come in stainless steel, so you should check these junctions before and after each use for any corrosion that may call for periodic replacement.

With all hardware in place, the hub unit is about 4×10×1 inches (plus U-bolts). It will remain a single unit from this point onward, so that your only field assembly requirements will be to extend tubing sections and install hitch pin clips. You are now ready to perform the initial 10 meter resonance tests on your field mast.

3. The drooping extensions for 12 through 20 meters. The drooping end sections consist of aluminum wire. Copper is usable, but aluminum is lighter and quite satisfactory for this application. Table 1 lists the approximate lengths of each extension below the element. Add 3 to 5 inches of wire—less for 12 meters, more for 20 meters—to each length listed.

Initially, I had hoped to use #14 aluminum wire. However, this material is becoming harder to find locally. A good substitute is common #17 aluminum electric fencing wire. Fence wire is stiffer than most wires of similar diameter, and it is cheap. Stiffness is the more important property, since we do not want the lower ends of the wire to wave excessively in the breeze, potentially changing the feed-point properties of the antenna while in use.

When stored, the lengths of wire extensions for 12 and 15 meters can be laid out without any bends. However, the longer extensions for 17 and 20 meters will require some coiling or folding to fit the same space as the tubing when nested. Fold or coil the wire around any kind of small spindle that has at least a 2 inch diameter. This measure prevents the wire from crimping and eventually breaking. Murphy dictates that a wire will break in the middle of an operating session. So carry some spare wire for replacement ends. The low cost of fencing wire (about $5 for 250 feet in my area) allows me to tote the entire spool with me to operating sites. All together, the ends require about 50 feet of wire.

Figure 6 shows the simple mounting scheme for the end wires. I push the straight wires through a pair of holes aligned vertically to the earth. I then bend the top portion slightly. To clamp the wire, I insert a hitch pin clip though holes parallel to the ground, pushing the wire slightly to one side to reach the far hole in the tube. The double bend holds the wire securely (for a short-term field operation), but allows the wire to be pulled out when the session is over or when I change bands.

Add a few inches to the lengths in Table 1 as an initial length for each band. Test the lengths and prune the wires until you obtain a smooth SWR curve below 2:1 at the ends of each band. Since an inverted-U antenna is full length, the SWR curves will be rather broad and suffer none of the narrow bandwidths associated with inductively loaded elements. Figure 7 shows modeled SWR curves for each band to guide your expectations. Unless your feed line is an exact multiple of a half-wavelength, the impedance that you might record on any of the antenna analyzer instruments may differ from the impedance at the feed point.

The impedance figures for a 20 foot height shown in Table 1 suggest that you should not require much, if any, adjustment once you have found satisfactory lengths for each band. However, leave enough excess so that you can adjust the lengths in the field, especially if you operate in an area where objects like trees and buildings are at different distances from the antenna than they are in your test setup. My initial tests required extension wires considerably longer than modeled, with the peak excess on 15 meters. Only during a second round of tests did I realize that my test mast is metal, ungrounded, and just about 22 feet long. The need for longer extensions likely resulted from coupling between the extensions and the mast, although nothing seemed amiss in reception tests. The pattern was clearly bi-directional.

Do not be too finicky with your SWR curves in the field. An initial test and possibly one adjustment should be all that you need to arrive at an SWR value that is satisfactory for your equipment. Spending half of your operating time adjusting the elements for as near to a 1:1 SWR curve as may be possible will rob you of valuable contacts without changing your signal strength is any manner that is detectable at the far end of the line.

Changing bands is now a simple matter. Remove the ends for the band you are using and install the ends for the new band. An SWR check and possibly one more adjustment of the end lengths will put you back on the air.

Some Final Notes

The inverted-U dipole with interchangeable end pieces provides a compact field antenna. Figure 8 shows the parts in their travel form, illustrating how compactly the antenna travels. When assembled and mounted at least 20 feet up (and higher is still better), the antenna will compete with just about any other dipole mounted at the same height. But the inverted-U is lighter than most dipoles at frequencies lower than 10 meters. It also rotates easily by hand—assuming that you can rotate the mast by hand. Being able to broadside the dipole to your target station gives the inverted-U a strong advantage over a fixed wire dipole.

You may experiment with other forms of construction according to your own skills and available materials. A wire version can be taped to bamboo poles, if you can obtain the material and find a means of connecting short transportable sections of bamboo. Try to avoid aluminum conduit and other heavy materials. Be certain

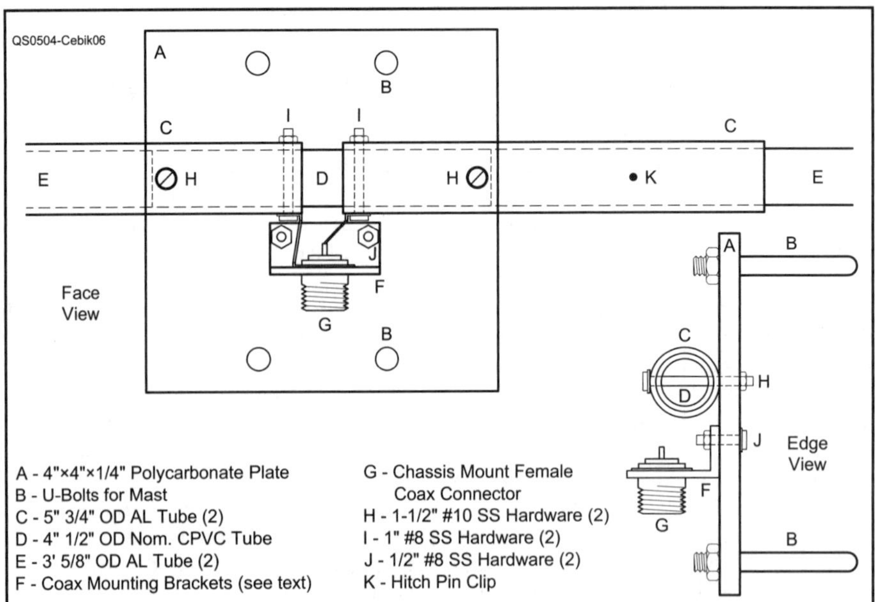

Figure 5—The element and feed-point mounting plate, with details of the construction used in the prototype.

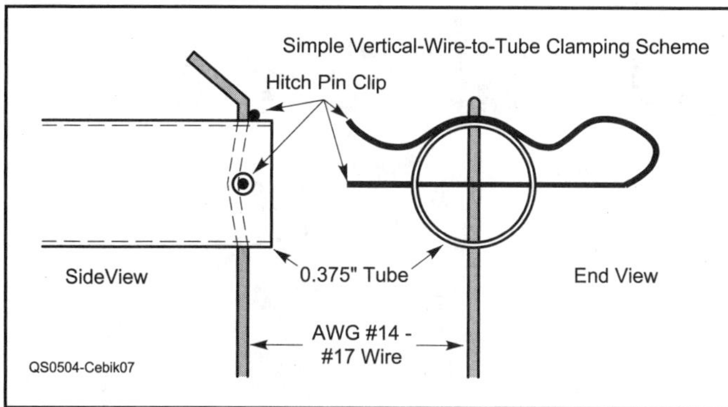

Figure 6—A simple method of clamping the end wires to the 3/8 inch tube end using a hitch pin clip.

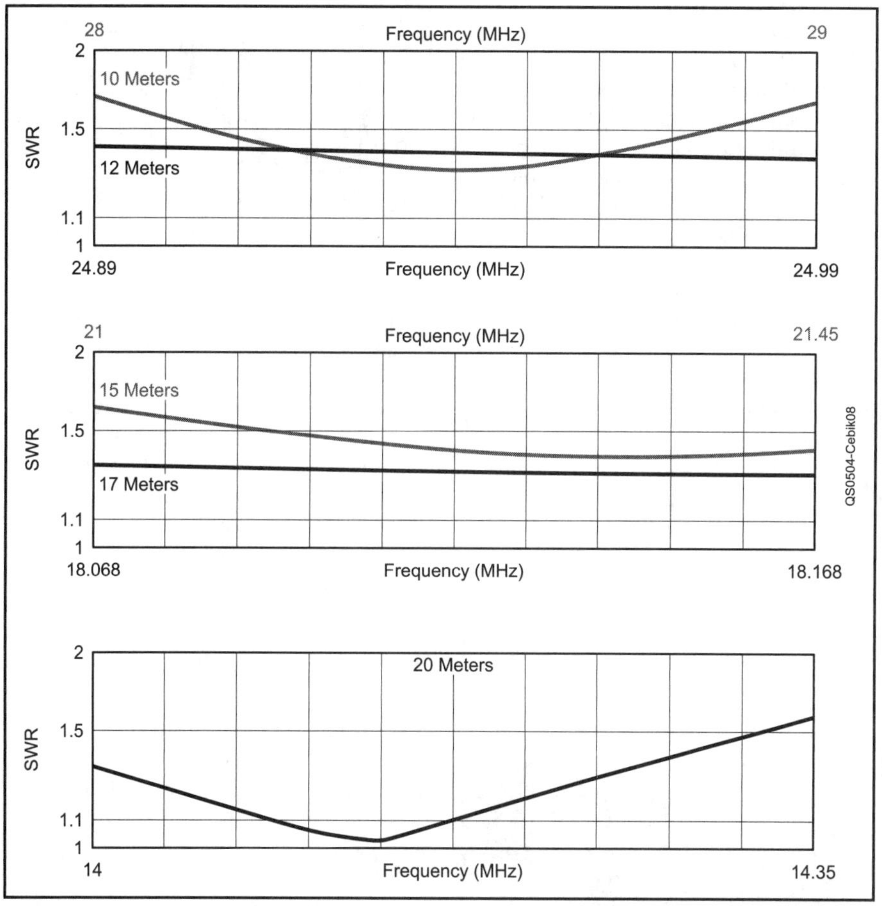

Figure 7—Typical 50 Ω SWR curves for the inverted-U antenna at a feed-point height of 20 feet.

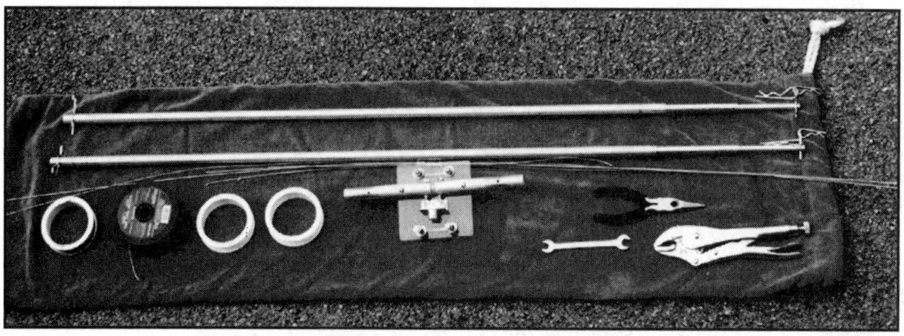

Figure 8—The entire inverted-U antenna parts collection in semi-nested form, with its carrying bag and the three dedicated tools for field assembly and disassembly.

that the drooping end wires are well secured to the 10 meter dipole ends both electrically and mechanically.

With a dipole having drooping ends, safety is very important. Do not use the antenna unless the wire ends for 20 meters are higher than any person can touch when the antenna is in use. Even with QRP power levels, the RF voltage on the wire ends can be dangerous. With the antenna at 20 feet at its center, the ends should be at least 10 feet above ground. For this reason, I have not tried to extend the inverted-U for use on 30 and 40 meters.

Equally important is the maintenance that you give the antenna before and after each use. Be sure that the aluminum tubing is clean—both inside and out—when you nest and unnest the sections. Grit can freeze the sections together, and dirty tubing can prevent good electrical continuity when the antenna is extended. Add a few extra hitch pin clips to the package to be sure you have spares in case you lose one.

You can carry all of the parts in any 3 foot long bag, either horizontal or vertical. A draw-string bag works very well. One of the photos shows the antenna pieces and their bag, along with the ribbon spools on which I store the wire extensions, the spool of extra wire, and the *field tools*. The tools that I store with the antenna include a wrench to tighten the U-bolts for the mast-to-plate mount and a pair of pliers to help me remove end wires from the tubing without cutting my fingers. The pliers have a wire-cutting feature in case I need to replace a broken end wire. A pair of Vise-Grip pliers makes a good removable handle for turning the mast. The combination of the Vise-Grip and regular pliers lets me uncoil the wire extensions for any band and give them a couple of sharp tugs to relax and straighten the wire.

Although the inverted-U is not the answer to every field antenna need, it will serve very well if it matches the kind of operating you do. Hitch pin clips simplify assembly and band changing in the field. At 20 feet and higher, the antenna will acquit itself very well. The entire antenna is inexpensive to build and easy to maintain. Those are reasonably good credentials for any antenna.

Licensed since 1954, L. B. Cebik, W4RNL, is a prolific writer on the subject of antennas. Since retiring from teaching at the University of Tennessee, LB has hosted a Web site (**www.cebik.com**) *discussing antennas—both theoretical and practical. He has written more than 15 books, including the ARRL course on antenna modeling. Serving both as a Technical and an Educational ARRL Advisor, he's also been inducted into both the QRP and QCWA Halls of Fame. LB can be reached at 1434 High Mesa Dr, Knoxville, TN 38938 or at* **cebik@cebik.com**. QST

The Ultimate Portable HF Vertical Antenna

Phil Salas, AD5X

I've had a tremendous response to the portable antenna project published in the July 2002 *QST*. Many folks had problems locating some of the parts, however. It seems that sprinkler-system parts are not as common in other sections of the country as they are in Texas! Also, the original article required that you drill and tap screw holes in brass couplings and the sprinkler-system risers. I wanted to eliminate the screw-hole tapping to simplify the construction.

Over the years, the antenna evolved from the original design through changes that used fiberglass, aluminum tubing, and then brass tubing. The most recent design is longer, lighter and more compact (when disassembled) than the original antenna. It is also easier to build, and easier to the find parts. Band coverage is also increased to include 60 meters, as well as 40 through 10 meters!

The Ultimate Portable Antenna is designed for easy transport. It breaks down into multiple mast sections, a whip section, an air-wound center loading coil section, and a small base support. No piece is longer than about 20 inches, so it will easily fit into most suitcases. The lead photo shows the disassembled antenna, including guys and radials. The fully assembled antenna has a length of almost 16 feet! See Figure 1. That's me, standing next to the antenna set up on my front lawn, in Figure 2.

The key to efficiency for short antennas (such as a loaded vertical less than a quarter wavelength) is the length. The longer the antenna, the greater the radiation resistance and therefore the less impact you have on efficiency due to ground and loading coil losses. This antenna is almost $^1/_4$ λ on 20 meters. For operation on 17 through 10 meters, you will shorten the antenna to $^1/_4$ λ. The length of the complete antenna minimizes the required loading coil for 60, 40 and 30 meters. So let's build it!

AD5X

The complete unassembled antenna.

Gathering the Parts

You can find most of the parts for this antenna at your local hardware store. The loading coil, coil taps, 10 foot telescoping whip, and SO-239 are available from MFJ (**www.mfjenterprises.com**). Be sure to mention this *QST* article when ordering the kit of parts (MFJ-1964-K) for a special price discount. Table 1 is the complete parts list.

A few notes about pipe sizes may be in order for anyone not familiar with plumbing fixtures. When you go to the hardware store to purchase the $^1/_8$ inch NPT nipples and couplers, don't expect to find anything measuring $^1/_8$ inch! The outside diameter of a $^1/_8$ inch pipe is 0.405 inches, or about $^{13}/_{32}$ inch. The standard threads are 27 turns per inch. (The $^1/_8$ inch designation comes from the approximate inside dimension of the pipe, although today you may find pipes with different wall thicknesses and the same outside dimensions. For this project you don't have to worry about the wall thickness of your fittings.) The NPT specification

AD5X

Figure 1—The complete antenna setup in the author's front yard.

AD5X

Figure 2—Standing next to the antenna to illustrate the height of the tubing sections and coil.

Table 1
Parts List

Qty	Item
1	5 inch long × 2.5 inch diameter × 10 TPI air wound coil (MFJ-404-008)*
1	10 foot telescoping whip (MFJ-1954)*
1	SO-239 chassis mount connector (MFJ-610-2005)*
5	Coil clips (MFJ-605-4001)*
2	3 foot pieces of 3/8 inch diameter brass tubing (ACE Hardware) (McMaster-Carr 8950K581 — 6 foot length)**
1	3/8 inch diameter wood dowel. (36 inch length at Home Depot—only 3 1/2 inch needed.)***
1	3/4 inch PVC T — All of the PVC fittings are white schedule 40 pipe (Home Depot)
1	3/4 inch slip × 1/2 inch female pipe thread PVC adapter (Home Depot)
1	3/4 inch slip × 1/2 inch slip PVC adapter (Home Depot)
1	3/4 inch slip PVC plug (ACE Hardware)
1	1/2 inch NPT (male thread) to 1/8 inch NPT (female thread) brass adapter bushing (ACE Hardware)(McMaster-Carr 50785K64)
8	1/8 inch NPT brass couplings (Home Depot)(McMaster-Carr 50785K91)
4	0.7 inch long 1/8 inch NPT all-thread nipples (these are also called "close nipples") (Home Depot) (McMaster-Carr 50785K151)
2	#8 brass wing-nuts (ACE Hardware)
2	#8-32 × 3/4 inch brass machine screws (ACE Hardware)
2	#8-32 brass nuts (pack of 6) (Home Depot)
2	#8 copper-plated steel (or brass) split lock washers (ACE Hardware)
1	36 inch length of 1/8 inch diameter brass rod (Home Depot)
1	3/8-16 × 1 1/4 inch hex head bolt, zinc plated. Choose the longest you can that is threaded all the way to the head. (Home Depot)
1	3/8-16 × 12 inch hex head or carriage bolt, zinc plated. Choose the longest bolt you can find in this size if you can't find a 12 inch bolt (Home Depot)
1	3/8-16 coupler, zinc plated (ACE Hardware)
2	3/8-16 nuts, zinc plated (pack of 6) (Home Depot)
1	3/8 inch lock washer, zinc plated (pack of 10) (Home Depot)
4	#6 stainless steel 3/8 inch sheet metal screws (ACE Hardware)
3	#8 solder lugs (Home Depot)
6	#8 × 1 1/2 inch brass wood screws (pack of 6) (Home Depot)
90 feet	Wire (any gauge, insulated or not, for six 15-foot ground radials.)
1	Alligator clip and miscellaneous short pieces of connecting wire.

*The total retail price for all the MFJ parts is $53.80, plus shipping/handling. Mention this article when you order and MFJ will sell a kit containing the telescopic whip, coil, coil clips, and SO-239 connector for $39.95 plus $6 s/h. The kit is MFJ-1964K. You can order additional coil clips for $2.95 each.

You can use 3/8 inch aluminum tubing if you prefer. Aluminum tubing is about half the cost of brass, but you will either need to drill the brass couplings and aluminum tubing so they can be connected with stainless steel sheet-metal screws or solder them together using aluminum solder and a torch (see **www.solder-it.com).

***You can use 3/8 inch fiberglass rod instead of the wood dowel if you prefer, but fiberglass rod is more difficult to find. Check out bicycle flags and driveway marker stakes as potential sources of fiberglass rod.

The brass plumbing items are also available from McMaster-Carr, (**www.mcmaster.com**) a mail-order supplier with no minimum order requirement. They stock 6 foot lengths of 3/8 inch brass tubing as well as the close nipples, couplings and brass bushings. The Table lists part numbers to help you look up those items on the Web site.

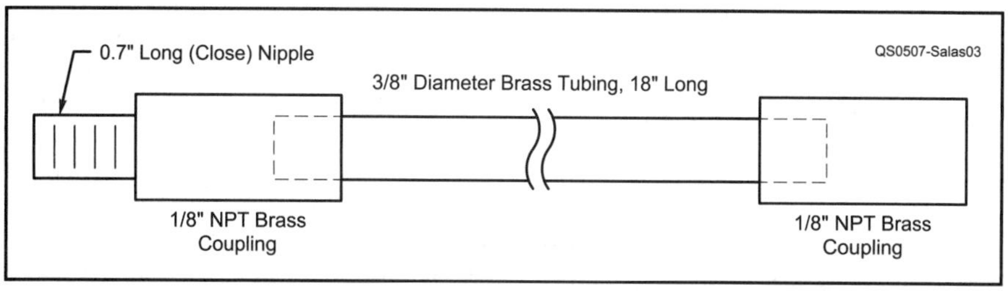

Figure 3—Assembly of the brass tubing sections.

means National Pipe Thread. There are several other thread specifications, but NPT is the most common.

I used 1/8 inch NPT close nipples for the antenna connectors. These are about 0.7 inches long, and are usually fully threaded over over the entire length. The PVC pipe fittings must be the white, schedule 40 PVC pipe. Do not use the thinner-walled pipe sometimes known as CPVC, which has a cream color or yellow tint. When fittings are designated as "slip," it means pieces are intended to be glued—they just slip together—rather than being threaded.

Brass Rod Preparation and Assembly

See Figure 3 for the assembly details. First cut three 18 inch sections of the 3/8 inch brass tubes with a hacksaw or tubing cutter and de-burr the tubing. The ends of the couplings that fit over the brass tubes must be reamed out with a 3/8 inch drill bit. Otherwise the couplings won't fit over the tubing. To do this, first screw a 1/8 inch NPT coupling on each end of a 1/8 inch NPT close nipple. Use wrenches to screw these together as tightly as possible. Next, clamp one of the couplings securely and ream out the opposite coupling with a 3/8 inch drill bit. (Use a drill press for this operation if at all possible.) Reverse, and ream out the other coupling. [See Figure 4 for one way to use a woodworker's clamp and a drill press.—*Ed.*] Now unscrew the couplings. One end of the nipples will break loose from one coupling, and the other end will stay tight in the remaining coupling. You'll now have a female and male end that will fit over each end of a section of brass tube, as shown in Figure 5. You will need four pair of these male/female brass connectors: three pair for the brass tubes and one pair

WR1B

Figure 4—A pipe coupling and nipple secured in a woodworker's clamp ready to drill out the end on a drill press.

AD5X

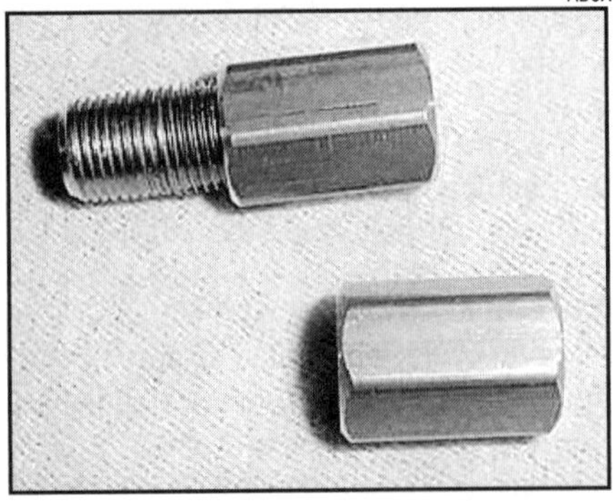

Figure 5—After the pipe couplings have been drilled out and one coupling removed from the nipple, the pair is ready to be installed on the ends of an 18 inch length of brass tubing.

for the loading coil assembly. If you'd like, you can solder the nipple/coupling assemblies together. The assembly tends to be very tightly secured even without soldering, however.

Now insert the male/female brass pairs just constructed over all three of the 18 inch brass tubes and solder the couplings directly to the tubes. This is easily done with a large soldering iron, or even better, with a torch and silver solder. Solder-It has a nice small butane torch that works well. See **www.solder-it.com**.

Loading Coil Assembly

Slide 1/8 inch NPT male/female coupling pairs over both ends of a 3/8 inch diameter, 3 1/2 inch long wood dowel. You will need to drill a 1/8 inch diameter hole completely through each of the 1/8 inch NPT brass couplings and dowel as shown in Figure 6. Next cut two 3 inch lengths of the 1/8 inch diameter brass rod. Insert one of these 3 inch sections through the holes on one brass coupling. Center the rod so that equal lengths are available on both sides of the coupling, and solder the rod to the coupling with a large soldering iron or torch. (Be careful not to burn the wood dowel with the torch!)

Now position a 3 inch length of the MFJ-404-008 coil such that the 3 inch brass rod just installed pokes through the last two turns on the coil. See Figure 7. Solder the coil turns to the rod. On the opposite end of the coil assembly, insert the remaining 3 inch brass rod through two adjacent turns on this end of the coil, through the brass coupling, and through the coil turns. Solder the coil turns to the brass rod, and then solder the brass rod to the coupling.

Next, indent every other turn on the coil with a small flat-head screwdriver. You should do this on opposite sides of the coil to give you plenty of adjustment capability.

Finally, on the end of the coil with the brass nipple (male end), solder a 6 inch piece of insulated wire terminated with an alligator clip. For extended outdoor use, you may wish to treat the wood dowel with varnish.

Top Whip

The MFJ-1954 10 foot telescopic whip comes with a standard 3/8-24 mounting thread. While the whip mounting stud is not a correct match for the 1/8 inch NPT coupling (27 turns per inch), I just thread the screw directly into the 1/8 inch NPT coupling. Because the 3/8 bolt is slightly smaller than the 1/8 inch pipe outside diameter, and the pipe coupling is tapered inside, they will go together. [See the "Alternative Construction Ideas" sidebar for a way to match the threads, if you are uncomfortable with the mismatch.—*Ed.*]

Base Assembly

For the base spike, I've used a 3/8-16 × 12 inch zinc-plated hex-head bolt. Only 1 1/2 inches of the bolt is threaded, and so I used the long smooth end of the bolt to go into the ground after cutting off the hex head. A damp cloth easily cleans the bolt after use. I also used a 3/8-16 × 1 1/4 inch zinc-plated hex-head bolt at the base of the PVC assembly, and a 3/8-16 zinc-plated coupler to attach the 1 1/4 inch bolt to the 12 inch bolt, as shown in Figure 8. This way you can leave the long bolt off if you want to bolt the base assembly directly to a metal plate or trailer mount, or screw on the long bolt for ground mounting.

AD5X

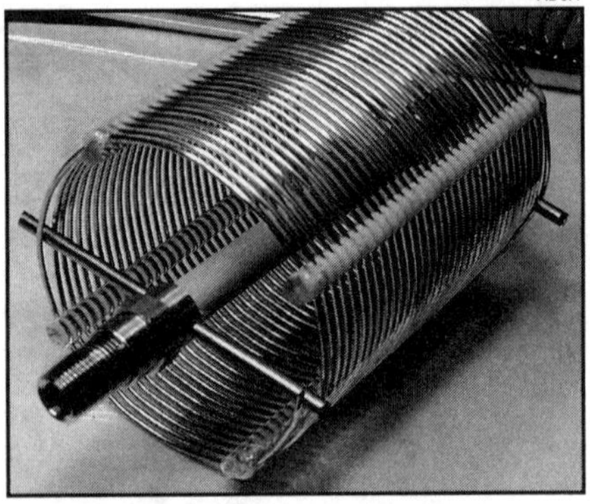

Figure 7—The loading coil assembly ready to be soldered.

Referring to Figure 8, drill a 3/8 inch diameter hole into the 3/4 inch PVC plug used for the base support 1 1/4 inch bolt. Cut off about half of the length of the 3/4 inch PVC plug to leave plenty of room inside the T for wiring. Solder a ground wire to the head of the 3/8-16 × 1 1/4 inch bolt as shown, or use a 3/8 inch solder lug. Insert the threaded end of bolt into the plug, and secure with a 3/8-16 nut and a lock washer. If you wish, you can glue the plug in place with PVC pipe cement instead of using the #6 stainless steel sheet metal screws shown. The screws make changing the support assembly easy in case you should ever want to, though. You might even consider using screws to attach the other two adapters into the T.

To prepare the 12 inch bolt, cut off the hex head and round this end with a file. Screw a 3/8-16 nut onto the threaded end, then add a lock washer and screw the 3/8-16 coupler onto the threads so the bolt is about halfway through the coupler. Tighten the nut against the coupler (with a lock washer between the nut and coupler). This 12 inch bolt assembly can now be easily screwed onto the 1 1/4 inch bolt on the base assembly for ground mounting.

Now place the SO-239 over the 1/2 inch hole in the 3/4 to 1/2 inch slip adapter and mark the location for the two #6-32 × 3/8 inch long stainless steel sheet metal screws that will hold it in place. (The adapter I used had a hex head on the outside lip, and by turning the SO-239 so diagonal mounting holes are over opposite points in the hex head, there should be enough material to hold the screws easily.) You'll see that these holes will be right in

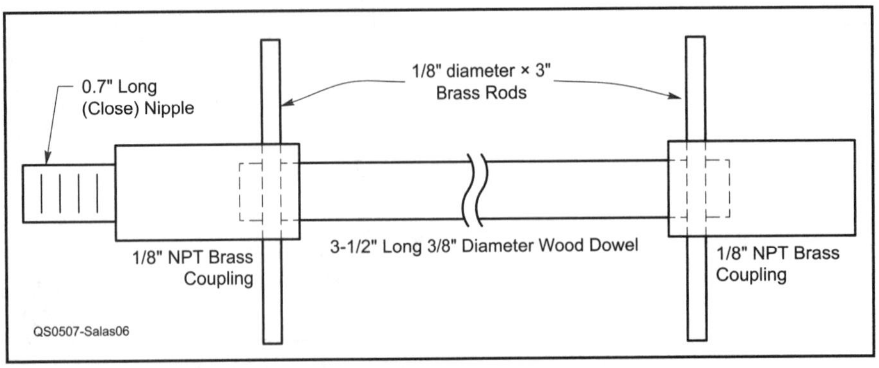

Figure 6—How brass couplings are installed on a length of 3/8 inch wood dowel and 1/8 inch brass rod in order to mount the coil.

Alternative Construction Ideas

As I began editing this article for publication in *QST*, I decided to build the antenna. After visiting three local Home Depot stores and most of the True Value and ACE Hardware stores in the area, I realized that builders may have difficulty finding some of the materials. For example, none of the stores I visited had any stock of brass or aluminum tubing in the size I wanted.

I contacted Phil Salas with a few questions about this article and the materials he specified. Phil informed me that since writing the article he had continued to try new ideas for building the antenna, and he found several alternative materials. His first suggestion was to use 1/2 inch aluminum tubing, with a wall thickness of 0.058 inch. He tapped the inside of that tubing for 1/8 inch pipe threads, and just turned the brass nipples into the pipe. This saves the step of drilling out the couplings to slip the 3/8 inch brass tubing inside. I was able to find some 1/2 inch aluminum tubing with a 0.050 inch wall thickness but not the thicker wall. Unfortunately, a 1/8 inch NPT nipple slips right inside that tubing, so I could not cut threads into the tubing. Perhaps 7/16 inch tubing with a 0.050 inch wall thickness would work, but I did not find any tubing in that size.

Phil also suggested McMaster-Carr (**www.mcmaster.com**) or similar mail-order/Internet stores as a good way to obtain the tubing and other plumbing parts. Their prices compare favorably with what I found in Connecticut, and with no minimum order they may save you some shopping time in a variety of hardware stores. Of course many of us probably enjoy browsing the aisles of our local hardware stores. You never know what new tool you may need or what other antenna projects may come to mind as you peruse the available materials. Phil has included McMaster-Carr part numbers for several of the pieces in Table 1.

I found a K & S Engineering display of hobby tubing at my local Do It Best hardware store. They ordered the 3 foot lengths I needed for a reasonable price. After building an antenna with that tubing, however, I realized that the 0.014 inch wall thickness was too thin to support the antenna without guy lines. Phil suggested that I could use that tubing by inserting a 1/4 inch dowel or 1/4 inch fiberglass rod, such as from a bicycle flag to stiffen the brass tubing. I ordered a 6 foot length of the brass tubing with a 0.032 inch wall from McMaster-Carr, as listed in Table 1. That proved to be a near ideal tradeoff between weight and strength.

I also discovered a much wider array of PVC pipe fittings than I had imagined. Many of the pieces appear to have similar specifications at first glance. For example, I found 3/4 inch × 1/2 inch adapters with male pipe threads on the 3/4 inch end and either male or female threads on the 1/2 inch end as well as one with both ends designed for slip fitting of the PVC pipe into the adapter.

There are many different ways to arrive at the end result you need. For example, if you can't find a 1/2 inch × 1/8 inch brass bushing for the top of the T, a 1/4 inch × 1/8 inch bushing will turn into a 1/2 inch × 1/4 inch bushing to achieve the same results.

All of my searching for parts shows that there may be a variety of materials that you could use to build a similar antenna. What you find in your area may not match what the author or I found. With a bit of creative shopping you will be able to find the parts you need to build the antenna.

The author suggests that you use #6 × 3/8 inch sheet metal screws to hold the pipe plug with the bolt into the bottom of the T. This will make it easier to disassemble that section if you have to fix a wiring problem inside the T. I decided that was a good approach for the fittings on the other two parts of the T as well. Photo A shows the T clamped in my woodworkers clamp, ready to drill the screw body holes.

WR1B

Photo A — The 3/4 inch PVC T is clamped in a woodworker's clamp, ready to drill the screw-body holes with a 9/64 inch drill bit. I had first drilled 5/64 inch pilot holes with the adapters and bottom plug installed in the T. Try to center the hole on the side of the T, and drill all the way through both sides. Then remove the adapters and plug, and drill the screw-body holes in the T.

Phil has found that threading the 3/8-24 stud on the bottom of the telescoping whip into the 1/8 inch NPT coupling is adequate. The perfectionist in me doesn't like the idea of mismatched threads. I decided to make a slightly different adapter than Phil suggests in his article. I used a 1/8 inch pipe coupling and nipple on the bottom of a 1 inch length of the brass tubing. (I drilled out the brass coupling with a 3/8 inch drill, the same as is done for the antenna tubing sections.) For the top section of my adapter I used a 3/8-24 coupling nut, which is about 7/8 inch long. I drilled out the threads on about half of that coupling, too, and pushed it onto the brass tubing. (I soldered these, and all other joints on my antenna.) Now all the threads mate properly. The 1/8 inch NPT to 3/8-24 adapter allows me to attach the whip to any antenna section, as desired. *—Larry Wolfgang, WR1B*

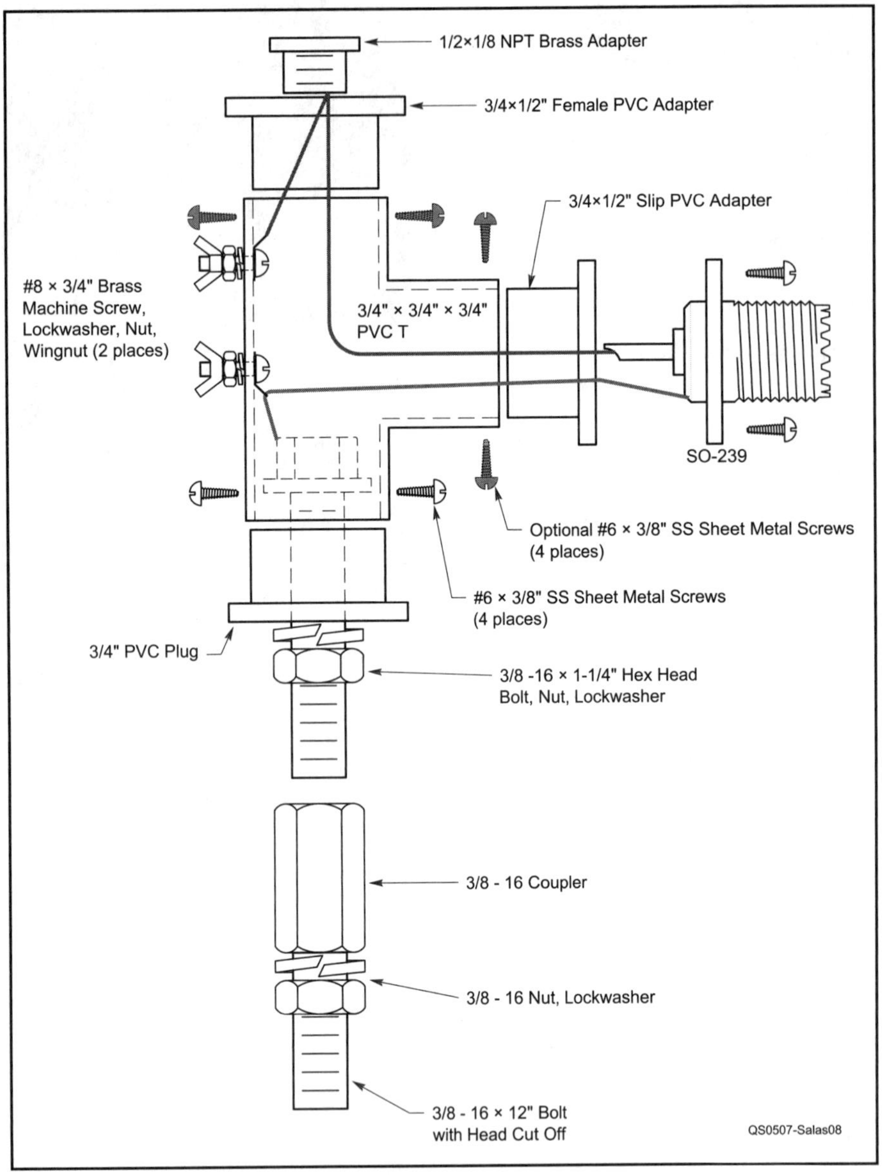

Figure 8—The antenna base and coaxial cable connector details.

the center of the PVC lip. Carefully drill two $^{5}/_{64}$ inch holes at these points. We will mount the SO-239 on the adapter later.

Place the $^{3}/_{4}$ inch PVC plug/spike assembly in the T and drill two $^{5}/_{64}$ inch diameter holes through the T and plug. [See sidebar Photo A for how I clamped the T in a woodworker's clamp to drill the holes. Try to drill straight through both sides in one pass to make alignment easier.—*Ed.*] Remove this assembly from the T and drill out these $^{5}/_{64}$ inch holes in the T to $^{9}/_{64}$ inch. Also drill out two holes in the SO-239 connector to $^{9}/_{64}$ inch, since the holes are not large enough to pass the #6 × $^{3}/_{8}$ inch sheet metal screws.

Use an $^{11}/_{64}$ inch drill bit to drill mounting holes for the #8-32 brass machine screws. I drilled the mounting holes close to the "top" and "bottom" of the T. Alternatively, if you align the bit with the inside edges of the side of the T and drill through the back of the T, you will be able to fit a screwdriver into the T to tighten this hardware.

Next we'll prepare the antenna interface at the top of the base. First, cut off part of the $^{3}/_{4}$ inch slip × $^{1}/_{2}$ inch female pipe thread PVC adapter so as to leave additional room in the T for wiring. Solder a piece of 14 gauge copper house wire directly to the inside lip of the $^{1}/_{2}$ × $^{1}/_{8}$ inch NPT brass adapter. You'll need a large soldering iron or a torch, since the brass adapter mass is pretty large. Screw this adapter tightly into the $^{3}/_{4}$ × $^{1}/_{2}$ inch PVC adapter.

Now solder a wire to the center conductor of the SO-239 connector as shown. This wire should be soldered to the wire stub on the $^{1}/_{2}$ × $^{1}/_{8}$ inch NPT brass adapter at the antenna interface, and then to the upper wing-nut assembly as shown. (Alternatively, you could use #8 solder lugs for these connections, and put the top #8-32 brass machine screw through the hole in the solder lug.) The $^{3}/_{4}$ × ½ inch PVC adapter can now be glued into place using PVC pipe glue. Solder a short piece of copper braid (from a piece of RG-58 cable) from the SO-239 ground (solder directly to the SO-239 body) to the brass ground screw, and finally to the wire soldered to the head of the 1$^{1}/_{4}$ inch bolt. (#8 solder lugs are a good alternative.) You can now complete the assembly of the base by inserting the PVC plug and 1$^{1}/_{4}$ inch bolt assembly into the T and installing the #6 stainless steel sheet metal screws as shown in Figure 8. Incidentally, the upper wing-nut assembly is used in case you need to add capacitive or inductive base matching should you want to improve the SWR on the lower bands. See Figure 9.

Ground Radial Network

The radial network is made up of six 15 foot radials, using 22 gauge insulated wire, though any gauge wire, insulated or not, can be used. I've found it best to make up three pairs of two wires each attached to a #8 spade lug on one end of each pair. This minimizes the hassle of deploying, and later rolling up, the radials. The three #8 lugs will attach to the ground screw on the base assembly. When the wires are rolled up, you should hold them together with twist-wraps. Solder a 1$^{1}/_{2}$ inch brass wood screw on the outer end of each radial. You can simply push these screws into the ground to hold the radials in place. Put a blob of hot glue on each wire/screw soldered interface to give it a little strain relief.

Guying

This antenna is self-supporting in a low breeze. In many cases, however, it will be necessary to guy the antenna because of its 16 foot length. For effective guying, I attached 9 foot lengths of nylon cord (3 pieces) just above the base of the 10 foot MFJ telescoping whip. Use a tie-wrap and close it just enough so that it won't slide over the base of the MFJ whip. Cut the 9 foot sections of nylon cord and heat the ends with a match to fuse the nylon so it won't unravel. Tie one end of a 9 foot section of nylon cord around each tie-wrap and secure with hot glue or epoxy. For the ground stakes, you can use the extra piece of brass tubing. (You only used 4$^{1}/_{2}$ feet of the 6 foot length.) Cut the remaining 18 inch piece of tubing into three 6 inch sections.

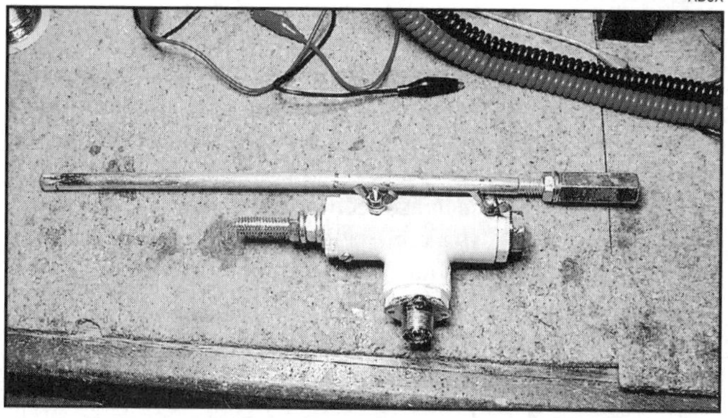

Figure 9—The completed antenna base and mounting spike.

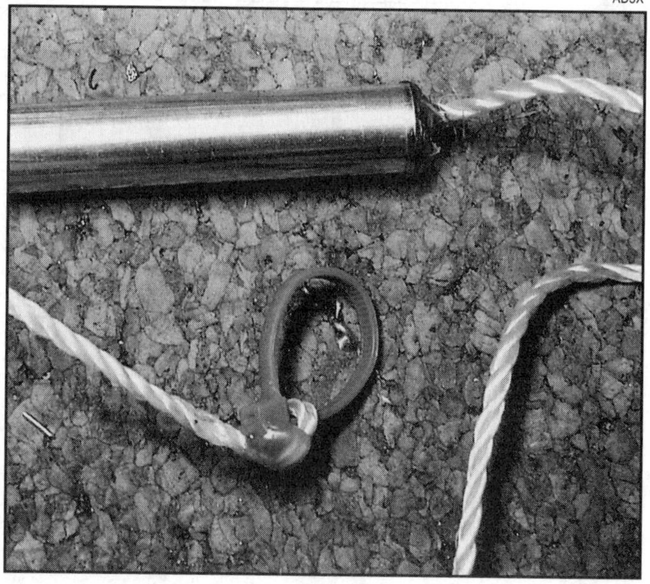

Figure 10—A set of 3 guys was made by using a wire tie (closed to slip over the top of the telescoping whip but not over the base of the whip). Hot glue was used to attach a nylon line to each wire tie. Hot glue was also used to attach the guy line to a left-over length of brass tubing.

Attach the end of the nylon cord without a tie-wrap to one end of each tube with hot glue. Also plug the open ends of the three tubes with hot glue. (You could also use some long spikes or bolts for stakes, if you are concerned that you won't be able to drive the brass tubing into hard ground.) For storage, wrap the nylon cord around each brass stake and hold it in place with masking tape. Figure 10 shows the details. See Figure 11 for a photo of the guys attached to the telescoping MFJ whip. When bolted to a trailer mount or plate, the antenna should really not need guying unless the wind is strong.

Antenna Assembly

To assemble the antenna, first screw the three brass-rod sections together, and then screw these into the top of the base assembly. Push the entire assembly firmly into the ground, keeping it as vertical as possible. Next, screw the loading coil and telescoping whip assemblies together. Slip the three guy tie-wrap/nylon cord assemblies over the whip and extend the telescoping whip. Screw this entire top assembly into the female end of the top brass tube. You only need to turn the brass fittings finger tight. Finally, push the guy rods in the ground and extend the six radials. Attach the common ends to the ground screw on the base assembly. Add coax, and you are ready to tune your antenna.

Antenna Tuning

Begin with 60 meters to determine the tap points on the coil. If your rig cannot tolerate a 2:1 SWR, you may need to add a 330 pF, 300 V, silver mica capacitor across the two wing-nut assemblies. See Figure 12 for a close-up view. This capacitor is fine for both 60 and 40 meters. The SWR on 30 meters will be closer to 1.7:1. If you need to improve this, use a 220 pF capacitor. You should not need any capacitors for 20 through 10 meters. Both my IC-706MKIIG and SG-2020 work fine into a 2:1 SWR.

To begin, use an antenna analyzer set to 5340 kHz to locate the coil tap point that gives the best SWR. Mark this tap point. Repeat the procedure at 5380 kHz. Move to 40 meters and repeat, again selecting two taps on 40 meters to give you the band coverage you desire. Repeat again for 30 meters—only a single tap is required for

Figure 11—A close-up of the guy lines attached to the telescoping whip. The coil and coil clips detail can also be seen.

Figure 12—The antenna base, with the ground radials and a matching capacitor attached to the brass screws on the PVC T.

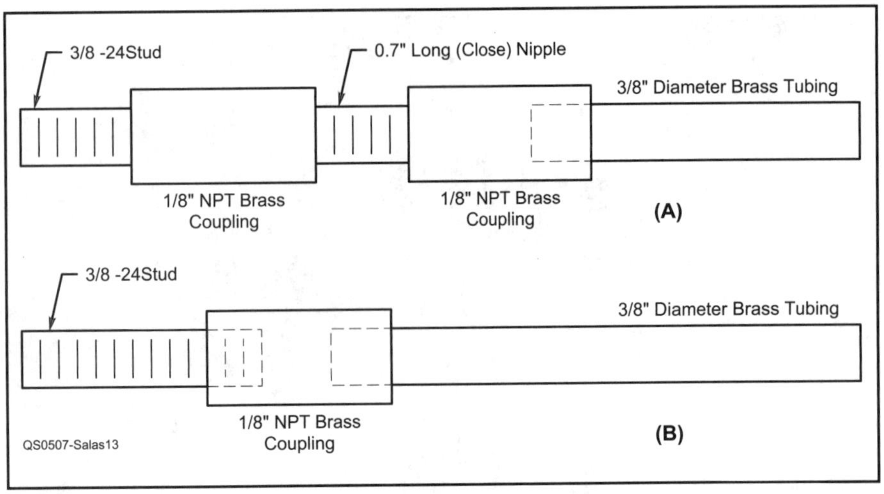

Figure 13—(A) shows how a 1/8 inch NPT brass coupling can be used to adapt the pipe threads to a 3/8-24 standard antenna mount. (B) shows an alternative attachment, if one only wants to attach the antenna to a standard mount, and not use the PVC base to mount the antenna on the ground.

this band. For 20 meters, you will find that only the top turn of the coil is necessary for resonance. The antenna is almost a quarter wavelength long on 20 meters.

For 17, 15, 12 and 10 meters, you will need to short out the loading coil and remove sections of the brass tube—then adjust the whip for resonance. On these bands, the antenna will be 1/4 λ long. Remove two brass sections for 17 and 15 meters, and all three sections for 12 and 10 meters. Use a permanent black marker pen to indicate the correct band positions on the telescoping whip.

Finally, attach the MFJ-605-4001 coil clips to the coil tap points determined above. You may wish to solder the clips in place to make the entire assembly a little more robust. From this point forward, you can just go back to these tap points, or re-adjust the top whip as necessary, and not have to worry about making SWR measurements.

Mounting Options

You can easily make a 3/8-24 threaded interface so that the antenna can be mounted on a standard 3/8-24 antenna mount. This would be useful for those with a standard ball mount on their car who want to use this extended-length antenna when stopped. As mentioned earlier, the 1/8 inch NPT thread is 27 turns per inch, and it is slightly larger than 3/8 inch diameter, with a slight taper. While the 3/8-24 standard stud will screw into an 1/8 inch NPT thread coupling, the 1/8 inch NPT nipple will not screw into a 3/8-24 threaded coupling. Therefore, an adapter is necessary if you want to mount this antenna to a standard 3/8-24 antenna mount.

One way to make an adapter is to purchase a 3/8-24 bolt and screw it tightly into a 1/8 inch NPT coupling. Cut off the head of the 3/8-24 bolt with a hacksaw and file carefully so that you don't damage the threads. Running a 3/8-24 die over the threads will clean up any damage you may have done. You can now either screw this assembly onto the 1/8 inch NPT nipple on the bottom brass tube section, as shown in Figure 13A, or screw the 3/8-24 bolt directly into the bottom 1/8 inch NPT coupling as shown in Figure 13B (if you'll never need the 1/8 inch NPT interface on the bottom antenna section).

Conclusion

Because of the interest in my original portable antenna, I've evolved that design into an antenna that is longer, lighter, more compact and easier to fabricate, and gives you more mounting options. You can also experiment with the antenna length. For example, you can remove a section or two, use more or fewer sections, decrease or increase section lengths, or place the loading coil in different positions. With the loading coil described in this article, you have quite a bit of latitude in the antenna length for a given band. For best efficiency though, try to keep the antenna as long as possible and the coil as high as possible.

Don't hesitate to make changes based on hardware availability. Try aluminum or copper tubing, or even wire wrapped 3/8 inch fiberglass or wood dowel. It's fun to design antennas "on the fly" while standing in the plumbing section of your hardware store. This makes for interesting discussions with the clerks, however!

Finally, I want to express my appreciation to Martin F. Jue, K5FLU, and Richard Stubbs, KC5NSZ, of MFJ, for working with me to provide a reasonably priced kit of parts to make this both an affordable and fun project to build.

Phil Salas, AD5X, is an ARRL Life Member. He's been licensed for 41 years, and enjoys HF operating (mostly CW). Phil's wife Debbie (N5UPT) and daughter Stephanie (AC5NF) are obviously very understanding of this hobby! Phil holds a BSEE from Virginia Tech, and an MSEE from Southern Methodist University, and is now fully retired after 33 years in the telecommunications industry. Phil can be reached at 1517 Creekside Dr, Richardson, TX 75081-2913 or at **ad5x@arrl.net** *if you have any questions or comments.*

By Robert Hollister, N7INK

A Portable NVIS Antenna

As the Amateur Radio Emergency Service (ARES) District Emergency Coordinator (DEC) and Radio Amateur Civil Emergency Service (RACES) officer for Cochise County, Arizona, I am always looking for simple solutions to complex problems. The problem in this case was to provide easy-to-use antennas for our county Emergency Response Vehicle (ERV) (Figure 1). Our ARES/RACES unit operates the ERV communications unit for the county sheriff's department in support of the Emergency Services Office, Search and Rescue Team and other emergency missions. Located in the southeast corner of Arizona, Cochise County consists of high desert, higher mountains and deep canyons. This presents a difficult environment to maintain reliable radio communications. Although the Amateur Radio clubs and the sheriff's department have a good network of VHF/UHF repeaters, there are many areas where repeaters or cell phones just do not provide the coverage we need to maintain contact with the sheriff's dispatch center in Bisbee.

A Solution

I recently read a Web article published by WØIPL that described the use of mobile monoband whip antennas in a near vertical incidence systems (NVIS) configuration (**www.w0ipl.com/ECom/NVIS/nvis.htm**). I decided, therefore, to experiment with this NVIS technique and build a portable kit for use with our ICOM IC-706 installed in the ERV.

The first step was to develop a shopping list (Table 1) and purchase the necessary items I thought I would need to make it work. At a recent hamfest, I purchased two monoband mobile whip antennas for each of the two bands I wanted to operate, and a dipole adapter that I had seen at the Dayton Hamvention.[1] I then broke out my trusty MFJ antenna analyzer to help with the trimming and SWR measurements.

First, I set up antenna number 1 (75 meters) on my Outpost tripod to start the tuning process.[2] The target was a 5 MHz frequency that I wanted to use to connect with the Fort Huachuca Military Affiliate Radio Service (MARS) station digital bulletin board. Many of our ARES/RACE members are also active in the Army MARS program and we have a TNC in our ERV for both HF and VHF digital communications. I knew that the normal range of the mobile whip antenna covered the 75 meter portion of the band effectively, but that it would require a much shorter stinger than the 48 inch standard length to reach the target 5 MHz frequency.

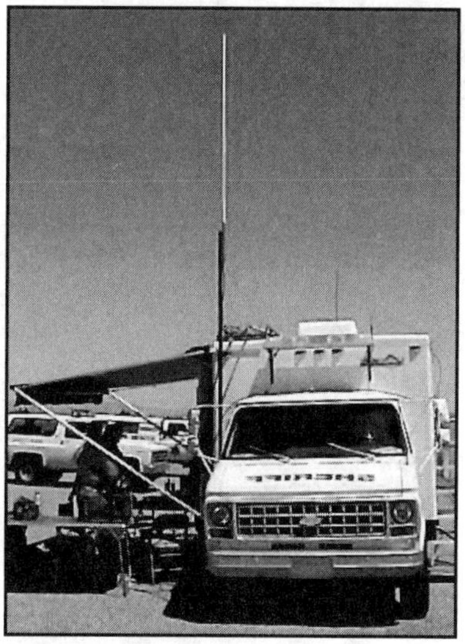

Figure 1—The Cochise County, Arizona emergency response vehicle (ERV).

[1]Notes appear on page 19.

Construction

Rather than trimming the stainless steel stinger that came with the antenna in steps, I started with a piece of wire coat hanger cut to 24 inches that fit into the collar on the top of the bottom section of the antenna. I scraped the varnish off the wire on the piece to be inserted in the collar for a good electrical connection (Figure 2). If you have ever tried to cut stainless stingers before, you will understand quickly why using the softer coat hanger wire is the preferred method. On the first test, the antenna was resonant at about 6.250 MHz. This told me that I was still much too long. I cut the wire stinger in half with my wire cutters to about a 12 inch stinger and tried again. I was now in the ballpark and only 500 Hz away. I continued cutting at half-inch intervals and tested the results.

I was soon at my desired resonant frequency. The stinger was now extended only 6 inches above the bottom section of the antenna. The antenna instruction sheet suggested that not more than 4 inches be inserted into the bottom section. The

Table 1
NVIS Antenna Parts

75 meter mobile monoband antenna	2 each = $30
Extra 48" stingers	2 each = $18
40 meter mobile monoband antenna	2 each = $30
Dipole adapter	1 each = $15
Stainless radiator clamps, 2"	2 each = $1
5 foot fiberglass pole, surplus	2 each = $15
RORO ("big foot") mount	1 each = $90
10" section, $^3/_8$" aluminum stock	1 each from junk box
10" section, 2" aluminum stock	1 each from junk box
	Total = $199

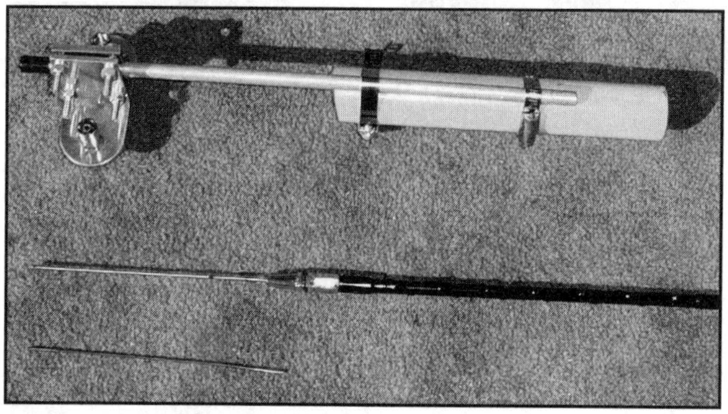

Figure 2—A coat hanger is used as a prototype "stinger" during testing and requires that the paint be scraped off its end. It is shown next to a new stainless steel tip. The dipole adapter is shown at the top.

total length of the now resonant stinger was 10 inches. I used this trimmed piece of wire as a template and laid it next to the original stainless stinger. After carefully marking the new length and allowing for adequate length to extend into the top of the lower section, I cut the stainless stinger to length with my hacksaw. After smoothing and rounding the cut end of the stainless steel whip with a diamond file, I rechecked the resonant frequency. It was right on. I took the new short stainless stinger and, using the same measurement as the first one, checked it with the second 75 meter antenna bottom section for resonance. It was close, but required a slight amount of adjustment to be brought to frequency. The second stinger was quickly cut and ready for testing (Figure 3).

NVIS Height

Since the dipole adapter I purchased did not come with any instruction sheet, I needed to figure out how to get the antenna to a suitable height above ground. I already had something we call our roll-on/roll-off (RORO) or "big foot" to use as an antenna support base plate (Figure 4).[3] The NVIS literature and field testing suggests that the optimum height above ground is not an exact science but that it should be approximately 10-12 feet high to reliably cover a range from 0 to 300 miles. Height above ground and the conductivity of the soil below the antenna will affect the optimum range and the resonant frequency. Minor adjustments to stinger lengths may be required at different locations, depending on ground conductivity. A height of 10 to12 feet above ground also ensures that it is high enough so that most pedestrians and NBA basketball players are not likely to run into the ends.

As I owned a number of 5 foot military surplus aluminum and fiberglass antenna mast sections, I decided to use two of the fiberglass sections as the main antenna support.[4] These are the same type we use to hold up our spare dual band (144/440) antenna when we need additional range over the vehicle-mounted antenna. I then had to decide how to easily and quickly attach the adapter to these mast sections. A short piece of a scrapped aluminum mast section cut to 10 inches, two each 2 inch stainless hose clamps, and another 10 inch piece cut from some $^1/_2$ inch aluminum tubing fit the bill. This allows a quick slip-on fit over the top of the fiberglass mast section (see Figure 3).

The next step was to attach the two modified mobile antennas and coax to the dipole adapter. I slipped the adapter over the top of one of the fiberglass masts, inserted it into the "big foot" and reattached the coax to my analyzer. It was too much to hope that it would still be exactly on frequency and, unfortunately, I was correct. There was enough interaction with the added metal of the adapter to require some additional tuning. Fortunately, it was minor and only required shortening both of the elements approximately another $^3/_4$ inch to bring it back to my desired frequency. The next step: On-the-air testing.

Testing

With the two fiberglass mast sections inserted into the "big foot" base, the completed horizontal dipole was now approximately 11 feet in the air (Figure 5). Hooking up the radio, the computer and the TNC (terminal node controller), my first attempt to connect to the Fort Huachuca MARS BBS (bulletin

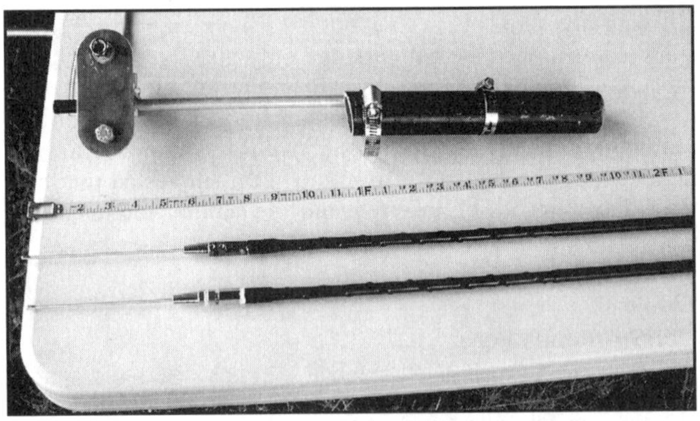

Figure 3—A top view of the assembled dipole adapter used to mount and secure both antennas, which are shown below it.

Figure 4—The RORO ("big foot") mount. This was custom-built for the author by a welding shop (see text).

Figure 5—The completed antenna in use in the Chiricahua mountains of Arizona.

board system) using PACTOR (a digital communications protocol) was a success. Solid S9 copy was obtained with approximately 40 W output. I quickly prepared a sample message to send, reconnected to the BBS, uploaded the message and disconnected. Again success! Before disassembly to move to the next frequency of interest, each stinger and bottom antenna section were color coded with a band of red and white paint to ensure the correct stinger could easily be reassembled with the correct bottom section.

I had purchased a second set of stingers (of normal 48 inch length) for the 75 meter antennas and a pair of 40 meter antennas at the time I made the original purchase. Setting up for the other two primary frequencies I wanted to use was straightforward and required no trimming. Using two standard-size 48 inch whips and collets, these two stingers were adjusted for the Arizona RACES frequency of 3.990 MHz on the Outpost tripod. These are adjusted for our Arizona state RACES net on 3.990 MHz, and the third set is set for our alternate frequency of 7.248 MHz. There was adequate flexibility using the standard stingers to adjust to the state MARS nets just outside the amateur band with no additional trimming required.

The state RACES net has stations representing most state counties and reaches all corners of the state. I am located in the southeast corner with the furthest station about 250 miles away. Again, attaching the antennas to the dipole adapter required some minor shortening to achieve resonance (1.2:1). During a Sunday morning RACES net, I conducted on-the-air A/B testing. I compared the NVIS configuration to a Hustler vertical (previously used for portable operation), attached to the Outpost tripod. The NVIS configuration demonstrated an approximate 1-2 S unit improvement for most signals compared to the vertical.

Switching to 40 meters, and using the same procedure with two 40 meter antennas on the Sunday morning Arizona MARS net, showed a similar improvement. Most signals showed a 1-2 S unit improvement in signal strength using the NVIS dipole compared to the vertical. The biggest audible improvement was noticed in interference/noise levels. Some stations were totally lost in the noise on the vertical and came back to Q5 on the NVIS dipole. This net included stations in Arizona and southern California. Similar distances to those of the Arizona RACES net were covered. At that time of day (one hour later than the RACES net), the 40 meter coverage was generally better on all stations.

Conclusion

Two weeks later, I was able to run a few more tests from a remote location in the Chiricahua Mountains. Over the Labor Day weekend, the Cochise Amateur Radio Association conducts a special event operation from the ghost town of Paradise. This provided the most realistic test yet, as we deployed the ERV and operated portable in the mountains, much like a normal search mission. I tested all configurations on both voice and digital frequencies. The antenna was quick to set up and provided reliable communication to the stations I needed to contact. The complete antenna kit requires less than 5 minutes to assemble and erect. This antenna has proven to be a reliable addition to our emergency communications needs and it meets all of my expectations.

Notes

[1]The antennas are available from the Lakeview Company (**www.hamstick.com/9106.htm**) and WBØW (**wb0w.com**) as well as other sources. The dipole adapter is available from a number of sources, including: Atoc Technologies (Iron Horse) (**www.atoctechnologies.com/**) and Quicksilver Radio Products (**www.qsradio.com/products.htm**).

[2]**www.alphadeltacom.com/pg6.htm**.

[3]The RORO mount was built to specification by a local welder for $90. A drawing can be provided for others who are interested in having one fabricated. I have seen similar designs available at several hamfests.

[4]The mast sections came from The Mast Company (K4TMC), PO Box 1932, Raleigh, NC 27602; **www.tmastco.com/**. Henry offers a variety of military surplus mast sections in aluminum or fiberglass. I prefer the lighter fiberglass but I have used both with equal success.

Photos by the author.

Robert Hollister, N7INK, was first licensed as N2BCY in 1978, while serving with the US Army in Germany. He's also held the calls DA2HO and GM5ELQ. Bob retired from the Army in 1986 as a communications intercept technician. He has a BA from The State University of New York and an MS in Applied Management from Lesley College, Cambridge, Massachusetts. Bob also has an Amateur Extra class license and is a life member of the ARRL. He can be contacted at 5457 S San Juan Ave, Sierra Vista, AZ 85650, or at **n7ink@cox.net**.

A 20 Meter Antenna for Sailboats

The Zepp antenna — not just for airships anymore.

Gary Davis, KD9SB

If you are one of the many sailing hams who has contemplated operating marine or maritime mobile from your sailboat, this antenna may be for you. For the past few years I have been searching for an easy to install, efficient 20 meter wire antenna that can be used on a fiberglass or wooden sailboat. This antenna eliminates the need for an RF ground system, or a trip up the mast to install the antenna. The antenna is usable on any point of sail without interfering with the sails either while tacking or under way.

Why the Zepp?

This antenna is ideal for a ham needing a quickly installed 20 meter antenna for use during a charter, delivery, leisure cruise, offshore passage or shipboard emergency. This Zepp antenna can be traced back to the early part of the twentieth century when it was developed for use on the Zeppelin airships. When simply hung from the airship gondola, it trailed behind or below the craft during flight. A similar configuration is often used on VHF and is called a "J" antenna.

The Zepp antenna consists of a λ/2 of wire end-fed by an electrical λ/4 of open-wire or ladder type transmission line. It is a very simple and inexpensive antenna to build. The antenna is matched to the 20 meter transceiver by a small antenna tuner. The configuration is shown in Figure 1. I have adapted the Zepp design for use on a fiberglass or wooden sailboat, where a proper sea water RF ground is often difficult to obtain.

Putting it Together

For the λ/2 antenna, I used 13-gauge 19-strand Variflex insulated wire. A lightweight plastic egg insulator is attached to each end of the antenna. To support the antenna, I used a monofilament plastic line of the type used on small grass trimmers. This line is 0.065 inches in diameter. The monofilament line is used to reduce antenna detuning, caused by wet support lines. A 6 foot line is used at the masthead end of the antenna to prevent detuning effects, due to the convergence of the stainless steel rigging at the masthead. A 2 foot line is used at the stern rail end. The grass trimmer lines are attached at both ends. The top end is attached to a spare masthead halyard and raised aloft The stern rail end is attached to the outermost corner of the stern rail, keeping it as far from the backstay and boom as possible. Figure 2 illustrates the arrangement.

I also added a lightweight Dacron line from the halyard to a deck level cleat. This forms a downhaul so I can retrieve the halyard in case the antenna wire or support line breaks aloft.

Matching an End-Fed λ/2

The λ/4 matching section is made from ladder line. I used 450 Ω line made from 18 gauge solid wire. Ladder line is also available in 14 gauge stranded wire, a more durable choice for offshore or longer term use. One wire of the ladder line is soldered to the stern rail end of the antenna, and covered with heat shrink tubing. The other wire is left unconnected and hanging free. At the antenna tuner end of the ladder line, both wires are connected to the balanced line output jacks via a dual banana plug. Remember, a jumper wire is often used to enable the balanced line output. I taped over the coax output jack, leaving only the transmitter jack free, to prevent any confusion during a rapid setup.

The λ/4 matching section serves not only as the antenna feed line, but is also used as the impedance matching device between the antenna coupler and the antenna itself. A λ/4 transmission line transforms the high impedance of the end of the λ/2 antenna to a low impedance. The impedance should be close enough to operate into a transceiver's internal tuner through a short section of coax, or an external antenna coupler may be needed, depending on the transceiver.

The antenna interface needs to make the transition from the balanced line to the coax connector at the radio. One way to do that is with directly connected coax with a common mode choke, another is with a balanced antenna tuner and a third is to use an unbalanced tuner with a balun at the output. The unit I use is a small MFJ Versa Tuner model MFJ-901B.[1] Another handy device is the MFJ model 212 Matchmaker. This device will allow you to approximately match the transceiver, coupler and antenna without radiating a signal. After the initial tune-up, connect the SWR meter in place of the MFJ 212, and adjust for as close to a 1:1 match as possible.

I cut my antenna and matching section to be centered at 14.300 MHz. For the antenna,

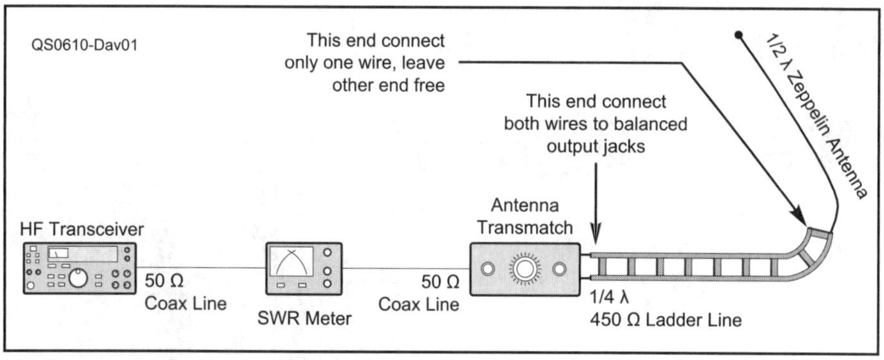

Figure 1 — Antenna and tuner configuration block diagram. A separate tuner may not be needed depending on the radio configuration.

Figure 2 — Photo of lower end configuration. Note that the balanced line is kept away from railings and other metal objects.

I used the usual formula, 468/frequency in MHz, resulting in a length of 32 feet 8 inches. The feed line length requires consideration of the relative propagation velocity, 0.91 for typical window type line. This results in a length of 15 feet 8 inches for the matching section. The antenna is broad enough to achieve a good match over the whole band.[2]

Making it Play

On my sailboat, a Catalina 30 sloop with the "tall rig" option, it was very easy to slip the thin feed line through the stern hatch, into the stern locker and through the aft berth bulkhead to the navigation station in the cabin. Try to keep the feed line away from large metal objects, such as the engine block and steering quadrant. Don't be tempted to tie the line to the stern rail or stanchions. Let it hang free.

If the feed line will be too short to reach the antenna tuner, a length of ¾ λ can also be used, but the bandwidth will be reduced. The bottom of the matching section can also be connected directly to coax. In that case, it's best to use a choke coil of about six turns about 6 to 10 inches in diameter in the coax adjacent to connection. The SWR will be low enough that a run to the transceiver or tuner should not incur significant losses on any reasonably sized craft.

On my boat, I use a 20 W Kantronics KT-120 transceiver. With any new RF-generating installation, it's a good idea to check for RF interference to the autopilot, fluxgate compass and any other sensitive electronic equipment. As always, make sure that no one will touch the antenna during transmissions.

My best DX has been to Russia and Armenia from Monroe Lake, a 10,000 acre fresh water reservoir in south central Indiana. Although I have not tested the boat and antenna on saltwater, I would expect performance to be even better.

You can also use the antenna in some other configurations. As an example, I once used a 12 foot boat hook as a center support for the antenna. This made it an end-fed inverted V. With this arrangement, I also made distant contacts.

Notes

[1]With this model tuner, an external SWR meter is also required. If your transceiver measures SWR, that indicator can be used.

[2]Your editor built a version to see how close it would match without a tuner. I used 14 gauge stranded window line and achieved best results with a matching section of 15 feet 2 inches. I found an inch or two either way didn't change much. Using an element length of 33 feet 10 inches I measured an SWR of less than 2.4:1 across the whole band. I had the feed line directly connected to RG-8X coax with a choke of 6 turns, 7 inches diameter, close to the ladder line. A 2.4:1 SWR may work directly with some radios, but should be within range of the internal tuner of any radio so equipped. This antenna will travel with me on vacation.

Photos by the author.

Gary F. Davis Jr, KD9SB, is a retired music business owner and radio broadcast engineer. He has had numerous construction projects published in various radio, electronics and sailing magazines over the past 37 years. Gary was first licensed in 1959 and holds an Amateur Extra class license, FCC Commercial General Radiotelephone license (GROL), and an FCC Global Maritime distress and safety system (GMDSS) operator/maintainer license. In addition, he has several US patents in electronics. Gary and his wife, Sandy, N9HKS, have owned sailboats for over 30 years. They have sailed their present boat, Shangri-La, *for over 18 years. Gary can be reached at 8475 E State Road 7, Columbus, IN 47203-9176, or* **gdaviskd9sb@sbcglobal.net**.

Compact 40 Meter HF Loop for Your Recreational Vehicle

Want something better than a mobile HF whip antenna while camping in an RV park? Try a tilt-up compact HF loop.

John Portune, W6NBC

Hams with a motor home or a travel trailer know that achieving high efficiency in an HF RV antenna isn't easy. Most often we settle for a bumper mounted whip or screwdriver antenna. That's fine for on the road operating, but leaves room for improvement for operation while parked.

A survey of campground HF RV antennas will turn up a number of potential solutions. These include tilt-up home station type trapped multiband verticals, long wires shot over a nearby tree or pairs of mobile HF whips mounted horizontally on a mast and used as short dipoles. A much better solution, in my opinion, is to use a compact transmitting loop. They can be small enough to be easily mounted on an RV, but are more efficient than mobile HF whips. My version of such an antenna for 40 meters is shown in the title photo. I've provided design details for other bands at the end of this article.

Compact loops can also be a good choice for a home station operation if aesthetics or local rules prohibit more obvious antennas. I have a 20 meter loop hidden in a tree, a 40 meter loop on a rose trellis and a 2 meter loop disguised as a weather vane.[1,2]

A compact loop is just a dipole made short (less than λ/10 long) then bent into an open-ended loop with a capacitor across the ends for tuning. Looking at it another way, a compact loop is a one turn parallel "tank circuit" big enough to radiate.

Why is it better than a typical electrical λ/4 mobile vertical HF whip? Simply stated, a compact loop or other horizontal antenna, suffers far less from ground losses. This one, modeled at 70% efficiency, is only a bit less efficient than a full-size 40 meter dipole. Comparatively, an electrical λ/4 mobile whip, such as a screwdriver antenna, for 40 meters may be only 5% efficient. Read a fuller explanation of this later in this article.

Radiation Pattern

In free space, the radiation pattern of a compact loop is like that of a λ/2 dipole lying along the axis of the loop with its polarization rotated 90°, in this case to vertical. As with all antennas, the ground changes the radiation pattern. Figure 1 is an *EZNEC* plot of the loop over average soil (RV on the X axis). As you can see, at this height, the loop is more or less omnidirectional, but this is also true for many 40 meters antennas at this height.

Unlike a dipole, however, there is no specified size for a compact loop. With the correct tuning capacitor across the ends, one can resonate a loop to a wide range of frequencies. This loop is roughly λ/5 in circumference. I intentionally made it larger than the classical λ/10, to improve the efficiency. In general, the bigger the loop, the lower the losses. On an RV, however, there is adequate room for a large loop. What's more, the extra size reduces the length of the coaxial tuning capacitor. The radiation pattern does skew a bit from the classical, but this is of no consequence at this height.

Do not mount the loop horizontally. If you do, losses may be severe. Loops do not like large objects, including the ground, broadside to the loop. Unless you can mount a compact loop high, keep it vertical.

Mechanical Details

In addition to nearby objects, the other main cause of loss in a compact loop is the resistance of its conductors and connections. Hams rarely consider skin effect but in small antennas its influence is appreciable. Compact loop designers should always pay careful attention to both the conductor size and the material used in construction in order to achieve high efficiency.

As an aside, you can estimate the real world loss in an antenna by measuring the bandwidth wherever you intend to use it and then again at the same height, but now clear of all nearby objects. In free space this loop has a theoretical bandwidth of roughly 10 kHz. On my RV it is slightly broader, indicating that the rig and the ground are to some degree lowering the efficiency.

Large diameter high conductivity tubing is mandatory. Copper is best. I use common hardware store rigid copper water pipe.

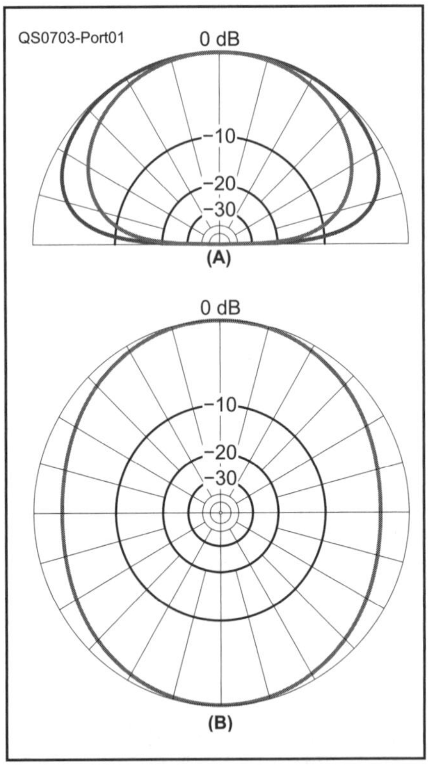

Figure 1 — EZNEC plots of the loop's pattern over average soil. At A, elevation plots in the plane of the loop (red) and at right angles (blue). At B the azimuth plot at 30° elevation.

[1]Notes appear on page 24.

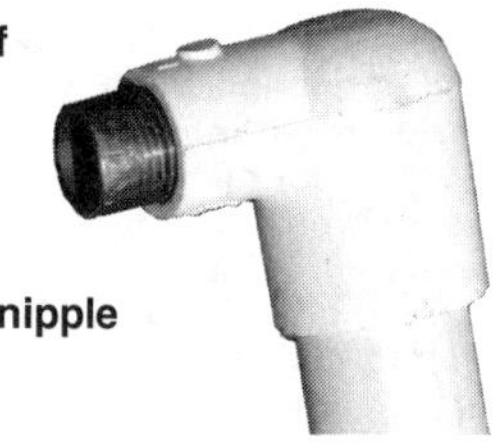

Figure 2 — View of the tilt-up hinges made from 1 inch PVC pipe, threaded elbow reducers and a ½ inch brass pipe nipple cut in half.

Aluminum is noticeably poorer at nearly double the loss (roughly half an S-unit). Steel and stainless steel are terrible. Also, don't use flat strap. Skin effect in a strap pushes the RF mostly to the edges, increasing conductor resistance even more. Copper pipe, ¾ inches in diameter, is the best compromise considering strength, cost and weight. It results in an antenna having reasonable efficiency. The use of available plumbing fittings and ordinary solder, applied with a propane torch, makes for a satisfactory assembly method.

Make the tilt-up hinges from 1 inch PVC pipe, threaded elbow reducers and a ½ inch brass pipe nipple cut in half as shown in Figure 2. The cut ends slip into copper elbows at the bottom of the loop to act as bearings. The normal location of this loop is on top of my RV, but I decided also to make it the same size as the bed of my ¾ ton long bed pickup as shown in Figure 3.

Remember, a compact loop doesn't have to be a particular size. At a pickup bed size, it can be sized to mount directly in the truck's rear stake pockets. This gave me a convenient place to develop the loop as well as a way to operate it at home without the RV.

For installation on my RV, I permanently attached two 1½ × 6 inch galvanized steel pipe nipples, as shown in Figure 4, on floor flanges to the roof. These perform the same function as the stake pockets of my pickup.

The Tuning Capacitor

I used 1¼ inch copper pipe and a 1¼ to ¾ inch reducer for the tuning capacitor as shown in Figure 5. This permits full legal power. You will need to drill through the small end of the reducer with a 29/32 inch drill so that the capacitor can slide on the loop. Also, drill a ¼ inch hole in the small side of the reducer and solder on a brass nut for a stainless steel locking screw. The tuning range with this capacitor is several megahertz.

The open ends of the main loop come within 3 inches of each other inside the capacitor. Maintain alignment by inserting a roughly 1 foot piece of ¾ inch acrylic rod into the open ends of the loop. Tape or heat shrink the rod to the ends of the loop. Several Internet providers supply acrylic rod.

You will also need to stabilize the open end of the capacitor with an assembly made from a PVC pipe cap. The cap requires a 7/8 inch hole drilled in the top. Glue the assembly together with ordinary PVC cement as shown in Figure 6. A PVC cap alone is not a good idea. If moisture gets in, a pathway for

Figure 3 — The loop mounted on the bed of the author's ¾ ton long bed pickup.

arcing will exist along the cap from the loop to the end of the capacitor. The voltage here is thousands of volts even at 100 W of transmit power. After tuning the loop, weatherproof both ends of the capacitor with tape or heat shrink tubing. The ½ to ¾ inch bushing helps you accomplish this. You should not need to readjust the capacitor after initial tuneup. Refer to the tuning section later in this article.

Feeding the Loop

There are several ways to connect to a compact loop. My preference is a tap match as shown in Figure 7. It resembles a gamma match, but is actually more like tapping an RF coil. It can be made of small diameter wire, but for rigidity I prefer ½ inch pipe and fittings.

By experiment, I discovered that the location of the matching section isn't critical. A good match can be achieved at many places on the loop. Changing the position mostly affects the loop's radiation null. I found by modeling with *EZNEC* that the best compromise is the bottom corner opposite the capacitor.

The matching section must be open at one end to permit connection to the feed coax. Use a CPVC plastic coupler as shown in Figure 8 for this. CPVC is not ordinary PVC. These special fittings are made of high strength, high temperature plastic to allow them to mate directly with copper water pipe. They are commonly available at hardware and plumbing supply stores.

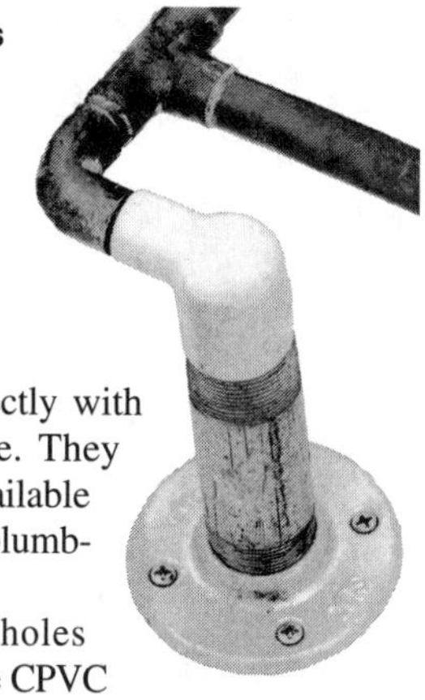

Figure 4 — Details of the galvanized steel pipe nipples mounted on floor flanges on the RV roof.

Drill 9/64 inch holes near the ends of the CPVC coupling through the matching section pipes for 6-32 stainless steel connection screws. Solder brass nuts inside the pipes. Use a stainless steel screw to hold the brass nut in place while soldering. Make the actual connection with a short coax pigtail fitted with ring terminals on one end and a PL-239 connector on the other. Weatherproof the pigtail.

You must also slip HF ferrite beads (roughly 50) over the feed coax near the loop to form a coaxial choke. Without the choke, currents on the outside of the coax make matching nearly impossible. Secure the beads with heat shrink tubing. I prefer a W2DU style ferrite choke balun for this. Several Internet vendors offer these.

Tuning the Loop

The best tool for tuning the loop is an MFJ-239 or similar SWR analyzer, but you can also tune it with an ordinary SWR bridge and your transceiver (at low power). Simply adjust the capacitor for the lowest SWR. The dimensions shown in Figure 5 for the matching section provide a good match over the entire 40 meter band. In a different location, such as an attic or a tree, you may need to lengthen the matching section. By experiment I have found that the closer the loop is to other objects the longer the matching section needs to be. The capacitance is changed by sliding the 1¼ inch tubing up and down.

Tuning While Operating

A big difference between a compact loop and a full-sized dipole is bandwidth. The loop is sharper. Most shortened antennas have inherently high Q. Therefore, some form of remote tuning while operating is desirable. My preference is an antenna tuner at the operating position, leaving the loop tuned to the center of the band. Some purists may object to this because the SWR on the feed line increases as one moves off center frequency. Actually, the loss is small. To be sure, I made signal strength tests to determine how much actual loss there is. With a tuner and 25 feet of RG-8X, I can not discern any appreciable difference in the transmit or receive signal strength over the entire 40 meter phone band. Subsequent on air experience has confirmed this.

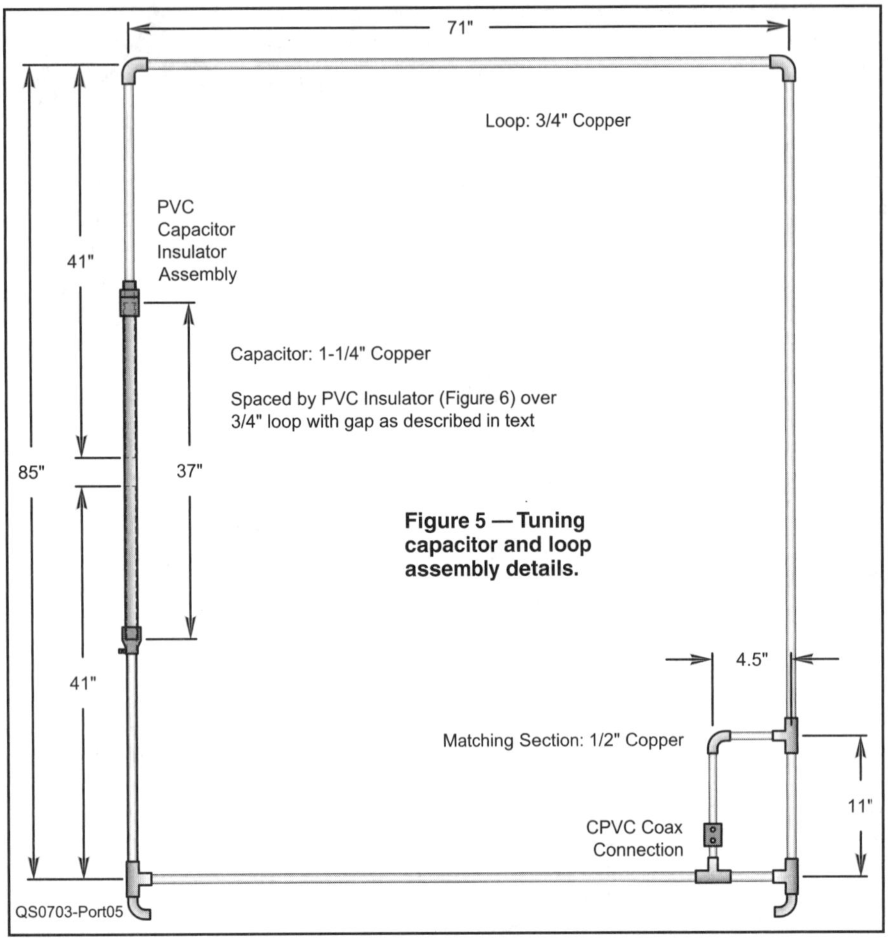

Figure 5 — Tuning capacitor and loop assembly details.

Figure 6 — Cover for open end of capacitor made by gluing together a 1¼ inch coupling, a 1½ inch pipe cap and a 1 × ¾ inch reducing bushing. Before gluing, drill a ⅞ inch hole through the top of the pipe cap and through the reducing bushing. A 1 inch coupling may be substituted for the reducing bushing.

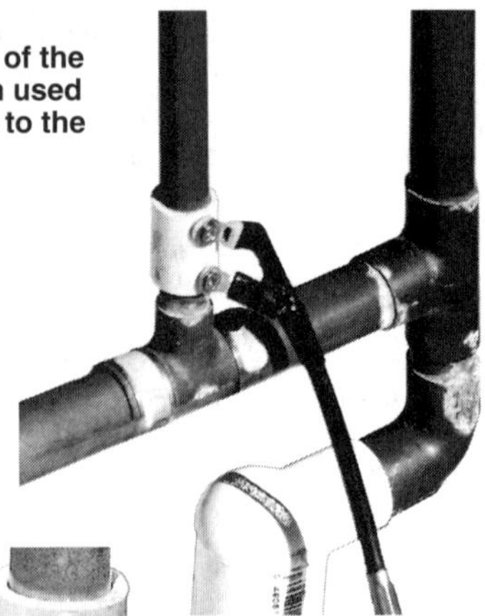

Figure 7 — Details of the tap match used to couple to the antenna.

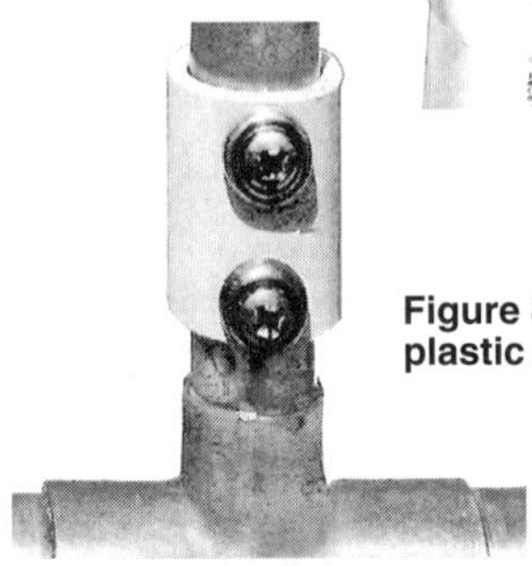

Figure 8 — CPVC plastic coupler.

Ground Plane Loss in λ/4 HF Whip Antennas

Let's now take a closer look at ground plane or soil loss. First of all, a λ/4 antenna is not a complete antenna. A resonant antenna needs to be at least λ/2 long. A λ/4 whip is, therefore, only half an antenna.

So how then does a λ/4 whip work? Visualize, if you will, a λ/4 whip mounted vertically on a metal mirror. What would you see looking at its reflection in the mirror? An upside down whip, of course. Well, that reflection acts as the other half of the antenna. It literally makes the whip into a λ/2 antenna. Therefore, a ground plane is essential for a λ/4 whip. What's more, that ground plane needs to be highly conductive — metal in most cases.

Not only this, but the ground plane should extend at least λ/4 in all directions from the base of the whip. When, for example, you use a 19 inch λ/4 2 meter whip on your car, the metal body is able to provide an adequate ground plane. It is large enough on 2 meters. On HF, however, it isn't. A vehicle's metal body on HF is usually too small to provide a complete ground plane. Instead the soil under the mobile rig carries significant current. The ground, and the resistance of any connections or wires in loading coils, adds significant resistance to the equivalent circuit. The result is that the resistive part of a typical HF vertical antenna is perhaps 35 Ω, in series with a radiation resistance, the part of the equivalent circuit that results in radiation, of only a few ohms, depending on the band in use.

To illustrate why this is a big problem, consider a full length 32 foot λ/4 vertical antenna on 40 meters. Theory predicts a radiation resistance of roughly 36 Ω in series with any soil resistance. The soil resistance can easily equal the radiation resistance; if so, about half of the transmitter's power is heating the soil. This is why we often use radials on a vertical antenna. They reduce the ground resistance.

Let's now shorten the whip to 8 feet, the average length of a mobile HF antenna. Now the radiation resistance will be roughly 2 Ω. In series with 35 Ω of soil resistance a typical 8 foot long mobile whip on 40 meters has an efficiency of roughly 5%. Worms get 95% of your transmitter's power.

In contrast, a compact loop has high efficiency because the soil resistance is no longer in series with its radiation resistance. That's why this loop consistently outperforms my conventional 40 meter HF mobile whip by as much as 2 S-units.

Loops for Other Bands

You can build a compact loop for almost any HF or VHF band. To do so, download a copy of the handy *Excel* compact loop spread sheet from AA5TB's Web site: **webpages.charter.net/aa5tb/loop.html** (scroll down for loop cal.xls). It dynamically calculates all the needed parameters. Using it, I have successfully constructed compact loops on several HF and VHF bands. For the matching section, you can essentially just scale the dimensions shown here by frequency. Try one — you won't be disappointed.

Notes

[1]J. Portune, W6NBC, and F. Adams, WD6ACJ, "A Weather Vane Antenna for 2 Meters," *QST*, Aug 2005, pp 35-38.

[2]Feedback, *QST*, Oct 2005, p 35.

John Portune, W6NBC, received a BSc in physics from Oregon State University 1960, his General Radiotelephone license in 1961 and his Advanced class amateur license in 1965. He spent five years in England as G5AJH and upgraded to Amateur Extra class in 1985 to become a volunteer examiner (VE). John retired as a broadcast television engineer and technical instructor at KNBC in Burbank and then from Sony Electronics in San Jose, California.

John is active on many bands and modes, predominantly from his HF RV mobile station. He has written various articles in ham radio and popular electronics magazines and remains active as a VE team leader, ham license teacher and Web site designer. You can reach John at 1095 W McCoy Ln #99, Santa Maria, CA 93455, or at **jportune@aol.com**. *More information on loops can be found on his Web site at* **www.w6nbc.com**.

Maximum Gain Portable HF Yagi

Here's a clever, portable beam, with gain on both 40 and 20 meters.

James Michener, K9JM

Inspired by VE7CA's November 2001 *QST* article describing a two element portable HF Yagi, I immediately concluded that such an antenna would be excellent for emergency communications as well as future ARRL Field Days. Given two supporting "boom" poles raised in the trees, I wanted to investigate how to achieve maximum gain from a 20 and 40 meter multiband version of this type of Yagi. Since it is easy to separate the two boom poles to support the Yagi wires, the question comes down to getting the most gain out of a rectangular area, while still keeping things simple.

Making It Coast to Coast

Field Day on the West Coast is especially challenging during a sunspot minimum. It's easy to work West Coast stations on the low bands, but the prize is having a solid signal to work the five times as many groups in the Midwest and the East. Breaking through the din all the way from California on 40 and 20 meters is a real challenge. This need gave birth to this unusual variant of a wire beam.

Gain and Directivity on Two Bands with a Single Antenna

Figure 1 shows the basic antenna. The antenna is a 40 meter two element beam consisting of an oversized driven element and a single reflector. On 20 meters the driven element becomes a double extended Zepp with a pair of reflectors, one behind each of the two current nodes. The antenna is also usable on 75 and 80 meters as a simple doublet. On 40 meters the performance is very much like a classic two element beam, with about 6.56 dBi gain in free space (see *EZNEC* plots in Figure 2).

On 20 meters there is an additional 2 dB of gain, or 8.46 dBi in free space, with gain nearly equal to that of a four element Yagi. In essence on 20 meters this antenna performs like two horizontally stacked two element 20 meter beams. While it nearly has the forward gain of a four element beam, its front to back is typical of a two element Yagi.

For best performance on 40 meters the antenna needs to be at least 40 feet in the air, although 50 to 60 feet is preferred. I tuned the antenna dimensions for the CW ends of each band and constructed my antenna out of 13 gauge insulated wire.[1]

The reflector spacing is not critical. The plots in Figure 2 are for 10 feet spacing for the 20 meter reflectors and 16 feet spacing for the 40 meter reflector. The spacing for the 40 meter reflector will most likely be determined by the length of the support poles you can find. There is a slight gain and bandwidth improvement with wider spacing, but any spacing between 15 and 20 feet will work.

[1] *EZNEC* model files are available at **www.arrl.org/files/qst-binaries**.

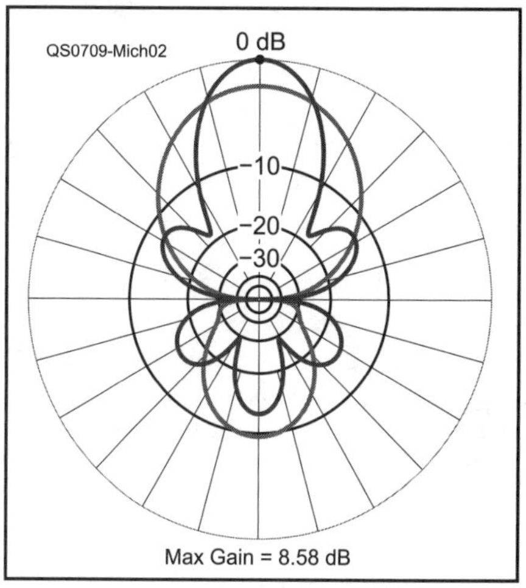

Figure 2 — Comparative azimuthal patterns for 20 (blue) and 40 meters (red) Yagis in free space.

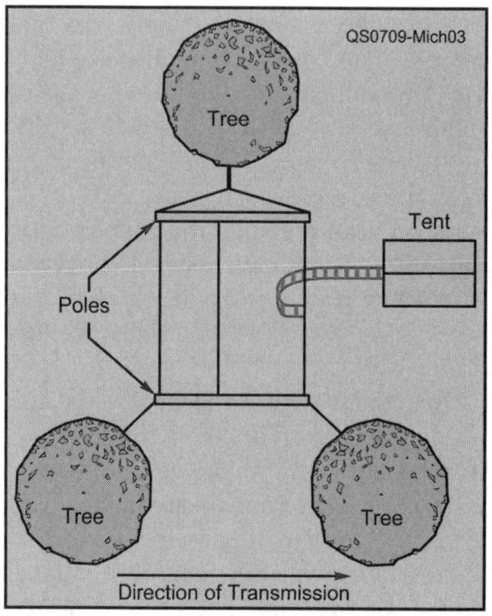

Figure 3 — Three-point technique for raising the K9JM 40 and 20 meter Field Day Yagi. This technique is flexible in deployment since the antenna can be aimed in a desired direction by tensioning the ropes properly.

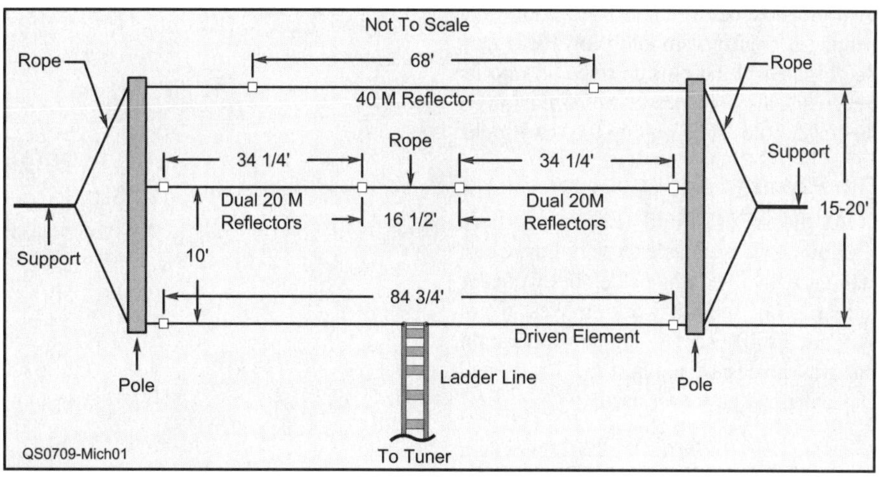

Figure 1 — Layout of K9JM 40 and 20 meter dual band Field Day Yagi. All wire is 13 gauge PVC insulated wire.

Figure 4 — One end of the duoband Yagi, showing the bamboo support boom. The window line feed is also shown. (Keen eyed observers will note that this version includes an experimental set of 15 meter wires. The experiment wasn't successful, however, mainly due to SWR changes when the wind blew.)

Figure 5 — Putting up the Yagi was easy. Keeping the rattlers out of the tent was more of a problem at the AE6C Field Day site in the wilds of northern California! Pete Murphy, W6WY (not to be confused with Murphy's Law) holds up a large California rattler, which perished from a brain injury.

Pulling it All Together

The input impedance is reactive on both bands. For simplicity, I fed the antenna with open wire window line, tuned with an old Johnson Matchbox balanced tuner.

While the antenna is mainly wire and rope — both commonly available — you may have difficulty finding poles for use as spreaders. For Field Day, I would suggest using bamboo poles. I found mine by doing a Google search using my city and the word "bamboo" and found a local supplier. You might find that talking with a local carpet distributor results in some free bamboo poles that they use at the centers of their carpet rolls.

Bamboo is great for Field Day because it's lightweight, strong and inexpensive, but it won't last many winters in a permanent installation. For year-round applications 1½ inch diameter fiberglass poles come in 20 foot lengths and generally sell for around $3 per foot. These poles can be found through major industrial distributors in major cities or mail order (MFJ or Max Gain Systems).

Most installations use a two point support. I found that it was easier and more stable to use a three point technique as shown in Figure 3. I suggest using a spacing between support trees of more than 110 feet. Finding two trees with the right heading and spacing may be difficult. The three point technique provides a degree of directional adjustment through varying the length of the ropes.

Although this antenna is a behemoth, as a test I was able to raise it by myself on my own property before Field Day. With three people, it went up smoothly on Field Day. See Figure 4. First pull up one end into its tree, while a second person holds tension on the lower pole. Then pull up the lower pole.

The Payoff

At the AE6C Field Day in northern California, we were able to work East Coast stations with ease, while sidelobes were sufficient to also permit working West Coast stations. In 40 years of amateur operation, this was the easiest and best low band Field Day antenna I have ever used!

James Michener, K9JM (ex-WB2SMD), has been a ham for 42 years. He has a BSEE, MSEE and MBA, with over 30 years of experience designing RF and video equipment for the television broadcast industry. K9JM is an ARRL Pacific Division Assistant Director and serves as net manager for the Northern California Net. He can be reached at 16851 Aileen Way, Grass Valley, CA 95949-7338 or at **k9jm@sbcglobal.net**.

A Super Duper Five Band Portable Antenna

Make camping and travel more fun with this easy to duplicate antenna.

Clarke Cooper, K8BP

One thing that makes Amateur Radio so wonderful is whether I am camping in the rugged Grand Canyon area or somewhere in the Great Northern Woods, I have the capability to work the world while operating low power (QRP).

As one who loves building and experimenting with antennas and operating QRP, it seems like every time I venture out into the backcountry I bring a new homebrew antenna to try out. Then in March 2006 while I was looking through an issue of *QST*, I noticed an MFJ advertisement for their 10 and 12 foot telescoping whip antennas. That's when my wheels started churning. After a few minutes of doing some calculations and sketches on paper, I came up with this rugged lightweight antenna.

After placing the order with MFJ for two of the MFJ-1954 10 foot whips, I ventured to my local Home Depot and purchased the rest of the parts. Surprisingly, this little project only took me about an hour to build. The fun really started several days later when I received the telescoping whips. The initial tune-ups with my MFJ antenna analyzer and Elecraft K2 indicated that this antenna loads up beautifully on 20, 17, 15, 12 and 10 meters without the need for a tuner. The best news is that it can be built for around $60.

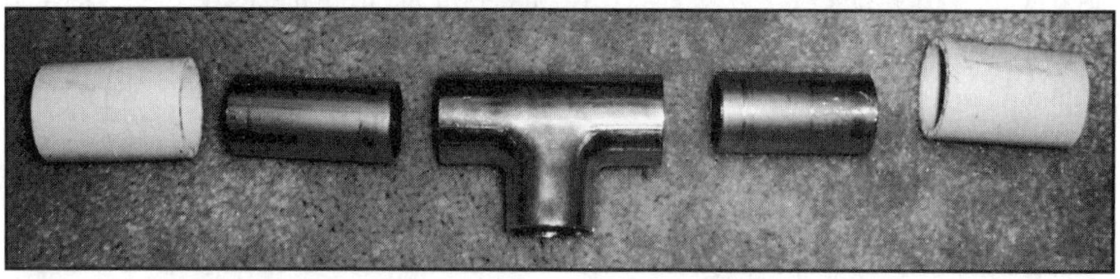

Figure 1 — Details of PVC and copper pipe pieces for antenna center.

Travel Friendly

When broken down for transport, this antenna, minus the mast, consists of two short 2 foot 7 inch long elements and two lightweight MFJ whip sections retracted to a length of 2 feet.

To operate on the 20 and 17 meter bands, loading coils are used to resonate the system, with precise tuning available using the telescoping whips. In order to operate on the 15, 12 or 10 meter bands, the loading coil is bypassed and with a short jumper across the coil terminals the antenna operates as a full sized unloaded half-wave ($\lambda/2$) dipole.

How Does It Perform?

Three days after completing my new antenna, I drove north from my winter residence in Phoenix to Grand Canyon National Park for a couple of days of camping. With this antenna 10 feet in the air and my Elecraft K2 at 5 W output, I had no trouble making SSB and CW contacts to all parts of the US as well as some long haul DX contacts. The DX contacts on that trip included Asiatic Russia, Australia, Belize, Lithuania, Italy, Japan and Hawaii on 20, 17 and 15 meters. My location was surrounded by high mountains to the north, east and west. I was quite pleased with my latest antenna's performance. Even though I found no activity at all on the 12 and 10 meter bands I still feel very confident that when they open up, this antenna will be a winner on these bands as well. I just can't wait until I start using this antenna on Pactor to get my e-mail.

Making it Happen

The parts required are listed in Table 1. Cut the copper and PVC pipe per Table 1 and Figure 1. Remove burrs from all cut edges using emery cloth. Assemble all the parts on a clean surface. At a small sacrifice in weight, a ⅝ inch by 2 foot dowel can be inserted into the copper T to provide additional strength. This is recommended for extended use.

Push two of the copper bushings into the center T and insert the two CPVC couplings into the copper bushings as far as possible.

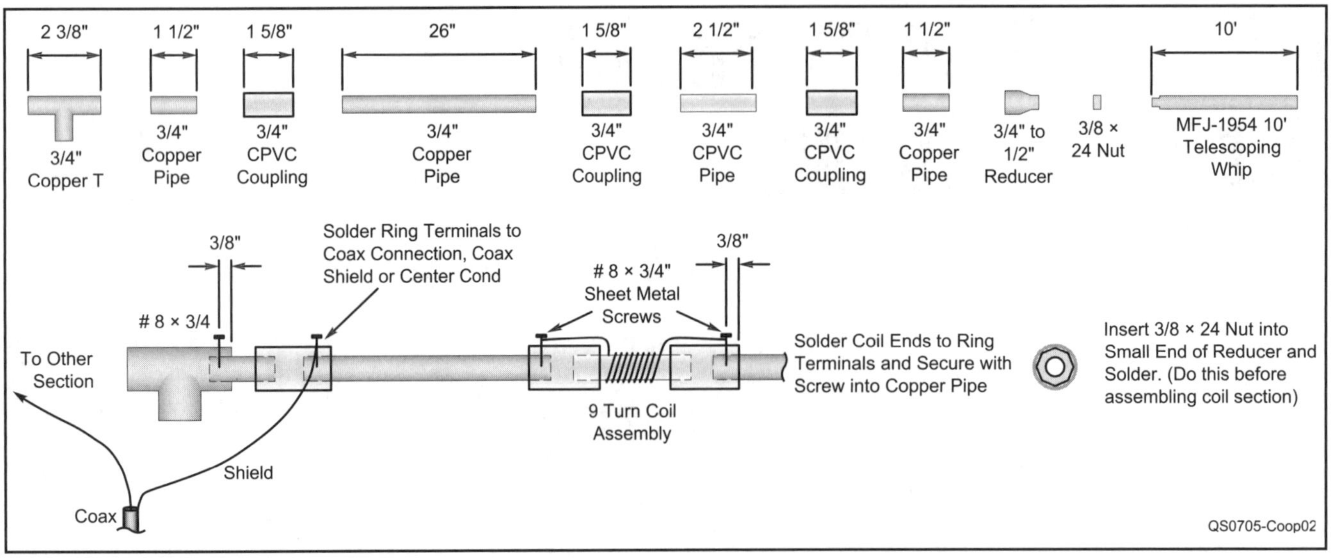

Figure 2 — Antenna assembly drawing.

Table 1
Parts Required to Assemble the Portable Dipole

Quantity one unless noted.

Alligator clips (4)
Copper pipe, ¾″ × 26 (2)
Copper pipe bushings (made from ¾″ pipe), ¾″ × 1½″ (4)
Copper pipe reducer, ¾″ × ½″ (2)
Copper T, ¾″
CPVC cement
CPVC gold pipe, ¾″ × 2½″
Fine thread stainless nuts, ⅜″ × 24 (2)
FSC hardwood dowel, ⅝″ × 4′
Insulated solid 20 gauge wire, 10′
MFJ -1954 (10 foot) telescoping whip (2)
Sheet metal screws, #8 × ¾ (package)
SXC CPVC couplings, ¾″
Ring terminals, 20 gauge size (6)

Table 2
Tools Required

Carpenter's saw	Hacksaw
Tubing cutter	Measuring tape
Electric drill and bits	Propane torch
Emery cloth	Soldering iron
Fine file	

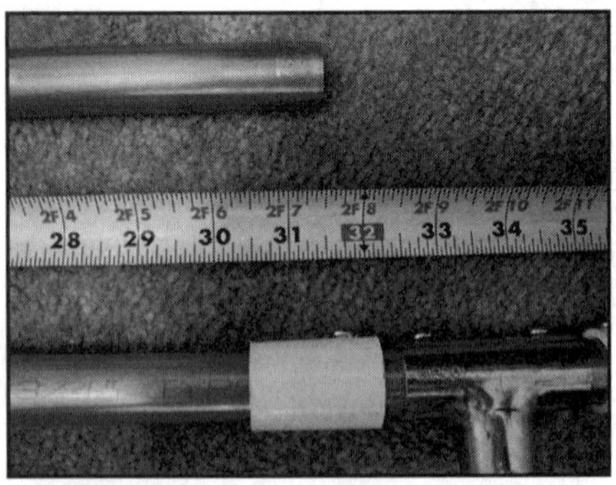

Figure 3 — Detailed view of inner element construction.

Drill four 3/32 inch holes on top of the center insulated T assembly and secure with four #8 × ¾ inch screws.

Next, insert both ¾ × 26 inch copper pipe elements into the CPVC couplings and drill four 3/32 inch holes. Then secure the four #8 × ¾ inch screws as shown in Figures 2 and 3.

Assemble the loading coil forms by lightly gluing two CPVC couplings to each end of a ¾ × 2½ inch CPVC pipe piece using CPVC cement. Make sure you apply glue only on the coupling half that will be used with the ¾ × 2½ inch CPVC pipe, as the other end of the couplings will be fastened later by #8 sheet metal screws. After assembly, wipe any excess glue from the outside of the form and from the inside of the free end of both of the coil CPVC couplings.

Drill two vertical holes in both coil assemblies, all the way through each end of the CPVC coil form assembly. Insert a length of 20 gauge wire all the way through both vertical holes of the coil form. On the short end, leave enough wire to connect to a ring terminal later. Carefully wind nine turns on the form and feed the remaining end through both holes. Adjust the coil windings by twisting the coil and pressing on the coils to remove any slack. When you are satisfied that the coil windings are tight, leave enough wire to connect to a ring terminal later. The completed coil assemblies are shown in Figure 4.

Insert the remaining ¾ × 1½ inch copper bushings into a ¾ × 1½ inch copper pipe reducer and carefully solder them together. Handle with extreme care until these items have cooled down.

Insert both ¾ × 24 nuts into the ½ inch ends of the pipe reducer until they are flush with the ends of the reducer. Carefully solder the nuts to the copper reducer. After they have cooled down, insert the ¾ inch ends of the reducers into the 20 meter coil form. Confirm that both coil windings run in the same direction. Then drill a hole in each outer section of the CPVC bushing, keeping both holes symmetrical.

Measure the coil wire tag ends and carefully strip off just enough insulation to make a loop that feeds into the terminals. Confirm that the center of the terminal is aligned with the mounting hole. Crimp and solder the wire connections.

Fasten each coil form to its respective pipe reducer and secure with a #8 × ¾ inch sheet metal screw. Slide the remaining end of each loading coil housing onto the end of each 2½ foot dipole element. Rotate the coil terminals so they are on top. Secure the coil assembly to each dipole element with a #8 × ¾ inch sheet metal screw. Attach the MFJ -1954 telescoping whips by screwing them into the end of each coil assembly as shown in Figure 6.

Attach a length of coax by separating the center conductor and outer shield and soldering a ring terminal on each. Secure each terminal to the outer side screws on the T assembly. Wind a 6-turn 7 inch diameter coil on the antenna end of the coax and secure with tape or wire straps. Connect the coax to the antenna by loosening up the side screws on each active element insulator. These screws are the ones on the outer edge of the CPVC coupler next to the element. Slide the coax terminals under each of these screws and retighten the screws. It doesn't matter which side the coax shield or center conductor is attached to. After these connections have been made, take an ohmmeter and hold one meter lead on the center T and with the other lead touch both elements. Both sides should indicate an open circuit.

If you wish, instead of the coiled coax

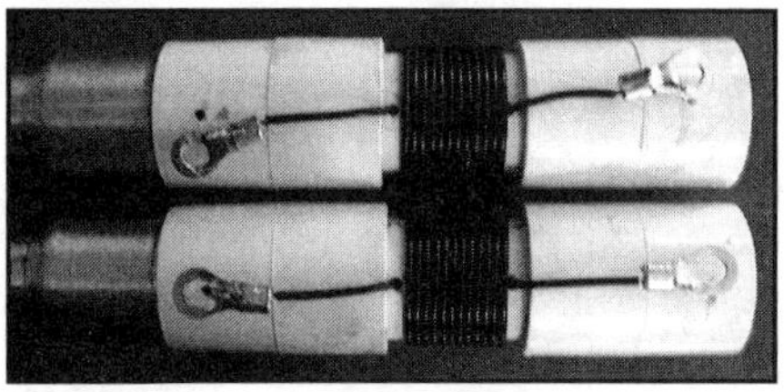

Figure 4 — Completed loading coil assemblies.

Table 3
Antenna Tuning Chart Settings and Resulting SWR

Band	Tune Frequency (MHz)	Max SWR	Coil	Whip Length
20 Meters	14.2	1.4:1	Yes	118″
17 Meters	18.11	1.0:1	Yes	78½″
15 Meters	21.15	1.1:1	No	99″
12 Meters	24.94	1.2:1	No	78″
10 Meters	28.8	1.2:1	No	64″

Figure 5 — One side of the antenna with retracted whip attached.

choke, you can use a 1:1 balun such as the one in Figure 6 that I made with a T200-2 toroid, a couple of feet of enameled wire and some odds and ends I had in my junk box.

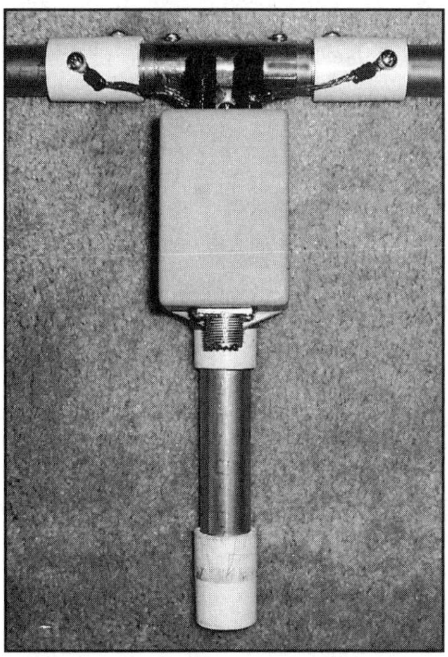

Figure 6 — Dipole feed connection — toroid balun option shown.

Tuning it Up

Table 3 provides the data required to adjust the antenna for all five bands. You will likely want to adjust the length to whip at each noted frequency and record the length that provides a 1:1 SWR, since your construction might be different than mine. With the antenna adjusted to the frequency shown it should provide an acceptable SWR across each band, no greater than the *Max SWR* shown in the table.

If you were going to operate most of the time close to 14,000 kHz on CW and the band edge SWR of 1:4 is excessive for your radio, it can be corrected very easily by slightly lengthening the whip length to obtain a 1:1 SWR in the CW portion. On the other hand if you want to adjust for a 1:1 SWR at around 14,350 you can make the same type of a slight adjustment, this time by shortening the MFJ whips. By using this same technique on any band there is no need to bring along any bulky heavy antenna tuner on your camping adventure.

Hoisting it Up

There are many ways of installing such an antenna in a portable environment. There are tripod mounts, improvised railing supports and many types of telescoping support poles available from a number of manufacturers. I have had good results using a Model S216 telescoping fiberglass pole manufactured by Hastings. These are available at electrical supply houses. There are other similar types sold by a number of Amateur Radio dealers.

In my usual installation, I have secured a short piece of CPVC tubing on the top telescoping section. On the other end of this short piece, I insert it into a CPVC coupling that is part of the T section on my antenna. I have also used short pieces of aluminum tubing fastened together made from old antenna beam elements. Even short pieces of inexpensive CPVC pipe will do for a 4 to 5 foot mast.

For the base, I often use a folding portable flood light base I picked up at a yard sale. If packing in by foot, I leave the base at home and use available supports such as tree stumps to hold up the mast by securing with a piece of small rope. With this method, one can set up this antenna within 5 minutes.

One final bit of advice — please be careful when setting up your antennas wherever you are. Before erecting any type antenna, always make it a habit to look for any electrical wires above and 360° around you.

Clarke Cooper, K8BP, has been a ham since 1961 and holds an Amateur Extra class license. Clarke has a degree in electrical engineering and has spent most of his career designing manufacturing plants and automated control systems. He retired in 2001 as Project Manager for Federal Mogul.

He is a Life Member and a four term president of the Muskegon Area Amateur Radio Council in Muskegon, Michigan. He is also a member of the Thunderbird Amateur Radio Club in Phoenix, Arizona. He especially enjoys mentoring new hams and operating QRP on both CW and SSB.

Over the years, Clarke has had the pleasure of sending out over 30,000 QSL cards. He is a member of ARRL, QCWA, ARES, RACES, Fists and Adventure Radio Society. During the summer you can reach Clarke at 4202 E Pontaluna Rd, Fruitport, MI and in the winter at 17237 North 15th Pl, Phoenix, AZ 85022 or by e-mail at **k8bp@earthlink.net**.

The Octopus — Four Band HF Antenna for Portable Use

An easy to make, easy to assemble and easy to pack travel antenna.

Geoff Haines, N1GY

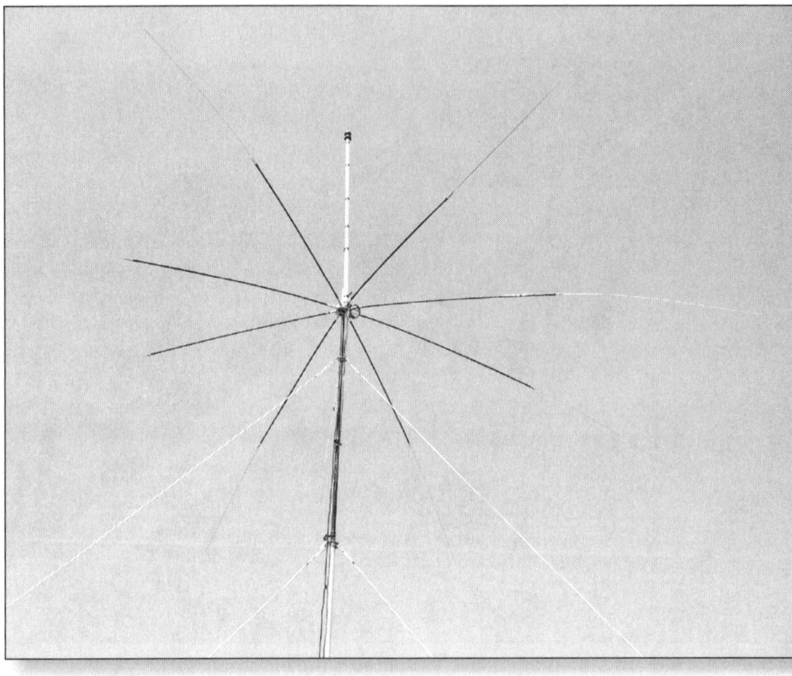

My participation in the world of Amateur Radio has led me in many directions. From DX to disaster recovery, I have used my meager skills to do many things. For me, the most enjoyable of these is the building of the simple gadgets and accessories that make my station work better.

One for the Road

I have a frequent need for an HF antenna to support portable operation on multiple bands. I recently thought of a simple way to provide one based on a *multiband fan dipole* in combination with a *helical mobile dipole*. Now those of you with 70 foot towers with multi-band Yagis on top who never leave their linear behind can leave the room. This idea is for those who like to go somewhere such as a beach or park where it's quiet and set up a modest station for an afternoon of radio fun. It also works for those of us who are Amateur Radio Emergency Service (ARES) members and have to set up a working HF station in the middle of a parking lot after a natural disaster such as a hurricane.

The Octopus is Born

Using the concept of a horizontal fan dipole, I drew up a design for a helical type dipole taken to extremes. The basic plan was to find an attachment mechanism that could be mounted on a mast and accept up to four helical dipoles, eight helical type antennas, radially around the mount in the horizontal plane.

Figure 1 — The hub of the array is a standard octagonal electrical box available at your local home improvement store. Stud mounts available from your local ham radio or CB shop are used to mount the mobile antenna pairs.

Finding the Pieces

Since I have often found the perfect parts for my projects at the local home improvement chain stores, I looked there again. Almost as soon as I got to the electrical department, my search was over. An eight sided electrical box, of the type used to mount overhead lighting, looked ideally suited for the task. See Figure 1. It had eight sides, a top and a bottom and would be relatively easy to modify. The punch-out slugs on the side were larger than I wanted for mounting the helicals, but could be adapted with fender washers. I drilled these out to ½ inch internal clearance.

The punch-out slugs in the top and bottom of the box are the right size "out of the box" to allow the entire mount to be slipped over the top of a ¾ inch diameter top section of telescoping mast. I used an angle bracket bolted to the bottom of the box to allow me to clamp it to the mast.

The next problem was how to mount the antennas. Checking the hardware aisle I found ⅜ × 24 bolts in the fasteners section. The long nuts for the other side of the mount were harder to find, but by going to truck stops (where they usually have a CB parts wall, visiting RadioShack and raiding my parts box, I eventually came up with eight double female stud mounts to attach to the central hub. The assembly details are shown in Figure 2.

Making it Happen

Four of these stud mounts were mounted directly to the hub, and four were mounted with insulating washers. The four insulated mounts were wired to a single SO-239 type connector mounted to the bottom of the hub box. Each insulated mount has a grounded mount 180° away from it. The connections from the insulated mounts to the SO-239 were made with ring terminals of appropriate size soldered to a wire harness that is connected to the SO-239, as shown in Figure 3.

Since this antenna system is intended for short term, temporary deployment, weatherproofing is not particularly significant for me. Should anyone want to use this idea for a more permanent installation, weatherproof-

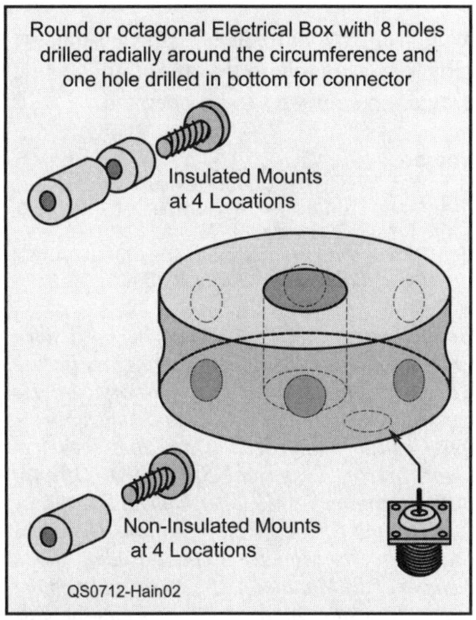

Figure 2 — A drawing of the electrical box showing the holes drilled for the stud mounts, four of which are insulated from the box and four of which are grounded to it.

Figure 4 — Here is a view of the octopus all set up for park-side DX or disaster recovery operations. The SUV, sitting on the mast base, keeps the whole array erect. The guy lines in the picture would be used only if the vehicle was needed elsewhere.

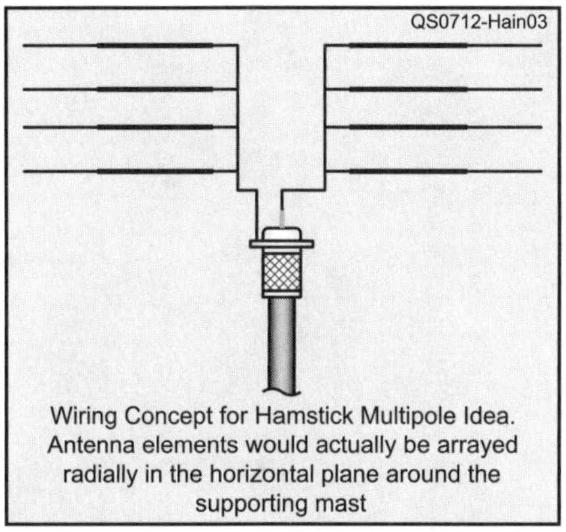

Figure 3 — This shows the schematic diagram of the wiring of the antenna elements.

ing the hub will require at least several coats of rust resistant paint and the sealing of any spots where water could gain entry.

The next step was to install four pairs of helical type mobile antennas cut for four different bands on the mount. With each pair, one is connected to an insulated mount and the other is installed in the grounded mount opposite to the first. I chose to use Valor Pro-Am helical whip antennas because the 3/8 × 24 connection of the whip to the helically wound lower shaft of the antenna makes them easier to store and assemble than the set-screw arrangement of the Lakeside Hamstick antennas. This is not to say one is better than the other. I have purchased both makes before and indeed use the Lakeside antennas on my mobile rig where storage is not an issue. The antenna, all ready to go, is shown in Figure 4.

How it Works

Since we now have four dipoles all connected to the same coax, a little explanation is in order. When the radio transmits RF up the coax to the antenna array, it sees all of the antennas as a high impedance load except the dipole that is resonant on the band being transmitted. Thus, almost all of the RF energy is directed automatically to the proper dipole. This is the way a multiband fan dipole works, and what we have here is such an antenna turned sideways from the usual configuration. See Figure 5.

Tuning it Up

There are two ways to make sure that the transceiver sees a 50 Ω load. One is to mount and raise one pair of antennas at a time, and tune each pair to present a 50 Ω load. It's very time-consuming, but it does work. Each pair will have to be retuned to some extent because of slight interactions between the pairs of antennas, but eventually all can be made to work. The second way is to get them approximately on frequency and use a manual or automatic antenna coupler or tuner such as my LDG Z-100.[1] This was my personal choice.

Once the antenna array goes up, I don't want to have to take it down until I am ready to go home. I chose to adjust each dipole to resonance at the middle of each band. To expand the bandwidth to the full scope of the available frequencies I rely on the automatic tuner, which also appears to handle the small mismatches caused by any interaction between sets of dipoles. This interaction is less than with a traditional fan dipole because the 40 and 75 meter antennas are at right angles to each other. The 20 and 15 meter antenna sets are also at right angles to each other and at a 45° angle to the lower band sets.

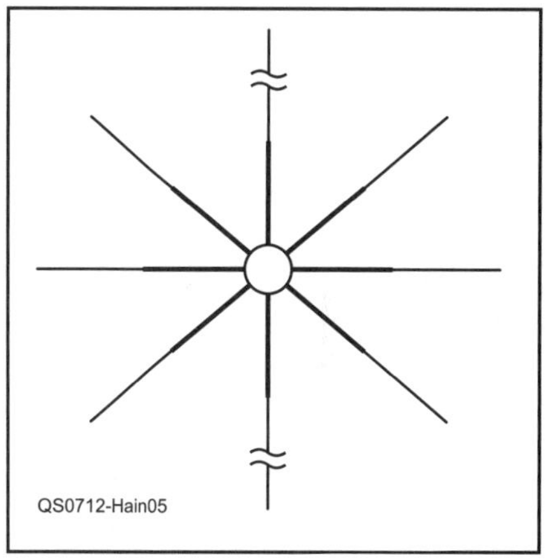

Figure 5 — Top view of the array.

Raising the Beast

The octopus array is mounted on a telescopic mast home-brewed of aluminum tubing available from several sources. The telescopic feature allows the array to be positioned from a low position for near vertical incidence skywave (NVIS) type operations or extended up to 26 feet for lower angle operation on the higher bands.[2] Any height

[1]Notes appear on page 32.

up to λ/4 will support NVIS operation, so the full height works well on 80 and 40 meters; the prime NVIS bands but lower heights can also be effectively used, if conditions warrant. I think that the low position should be above 8 or 9 feet off the ground for safety's sake. No one likes to run into the sharp end of an antenna.

Don't Expect Miracles — but Good Results for its Size

Any short dipole array, including the octopus, is a compromise. If time were no object and funds unlimited, there are antenna systems that work better. However, if your antenna system has to ride around in the trunk of your car waiting for the next ARES callout, or you just have one afternoon to devote to catching some fun DX at the beach, this antenna will do the job well enough. A decent enough antenna is better than no antenna at all.

To store the array, just remove the helicals and put them back into whatever bag they are normally stored in. The mast is telescoped to its minimum length and stored the same way. You can leave the hub on the mast or store it separately — your choice. The initial setup only takes about 15 to 20 minutes and once done, you are finished with antenna work until it's time to go home. There are no radials to lay out and trip over. When disassembled, it all fits in the trunk of your car.

But Wait — There's More!

This antenna array also allows the use of a VHF or UHF antenna mounted above the hub on the very tip of the mast as shown in the lead photo. Therefore, with just one mast you can provide communications over a broad range of amateur bands and ranges with a minimum of effort. Sounds like the Amateur Radio we all know and love.

How's it Play?

I raised the octopus about 11 feet above ground for my initial testing. The radio and autotuner combination had no trouble getting a good match on any of the four bands, 75, 40, 20 and 15 meters. During an all-day display of Amateur Radio at a local American Red Cross center, I set up the mast and antenna to its full height of 26 feet. I had no trouble at all checking into the Maritime Mobile Service Net on 20 meters and similar success in making contacts on all of the bands except 75 meters, which is usually pretty dead here during the daytime. I was able to get a good match on all four bands, so I have no doubt that it will work on 75 meters when the band is open. To take advantage of the directional effects of the dipoles, particularly on the higher bands, the antenna can be manually rotated as shown in Figure 6.

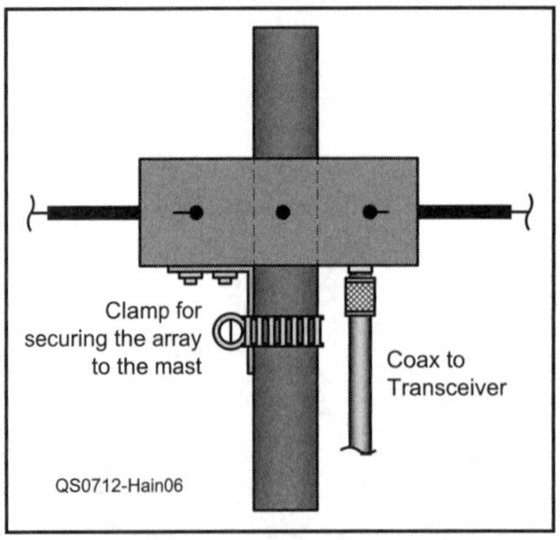

Figure 6 — This diagram shows clearly how the hub is secured to the mast to allow "armstrong" rotation of the array, if needed.

One cautionary note: This antenna mast should never be tilted up with the antenna on board. It should always be extended upward with the mast already vertical. I discovered this when I attempted the tilt-up method, with disastrous results. The mast bent under the weight of the antenna and had to be repaired. As long as it is extended vertically, and then guyed, there should be no problem in any reasonable wind.

If you decide to try this approach, you can substitute your favorite band or bands of choice. You could build a one, two or three band version just as easily. In terms of overall performance, the array appears to work just as well as any single band helical type dipole. Note that any helical dipole will be more efficient at 20 meters and higher than on the lower bands, but will work as well as other antennas their size. The main advantage of the multiband configuration is that when you want to change bands you just switch the radio to the new band and give the TUNE button a jab. Two seconds later you are ready. No going out to the mast, lowering it down to 6 feet and removing one set of antennas to substitute another pair. All of that could easily soak up 15 or 20 minutes, and you still have to raise the mast back up to the proper height.

Time to Hit the Road

The octopus resides in the back of my SUV, along with the telescopic mast and its base, also home-brewed. When deployed, the base sits under one of the tires on the SUV and keeps the mast vertical until the guy ropes are deployed. Then I can drive the car away and set up the rest of the station in a more convenient spot. Total time from the initial setup of the mast and antenna to being on the air ready to catch some DX or assist after a disaster is about 30 minutes. It can be less if I use the radio mounted in the car instead of a separate station set up as in the "Radio in a Box" pictured in *Up Front* in the October 2005 issue of *QST*. On the other hand, using the radio in a box means that the vehicle is available to do other things while the station continues to be operated.

Notes

[1]J. Hallas, W1ZR, "Product Review — Automatic Antenna Tuners — A Sample of the Field," *QST*, May 2004, pp 71-76.

[2]R. Straw, N6BV, "What's the Deal About 'NVIS'?" *QST*, Dec 2005, pp 38-43.

Geoff Haines, N1GY, has been licensed since 1992 and holds an Amateur Extra class license. He retired after a career in respiratory care. He currently holds several ARRL appointments in the West Central Florida Section, including Technical Coordinator, Technical Specialist, Official Bulletin Station, Net Manager, Official Emergency Station and Official Relay Station. He is the President of the Manatee Amateur Radio Club, a member of the Manatee ARES group and member of the Bradenton Amateur Radio Club, the Yale University Amateur Radio Club and the Meriden (CT) Amateur Radio Club. In his spare time, he enjoys homebrewing antennas and accessories for his Amateur Radio operations. Geoff can be reached at 708 52nd Av Ln W, Bradenton, FL 34207; **n1gy@arrl.net**.

Portable Two Element 15 Meter Yagi

Jack B. Morgan, KF6T

ARRL Field Day 2007 was on the horizon and our local club (MLDXCC) did not have an antenna lined up for 15 meters. We were worried that the band might open up during the event and important opportunities for contacts would be missed. It happened in 2006, and only a few contacts were made using our 40 meter antenna. So we needed a better solution.

Some of the requirements established were that this antenna be compact, light weight, easily assembled, no fancy matching required and it needed to cover the entire band.

Description

I selected a two element design to meet these requirements. I have had experience with this approach in the past and have built similar antennas from 40 to 6 meters with good results.[1] With a small sacrifice in gain and front to back ratio (F/B), you can always achieve a direct 50 Ω match with two elements on a short boom. I used *EZNEC* to confirm the element dimensions and spacing.[2] The elements are isolated from their supports so the clamps have little effect on dimensions or performance.

[1]Notes appear on page 35.

I was so confident that the computer generated dimensions were correct that I didn't test the antenna until we raised it on Field Day weekend. I was relieved that the VSWR was exactly as predicted (see Figure 1). The new antenna played very well during the 15 meter openings. The main lobe of propagation is at 30° elevation at 20 feet above flat land (see Figure 2) and even lower at our cliff-like location. This provides good propagation over the intermediate distances important to Field Day.

Construction

The most unusual aspect of this antenna is that it uses a boom with a square cross section. I used an 8 foot piece of 1.25 × 1.25 inch 6061-T6 aluminum with ⅛ inch wall thickness. This boom is very strong and offers some interesting advantages. First of all, no U bolts are required except for

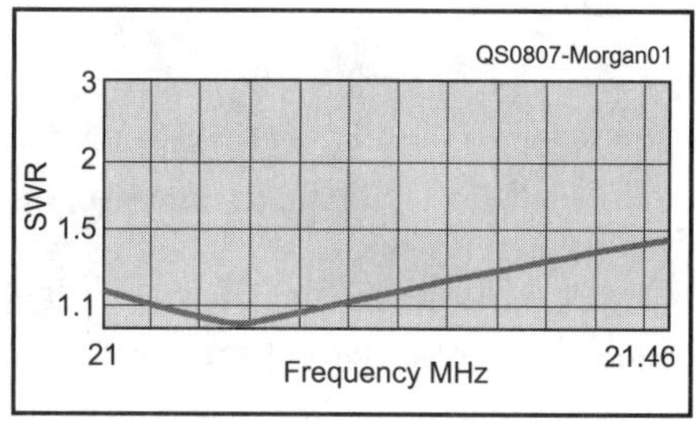

Figure 1 — SWR curve of the 15 meter two element Yagi antenna.

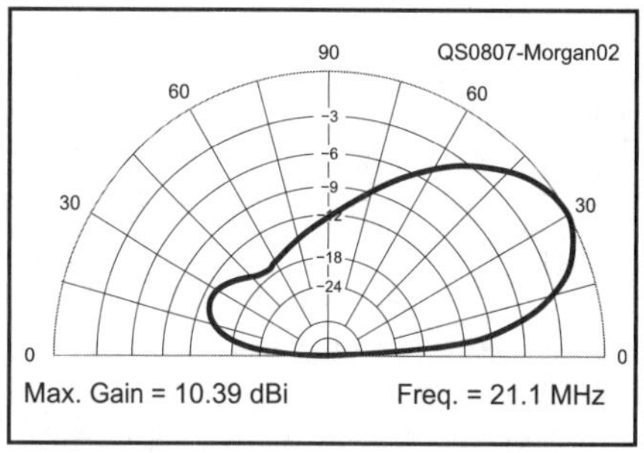

Figure 2 — Elevation plot of the 15 meter antenna at 20 feet, in the peak of the main beam.

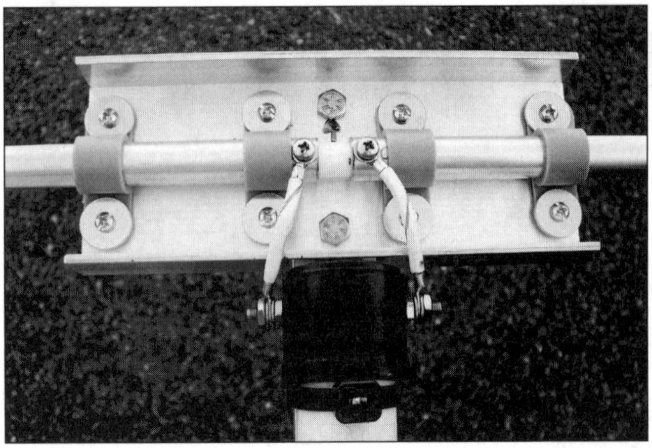

Figure 3 — Driven element assembly.

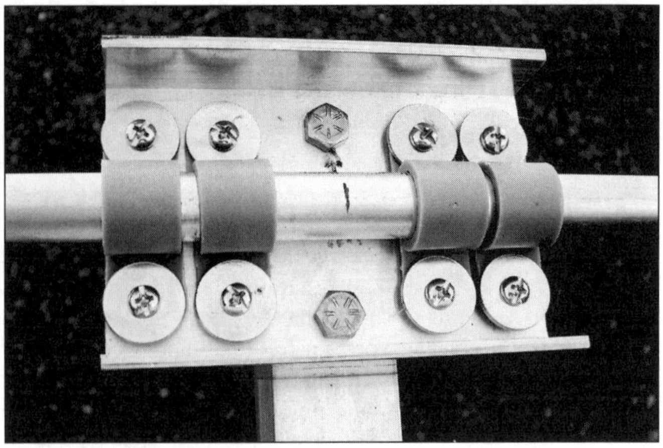

Figure 4 — Ref ector assembly.

Figure 5 — Oatey 33916 ½ inch EMT clamp used to secure elements.

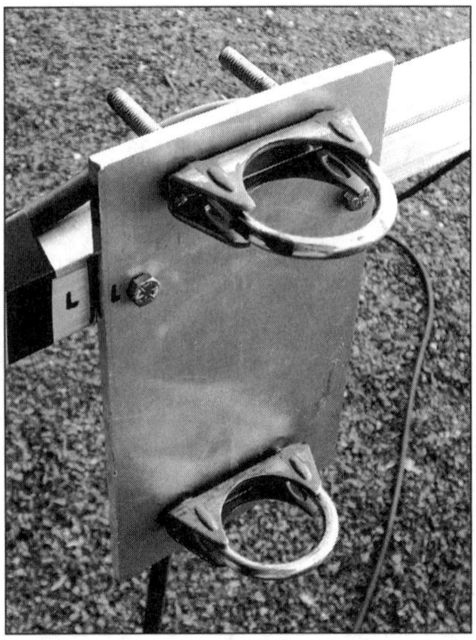

Figure 6 — Plate installed on square boom.

mast attachment. Everything else just bolts together. Secondly, because the mounting surfaces are square, element to mast alignment is guaranteed! We assembled this antenna in ten minutes at the Field Day site. All components are 8 feet or less, so this antenna is easy to transport or store.

The two elements are made from ⅝ (0.625) and ½ (0.500) inch 6061-T6 aluminum tubing with 0.058 inch wall thickness. These tubes will usually telescope together but sometimes need a little adjustment with fine sandpaper. If the fit is too tight, you can get into a situation in which the elements will gall and disassembly will be difficult. If you use any sort of lubricant, apply it just before assembly to keep it from acting like fly paper to dirt or sand.

The elements are connected together at a 2 inch overlap point using a single 8-32 stainless steel screw and Nyloc lock nut.

The two element clamps are fabricated from a one foot section of 3 × 1 inch aluminum U channel with 0.125 inch wall thickness. The channel is cut, then drilled and tapped as shown in a detailed drawing on the *QST* binaries Web site.[3] Note that the boom attachment holes are initially made smaller as pilot holes. At final assembly the finish drilling is accomplished.

I used three gray polypropylene conduit clamps, as shown in Figures 3, 4 and 5 to support the elements on the boom.[4] These clamps are intended to be used to mount ½ inch electrical conduit to a wall. They work very well in this application. Fender washers (0.75 inch diameter) are used under the 8-32 screws to better support the poly clamps. The U channel plates are tapped to accept these screws and are secured on the far side with stainless steel lock nuts.

The element dimensions are shown in Table 1. I have determined the dimensions for adapting the design to 12, 10 and 6 meters and these are presented in Tables 2 through 4. This dimensions have been verified through modeling, but these versions have not been made and measured.

The driven element has a 0.500 inch wide gap that is filled with a ½ inch diameter insulating rod made of fiberglass, Teflon or Delrin. The 2.5 inch long insulator is inserted 1 inch into each split elements. I drilled and tapped (8-32) holes through the tubing wall and into the center of the insulator. By not drilling all the way through, we prevent an oversized screw from shorting on the far side at the point of balun attachment later. It is a lot easier to drill these holes with the elements and insulator mounted on the support clamp beforehand.

Each element clamp is mounted to the boom using two ¼-28 (UNF fine thread) stainless bolts. I clamped the parts together, got them square, then drilled (0.250 diameter) through the clamp pilot holes continuing through the boom in two places. Mark the mating parts with a permanent marker pen so they will exactly match up at a later time.

A current balun is used to feed the driven element. This keeps the coax from radiating unwanted common mode signals, an important consideration in a Field Day environment. It is mounted as close as possible to the driven element as shown in Figure 3 and held in place with tie wraps. The reflector assembly is shown in Figure 4.

The mast to boom plate is fabricated from a scrap piece of 0.25 inch thick aluminum plate. The suggested mounting dimensions are shown on the binaries Web site. The U bolt hole dimensions will depend on the actual mast diameter selected. This design used 2 inch U bolts and is shown mounted on the square boom in Figure 7. I located this plate at the center of balance of the antenna, with the balun mounted.

Summary

This antenna was mounted on a windy ridge top 7000 feet up in the Sierra Nevadas during the 2007 ARRL Field Day. The small diameter elements shrugged off the wind and displayed an acceptable amount of droop. Analysis using Dave Leeson's spreadsheet program (element strength calculator by W6QHS/W6NL) shows that these elements should survive 72 mile an hour winds.[5] A heavier duty version of this antenna could be made starting with larger diameter aluminum and heftier clamps, but this design fit our needs perfectly.

Table 1
15 Meter Antenna Dimensions

Item	*Overall Length (inches)*	*Cut Length (inches)*	*Notes*
Driven Element	259¼		Check tip to tip dimensions.
2 × ⅝" inner sections	47¾	47¾	Cut from one 8' section.
2 × ½" element tips	81⅝	83⅝	Allows 2" overlap.
Reflector	280¾		Check tip to tip dimensions.
1 × ⅝" center section	96	Not cut	Check stock is 96" long.
2 × ½" element tips	92⅜	94⅜	Allows 2" overlap.
Boom	96	Not cut	Use 8' section of 1.25 × 1.25".
Driven element to reflector spacing	93		Clamps flush at boom ends.

Table 2
12 Meter Antenna Dimensions

Item	*Overall Length (inches)*	*Cut Length (inches)*	*Notes*
Driven Element	220		Check tip to tip dimensions.
2 × ⅝" inner sections	47¾	47¾	Cut from one 8' section
2 × ½" element tips	62	64	Allows 2" overlap.
Reflector	238		Check tip to tip dimensions.
1 × ⅝" center section	96	Not cut	Check stock is 96" long
2 × ½" element tips	71	73	Allows 2" overlap.
Boom	83		Use 8' section of 1.25 × 1.25".
Driven element to reflector spacing	80		Clamps flush at boom ends.

Table 3
10 Meter Antenna Dimensions

Item	*Overall Length (inches)*	*Cut Length (inches)*	*Notes*
Driven Element	194½		Check tip to tip dimensions.
2 × ⅝" inner sections	47¾	47¾	Cut from one 8' section.
2 × ½" element tips	49¼	51¼	Allows 2" overlap.
Reflector	206¼		Check tip to tip dimensions.
1 × ⅝" center section	96	Not cut	Check stock is 96" long.
2 × ½" element tips	55⅛	57⅛	Allows 2" overlap.
Boom	79¾		Use 8' section of 1.25 × 1.25".
Driven element to reflector spacing	76¾		Clamps flush at boom ends.

Table 4
6 Meter Antenna Dimensions

Item	*Overall Length (inches)*	*Cut Length (inches)*	*Notes*
Driven Element	105½		Check tip to tip dimensions.
2 × ⅝" inner sections	47¾	47¾	Cut from one 8' section.
2 × ½" element tips	4¾	6¾	Allows 2" overlap.
Reflector	117		Check tip to tip dimensions.
1 × ⅝" center section	96	Not cut	Check stock is 96 inches long.
2 × ½" element tips	10½	12½	Allows 2" overlap.
Boom	43		Use 4' section of 1.25 × 1.25".
Driven element to reflector spacing	40		Clamps flush at boom ends.

Notes

[1]J. Morgan, W1FEA, "The Building-Supply Yagi," *QST*, Mar 1991, pp 22-24.

[2]Several versions of *EZNEC* antenna modeling software are available from developer Roy Lewallen, W7EL, at **www.eznec.com**.

[3]**www.arrl.org/f les/qst-binaries**.

[4]Oatey 33940 ½ inch EMT clamp.

[5]**www.realhamradio.com/Download.htm**. Unzip and use the "element" file.

Jack Morgan, KF6T, was first licensed in 1955 as W1FEA and holds an Amateur Extra class license. Jack started working at the ARRL designing many projects for The ARRL Handbook *while working toward his BS and MSEE degrees. He subsequently worked at Eimac and Varian before starting his own company in Silicon Valley. Retired from that, he started teaching Computer Technology at a local high school. He also pioneered competitive snowboarding as a high school varsity sport in California. He currently coaches a snowboard team and has been a gold medalist at national snowboard competitions. You can reach Jack at 2040 Pheasant Hill Ln, Auburn, CA 95602 or by e-mail at* **jack@racetimesystems.com**.

The Tee Pee V 20 and 40 Meter Antenna

This simple to prepare and erect dual-band HF antenna is great for Field Day or emergency use.

Robert Giuliano, KB8RCO

The Need

Our local radio group enjoys setting up portable stations for special events and public service outings. Typically, we have no problem locating a couple of trees or other antenna supports. On one occasion, we found ourselves in a field with no available supports. Fortunately we had brought a Mosley ground mounted, trapped vertical and we were on the air as usual. This incident led to a discussion about antennas for portable work. We decided to find a simple, self-supporting, but effective antenna that could be put up about anywhere and be carried easily.

The Idea

During a local hamfest, I saw many telescoping fiberglass poles advertised as antenna supports. An Internet search turned up a 22 foot telescoping pole with a ⅝ inch diameter tip that looked sturdy.[1] A couple of e-mails and phone calls (researching and ordering) later, the 22 foot pole was on its way.

I strapped the pole to a support on my deck, hooked a pulley at the top for swapping antennas during experimenting, strung a dipole as an inverted V, trimmed it to 14.1 MHz and gave it a try. Simple, portable and practical — this pole can do the job.

The Tee Pee V concept came from Robert, N8RGF, as he looked up at our inverted V and asked, "Why do we need four guys *and* the dipole wires?" The next comment was something about parallel dipoles with a common feed point and whether that concept would work for two inverted Vs. Can we combine the concepts and get two band operation from one pole using the antenna as the guys?

The Development

To me that sounded like a challenge, and I was determined to check the concept with

[1]Notes appear on page 38.

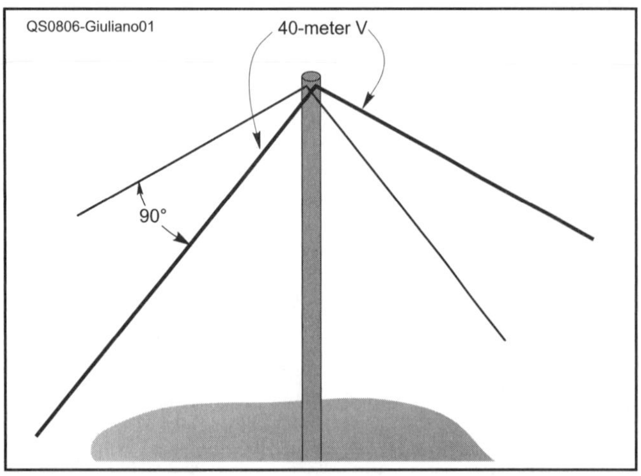

Figure 1 — Isometric view of the antenna configuration.

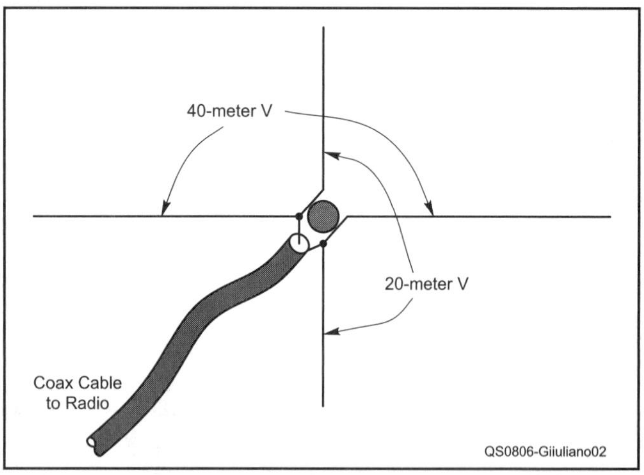

Figure 2 — Details of the feed point connection arrangements.

Table 1

SWR Bandwidth of the Tee Pee V on 40 and 20 Meters

Band (meters)	*Frequency Range (MHz)*	*Maximum SWR*
40	7.0- 7.30	1.9:1
20	14.0-14.35	1.4:1

some antenna modeling. The development process would be:

- Rough model to determine if the concept was feasible.
- Build a prototype based on the model results.
- Verify the prototype for match and hopefully put it on the air.

The Modeling

I remembered reading about *EZNEC* as an antenna modeling application.[2] I hoped that *EZNEC* could be used to verify the concept and provide starting dimensions for the antenna. I had heard good things about it and decided to give it a try.

EZNEC was able to provide all that I needed to know about two antenna aspects of the Tee Pee V. I first determined the modeled SWR bandwidth of the antenna made from 16 gauge wire. The results are shown in Table 1. Because the two antennas are perpendicular to each other, the coupling that often causes problems with multiband parallel wire dipoles is minimized, and *EZNEC* predicts that operation on each band is much as a separate inverted V.

An isometric view of the antenna is shown in Figure 1, while the details of the connections are shown in Figure 2. The origin of the name should be clear from the figure. I didn't use a balun at the feed point. If you experience the effects of common mode currents on the outside of the coax (I had none) you may want to add a balun. A choke as simple as half a dozen turns of coax, about 8 inches in diameter, might be all you need.

Real Ground Model

The predicted performance over typical ground is shown in the azimuth patterns for both bands in Figures 3 and 4. As seen, the patterns are just what would be expected of any relatively low inverted V, largely omnidirectional. The other information I wanted from modeling the Tee Pee V was the approximate wire lengths, and of course, some indication that the antenna would work. The wire lengths were needed to determine how much wire and how much rope would be required. It would also provide the "clear area" needed for antenna erection.

Sizing the Antenna

The antenna model determined the element lengths to be approximately 17 feet for 20 meters and 33 feet for 40 meters. The total wire required would thus be about 100 feet. A little geometry was used to keep the angle with respect to the pole at 56.75°.[3] This allows each guy stake to be the same distance from the pole and makes setup easy.

The distance the stakes need to be from the pole was calculated by multiplying the height by the tangent of the included angle (56.75°) and that came out to be about 33.5 feet. The 56.75° angle of each wire to the pole results in the angle between the antenna wires of 113.5°, nicely larger than the design goal for inverted Vs of being greater than 90° according to most dipole documentation I've read.[3]

The Materials

What is needed for the Tee Pee V? I already had the pole, but that is the first item needed. Next was an antenna to coax connector. I used the good old reliable Budwig HQ-1 dipole center insulator (available at radio dealers) because of its light weight and small size. Their HQ-2 end insulators are a good choice to go between the element ends and the support ropes. Any lightweight connection arrangement will do, even direct connection of the wires to the coax.

Next was wire. For my concept prove-out, I purchased cheap 16 gauge speaker wire (largest diameter where I shopped) made from stranded copper, one side tinned. Last, I added some antenna rope to extend the additional distance to reach the ground and I was ready to build. Again, geometry was used to determine the exact length at 40.1 feet. Subtract the length of wire and the rest is rope. You will need enough extra rope to tie around the stake and any needed antenna wire trimming.

Build the Prototype

I cut the 50 feet of two conductor wire into a 17 foot section and a 33 ft section. Then, I separated the cord into 2 single wires, one for each leg. The next step was to attach the wire to the center insulator such that the legs are arranged with the proper lengths 180° from each other.

In an effort to make an easy connection to the rope, and with enough wire to trim the antenna for a good match, I measured 1.5 feet from the tip of each wire, slit the wire, bunched it out into a loop with part of the wire on each side, and soldered the wire into a ring. The extra 1.5 feet can be trimmed to get the desired match.

Next was a Saturday morning trip to the field with an MFJ analyzer and the makings for the Tee Pee V. With a helper to hold the pole, I trimmed the 20 meter antenna

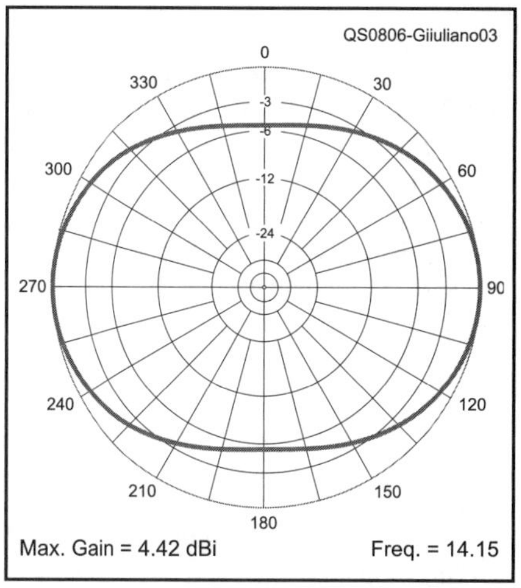

Figure 3 — Azimuth plot of 20 meter Tee Pee V at elevation angle of 35°.

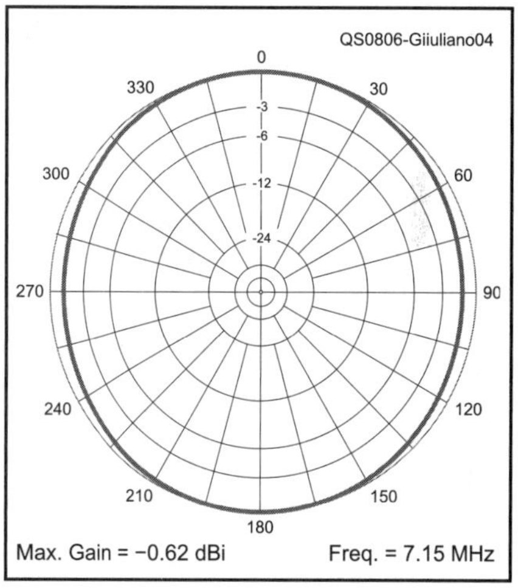

Figure 4 — Azimuth plot of 40 meter Tee Pee V at elevation angle of 35°.

first. I trimmed and trimmed and just before reaching the solder loop, I got the match I was looking for. My guess of 1.5 feet was cutting things too close — only 1.5 inches remained.

Next was the 40 meter section. This time I cut off less than a foot to achieve a good match. Wow, close call, but 50 feet and the 17/33 foot split worked out fine.

On the Air

A couple of checks back and forth between bands with the analyzer and the antenna measured at the predicted flat SWR across both bands. Next I connected my Yaesu FT-100D transceiver to the antenna and gave it a try. I check into a local net on 40 meters and received a good signal report (59) and I confirmed 40 was working.

Now for the 20 meter side. I usually ask the Maritime Mobile Net operator, but the Coast Guard net had just started. I made a quick swing through the band to find a contact and got a 59 report from Georgia.

This made me feel the antenna was usable. No DX, but this is the bottom of the sunspot cycle. The FT-100D without the tuner fed the antenna with 100 W (no roll back in power) and the meter showed low SWR.

The Mosley vertical was also set up, so we checked it against that. Swapping the antennas back and forth, the signals were pretty close. Overall, I'd say the antenna is capable.

Repeatable?

Another Saturday trip to a different spot in the park would determine if everything will stay in tune.

I picked a spot for the pole and stepped off 11 paces (about 33 feet) in each direction where I pounded in a tent stake for each of the support ropes and wires. I tied slip knots around the stakes (adding rope slide stops would make this even easier), and raised the pole. The pole stayed up with the loose guy ropes and I was able to step out to adjust the ropes for a touch more tension to keep everything straight. Wow — this antenna can go up with one person.

Conclusion

The Tee Pee V is a simple dual band antenna that can be setup by a single person and will get you a decent signal on the design bands. Best of all, if you trim it properly, it doesn't need a tuner. I have set up this antenna in many locations now, and it appears to give a good consistent SWR reading in each location.

I hope you build and experiment with the Tee Pee V and I hope it fills a need in your antenna pack. You never know when you will need this type of antenna for field work.

I'd like to thank Robert Laundra, N8RGF, for the inspiration and his direction that resulted in this antenna.

Notes

[1]**www.tmastco.com**.

[2]Several versions of *EZNEC* antenna modeling software are available from developer Roy Lewallen, W7EL, at **www.eznec.com**.

[3]For a discussion and example of using geometry and trigonometry to determine antenna dimensions, see **www.arrl.org/files/qst-binaries**.

Robert Giuliano, KB8RCO, is an engineer for an automotive supplier that tests electro-hydraulic brake systems. He was first licensed in January 1994 as a Technician Plus and currently holds an Amateur Extra class license. Robert is a member of the KC8KGA special event radio group, which was the inspiration behind this project and article. You may contact him at 1069 Arboretum Ct, Saline, MI 48176 or at **kb8rco@yahoo.com**.

Zip Cord Antennas and Feed Lines For Portable Applications

Don't let lack of real transmission line keep you off the air!

William A. Parmley, KR8L

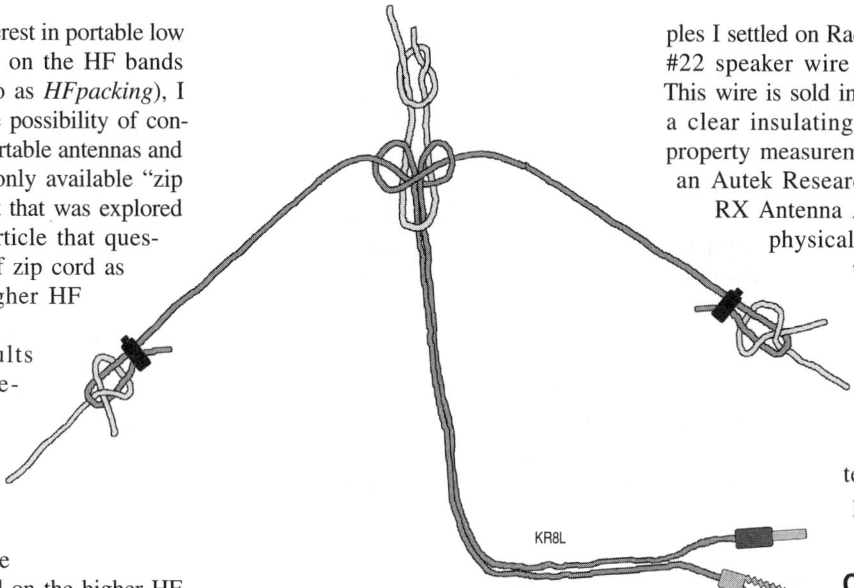

Because of my interest in portable low power operation on the HF bands (often referred to as *HFpacking*), I decided to look into the possibility of constructing lightweight, portable antennas and feed lines using commonly available "zip cord."[1] This is a subject that was explored previously in a *QST* article that questioned the usefulness of zip cord as a feed line for the higher HF bands.[2]

Although my results are in general agreement with the earlier tests, I believe I can show that with the proper selection of material and careful deployment, feed line losses can be minimized on the higher HF bands, making zip cord dipole and feed line combinations satisfactory performers. This is especially true for portable, low-power operations in which some compromise may be acceptable in the interest of saving weight and bulk. Other potential uses for this type of antenna and feed line combination might include ARRL Field Day or a "stealth" situation in which the operator needed an easily stowed antenna that could be erected quickly whenever she wanted to operate. Of course, this would also make an ideal addition to an emergency operations *Go Kit*.

My plan was to use zip cord as a half-wavelength feed line so that the dipole's impedance would be repeated at the transmitter end of the feed line. The impedance terminating a feed line is repeated every half wavelength. Please see any edition of *The ARRL Antenna Book* for more discussion.[3] Because of this, the characteristic impedance of the feed line would be of secondary importance. Also, because the resulting feed line would be relatively short, losses would be minimized. This meant that I would not be able to erect the dipole at "optimum" height, but since the goal of this project was to create light weight, compact, pocket sized antennas and feed lines that could be used in the field with temporary supports, antenna height was not a primary consideration.

After testing a couple of different samples I settled on RadioShack No. 278-1385, #22 speaker wire as a likely candidate. This wire is sold in 100 foot rolls and has a clear insulating jacket. All electrical property measurements were made using an Autek Research Model VA1 Vector RX Antenna Analyst. Formulas and physical property information were taken from *The ARRL Antenna Book*, *The ARRL Electronics Data Book*[4] and the *VA1 Instruction Manual*. I made an effort to characterize the wire as completely as possible, as discussed below.

Table 1
Measured Velocity Factor

Frequency (MHz)	*Velocity Factor (VF)*
3.31	0.68
6.75	0.69
13.67	0.70
27.77	0.71

Table 2
Calculated Attenuation of Zip Cord Compared to Small Coax, dB/100 feet

Frequency (MHz)	*RS 278-1385*	*RG-174*	*RG-58*
3.31	0.97	2.7	0.8
6.75	1.48	3.3	1.2
13.67	2.39	4.0	1.6
27.77	3.41	5.3	2.4

Characteristic Impedance

This is a property that I could not measure directly with the VA1, although I was able to make some calculations and educated guesses.

To begin I measured the center-to-center distance (S) between the conductors of RadioShack No. 278-1385 speaker wire as 0.082 inches, and found the conductor diameter (d) of #22 wire listed in *The ARRL Electronics Data Book* as 0.0253 inch. For parallel conductors with air dielectric the characteristic impedance is given by:

$$Z = 276 \times \log(2S/d)$$

or

$$Z = 276 \times \log(2 \times 0.082 / 0.0253) = 224\ \Omega$$

Again, this is for an air dielectric. The plastic insulation on the wire should reduce the characteristic impedance by some amount, and although I didn't have a way to measure this effect I did come up with a way to estimate it. Here's where the "educated guess" part comes in.

First I did a similar calculation for nominal 300 Ω twin lead, yielding a result of about 400 Ω, meaning that the polyethylene insulation must reduce the calculated

[1]Notes appear on page 41.

value by a factor of about 0.75. *The ARRL Antenna Book* says that for an insulated line, the characteristic impedance can be calculated by multiplying the "in air" value by the inverse square root of the dielectric constant for the particular insulation being used. For polyethylene the dielectric constant is 2.3, so the adjustment factor should be about 0.65. However, with typical 300 Ω line the insulator is very thin so that the field between the conductors is only partially in the insulation and partially in air. It seemed reasonable that the adjustment factor (0.75 in this case) would fall somewhere between pure insulation (0.65) and pure air (1.0).

The type of insulation used on the RadioShack speaker wire isn't specified, but I made the assumption that it is polyethylene or a similar material (this seemed safe since many plastics have a similar dielectric constant of about 2.3). In addition, since the insulation between the conductors of the speaker wire is thicker (perpendicular to the plane containing the two wires) than the insulation for 300 Ω twin lead, it seemed reasonable that it must have a greater influence on the characteristic impedance since more of the field between the conductors will pass through the insulation and less through the air. Based on all of this I assumed that the adjustment factor for the characteristic impedance would probably be closer to 0.65 than to 0.75, so the characteristic impedance for this line might be:

$Z = 224 \times 0.65 = 145.6$, or about 150 Ω.

Again, this value is not particularly important, but keep it in mind and we will use it later to estimate line loss.

Velocity Factor

The velocity factor (VF) of the line was measured using the VA1. The technique involves shorting the end of the line, then sweeping the instrument over a range of frequencies to find the lowest impedance at several points. The lowest frequency will be the frequency at which the line is a half wavelength, the next will be two half wavelengths, etc. The velocity factor can then be calculated by the ratio of the physical line length to the value of a half wavelength in vacuum at the particular frequency (given by the formula $L = 492/f$, where L is in feet and f is in MHz). The results for my roll of No. 278-1385 speaker wire near four amateur bands are shown in Table 1.

Since I planned to build my antennas mostly for the 20 meter band and above, I chose to use 0.70 as the velocity factor of this line. (I could just as well have said that I picked 0.70 because it is a round number or because it is approximately the average of the four readings. I think these results are amazingly consistent considering the use of a consumer grade handheld instrument. There is only a 4% variation in measured velocity factor over a frequency range of about an order of magnitude.)

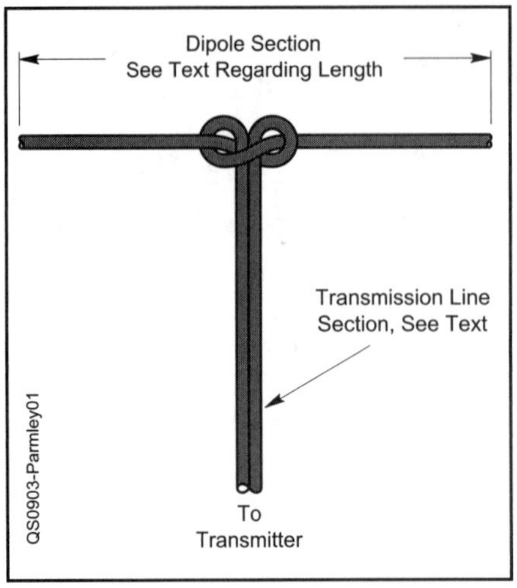

Figure 1 — The center of the antenna section can be secured with an electrician's (or underwriter's) knot as shown.

Attenuation

The attenuation or line loss was also measured using the VA1. For this calculation the series of minimum impedance measurements taken during the frequency sweep (see Velocity Factor, above) were applied to the following formula from the *VA1 Instruction Manual*:

Loss = $8.69 \times Z_{MIN}/Z_0$, where Z_0 is the characteristic impedance of the line.

The calculated loss is given in Table 2. Values for RG-174 and RG-58 (as read from the log-log graph in *The ARRL Antenna Book*) are listed for comparison.

Now, let's think again about how I planned to use this feed line (a single half wavelength between transceiver and antenna). As the frequency increases so does the loss, but the length of a half wavelength of feed line decreases. As a result, the feed line loss remains less than 1 dB as we go up in frequency. In fact, the loss of a half wavelength line decreases from about 1 dB at 80 meters to about 0.5 dB at 10 meters.

If the estimate of a 150 Ω characteristic impedance is correct, then the SWR on the line will be about 3:1, which introduces at most an additional 0.7 dB of loss (as read from the graph of mismatched line loss in *The ARRL Antenna Book*). Total feed line loss would then be about 1.5 dB, and certainly less than 2 dB. As you can see from the table, the line is slightly more lossy than RG-58. It is closer in size and flexibility to RG-174 and is much less susceptible to damage from bending and rough handling than is coaxial cable, an important consideration given the intended use. It might be reasonable to make the feed line a full wavelength long on the higher bands in order to increase the antenna height. Whether the increase in height would offset the increase in feed line loss probably depends on the individual installation.

Construction

Having characterized the wire as completely as possible I next wanted to check my calculations by building and testing some examples. First I calculated the antenna length using the formula $L_A = 234/f$ and then calculated the feed line length using the formula $L_F = (VF) \times 492/f$. I used the quarter wavelength formula for the dipole since the wire was going to be "unzipped" to form the two halves of the antenna and applied the standard 5% reduction in length for "end effects." I used the half wavelength formula for the feed line multiplied by the measured velocity factor. A single piece of speaker wire of length $L = L_F + L_A + X$ was then cut, where X had a couple of feet added to the length for margin of error. The feed line length (L_F) was measured and marked, and the remainder of the wire, including the extra length, was unzipped to make the dipole section. An electricians' knot (see Figure 1) was tied at the junction of the dipole and the feed line.

At the transmitter end of the feed line I unzipped the wire a couple of inches and attached a banana plug to one side and an alligator clip to the other. The banana plug fits perfectly in the center conductor of a transceiver's SO-239 coax connector, while the alligator clip makes a convenient way to attach to the transceiver's ground connection (as shown in Figure 2). I completely ignored the issue of feeding a balanced load with an unbalanced source here, but if you think about it, this is not an uncommon practice for simple wire antennas, and hams have been doing this kind of thing for ages. We're just more accustomed to seeing this where the feed line (coax) meets the antenna (dipole) than where the transceiver meets the feed line. For low power (QRP) applications I have not found this arrangement to be a problem. For those who are concerned about this, a few turns of the transceiver end of the feed line can be wound through an appropriate toroidal core to make a 1:1 transmission line transformer. I did test the transformer configuration briefly and found no difference in performance, feed point impedance or SWR.

For the initial setup of the dipole I measured and marked the calculated antenna length on each leg and folded the wire back on itself at this point and taped it in place. Later I found that a spring compression cord

Figure 2 — Rear of radio showing banana plug and clip lead connections.

Figure 3 — Self contained station using a zip cord antenna and feed line.

stop of the type found on drawstrings on jackets and other items of clothing worked much better than tape, especially when the length of the dipole legs needed to be adjusted. These should be available in most fabric stores. In addition to shortening the dipole while leaving the option of lengthening it without subsequent splicing and soldering, I found that this technique created a convenient attachment loop at the end of the wire. Finally, I tied a piece of light nylon line to the loop on each end of the dipole (no other insulator was used) and proceeded with deployment and testing.

Deployment and Testing

After building antennas and feed lines for 30, 20 and 17 meters, the initial testing was done by installing the antennas in an inverted V configuration with the apex at about 20 feet. This was done using either a telescoping fishing pole, or by tossing a line over a tree branch and pulling the dipole up with that. The ends of the dipole were brought down to 6 to 8 feet off the ground and tied off with nylon line that was then tied to tent stakes. The dipole was pruned to resonance using the VA1 by changing the fold point at the end. The extra wire was left in place and was not trimmed off. The 20 meter and 17 meter antennas were also tested as indoor dipoles by attaching the apex to a ceiling lamp and taping the ends to the walls with masking tape. In this configuration they were easily tuned to resonance.

In practice I found that once the antenna was tuned to resonance it was possible to adjust and optimize the feed point impedance by changing both the horizontal and vertical angles between the two legs. In my particular outdoor installation the best match was found with the dipole legs arranged at a horizontal (azimuthal) angle of between 90 and 120°. For indoor applications the feed point impedance was found to be adjustable by changing the amount of droop in the legs, proximity to walls or floors, and the angle between the legs. As should always be done with parallel wire feeders, I made an effort to keep the feed line clear of other objects and equidistant from both legs of the dipole to the maximum extent practicable.

Conclusion

I have used these antennas and feed lines for SSB, CW and PSK31 at QRP power levels in indoor, backyard and backpack portable situations. Figure 3 shows a self contained portable station. Portability is excellent, deployment is simple and on-air performance seems to be very good. As noted in the original article by K1TD, some zip cord may be significantly more lossy than the type that I used, so it is important to make measurements before committing to a particular type. (One sample I tested, an example of #24 speaker wire, was at least as lossy as the wire that Jerry used.) If you want to try building your own zip cord antennas and feed lines I suggest spending some time with *The ARRL Antenna Book,* particularly the chapters on transmission lines and Smith charts. This will enhance not only your understanding but also your enjoyment of this project.

Notes

[1]Information about HFpacking is available on the Internet at **hfpack.com**.

[2]J. Hall, K1TD, "Zip Cord Antennas — Do They Work?" *QST,* Mar 1979, pp 31-32. This was reprinted in C. Hutchinson, K8CH, Editor, *More Wire Antenna Classics Volume 2.* Available from your ARRL dealer or the ARRL Bookstore, ARRL order no. 7709. Telephone 860-594-0355, or toll-free in the US 888-277-5289; **www.arrl.org/shop**; **pubsales@arrl.org**.

[3]R. D. Straw, Editor, *The ARRL Antenna Book,* 21st Edition. Available from your ARRL dealer or the ARRL Bookstore, ARRL order no. 9876. Telephone 860-594-0355, or toll-free in the US 888-277-5289; **www.arrl.org/shop**; **pubsales@arrl.org**.

[4]*The ARRL Electronics Data Book* is out of print.

Photos by the author.

Bill Parmley, KR8L, was first licensed in high school, then dropped out of Amateur Radio while in college and the military. He was licensed again in 1979 and upgraded to Amateur Extra in 1981. He is an ARRL member. His interests range from QRP on 160 meters to meteor scatter on 222 MHz and from CW to digital voice. Bill has a Master's degree in physics from Michigan State University. He served as a nuclear submarine officer in the US Navy, and has worked as a nuclear engineer for several electric utilities, and as a safety engineer, project manager and program manager for the US Department of Energy. Bill and his wife Anne, KA8TER, are now retired and enjoy living on the small southern Illinois farm where he grew up. You can read more about his Amateur Radio interests and activities at **kr8l.us***, and you can contact him at 1123 Country Club Rd, Metropolis, IL 62960 or at* **kr8l@kr8l.us**.

A 3½ Band Yagi

KC6T builds an elegant 10/15/20 meter triband beam, with bonus "backwards" coverage of 12 meters.

Bill Stein, KC6T

Some years ago I had a great time constructing a five band, 2 element cubical quad antenna.[1] But I have always wanted to try my hand at a competitive multiband Yagi, one that could be constructed with commonly available materials. Because of my space limitations, a 14 foot maximum boom length was placed on my list of wants, along with the ability to operate on at least 10, 15 and 20 meters. In addition, I expanded the list to include full legal-limit operation, a wide SWR bandwidth and low losses consistent with moderate antenna weight and wind loading.

The usual suspects were considered, including log-Yagi designs, the Log Periodic Dipole Array (LPDA), the traditional trapped antenna and open-sleeve designs—each having its own advantages. I modeled the antenna configurations using *EZNEC*,[2] and ended up combining features found in two basic antenna types: a trapped tribander and an open-sleeve design.

- It is a tribander, covering all of the 20 and 15 meter bands, and 1 MHz of the 10 meter band, with a wide SWR operating bandwidth, low losses and a competitive gain and front-to-back ratio.
- It also operates on 12 meters, but with a level of SWR across the band that is high enough to require a tuner. Here it also has a competitive gain and front-to-back ratio—but the pattern is *reversed*. The front for this portion of the antenna is the back of the triband portion. So I call this my "3½ band antenna."
- It weighs about 65 pounds, has a calculated wind loading (at 45° to the boom) of 121 pounds at 70 mph, and has a center of weight balance and a center of wind balance that are very close to each other.
- Boom length is 14 feet, with a turning radius of 18 feet.
- The antenna features 3 elements with a single, low-loss 15 meter trap in each element half (of the 15 and 20 meter elements), plus an open-sleeve, parasitically driven 4 element, 10 meter array—this is effectively a 3 element array as well.
- It has a single coaxial feed line, coupled to the driven element through a commercially available choke balun. It needs no feed-point matching system, such as a beta or hairpin match.
- A wide SWR operating bandwidth is achieved without trap losses usually encountered with conventional trapped antennas. More on this later.
- The number and length of the individual element parts takes into consideration that 6 foot lengths of aluminum tubing may be more economical in the quantities required.
- The antenna is shown schematically in Figure 1.

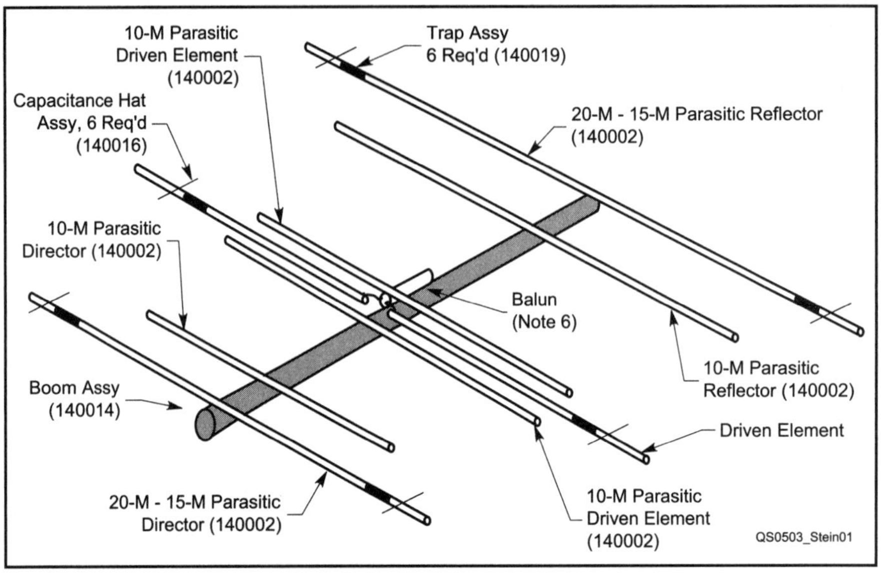

Figure 1—Schematic layout of KC6T 3½ band beam. Numbers in parentheses refer to drawings available on the ARRLWeb.[3]

Why Use Traps?

There is much discussion about lossy traps in today's ham community and about the advantages of a trapless antenna. There is an excellent selection of commercial antennas available that include arrays of 2 element antennas on a common boom, log periodic dipole arrays (LPDA), hybrid log-Yagi systems, the familiar triband trapped array and others. All choices incorporate trade-offs—whether they be mechanical sturdiness versus weight, gain versus boom length, operating SWR bandwidth versus gain bandwidth, the number of bands in a single antenna array versus other performance figures—and this antenna is no different. Traps can be lossy—but even the mechanical joints where antenna sections are joined together can be lossy, especially after some weathering. What is interesting is just how lossy a system might be and what, if anything, can be done about the losses.

With a conventional trapped triband Yagi each element half has two traps. The trap closest to the boom is a parallel L-C circuit that resonates in the 10 meter band. It acts to *disconnect* the portion of the element extending beyond the proper length for 10 meter operation. When I operate on 15 meters, the 15 meter trap

[1]Notes appear on page 45.

acts in the same way but the 10 meter trap (which is still in the circuit) acts as a loading coil, shortening the required physical length of the 15 meter section for resonance. The loading also causes the SWR bandwidth of the resulting 15 meter antenna to be somewhat narrower compared to an unloaded antenna. When I operate on 20 meters, both traps act like loading coils, resulting in more physical shortening of the required 20 meter elements and further narrowing of the SWR bandwidth on 20 meters. Trap unloaded Q has an effect on these properties, but regardless of the Q, the trade-offs tend to make the operating SWR bandwidth more limited compared to a full-size antenna system. Adding more bands by adding more parallel-resonant traps in series compounds the problem further.

This antenna uses only a single, high-Q trap in each element half for the 15 and 20 meter bands. This improves SWR bandwidth on these bands significantly, while minimizing the total losses attributable to traps. The calculated total trap losses for all six traps is about 0.1 dB, and the operating SWR bandwidth is the whole 20 meter band and most of the 15 meter band.[4,5] The 10 meter portion of the antenna is an open-sleeve driven array with no traps that yields an operating SWR bandwidth of 1 MHz.

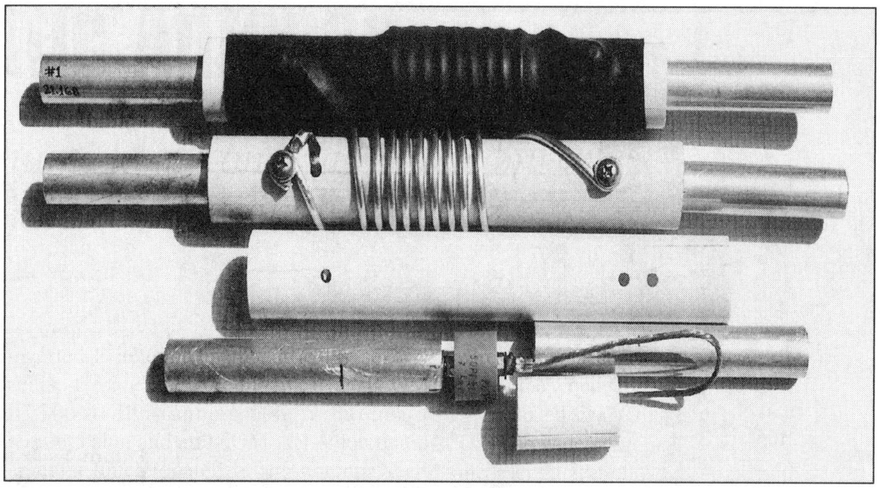

Figure 2—The trap assembly is shown in various stages of fabrication. The mechanical components are detailed in the drawing package available on the ARRLWeb site. This photo shows the 50 pF NPO capacitor attached to its trap terminal, complete with a flexible wire. The coil form and insert are also shown. Although this particular design used a shrink-sleeve cover, the final design used two coats of gray epoxy paint instead.

Figure 3—The area around the boom-to-mast plate for the KC6T beam. The Cal-Av Labs balun is mounted under the rearward 10 meter open-sleeve driver. All elements are mounted on the boom-to-element plates using cast acrylic-insulating blocks.

Trap Construction

Careful attention to the step-by-step instructions will permit trap construction with a minimum of difficulty, since all six traps are identical. The completed trap consists of a 10 turn inductor[6] wound on a 3/4 inch schedule 40 PVC water pipe coil form and a parallel-connected 50 pF, 7500 V NP0 transmitting capacitor.[7] PVC is not the ideal material from which to make the coil form in terms of electrical losses, but it is readily available and more than strong enough.

Figure 2 shows a trap assembly in various phases of construction. In this early model I used shrink sleeving over the completed assembly, but this was not included for the final design. Heat from the shrinking operation caused the PVC to take a slight set, resulting in loosened electrical connections.

A complete set of fabrication drawings for this antenna can be downloaded from the ARRLWeb.[8] Preparation of the PVC forms and other trap components is detailed in these drawings. The basic idea is to make the coil form from 3/4 inch PVC and the inserts from 1/2 inch PVC. The inserts are split axially with a single-width hacksaw blade, then placed in a fixture and drilled for 5/8 inch ID holes. The fixture is simply a 3/4 inch diameter piece of PVC that is also split axially with a single-width hacksaw blade cut and then held together with a pair of hose clamps until the insert has been drilled.

After you have fabricated the trap components, start construction by assembling the transmitting capacitor to the trap terminal/capacitor mount. Attach a flexible multi-strand copper lead, about 8 inches long, to the second capacitor terminal. Next, install one of the coil-form inserts into the end of the coil form that has a single hole through the sidewall. Press the coil form in until it is flush with the end of the coil form. Insert the capacitor/terminal tube assembly into the opposite end of the coil form, terminal tube first. Gently press the assembly terminal tube through the installed coil form insert until the terminal tube extends beyond the coil form by 2 1/2 inches. Do not force the terminal through the insert, since damage to the capacitor may result. If required, redrill the bore of the installed insert until the capacitor terminal tube can be inserted with moderate pressure. Feed the capacitor flexible wire lead through the inside hole at the open end of the coil form.

Now install the remaining coil form insert into the open end of the coil form and press in the plain terminal tube until it extends beyond the coil form by 2 1/2 inches. Using 5 minute epoxy and a small screwdriver, apply the glue to the aluminum reinforcement pieces and insert them until they are flush with the inside end of the terminal tubes. Allow the glue to set. Using the end holes in the coil form as a guide, drill the appropriate size hole through the coil form insert, the terminal tube and the reinforcement tube. This gives a screw on each end of the coil form to electrically connect the terminal tubes, the capacitor and the coil. A #8 pan-head self-tapping screw could be used, although I prefer to use #8-32 stainless-steel machine

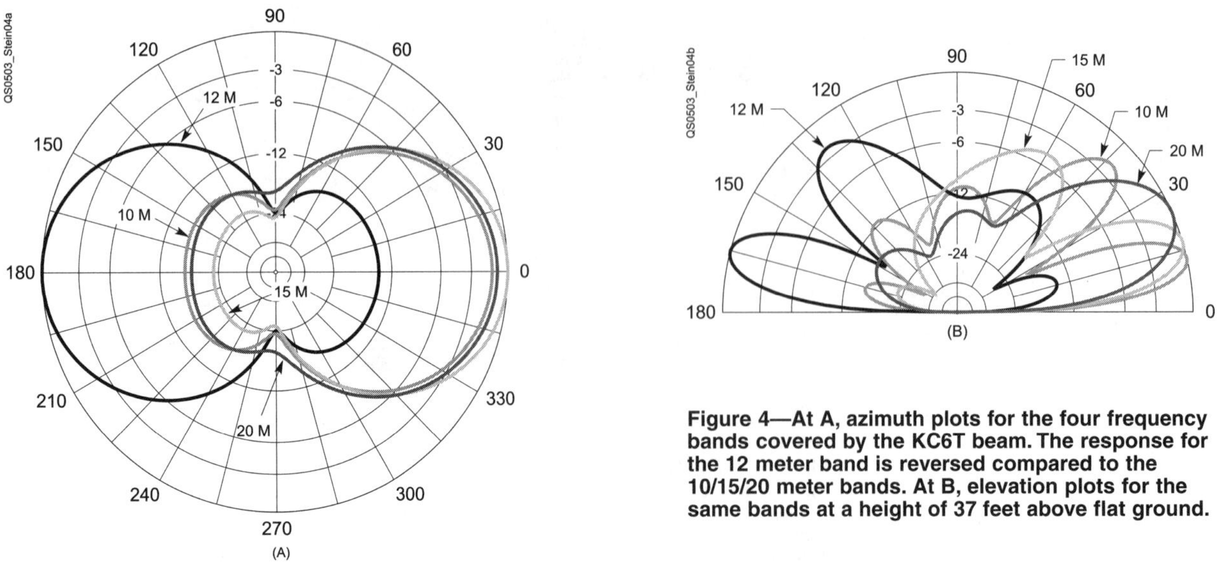

Figure 4—At A, azimuth plots for the four frequency bands covered by the KC6T beam. The response for the 12 meter band is reversed compared to the 10/15/20 meter bands. At B, elevation plots for the same bands at a height of 37 feet above flat ground.

screws and washers, requiring me to use the correct size drill and tap. These screws also act to secure the terminal tubes to the coil form assembly.

Form an "eye" in one end of the coil wire and attach to the terminal screw on the capacitor side of the coil form. Carefully route the first turn to provide for the main body of coil turns to be away from the internal capacitor body. Wind 10 turns as evenly as possible, forming another eye to terminate the coil on the opposite end of the coil form. Trim off excess length from the flexible wire from the internal capacitor and attach a terminal lug. Attach the lug to the adjacent coil terminal.

Using a grid dip oscillator, carefully adjust the trap windings until the assemblies resonate between 21.175 and 21.200 MHz. This is done by either squeezing or prying apart the coil turns as evenly as possible. In this process, a coil should be as far removed from metallic working surfaces or other electrical components as possible. Make sure that the coil forms and coils are free of grease and dirt. Then use masking tape to cover the exposed trap terminals. Paint the assemblies with two coats of epoxy spray paint, allowing the first coat to dry thoroughly before applying the second. Remove the masking tape when the trap assemblies are completely dry.

When installing the trap assemblies to their respective elements, the screw terminals should be oriented toward the ground (facing down). The end of the coil form through which the capacitor flexible wire protrudes should also be oriented toward the antenna element tip. This arrangement provides a convenient drain for any moisture that might appear inside the trap coil form.

Remaining Antenna Components

I made the boom-to-mast plate from a T-6061 $^{5}/_{16}$ inch thick aluminum blank, 17$^{3}/_{4}$ inches long by 9 inches wide. It is a complex part because the center of weight and wind load is where the driven element and the close-spaced 10 meter parasitic driven elements are all located. This plate does not permit the mast to extend above the boom of the antenna. See Figure 3 for a close-up of the center of the boom.

All U bolts are 2 inch by $^{5}/_{16}$ inch. I fabricated and installed a pair of metal strips for each boom-to element plate as insurance that they will remain aligned perpendicular to the boom, even though this was probably overkill.

I made the element-support insulators from cast acrylic material that has a UV inhibitor. Figure 3 shows how each element is supported on an insulator. Notice how the balun[9] is fitted underneath the rear 10 meter parasitic driven element. The insulators are attached to the element-to-boom plates with #10-32 stainless-steel screws, as are the elements attached to the insulators.

A generous amount of overlap between the telescoping aluminum tubes is used for mechanical strength. The trapped driven element is split at the center for the antenna feed point. An insulating coupler is inserted inside the two halves. I used a fiberglass tube but you could also use a solid acrylic plastic rod with a 1 inch OD. The two center holes (#18 drill through) are provided for two #8-32 stainless-steel machine screws, nuts and washers to connect the two feed wires from the balun.

The center sections of the parasitic elements are reinforced with smaller OD aluminum tubing, as is the center of the boom. Reinforcement pieces ("BE" in the drawing package) provide added strength to the boom wall where the four outer elements are attached. An additional 8 inches of 2 inch OD tubing will be required to make the four pieces needed.

All boom-to-element support plates are made from $^{3}/_{8}$ inch thick, 2 inch wide 6061-T6 aluminum. A different boom-to-element plate is used for the trapped driven element, while the remaining six elements use the same design for their plates.

The two outer boom sections are each attached to the boom's center section with a $^{1}/_{4}$-20 stainless-steel machine bolt, lock washer and nut. After placing the two internal boom reinforcement sections ("BE"), drill a small hole through the outer boom section and reinforcement, add a small sheet-metal screw to hold each of the reinforcement sections in place during assembly of the boom to the boom-to-mast plate.

I used a capacitance-hat assembly shown in the drawing package. An even simpler capacitance hat is also suggested in that package. The design of the balun mount depends on the balun you choose to use. To keep feed lines short at the attachment point to the trapped driven element, choose a balun that will fit along the top of the boom. Mine was mounted underneath the rearmost 10 meter parasitic driven element.

Buying the Aluminum Tubing

The required quantity of tubing de-

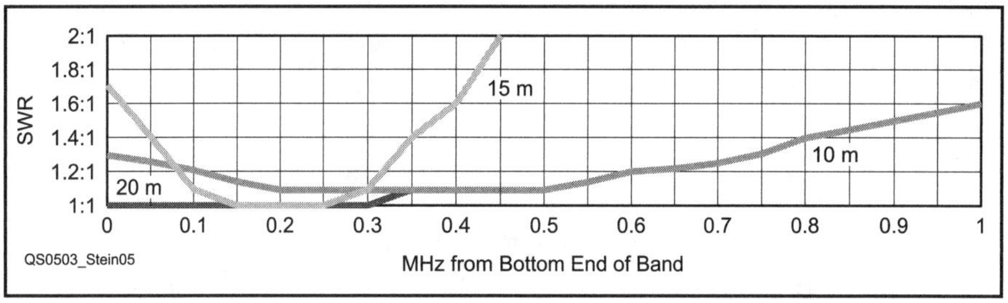

Figure 5—Measured SWR plots across the 10/15/20 meter bands at a height of 37 feet.

pends on the number of antennas being fabricated but also on the available lengths of each piece. In general, longer tubing lengths will result in less scrap. But shipping costs and minimum-order requirements may more than offset any scrap material savings. Six-foot lengths of tubing can be shipped for modest cost by UPS. I chose to buy tubing in these lengths as a reasonable trade-off between (large) minimum quantities, shipping costs and tubing prices.[10]

Antenna Assembly

When you assemble each element, remember to use conductive compound at all the tubing interfaces. The trap assemblies are inserted in the appropriate element section until the trap body is flush against the mating antenna element sections. Place the capacitance-hat assemblies on the ends of the trapped elements where shown.

The boom center section (2 inch diameter by 72 inches long) has internal reinforcement pieces ("BC"). Note that these are *not* placed in the center of the 2 inch diameter center portion of the boom. Assemble the element insulators to the element-to-boom plates and then mount the elements to their respective element plates. Then install the completed element assemblies to the boom. Install the boom-to-element plate reinforcement parts ("BE"). These help keep the boom from being crushed by the element plate-to-boom U bolts. Assemble the boom-to-mast clamp to the boom and double-check the element-to-element spacings. Finally, install the balun to the boom such that the balun body passes under the rear 10 meter parasitic driver. Connect the short balun leads to each side of the split driven element.

Antenna Adjustment

Figure 5 shows the measured SWR curves across the frequency bands covered. This shows a slightly narrower SWR operating bandwidth for the 15 meter band than the computer model, but the 20 and 10 meter SWR data are in close agreement with the calculated values. The 12 meter SWR performance shows the need for an antenna tuner. Most transceiver tuners will do an adequate job.

If you wish to move the frequency at which the lowest SWR occurs, you may do so, using the information in the drawing package. Because the 10 meter portion of this antenna is parasitically driven, changes in the lengths of both the 15 and 20 meter elements will have minimal effect on the 10 meter performance. But changes to the 15 meter section will affect the SWR characteristics of the 20 meter section.

Calculated azimuth and elevation patterns are shown in Figure 5. The performance of the Yagi on 12 meters is quite respectable, even though the forward direction is off the backside of the antenna.

On 10 meters the spacing between the trapped driven element and the two 10 meter open-sleeve driven elements is quite sensitive. The SWR on 10 meters can be shifted to another frequency by altering the length of the two 10 meter parasitic driven elements equally, and altering the 10 meter director and reflector proportionally. For example, the lengths of each 10 meter element half-length are:

- Parasitic reflector: 111 inches
- Parasitic driven: 100 inches
- Parasitic director: 96 inches

If you decide to shorten the driven element by 1 inch (99/100 = 99% of the original length):

- New parasitic reflector: 0.99 × 111 inches = 110 inches
- New parasitic driven: 99 inches
- New parasitic director: 0.99 × 96 inches = 95 inches

The SWR characteristics for the 15 meter band can be shifted in frequency by the same means suggested for 10 meters. A shift in the SWR characteristics for 20 meters section can best be accomplished for small changes by changing each of the 20 meter element lengths after the trap by the same length. Thus if you shorten each tip of the 20 meter driven element by 1 inch, you should also shorten each tip of the 20 meter reflector and director by 1 inch.

The Final Installation

I mounted my antenna at 37 feet, using a guyed tubular crank-up tower—a vintage Tristao model. You may find this antenna to be a mechanical challenge to fabricate, but for your efforts you will be rewarded with electrical stability and good performance. So far my antenna has been in place over 3 years. It has survived a 75 mph windstorm and has worked some DX toughies. It has not shown any changes in its SWR characteristics.

Acknowledgments

My thanks to the many ham antenna enthusiasts and experimenters I have contacted, each willing to exchange their views and antenna experiences during many enjoyable QSOs. My special thanks to my wife, Diane, who encouraged me to design and build this antenna. She put up with a huge number of hours spent over a hot computer and a similarly large number of hours spent fabricating and installing this beauty on my tower.

Notes

[1] "A Five-Band, Two Element Quad for 20 through 10 Meters," W. Stein, *QST*, Apr 1992, p 52.

[2] *EZNEC 3.0*, Roy Lewallen, W7EL, PO Box 6658, Beaverton, OR 97007; **www.eznec.com**.

[3] See **www.arrl.org/files/qst-binaries/stein_yagi.zip**.

[4] See Note 2.

[5] *Reference Data for Radio Engineers*, Howard W. Sams & Co., 6th edition, 1975, pp 6-4 to 6-8.

[6] #8 solid aluminum ground wire is available in 40 foot lengths from RadioShack, catalog no.15-035.

[7] Available from Surplus Sales of Nebraska, 1502 Jones St, Omaha, NE 68102. Their Web site is **www.surplussales.com**.

[8] See Note 3.

[9] The balun I used was a Model EB-1, manufactured by CAL-AV Labs Inc, 1802 W Grant Rd, #116, Tucson, AZ 85745, purchased at HRO.

[10] I purchased aluminum tubing in 6 foot lengths from Texas Towers, 1108 Summit Ave #4, Plano, TX 75074.

Photos by the author.

Bill Stein, KC6T, was first licensed in 1954, and enjoys both ragchewing and chasing DX, operating CW, SSB and PSK31. He is an active designer and homebrewer, having built not only antennas, but also a microcontroller based SWR/power meter, a solid-state audio record/playback device integrated with his HF transceiver, and the usual solid-state keyer projects. He is especially fond of restoring and operating vintage ham equipment. Bill received his BSE from UCLA in 1958, and spent most of his career working in the computer peripherals field. For the last 7 years, Bill has been self-employed as a consultant to management and is involved with industrial electronic products. You can reach the author at **steinwa@earthlink.net**.

A Dual-Band Wire Beam for 17 and 12 Meters

Allen Baker, KG4JJH

This wire beam Moxon antenna offers double duty on 17 and 12 meters with one 50 Ω feed line.

The enthusiasm generated by the Moxon[1] has left many amateurs wondering if a multi-band version is possible. Interaction between elements, however, has denied the builder an easy answer. Nevertheless, a dual-band version *is* possible for nonconsecutive bands.

The Design

In one of his articles, W4RNL describes three methods of nesting Moxon rectangles that allows the antenna to retain its desirable characteristics.[2] (For a detailed explanation, I encourage the reader to study this enlightening piece.) The method presented here is the first and simplest of the three versions to build. The antenna's 50 Ω feed line attaches to the 12 meter driven element, and from there a 75 Ω transmission line feeds the 17 meter driver.

The original antenna designs were based on the use of 3/4 and 1/2 inch aluminum tubing for construction, but I wanted to use wire elements, in order to allow for portable operation. Reducing the element diameter will reduce the SWR bandwidth, but this does not become an issue

[1]Notes appear on page 50.

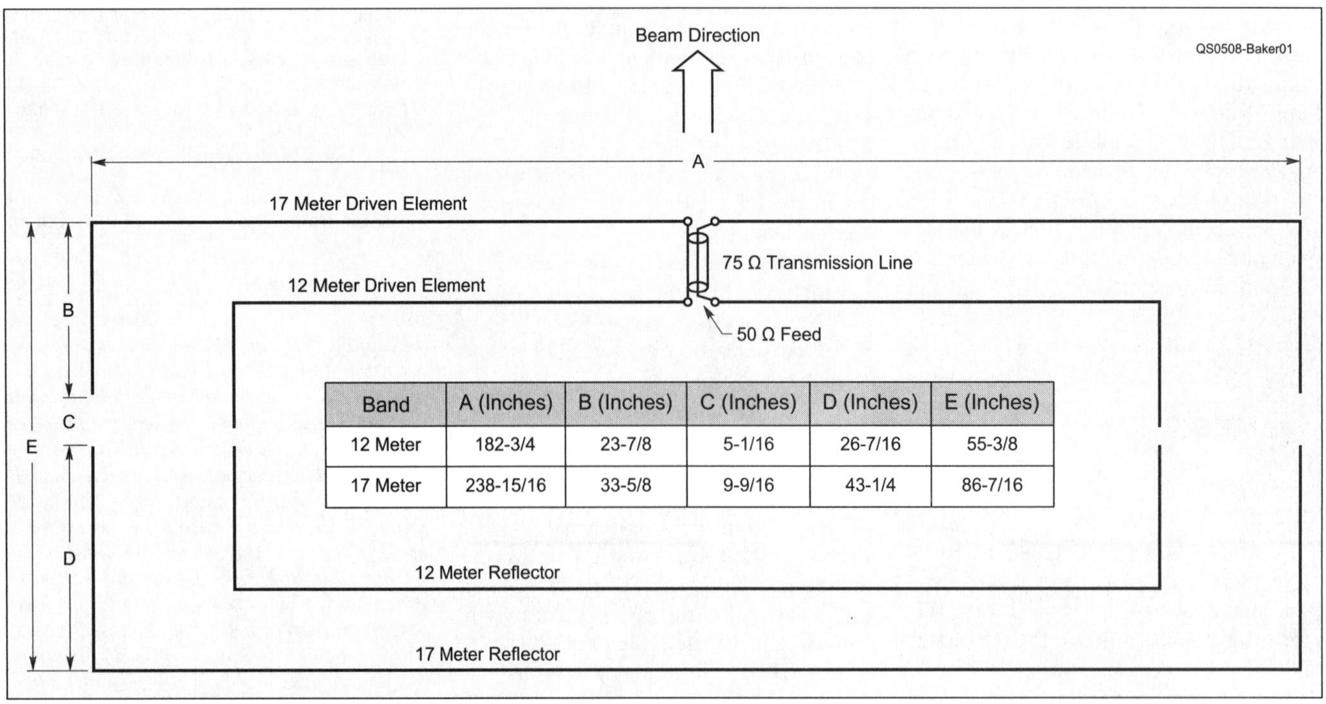

Band	A (Inches)	B (Inches)	C (Inches)	D (Inches)	E (Inches)
12 Meter	182-3/4	23-7/8	5-1/16	26-7/16	55-3/8
17 Meter	238-15/16	33-5/8	9-9/16	43-1/4	86-7/16

Figure 1—The dual-band Moxon antenna layout, with basic element dimensions.

for the narrow 12 and 17 meter bands. Numerous *EZNEC* modeling sessions yielded a dual-band design that appeared to equal Moxon wire monobander parameters.[3] Figure 1 shows the basic antenna layout and Table 1 shows the modeled data for both free-space and 30 foot elevations. Detailed construction drawings are available at the ARRL Web site.[4]

The Hub

To ensure a rugged and weatherproof design, fiberglass spreaders are mounted on an aluminum plate using stainless steel U-bolts. A view of the center hub, together with the top guy anchoring post can be seen in Figure 2. Two die-cast aluminum flanges secure the mast to the hub on the bottom, and the guy line post on the top. The hub is made from a one-foot square of aluminum plate of 0.190 inch thickness. Using a paper template taped to the aluminum plate, center punch and drill all holes. Next, cut the plate into a circle to reduce the weight. On the top of the plate, draw a 4 inch diameter circle, in the center, with a permanent marker. This circle will be used later to align the spreaders. Four 3/8-16 × 1 1/2 inch stainless steel hex bolts and stop nuts are used to attach the flanges to the top and bottom sides of the plate. Figure 3 shows the hub bottom.

The Spreaders

The four main spreaders are made of 3/4 and 1/2 inch OD × 1/8 inch wall fiberglass tubing. A feed line spreader supports the coax and transmission line on one side, while a balance spreader provides balance and element support on the opposite side. A 12 inch piece of 1/2 inch OD tubing is inserted into one end of each 3/4 inch tubing at the hub for reinforcement. A cutting schedule is shown in Table 2.

To keep from drilling holes and weakening the spreaders, the wire elements and guy lines are attached using nylon collars.[5] These "collars" slide over the fiberglass tubing and tighten with a setscrew. Four 1/2 inch ID collars are drilled and tapped (opposite the setscrew) to accept 1/4-20 nylon screws which are used as the 17 meter element corner anchors. Four 3/4 inch ID collars are drilled and tapped (opposite the setscrew) to accept 1/4-20 nylon screws, which are used as the 12 meter element corner anchors. The nylon screws and collars provide a weatherproof method of securing the elements without adding metal that might affect resonance. The collars also provide an easy way to adjust the elements. A partial view of the inner element (12 meter) collars can be seen in Figure 4.

The Feed Line/Balance Spreaders

Two type-C, 1/2 inch PVC conduit bodies are mounted on the feed line spreader

Table 1
Modeled Antenna Data

Elevation	*Free-Space*		*30 Feet*	
Center Frequency (MHz)	18.118	24.94	18.118	24.94
Gain (dBi)	5.97	5.81	10.94	10.84
Front-to-Back Ratio (dB)	27.99	30.64	21.35	25.46
Complex Impedance (Ω)	45.1 – *j*5.3	44.4 + *j*12.9	48.3 – *j*8.9	41.6 + *j*12.8
SWR	1.16:1	1.35:1	1.2:1	1.4:1

Table 2
Cutting Schedule

Description	*Quantity*	*Total Length (Inches)*	*Materials*
Spreader reinforcement	6	12	1/2" OD fiberglass tubing*
Main spreader extension	4	48	1/2" OD fiberglass tubing*
Main spreader	4	96	3/4" OD fiberglass tubing*
Feed line spreader	1	48	3/4" OD fiberglass tubing*
Balance spreader	1	48	3/4" OD fiberglass tubing*
17 meter driver	2	153	14 gauge Flexweave wire and ring terminals
17 meter reflector	1	325 1/4	14 gauge Flexweave wire and ring terminals
17 meter insulator	2	10 5/8	Screen base (1/2" wide)
12 meter driver	2	115	14 gauge Flexweave wire and ring terminals
12 meter reflector	1	235 5/8	14 gauge Flexweave wire and ring terminals
12 meter insulator	2	6 1/8	Screen base (1/2" wide)

*All fiberglass tubing is 1/8" wall.

Figure 2—A top view of the antenna center hub during construction, together with the guy anchor post.

Figure 3—A bottom view of the antenna center hub.

Figure 4—Element and guy anchor collars. The inner (12 meter) element collars are shown.

Figure 5—One of the conduit bodies. These house the transmission and phasing line junctions to the driven elements.

and provide access to the feed line, transmission line, and driven element terminal posts. The conduit bodies slide over the 3/4 inch tubing and are secured with stainless steel setscrews. After you have determined the location of each conduit body, cut slots (inside the conduit bodies) on the top of the 3/4 inch fiberglass feed line spreader to allow coax cable entry and exit. Two 3/4 inch ID collars are mounted on the balance spreader for element support. Once again, drill and tap both collars (opposite the setscrew) to accept a 1/4-20 nylon screw. Also, drill and tap two 1/4-20 holes in the bottom of both conduit bodies for the setscrews. Figure 5 shows a close-up of one of the conduit bodies.

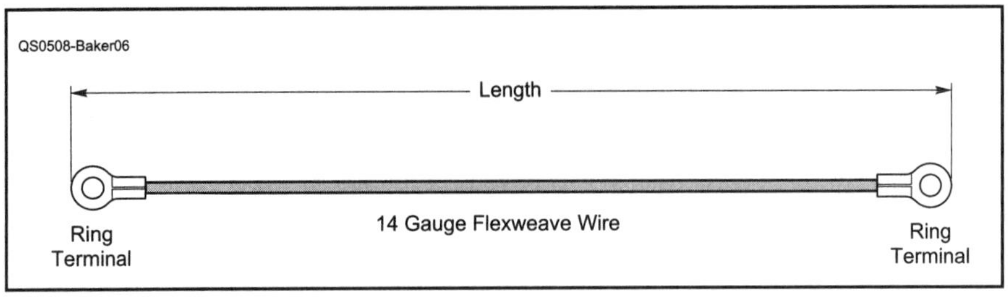

Figure 6—Be sure to include the ring terminals when measuring element length.

The Elements

The elements are cut from 14 gauge Flexweave wire, according to the cutting schedule of Table 2. One hundred feet of wire will provide enough length for both bands. The insulators are made from 1/2 inch wide strips, cut from screen base (used for screened-in porches). The elements are attached to the insulators and feedpoints by crimping and soldering ring terminals to the wire ends—securing them to the insulators with #6 thread-cutting screws. You should be careful to include the lengths of the ring terminals when cutting the wire to get the correct lengths (Figure 6).

Figure 7—The guy wire center post, turnbuckles, and hub. Note the conduit boxes at the lower right. These serve as feed line junctions to the driven elements. The feed line and phasing line pass through the fiberglass support arm.

The Guy Lines

Nylon collars are drilled and tapped to accept 3/16 inch stainless steel eyebolts that are used for the guy line tie points. Four 1/2 inch ID collars are used for the four main spreaders and two 3/4 inch ID collars for the feed line and balance spreaders. Using a combination of forces, that are provided by the guy lines and elements, the main spreaders are stressed into an "S" curve to supply the tension needed to stretch the elements taut.

Prepare the guy line post by cutting a 14 inch piece of 1 1/4 inch schedule 40 PVC pipe. Measure 2 inches down from the end, and drill a 1/4 inch OD hole all the way through the center of the pipe. Next, measure 1 5/8 inches down from the same end and 90° away from the first hole, and drill a 3/16 inch hole all the way through the center of the pipe. Install two 1/4 inch and two 3/16 inch stainless steel eyebolts with hex nuts. With the eyebolts on top, secure the pipe into the top hub-mounted flange by tightening the setscrew.

Prepare six guy lines using Kevlar twine, fishing swivels, and clips. The Kevlar is stronger than fishing line and it will not stretch as much as nylon or polyester. The total length for the four main spreader assemblies (Kevlar length plus swivel/clips on each end) should be 96 inches. Make the feed line and balance spreader guy line assemblies 46 inches long. I used standard fishing line knots on each end and, with a little practice, I was able to get the lengths correct.

Each guy line is attached to an aluminum/stainless steel turnbuckle at the guy line post to make adjustments easier, as shown in Figure 7. The other ends of the long guy lines attach to the four 1/2 inch OD collars with eyebolts mounted on the main spreaders, and the short guy lines attach to the two feed line/balance spreader 3/4 inch OD collars. Figure 8 shows a closeup of a typical guy line anchoring collar, together with an element support collar.

Figure 8—A typical element support collar and guy-anchoring collar.

Assembly

Since I wanted to be able to disassemble the antenna, I fastened the spreaders together with screws and nuts. Insert a 48 inch long piece of 1/2 inch OD fiberglass tubing into one end of the 3/4 inch OD × 96 inch long main spreader with a 12 inch overlap. Drill two 9/64 inch OD holes through both tubes—3 inches from each overlapped end—being careful to drill straight through the centerline of the tubes and perpendicular to its surface. Secure using two 6-32 × 1 inch stainless steel screws and stop nuts at each overlap, but do not over tighten. Repeat this process for all four main spreaders.

Spread a thin bead of *Liquid Nails* on one side only (to allow moisture to drain) of each 12 inch piece of 1/2 inch OD fiberglass tubing doubler. Insert each doubler flush into one end of the four 3/4 inch OD × 96 inch long main spreaders and the two 3/4 inch OD × 48 inch long feed line/balance spreaders. Allow several days for these joints to cure because little drying air can get to the joint. Insert each reinforced spreader under a pair of hub-mounted U-bolts and align the ends with the 4 inch diameter circle marked earlier. Add stainless steel lockwashers and tighten each U-bolt until snug, being careful not to crack the fiberglass.

For convenience, mount a temporary short mast that places the hub at a comfortable working height. Slide all collars onto the spreaders and tighten the setscrews, using the dimensions on drawing sheets 3 and 5 as starting points.[6] (These starting points are applicable only if you used identical materials as listed in the materials list on drawing sheet 5.) Remove the conduit-body top plates, and drill and mount 8-32 × 1 inch stainless steel screws, nuts, and lock washers, spaced 1/2 inch apart. You will have to remove some of the lip on the bottom of the top plate with a Dremel tool to make room for the screw heads.

Mount an SO-239 connector on the hub and route a length of 50 Ω coax (RG-8X) from there to the 12 meter conduit-body inside the feed line spreader. (The feed line spreader ID is 1/4 inch, so make sure your coax OD is slightly smaller.) Solder the coax to the SO-239 and, using ring terminals, attach the other side to the 12 meter feed point screws. Route a length of 75 Ω coax (RG-59, VF = 0.8) from the 12 meter feed to the 17 meter feed, inside the feed line spreader.

Using ring terminals, attach the center conductor of the 75 Ω phasing line to the left 12 meter/17 meter drivers and the braid to the right 12 meter/17 meter drivers (or vice versa). This is an important step, as reversing the phase of the transmission line between elements will negate its purpose. I added a plastic cap (fastened with RTV silicone glue) to weatherproof the back of the SO-239 connector.[7] I recommend the use of stranded wire center conductors for both coax cables to reduce the possibility of breakage when assembling the feed lines inside the conduit bodies.

At this point you will notice that the main spreaders have started to sag. Adjust the tension in the guy lines by adjusting the turnbuckles until the tips of the main spreaders are a few inches higher than the hub. Cut the wire elements and mark each corner with a permanent marker according to the lengths in Figure 1 and Table 2. Using a soldering iron, tin a 1/8 inch wide band around each mark with solder, to permanently mark the wire element corners.

Attach the 12 meter driven element to the right feed point and route the wire around the collar-mounted nylon screw, placing the solder mark under the nylon washer and screw and tighten. (Do not over-tighten the nylon screws—I learned that the hard way. If you do twist the head off, drill a 1/16 inch hole in the remaining threaded portion, insert a #1 screw extractor, and turn counterclockwise.) The nylon washer is important—it prevents the wire element from moving as you tighten the screw. Continue, in similar fashion, to the 12 meter insulator, the 12 meter reflector, another 12 meter insulator, and then attach the 12 meter driver to the left 12 meter feed point. The same procedure is used for the 17 meter elements.

Final collar and turnbuckle adjustments should be made in small increments—this will keep the spreaders stressed evenly and the elements level and equidistant from each other. When you are satisfied, line up the center of the feedpoints and tighten the conduit-body setscrews. If desired, trim the spreaders to within 2 inches of the collars, using a hacksaw.

I highly recommend assembling all the components, making *all* element and feed point adjustments, and then tackling the feed point wiring (and associated slot-cutting) last. Once the feed and transmission line coax has been installed in the conduit boxes, it becomes difficult to move them.

The fiberglass-tubing supplier recommends spray-painting the spreaders flat black to increase their life. First, clean them with a solvent, such as acetone. Then, use a good primer, allowing it to cure for two days. Finally, finish with a topcoat of flat black.

A 5 inch long, 1 1/4 inch pipe nipple was used to make the transition from the four-hole flange on the bottom of the antenna to my mast. Before you mount the antenna on the mast, check all hardware to make sure it is secure. The finished antenna—up and ready to go—can be seen in Figure 9.

Results

The complete antenna weighed in at 15 pounds, which was a bit too much for a push-up mast. I fastened the antenna to my recently acquired AB-952 military surplus mast, and raised it to a height of 30 feet. This mast is a bit cumbersome to move, but it provides a rock-steady platform for supporting antennas as high as 35 feet using 4 inch OD aluminum tubing. At this elevation, my MFJ-259B

Figure 9—Are the bands open? The completed antenna ready to snag some elusive DX!

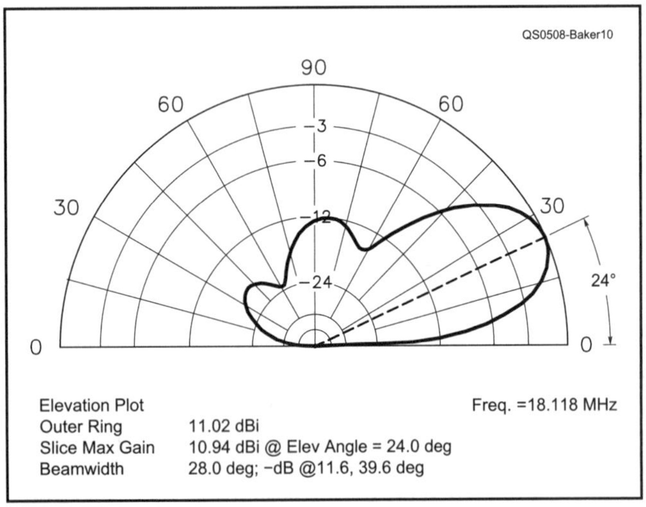

Figure 10—A modeled elevation plot of the antenna at 30 feet on 17 meters using *EZNEC*. The gain at 12 meters is essentially the same, except for a lower take-off angle. [Bear in mind that the antenna's gain and launch angle will be seriously affected by the local ground characteristics and the antenna's height above that ground.—*Ed.*]

antenna analyzer indicated that both bands were within the modeled SWR, and never exceeded 1.4:1. An elevation plot (*EZNEC*) of the antenna on 17 meters is shown in Figure 10. The 12 meter gain model is essentially the same, save for a lower take-off angle (18° versus 24°) at the same height. Modeled SWR at 17 meters is about 1.2:1 at the band center (18.118 MHz).

PJ2/K9LZJ on DXpedition in Curacao supplied my first contact. Using only 5 W SSB, this 1900 mile contact gave me a 5-5 report and congratulated me on my setup. KF6IWW provided the second QSO and gave me a chance to work both bands using 5 W on PSK31 and PSK63 modes. The contact ended up on 12 meter SSB using 50 W, where we had a very nice chat for over an hour. As a quick front-to-back ratio test, I pointed the beam toward a station in Ireland and observed a signal strength of S-9. After rotating the antenna 180°, the meter read S-6. Using the traditional 6 dB/S-unit approximation, this worked out to 18 dB, which was 3 dB below the modeled F/B of 21 dB. With the assistance of NG4T, additional testing on both bands was performed using CW, digital, and SSB modes at power levels of 5 to 100 W.

I am delighted that the antenna is performing as well as a single-band Moxon, with excellent gain, directivity, and front-to-back ratio. The Moxon rectangle is, in my opinion, the best performer among the family of dual-coupled 2-element parasitic beams in terms of gain, front-to-back ratio, and SWR bandwidth. This dual-band version retains these characteristics in an antenna that is less than half the footprint (in square feet) and within 1 dBi (modeled) gain of a commercial dual-band 4-element Yagi.[8]

Notes

[1]Moxon Antenna Project (**www.moxonantennaproject.com**).

[2]"The Elusive Moxon Nest," L.B. Cebik, W4RNL (**www.cebik.com/moxon/mox1712.html**).

[3]*EZNEC*, W7EL (**www.eznec.com/index.shtml**).

[4]**www.arrl.org/files/qst-binaries/baker0805.zip**.

[5]Quad guy ties (**www.antenna-magic.com/antenna/Product.htm**).

[6]See Note 4.

[7]Both the SO-239 socket and its mating connector, the PL-259, are not waterproof. Remember to liberally wrap the plug and socket assembly and the plug cable entrance with vinyl electrical tape (Scotch 33+) or coaxial sealing tape (CoaxSeal) for any outdoor application.—*Ed.*

[8]*OB4-2WARC,* Optibeam (**www.optibeam.de/**).

All photos by the author.

Allen Baker, KG4JJH, became licensed in September 2000, after a lifelong dream of becoming a ham. Active on SSB and the digital modes, Allen enjoys experimenting with antenna designs. He has a BS in Industrial Engineering from Tennessee Technological University and is an Instrument and Controls Engineer at the BWXT Y-12 National Security Complex in Oak Ridge, Tennessee. You can reach him at **kg4jjh@arrl.net**.

A Stub Directed 80 Meter Beam

This antenna uses a pair of relays and a selector switch to pick one of three different radiation patterns.

Kenneth Carr, WØKUS

Most hams have worked the 80 meter band at one time or another. Many nights the white noise or other QRN is S9 or more. Add to that the QRM, and we often wish for a directional antenna. Wouldn't it be nice to lower that noise floor a couple of S units or have some gain? While we are making up that wish list, let's add a front to back ratio of about 25 dB.

Well, if you are willing (and able) to build this stub directed beam you can do all of that and more. The antenna consists of three identical inverted V antennas, configured as a three-element beam. The operation is a bit out of the ordinary. The radiation pattern can be switched between two directions with a gain of over 10 dB. The third pattern is a lot like an ordinary dipole, which can be used for general coverage receiving or transmitting.

You can instantly switch to any of the three patterns. Perhaps it would be prudent to stop talking for the instant it takes to turn the switch, especially if you run an amplifier. This is not a project for the beginner, but is well within the reach of many hams.

The configuration works well, but it does take up some room. I have a small lot and this antenna is 106 ft long and about 120 ft wide. So much for putting up my dream antenna at my home! Enter "Radio Ray"—Ray Galep, WA9SXJ. He has lots of towers and we have built several big antennas together in the past. Ray had two towers in place that aligned east and west. We paced off his back lot and went to work on it.

Several weeks on the computer led to what looked like a workable antenna design. I used Roy Lewallen's *EZNEC* program to plot and model the antenna. After a few trial runs the gain turned out to be 10.54 dBi at 35° above the horizon. (This was for both directional patterns but not the general coverage pattern. See Figure 1). The front-to-back ratio was 26.56 dB at 45° above the horizon. From 5° to 60° it never fell below 20 dB. See Figure 2. This turned out to be a real plus when we found it reduced the noise floor 1 to 2 S units on receive.

We connected the center inverted V directly to a 50 Ω coaxial cable. That is always the driven element. In order to get a directive pattern we had to load the center of the parasitic elements. We accomplished this with tuning stubs. The stub we wanted for a director was cut to reflect $-j115$ Ω. The one we used for a reflector was cut for $+j60$ Ω. The effect of the center loading made the elements look longer or shorter electrically, forming a 3-element antenna. The director and reflector Vs are equally spaced from the driven element so that we can reverse the beam direction. By switching the loads from front to back we could reverse the pattern.

This looked good on the computer, but how do you do it in practice? I got out the Smith charts and looked up the stub lengths needed to build the antenna. The capacitive element is the shortest, so we placed one capacitive stub on each parasitic V first. Using the Smith chart, we

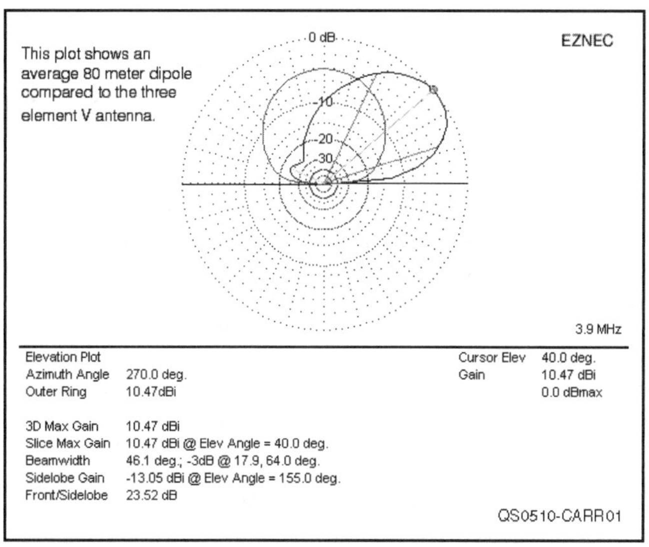

Figure 1—This *EZNEC* plot compares the stub-directed 3-element V antenna with an average 80 meter dipole.

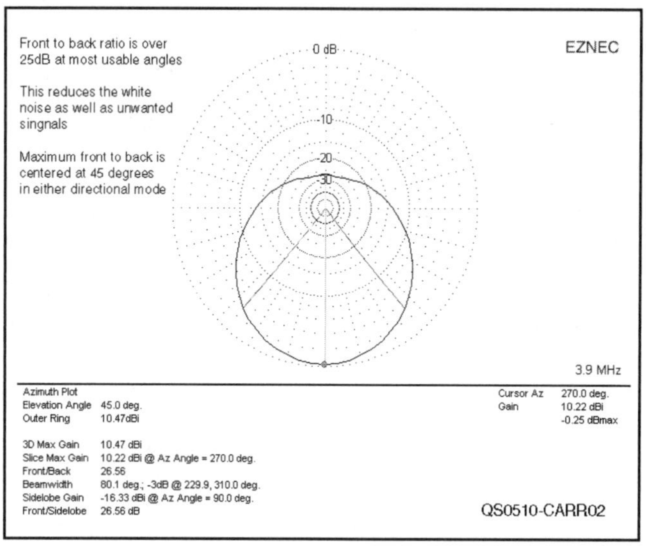

Figure 2—*EZNEC* shows that the antenna has a front-to-back ratio of 26 dB at a take-off angle of 45°.

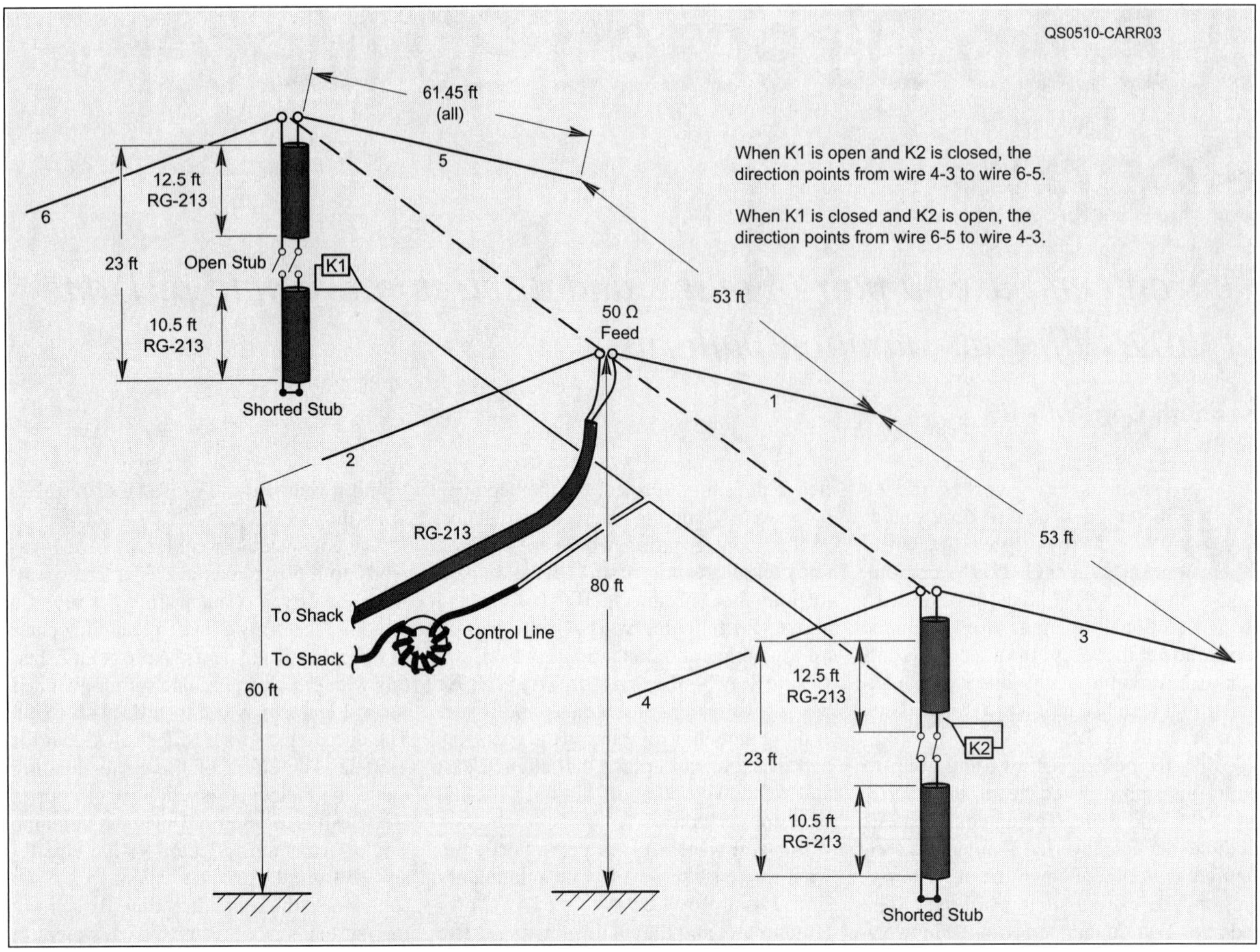

Figure 3—This drawing illustrates the construction of the stub-directed V antenna. Sections 1 and 2 are the driven element. Either of the other two V antennas can be switched to form a reflector or a director.

calculated the length of the capacitive stub to be 18.94 ft. We used RG-213 coax with a velocity factor of 0.66. You have to multiply the free space length by the velocity factor of the coax to get a physical length for the stub. This turned out to be 12.5 ft. We then calculated the $+j60\ \Omega$ inductive stub for the reflector element. This was longer than the capacitive stub (23 ft), and shorted at the end. We then checked our figures by measuring the capacitance using an RF impedance bridge. (Not all coax is created equal, it seems.) I used a circuit I built years ago from a *QST* project.[1] That project appeared in *The ARRL Handbook* for a number of years in the 1970s. You can make similar measurements with today's antenna analyzers, such as the MFJ-259B or the Autek RF1. Smith charts will give you a good starting point but if you can measure the impedance at the frequency you are going to use, the results will be more accurate.

We used relays to select the capacitive and inductive stubs. First, we subtracted the length of the shorter capacitive stub from the length of the inductive stub. Then we placed a relay at the end of each capacitive stub, to add the extra length for the inductive stub. With the relay open, the stub is 12.5 ft long and open at the end. This places a load of $-j115\ \Omega$ on the dipole to make it a director. With the relay closed, the length of the inductive stub is 23 ft, shorted at the end. This places a load of $+j60\ \Omega$ on the dipole and it becomes a reflector. We built both parasitic elements this way. Figure 3 shows a drawing of the antenna configuration.

The relays must be double pole, to break both the braid and the center conductor of the matching stubs. We connected the relay contacts to SO-239 sockets mounted into PVC pipe to keep out the weather. This also insulates the two stub lengths from each other when the relays are open.

We used dc relays to minimize heating in the PVC pipe insulators. I tried one antenna with ac relays and found they ran very hot because the PVC pipe would not dissipate the heat well. DC relays don't seem to heat up. On one unit I put diodes on the power transformer to supply pulsating dc to the ac relays. This seemed to cool them down as well. I think that reduced the losses enough to let the relays run cooler.

Connect the relay contacts braid to braid so they don't transpose the coax. When both relays are turned off, the antenna goes into the general coverage mode. This puts a $-j115\ \Omega$ load on both sides of the parasitic elements.

Power for the relays comes from a small control box in the ham shack. (See Figures 4 and 5.) Voltage from a 12 V transformer is rectified and fed to a wafer switch that selects the proper relay and lights up a display on the control box. Simple, but effective.

The control box selects a direction by switching in one relay. That element then

[1]J. Hall, K1PLP (now K1TD) and J. Kaufmann, WA1CQW, "The Macromatcher: An RF Impedance Bridge for Coax Lines," *QST*, Jan 1972, pp 14-20.

Figure 4—Relays are sealed in sections of PVC pipe, as shown in this photo. SO-239 coaxial connectors allow the relays to be connected at the end of a capacitive tuning stub to add the extra length of coax to make the inductive tuning stub. The control box is mounted in the shack and allows selection of one of three radiation patterns.

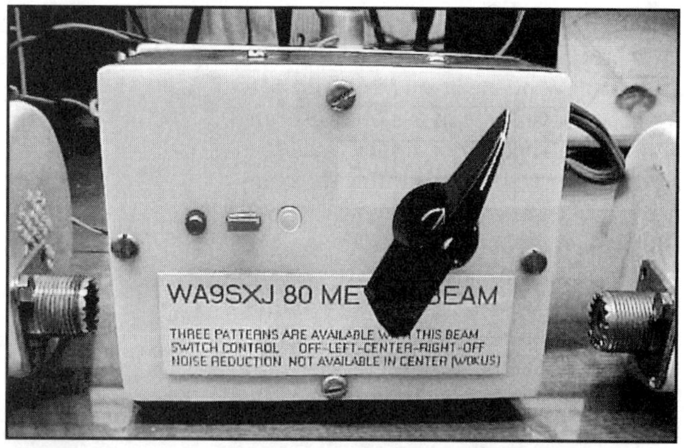

Figure 5—The control box in the shack includes a 12 V ac transformer and a diode to rectify the ac, which goes to a wafer switch. The switch selects either the right or left relay to change the direction of the antenna main lobe. The center-off position provides a pattern similar to a single dipole. LEDs on the control box indicate right, left or center-off positions. Additional wafer positions beyond the left and right selector positions turn off the unit.

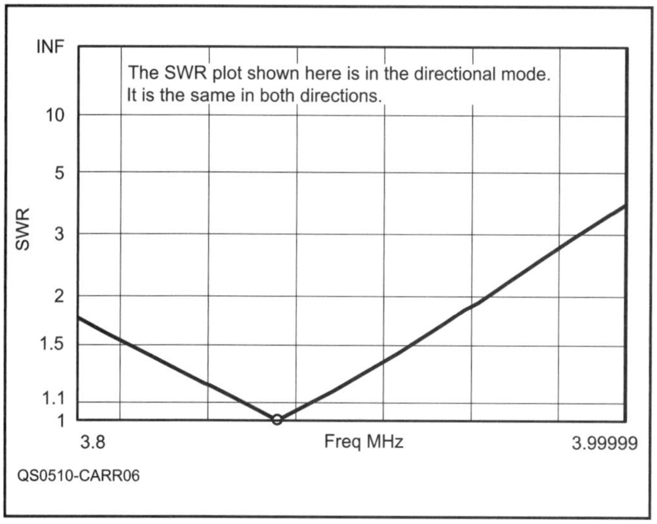

Figure 6—The antenna exhibits an SWR of less than 2:1 over much of the 75 meter phone band.

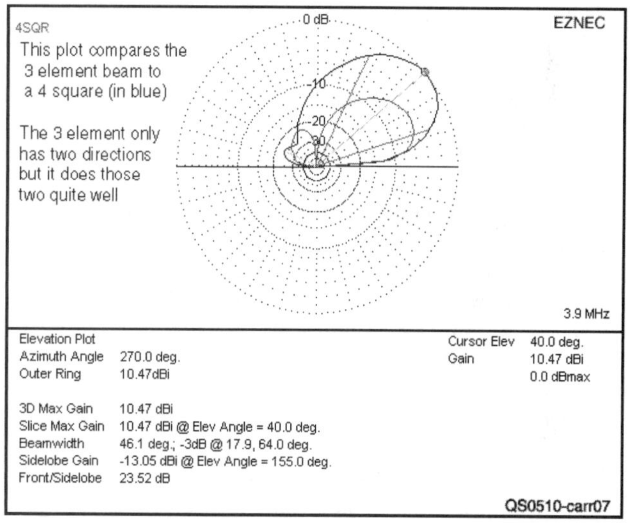

Figure 7—We used *EZNEC* to compare the stub-directed V antenna with a 4-square array.

goes to the inductive mode and becomes a reflector. The other is already set to be a director so you now have a 3-element beam. As you switch to the other direction it drops the reflector relay open and it then becomes a director. The opposite side then closes and that one becomes a reflector. The beam now points in the opposite direction.

Now for some checks. SWR was not bad so we set up a 100 W rig in a trailer at the antenna site to check it out. Figure 6 shows the SWR plot for the antenna. On the air checks were nothing short of spectacular. We had quite a group on the frequency and I was a hoping it would work as well as the computer said it would. From River Falls in Wisconsin we

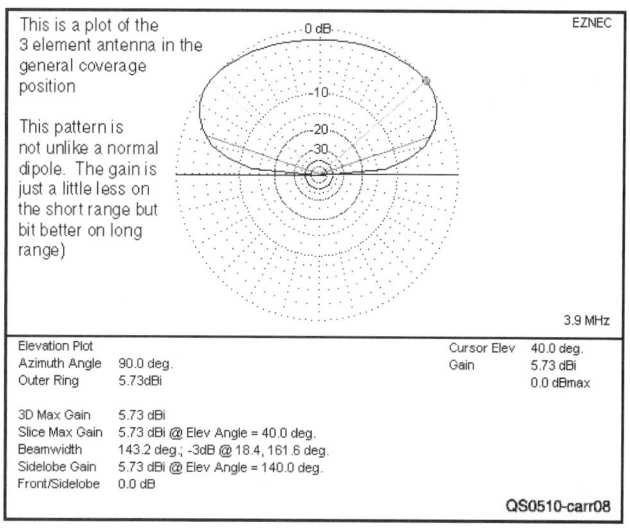

Figure 8—This *EZNEC* plot shows the general coverage pattern of the stub-directed V antenna. This is similar to a typical dipole antenna on 75 or 80 meters.

worked two mobile stations out west, one in Utah and another in California. In both cases, 100 W on the beam easily out-performed stations here in the Midwest running full legal power to dipoles and other antennas as well. In addition we compared it to a known good dipole that Ray already had in the air. It was always better on transmit and spectacular on receive. Signals that were barely audible on the dipole came in as much as 20 dB above the noise on the beam.

We then asked a ham friend in the Chicago area (WS9D) to check the front-to-back ratio. We were delighted to hear Kirby say that it was 25 dB. Later checks confirmed that it was not just that area but the same in the opposite direction as well. At this point we knew this antenna was a winner! It covered stations east and west better than anything we had ever heard or seen.

Results were almost exactly as *EZNEC* had predicted. In the general coverage mode it was still good, and was similar to Ray's dipole antenna. Using *EZNEC* we then compared the new 3-element beam with other antennas. Figure 1 includes the plot of a dipole antenna for comparison. Figure 7 compares it with a 4-square antenna. The pattern shown in Figure 8 is in effect when no relays are active in the stubs.

For Ray's installation we strung Phillystran between the two east-west towers and taped the tuning stubs to the catenary, along with the relays and their control wires. The control wires then went on to the center V, where they were taped to the feed line. We also put a ferrite choke in the control wires at that point just to keep any RF from getting into the control circuit wires.

Shortwave broadcast stations have used a similar technique to change the directional pattern of their antennas for years. It may be a bit of a challenge to build an antenna like this, but if you have the space and trees or other antenna supports, it can be worth the effort. You can also adapt the same idea to other bands. Have fun!

Ken Carr, WØKUS, was first licensed in 1952 as KZ5KC, while serving with the US Army in Panama, in the Canal Zone. Ken later served as a technical writer, a service representative, and held other positions with the Oliver Tractor Company. He was also an instructor at North Iowa Area Community College and held various positions with several construction equipment companies. Those jobs took him to every state in the union and many countries. And, obviously enough, Ken enjoys experimenting with antennas! You can reach the author at **w0kus@fiai.net**.

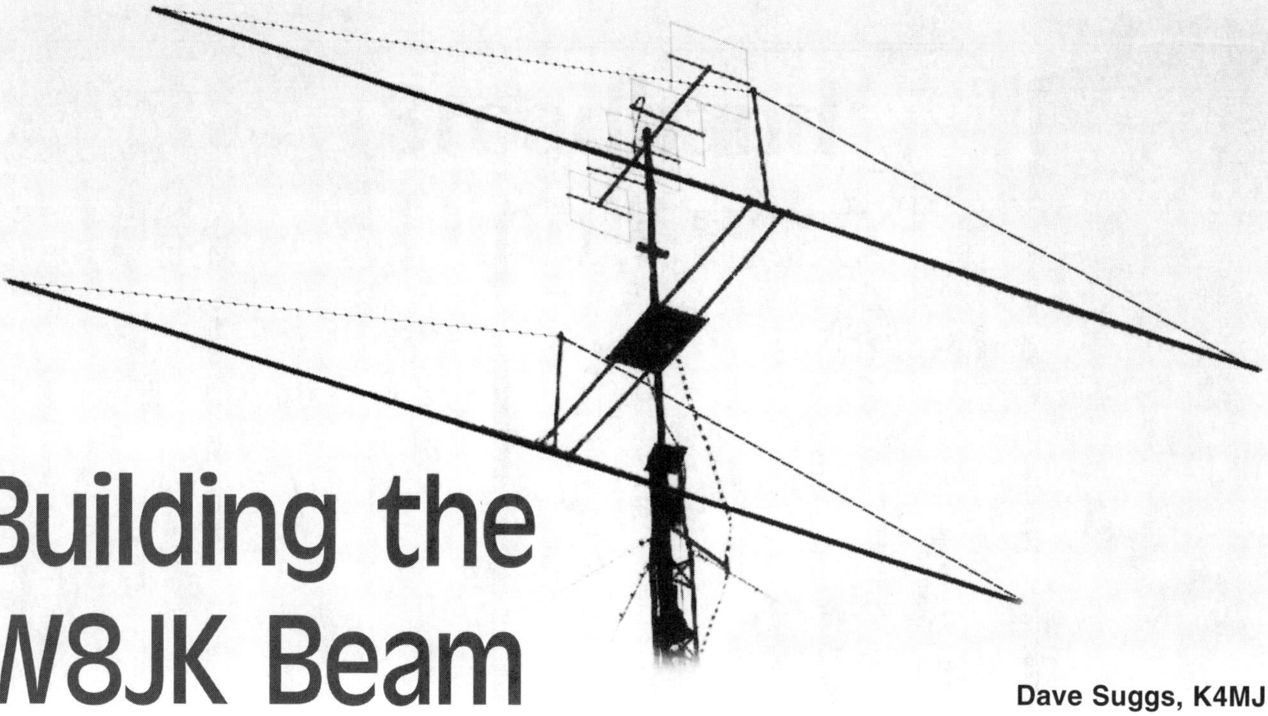

Building the W8JK Beam

Dave Suggs, K4MJN

Employing a mechanical design using readily available materials, here's a classic beam with gain on six bands.

Homebrew antennas have been a special interest of mine over my 47 years of ham radio. It began with wire contrivances and feed lines that drove my parents batty during my pre-teen SWL days. Now my dear wife's mandate is "Paint the tower dark green and I don't want to see any antennas out any window. *Period*!"

With those parameters established, I've fashioned dipoles, V beams, loops, Zepps and other wire arrays over our small acreage in Blythewood, South Carolina, for over 31 years. They're always out of sight from the windows!

Antennas fed with open wire line and a tuner have been *de rigueur* for many years. You can load up most anything with a tuner, but getting a decent signal to radiate is something else. For design, I defer to the antenna experts. All of my antenna projects have, unabashedly, been based on articles by those amateurs whose knowledge of RF current, takeoff angles and patterns has been held in high esteem.

After acquiring a well-weathered 40 foot piece of tower, I decided to build a beam for the first time. I searched my collections of past *QST* magazines, old *ARRL Handbooks* and the League's antenna books. Anything needing coax, traps or heavy duty rotators lost my interest. I looked for something lightweight, with modest gain, multiband capability and a low radiation angle for DX. It also had to be forgiving if my measurements might be a bit off.

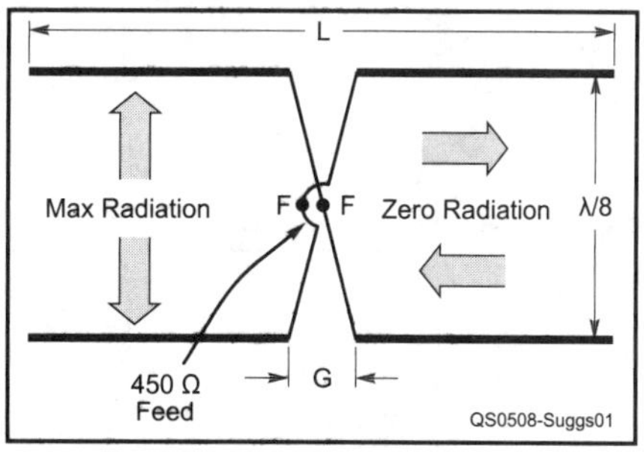

Figure 1—The classic W8JK antenna configuration uses two elements fed out of phase and spaced about $^{1}/_{8}\,\lambda$ apart at the lowed frequency of operation. The 450 Ω twin lead feed connects in the center at FF.

The W8JK Antenna

Thumbing through my treasured 1948 *ARRL Handbook*, I became increasingly interested in the end-fire array designed by Dr John Kraus, W8JK, in 1937. The design was nearly as old as I and the fit was perfect! Searching more, I found his article in the July 1970 *QST*.[1] Then the June 1982 *QST* had an updated article.[2] It was definitive information I could digest.

Here was a simple, compact *six band* antenna that would cover 20 through 6 meters, including 17 and 12 meters, with one feed line, two elements and no traps or loading coils! Not only that, it didn't have to be too high, nicely fitting my 40 footer. The gain ranges from about 3.75 dBi on 20 meters to 6.75 dBi on 6 meters.[3] Not bad for a small multiband beam!

Figure 1 shows the antenna configuration described in W8JK's 1982 *QST* article. The W8JK beam consists of two dipoles spaced $^{1}/_{8}\,\lambda$ apart at the lowest frequency used and fed 180° out of phase. The total length (L) of the dipoles can range from less than $^{1}/_{2}\,\lambda$ to more than $^{3}/_{2}\,\lambda$. I was interested in using the antenna for 20 through 10 meters, so I followed the dimensions in the 1982 article. W8JK determined that if the L dimension is about 24 feet (somewhat less than $^{1}/_{2}\,\lambda$ on 20 meters), the antenna can be used on all bands from 20 meters through 6 meters. This antenna has a clean bidirectional pattern over a 2:1 frequency range (14 to 28 MHz). Coverage can be stretched to 3:1 (14 to 50 MHz), but the

Title photo: Flying high and chasing DX! Topped with a 2 meter quad, this array was originally installed in 1985 and survived Hurricane Hugo in September 1989 with no damage. The flexibility of PVC construction and dacron guys on the elements seem to add longevity to this array in high winds—18 years of service with no problems.

[1]Notes appear on page 59.

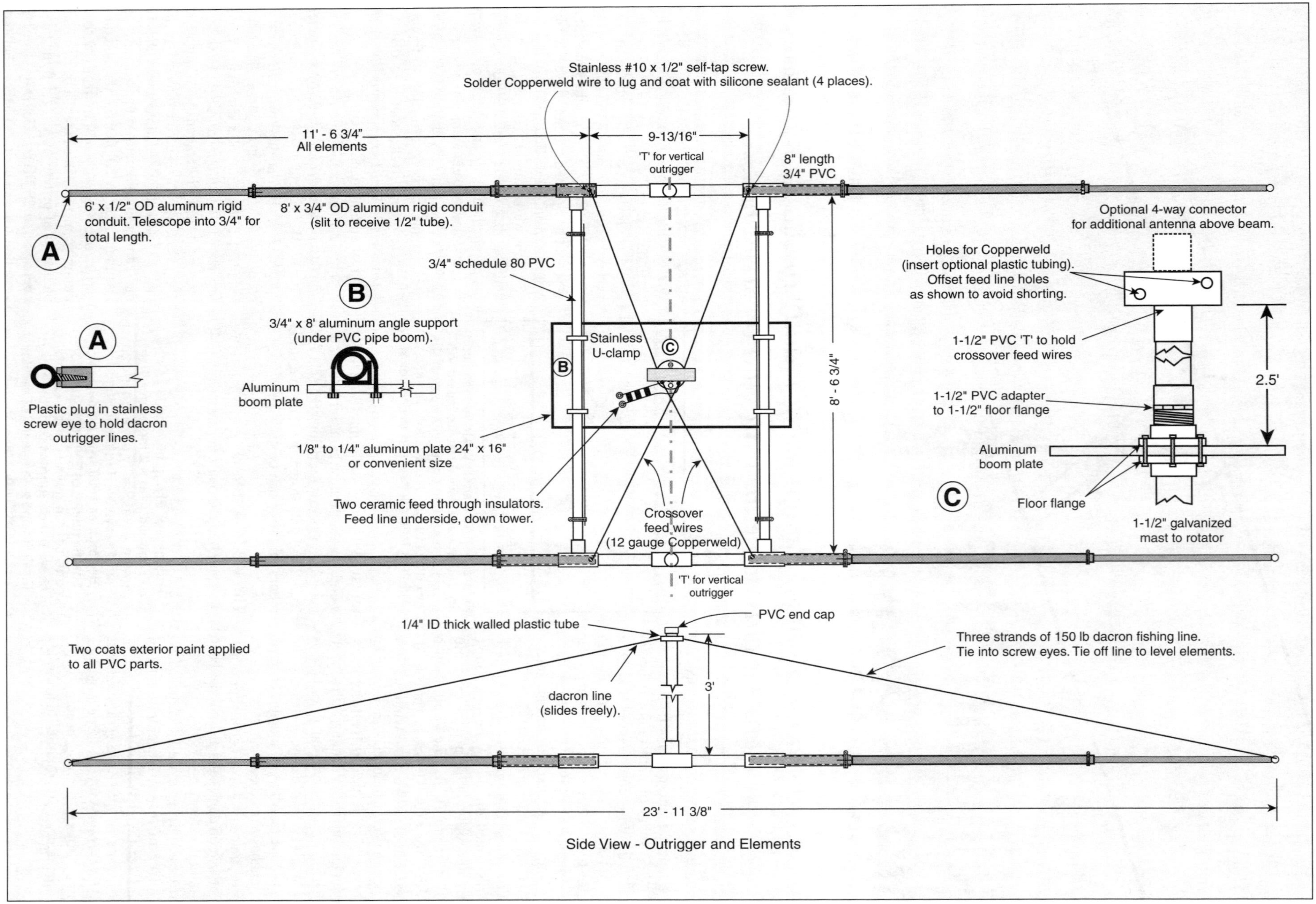

Figure 2—Construction details of the author's W8JK array. Inexpensive PVC and aluminum conduit make for an easy to build DX antenna from the early days of Amateur Radio.

pattern starts to deteriorate at the high end. The gap between element ends (G) isn't critical. I found I could reasonably construct a beam with roughly 24 foot overall elements and a 9 foot boom. The rotator could be lightweight and, with a thrust bearing, very manageable on the tower.

I decided to use standard PVC pipe and aluminum conduit readily available at most home centers and hardware stores for building the array. The PVC material on the first beam I constructed was schedule 40. My second version used schedule 80 pipe, which is a bit sturdier. Wherever possible, I used stainless hardware throughout the construction. Figure 2 shows the construction details.

Boom Construction

A single boom, even one made from large diameter PVC, seemed out of the question. What kind of boom-to-mast bracket could I fashion with 2 inch or larger PVC? Rigidity was required and there's too much wiggle and deflection with any length of PVC. I decided to use two $\frac{3}{4}$ inch parallel PVC sticks that U clamped to an aluminum plate.

I had a plate in the junk box that was 24 × 16 inches, $\frac{1}{8}$ inch thick, with a 1 inch, 90° lengthwise lip that added strength. It once served as a real estate sign. You can be flexible here by making it longer, wider or thicker. Quarter inch plate would be even better.

Two stainless steel U clamps secure each PVC boom pipe to the aluminum boom plate. As shown in detail C in Figure 2 and in Figure 3, two $1\frac{1}{2}$ inch galvanized floor flanges are centered on either side of the plate for mast attachments. I secured these flanges with stainless machine screws and lock washers. The bottom flange threads nicely into the $1\frac{1}{2}$ inch galvanized pipe mast, which mounts to the rotator through the thrust bearing. A $1\frac{1}{2}$ inch PVC adapter screwed into the top flange holds a vertical $1\frac{1}{2}$ inch PVC pipe 30 inches long to support the crossover wires and feed point.

I used 12 gauge Copperweld wire for the element's crossover feed. The crossover wires also serve as a guy wire structure, adding to boom rigidity. You could also use stranded wire here, but I like the Copperweld because it's strong and won't stretch over time. After some experimentation, I came up with the idea of clamping $\frac{3}{4}$ inch aluminum angle underneath the PVC boom pieces for additional strength. The aluminum angles have no effect on antenna performance.

Next, use more $\frac{3}{4}$ inch PVC pipe and T adapters to make the element mounts at the ends of the boom. These pieces also allow for element support outriggers and add rigidity to the boom structure. It's a good idea to get everything cut and lined up before using PVC cement to lock it all in place. Use a hacksaw to cut a slit at the end of each 8 inch PVC element mount and make sure the inside is smooth.

The Feed Point

The 30 inch PVC pipe glued into the adapter in the top flange is topped off with a PVC T or 4-way joint to support the crossover feed wires (Figure 4). I used a 4-way joint and added a short mast above it to support a small homebrew 2-meter quad. This arrangement will support any short, lightweight VHF/UHF antenna or vertical.

Place end caps on the T and drill $\frac{3}{16}$ inch holes as shown. I inserted small poly tubing through the holes to help with abrasion from the crossover wires. It may be necessary to add a length of plexiglass or a PVC spreader here and there to keep the crossover wires from shorting together.

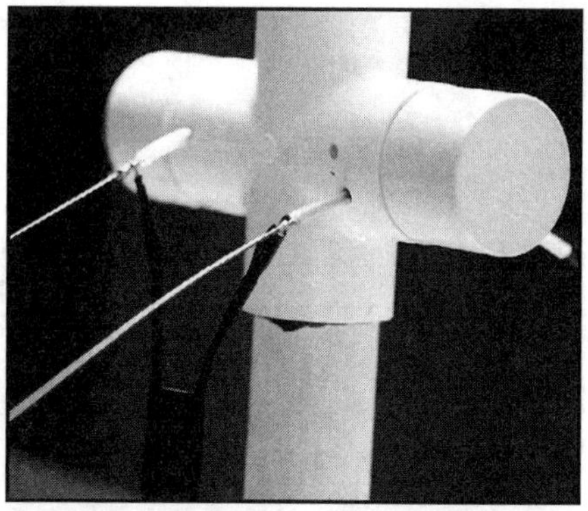

Figure 4—A PVC T or 4-way adapter supports two element crossover wires above the boom and provides a convenient attachment point for the 450 Ω feed line. The crossover wires are made from no. 12 Copperweld wire and double as guy wires for the twin boom.

Figure 3—Galvanized pipe floor flanges are used top and bottom to connect the mast and crossover feed support. This is the top flange with a PVC adapter. The hole to the right of the flange is for a coaxial feed line for the 2-meter quad mounted above the W8JK array. Ceramic feedthrough insulators on left are for the 450 Ω ladder line feed.

Next drill two holes in the boom support plate and install a couple of ceramic feedthrough insulators. Location is not critical as long as you clear the mast. Once the elements are finished and everything is in place, you will solder a piece of 450 Ω ladder line at the midpoints of the crossover feed wires and feed the ladder line over to the feedthrough insulators on top. The other end of the feedthroughs will connect to the feed line to your station.

Building the Elements

The elements are made from $\frac{3}{4}$ inch PVC pipe and $\frac{3}{4}$ and $\frac{1}{2}$ inch rigid aluminum conduit. You could use aluminum tubing here and save a little weight, but the conduit is easily available and has worked just fine for me.

As shown in Figure 2, carefully hacksaw a 1 inch slit at one end of an 8 foot piece of $\frac{3}{4}$ inch conduit to receive the $\frac{1}{2}$ inch element section. Good electrical contact between the two pieces of conduit is a must! Burnish the outside of the $\frac{1}{2}$ inch piece to a high shine with 000 steel wool. Remove any rough edges and burnish the inside of the $\frac{3}{4}$ inch piece as well. This removes any aluminum oxide that is present. A good conductive grease is appropriate here too. Slide the elements together to measure out to 11 feet, $6\frac{3}{4}$ inches as shown in Figure 2. Clamp the element joint securely with a stainless $\frac{3}{4}$ inch hose clamp and check continuity with ohmmeter.

Lay the completed aluminum element sections alongside the slit 8 inch PVC element mounts and position them so they will end at the far side of the PVC T section on the boom. Mark each aluminum element at the slit end of the PVC and then slide the elements into the PVC section up to the mark. Hose clamp securely.

Figure 5—The crossover feed wires are attached to the element ends with large solder lugs and stainless sheet metal screws. Weatherproof the connection with a blob of silicone sealant.

Figure 6—Detail of the outrigger support tube and dacron element guy lines. The plastic tubing reduces wear and allows the dacron to slip as needed when the elements flex in strong winds.

The crossover feed wires are connected to the elements with 1/2 inch long #10 stainless self-tapping sheet metal screws and heavy duty, ring-type solder lugs. Make sure the screws are long enough to pass through the lug and PVC and get a good bite in the conduit. Drill a hole smaller than the screw into both the PVC T and the element tubing (Figure 5). You want to make the contact right at the end of the element.

Feed the crossover wires through the holes in the T at the top of the mast and cut them a bit longer than needed to get to the element mounts. Attach solder lugs to the ends of each wire on one side of the boom and wrap a couple of turns for strength. Screw these lugs into the elements.

Pull the other ends of the wires until there is no slack. Then wire up the solder lugs and screw them to the elements. Keep the crossover feed wires taut and be sure to wire them up out-of-phase as shown in the drawings. If you encounter shorting of the crossover feed, add a dog bone insulator or suitable piece of PVC to separate them. Once everything is in place, solder the lugs. Solder the ladder line to the crossover wires as described earlier, check continuity again and cover all connections with clear silicone sealant.

Element Outriggers

The aluminum elements need support, as the PVC element holders are not strong enough by themselves. I used more PVC and some fishing line to make outriggers, as shown in Figures 2 and 6. Each element has a center support made from a 3 foot piece of 3/4 inch PVC, mounted vertically in the T section at the element center. I ran a support line from one element tip, through the center support, to the element tip at the other end. The support lines are made from four strands of 150 pound test dacron fishing line (dacron will weather better than monofilament).

The dacron lines are not tied to the center supports. I drilled a hole near the top of each support and inserted a piece of 1/4 inch poly tubing. The dacron line runs through the tubing and can slip back and forth in windy conditions. The ends of the dacron lines are tied to stainless screw eyes in plastic plugs inserted into the element ends. Caps seal the outrigger tops.

This completes the antenna assembly. Make sure that all PVC joints are glued and the hardware is tight. I applied two coats of dark green Rustoleum exterior paint to all of the PVC pieces to help retard the effects of sun and weather.

Tower Feed

Some care is required in feeding this antenna because of the very low radiation resistance. See the sidebar "Feeding the W8JK Beam" for more information on various approaches. For convenience, I opted to run 450 Ω ladder line from the antenna to my house. Then I put a 4:1 current balun on the outside wall and ran RG-8 coax into the shack to a tuner.

Ladder line *cannot* be cable-tied along a steel tower! I made standoffs from 3/4 inch PVC T adapters, with a pipe length that will hold the ladder line 3-4 feet away from the tower. See Figure 7. The 3/4 inch T sections are perfect for threading and holding 450 Ω ladder line. You may wish to add a black UV resistant cable tie and loop it through the T and a window in the ladder line to keep it from slipping through. I use the same system to bring the feed line along the house to the balun.

At the top of the tower, tie off a standoff long enough (at least 5 to 6 feet) to leave a sufficient loop of ladder line to allow element rotation of 180°. (The pattern is bidirectional, so you don't need 360° rotation like a conventional beam.) You don't want it wrapping around the mast, so allow plenty of slack to keep it away. Attach the feed line to the feedthrough insulators under the boom plate. I used ring lugs with lock washers soldered to the feed line on the feedthroughs. Coat both upper and lower connections with silicone sealant.

If you use another antenna with coax feed above the beam, build another long standoff at 90° to the ladder line standoff and allow

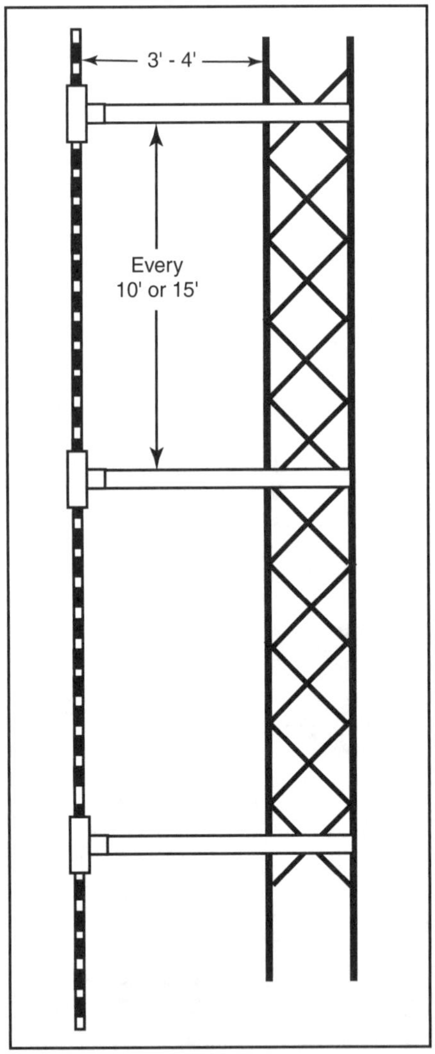

Figure 7—The 450 Ω ladder line runs through 3/4 inch PVC standoffs spaced down the tower. Attach the feed line to the standoffs with cable ties.

Feeding the W8JK Beam

The short elements and close spacing of this classic W8JK beam make it an attractive multiband antenna for low heights and limited space. The one drawback of the design is its very low radiation resistance—on the order of 2 Ω at 14 MHz. This makes feeding the antenna a bit more challenging than other designs, particularly on 20 meters.

In his 1982 article, John Kraus, W8JK, described a "trombone" matching system to short the feed line at a current maximum (any multiple of $^1/_2\ \lambda$ from the feed point) and couple a 50 Ω feed line to the shack at this point. The trombone is essentially a section of 300 Ω twin lead made from telescoping aluminum tubing. Its length can be adjusted to find the matching point for the various bands. Although this arrangement provides a good low-loss match to 50 Ω, you need to go outside and adjust the trombone section length to change bands.

Kraus also suggested that the antenna could be fed with twin lead, with tuning done at the station end. This approach makes for convenient band changes, but, as Kraus pointed out, brings high-voltage points on the feed line into the station. Also, because of the high SWR, losses in the tuner and feed line—even with low-loss twin lead—can negate any gain the antenna offers at the low end of its frequency range. For example, line loss with 100 feet of ladder line is about 3.6 dB at 14 MHz. Losses drop considerably as the SWR improves at the higher frequencies.

Initially, I decided to run ladder line all the way into my shack and use a tuner for quick band changes. I had some trouble with RF floating around the shack, so I brought the ladder line to the outside wall of the shack, fed it into a 4:1 current balun rated for 2 kW and ran about 15 feet of RG-8 coax to the tuner. I've been happy with this feed arrangement over the years and have had many enjoyable QSOs with the antenna. In preparing this article for publication, the ARRL technical folks pointed out that computer modeling shows that losses in the feed line (twin lead and coax), balun and tuner add up quickly to 8 or 9 dB on 20 meters. In addition, the high voltage present at some points on the feed line can stress a conventional 4:1 balun. The loss drops quickly on the higher bands and hasn't bothered me, as I operate mostly on 15 meters and like the convenience and other features of the antenna. If you decide to try one of these W8JK beams, you may want to consider a different feed arrangement.

Thanks to ARRL Technical Advisor John Stanley, K4ERO, and *QST* Assistant Technical Editor Joel Hallas, W1ZR, for modeling the antenna and feed options and providing information for this article.

Figure 8—Here's the author's completed W8JK beam before painting the PVC with green paint for weather and sun protection. The green also helps the antenna blend in with the trees.

enough slack for rotation. Run the rotator several times to make sure you can turn at least 180°. This does limit a VHF/UHF quad. My "druthers" should have told me to use a VHF vertical on top rather than a quad. No more rotation problem!

On the Air!

After many summer weekends building (and often modifying midstream), the beam was ready to go up the tower (see Figure 8). The final mounted height of the antenna measured out at 47 feet (about $^3/_4\ \lambda$), so peak gain is at about 22° according to W8JK's article. The half-power beamwidth is 62°.

My antenna tuner matched the rig perfectly on all bands from 20 to 10 meters. Over the years, I've had lots of great US and DX contacts. Normally I run barefoot or an 800 W amplifier and can work anyone I can hear. I've found the W8JK to be a very effective DX contest antenna. I've not tried it on 6 meters, but it would be interesting to hear how it would work there.

The W8JK is bidirectional. This pattern has its pros and cons, but I've found it to be an asset. I've had numerous contacts on 15 meters with old friends, one in Boston and the other in the Virgin Islands. We would work each other on Sunday afternoon schedules with regularity and strong signals for all. On 20 meter nets, I've heard many a net control say he's turning the beam around to pick up someone on his backside. This bidirectional feature is a great antenna for a net control station, saving time for rotation.

Hope "you alls" like this "Bubba Engineering." I only regret I didn't find any use for duct tape in the construction. Build one and enjoy!

Notes

[1]J. Kraus, "W8JK 5-Band Rotary Beam Antenna," *QST*, Jul 1970, pp 11-14.

[2]J. Kraus, "The W8JK Antenna: Recap and Update," *QST*, Jun 1982, pp 11-14.

[3]John Kraus's 1982 *QST* article shows gain for this design ranging from 5.7 dBi on 20 meters to 8.2 dBi on 6 meters. Modeling with modern computer tools indicates that more realistic figures are 14 MHz, 3.75 dBi; 18 MHz, 4.3 dBi; 21 MHz, 4.5 dBi; 24 MHz, 4.6 dBi; 28 MHz, 4.9 dBi; and 50 MHz, 6.75 dBi. The computer model takes into account the very low radiation resistance of the design and real-world losses in the feed system and elements. Performance could be improved by lengthening the elements and optimizing the feed system.—*Ed.*

Dave Suggs was first licensed as K4MJN in 1958 and holds an Extra class license. He is a graduate of the US Air Force Radio Operator School and earned a BS degree from Clemson University. Dave has been self-employed as a graphic designer since 1966. You can reach him at 5501 Hardscrabble Rd, Blythewood, SC 29016 or by e-mail at **k4mjn@arrl.net**.

A Wire Antenna Combination for DX

KTØNY takes two vintage flat-tops with tuned feeders, adds a wire Yagi and gets DX performance suitable for the 21st century. Will it work at your location?

Tony Estep, KTØNY

Many hams are faced with cold realities that force a compromise antenna installation. This article is dedicated to amateurs who are in that situation, but still want to be competitive in DX pileups. It describes one amateur's approach to a low-visibility antenna setup that covers a lot of directions, gives acceptable gain, and works on a wide range of frequencies. As you'll see, my scheme involves switching among three inconspicuous antennas. Two are combined into an "over-and-under" two-radiator long wire-dipole combination and the third is a Yagi that lies flat on the roof.

In my neighborhood, a tower is out of the question. Even an end-fed vertical won't do, since there's no way to have a radial system in my yard. Fortunately, however, antenna supports in the form of trees are conveniently provided. Looking at the two trees that are close enough to my shack, I tried to sketch out something that would give me a chance of a reasonable signal to DX locations. I wanted to get decent performance in as many directions as possible, with a low-visibility installation. Moreover, I want to work all bands from 80 to 10 meters.

I could readily see how I could put up at least one pretty good wire antenna between trees, but it appeared that my geographical coverage would be limited. The wire would run North-South, so I'd be shooting a good signal to Africa and Australia, but Europe would be 45° off of the perpendicular, and Asia would be right off the end of the wire. I didn't want an antenna that would give good gain in one or two directions while leaving the rest of the world unreachable. After scratching my head over this dilemma, I invested in Roy Lewallen's *EZNEC*[1] antenna analysis software. The $89 was some of the best money I've ever spent in ham radio.

The KTØNY Solution

I started with the most obvious choice, a center fed dipole running from the north tree to the south tree. I chose a length of 56 feet, equal to an extended double Zepp on 15 meters. This antenna gives a figure-8 pattern pointing East and West. Its 20 meter pattern is shown in Figure 1.

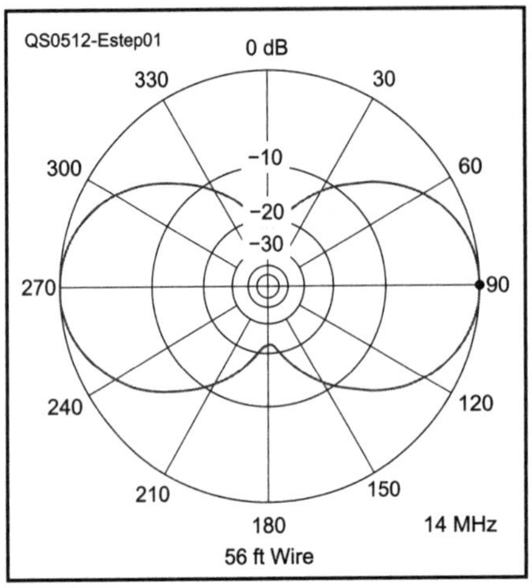

Figure 1—20 meter pattern of a 56 foot wire, oriented N/S and center-fed, at 40 feet; the low-angle radiation is E/W. (Important note: All patterns shown are for 8° take-off angle.) The length is chosen because this forms an extended double Zepp on 15 meters. This dipole will give limited directional coverage at low angles on 40, 30, 20, 17 and 15 meters.

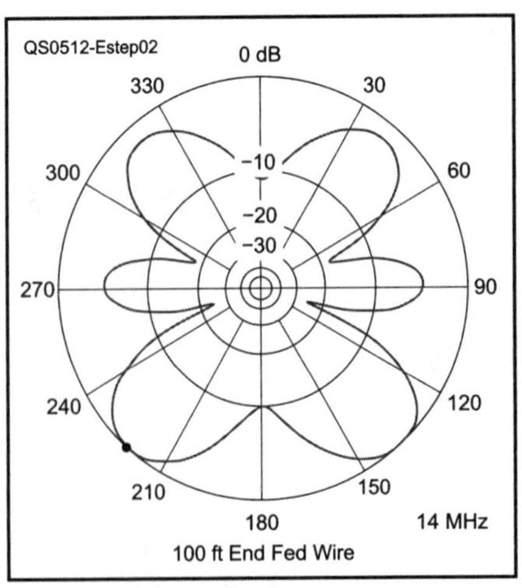

Figure 2—20 meter pattern of a 100 foot N/S oriented long wire, fed 13% from the north end, at 40 feet. This gives coverage that nicely complements that of the center-fed dipole.

Some tinkering with *EZNEC* gave me an idea for a way to supplement this pattern. Consider Figure 2. But the two antennas can't occupy the same space—or can they? I quickly concluded that they could. I sketched out the arrangement shown in Figure 3.

Both antennas are fed with tuned feeders. Figure 4 shows the support details. At first I brought the feeders into the shack and tuned them with a balanced tuner. At present, the lines lead to 4:1 current baluns right outside the shack window; a 6 foot length of coax brings them in to a standard unbalanced tuner. Both systems work, but I prefer having only coax inside the shack. My Ten-Tec 238A antenna tuner has a convenient antenna switch that lets me select my directional pattern—the poor man's rotator!

The idea is that I can switch back and forth between antennas, so that one of the major radiation lobes will be pointed (more or less) at the DX I'm trying to work. When you put the two patterns together, the coverage is as shown in Figure 5.

This looks good. The gain is a little higher to SE and SW than to NE and NW, which is the opposite of what I would like, but that's an unavoidable consequence of feeding the longwire 13½ feet from the north end. Of course, at your station the wires may have to run in a different direction. No matter. You should be able to cover many compass points, however your house and lot might be oriented.

The story is not quite as favorable as Figure 5 depicts, however. Using *EZNEC*, I modeled the actual shape of the system, including the droop in the center, the effects of transmission lines in the radiation field, and the interaction between the two antennas. These effects reduce the gain somewhat.

This is unfortunate, but can't be helped. Figures 6 and 7 give a realistic appraisal of

[1]*EZNEC* software available from **www.eznec.com**.

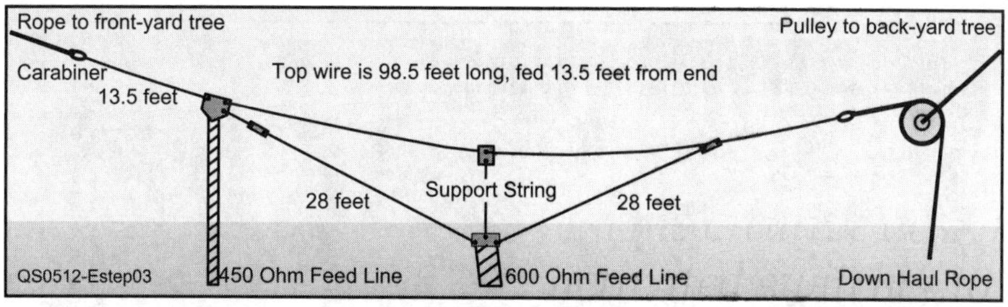

Figure 3—As constructed, the KTØNY "over-and-under" antenna consists of an off-center-fed 98.5 foot wire with a 56 foot center-fed dipole hanging below it. The ends are at about 42 feet, the low point (at the center of the dipole) is at about 32 feet.

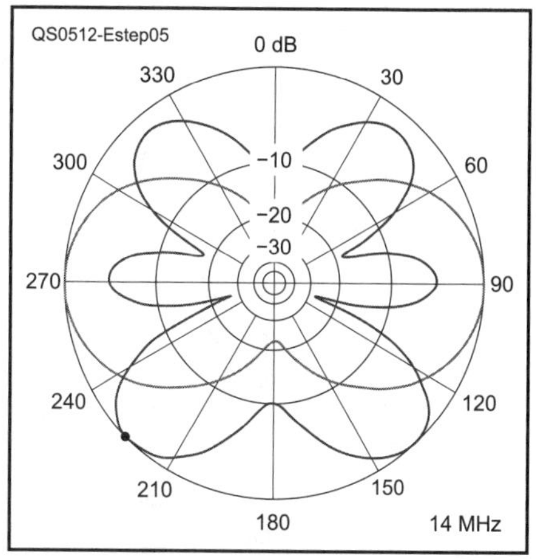

Figure 5—Total coverage of theoretical wire combination on 20 meters. Patterns on 15, 17 and 30 meters have generally similar shape.

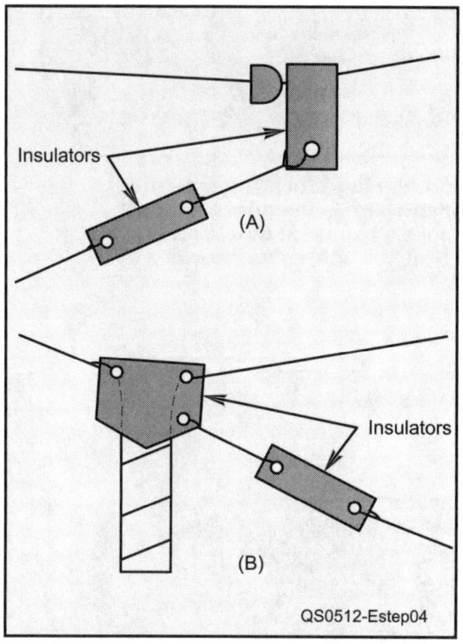

Figure 4—Details of the fittings for hanging the dipole below the long wire. Parts are made of ⅜ inch thick Delrin. At (A) right end support, at (B) left end support.

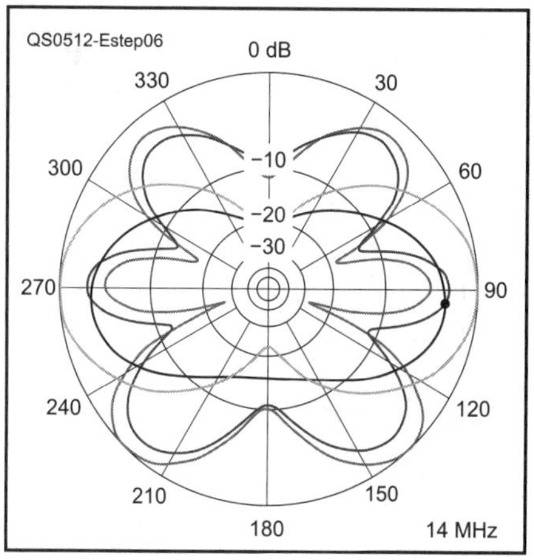

Figure 6—Ideal 20 meter patterns (green, red) and predicted pattern (black, blue) superimposed. The droop in the middle and the effect of interactions degrade the results, especially for the center-fed dipole.

the gain achieved. Very roughly speaking, on both 20 and 15 meters, the gain of the antenna system in the direction of its favored lobes is almost as good as that of a 2 element Yagi at 25 feet as shown in Figure 8.

Due North

The composite pattern figures reveal that the antenna so far puts little signal directly north. This bugged me, because some truly juicy DX (4S7, XW, XX9, and others) lie right in that northerly null. So I hit on a desperate idea for one last DX antenna. It is good for only one direction, on one band, but on the plus side it is almost completely invisible.

I built a single-band, single-direction wire Yagi, for 20 meters only, consisting of two wires lying right on the roof. They are about 20 feet above the ground, one on either side of the peak of the roof (in other words, the interaction between the elements takes place through the roof). Because the feed line has to exit to the south, I had to use a driven element/director combination, which is not as good as a driven element plus reflector. But it does work, and the resulting 20 meter pattern is shown in Figure 9.

A little experimentation with *EZNEC* will reveal that the pattern and gain deteriorate rapidly as you go above the optimal frequency. Therefore, the best setup is to tune the antenna for the high end of the 20 meter band. The SWR and gain will be acceptable across the whole band. I discovered that there are some tricks involved in actually constructing this antenna. After some trial and error, I found a simple procedure that worked well. (It will be much easier if you have an antenna analyzer.)

Make the two elements as shown in Figure 10, and cut them a bit oversize (note that dimensions are for insulated 14 gauge wire). Take them up to the roof, along with a short length of coax and your analyzer. Lay out the driven element, hook it to the analyzer, and find the frequency at which it displays lowest SWR. Trim it until the lowest SWR is at 14.300, even if you plan to operate mostly CW.

Now lay out the director right next to the driven element, and cut it to be 15 inches shorter. Then install it 13½ feet in front of the driven element. You now have a full-size, wide-spaced 20 meter Yagi.

The SWR curve will probably be pretty flat across the band. The *EZNEC* modeled curve has a fairly pronounced dip, so the "improvement" is due to losses caused by proximity to stuff in the near field. If you're going to put the antenna right on the roof, this can't be avoided. Coupling to the household wiring is a bad thing, of course, and the more you can elevate the wire Yagi, the better it is likely to perform. But low and lossy as it is, mine gives a boost of up to 10 dB over my long wire on signals coming from due north.

DX Results

The total setup is unobtrusive. Few notice it. From the viewpoint of the neighbors, all that is really visible is some window line running up into the front yard tree. The rooftop Yagi is almost impossible to see.

And, yes, it does work DX. After a long period of inactivity, I put up my first version of this antenna and became active in February 2004. In the ensuing 12 months, operating after work in the evenings, usually with limited power, I've worked 251 countries. I have snagged DXCC on phone, cw and digital, and I just completed DXCC on QRP. My wires have propelled my signal through the pileups on just about every DXpedition that has been on since I've been active. And best of all, I have managed that most elusive of my personal goals, WAZ.

The complete configuration is shown in Figure 11. There are limitations and frustrations, of course, but they are a lot less than I expected when I first contemplated a wire antenna. I hope that this article contains some ideas that may be helpful to others in similar situations.

Tony Estep, KTØNY, was first licensed in

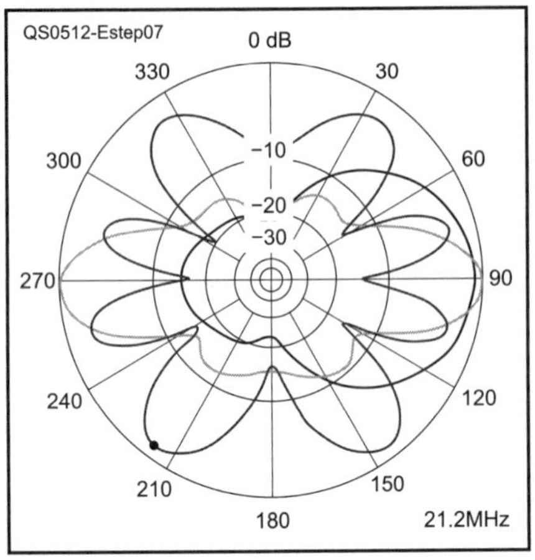

Figure 7—On 15 meters, the radiation in favored directions is roughly comparable to a 2 element Yagi (black) at 25 feet.

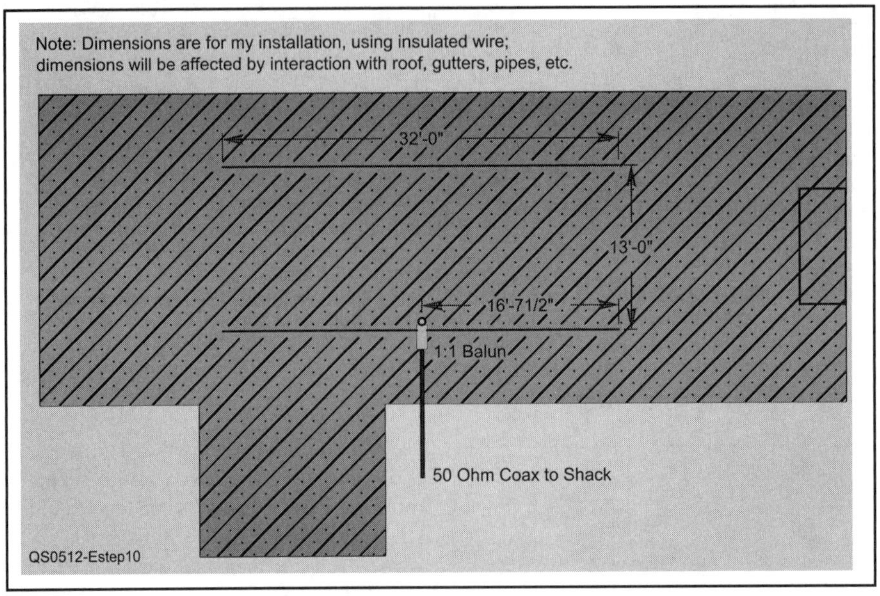

Figure 10—Two wires lying right on the roof, and running E/W at a height of 22 feet, form a fixed, single-band Yagi. The southern wire is the driven element, and the northern one is a director. To build it, make a balun-fed dipole resonant at the high end of the band, then add a director 15 inches shorter as shown. Dimensions shown are for PVC insulated wire.

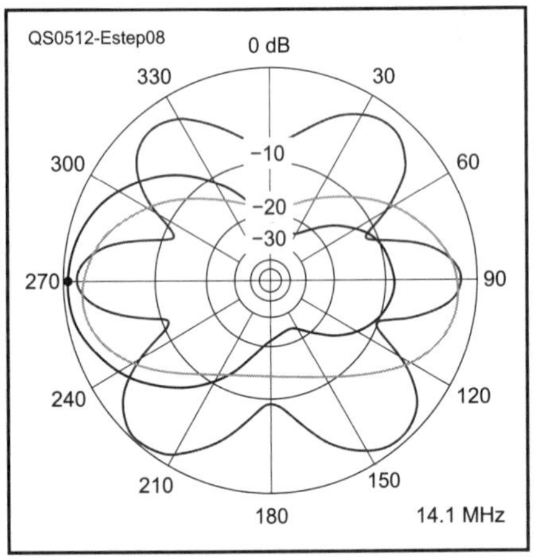

Figure 8—On 20 meters, the best gain is again similar to a low 2 element Yagi. It doesn't have much gain over a dipole, but covers more territory. On 40 and 80 meters, coverage is essentially omnidirectional. On 12 and 10 meters, it's mainly NE/SW and NW/SE.

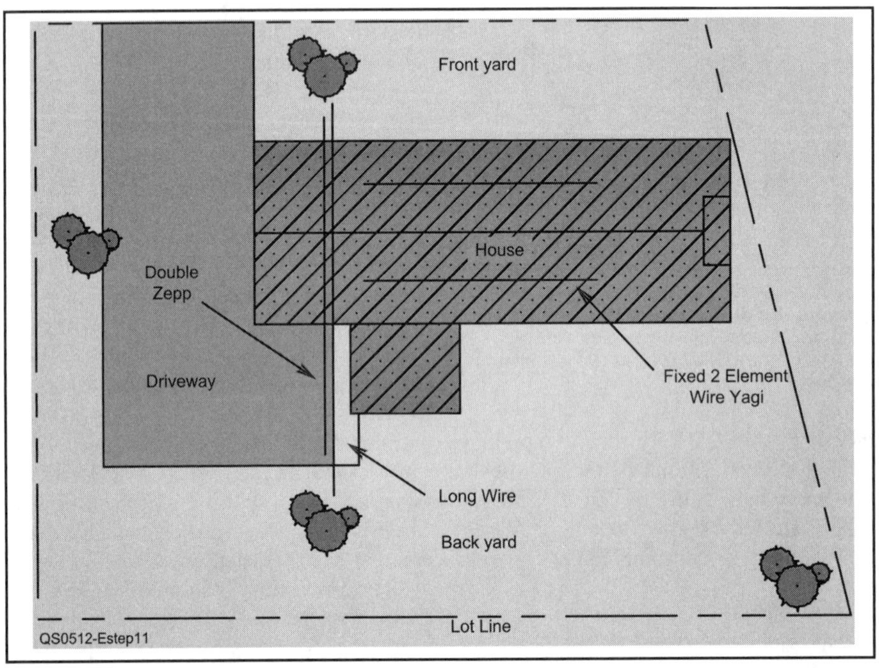

Figure 11—My house and yard are in a typical suburban configuration with neighbors on both sides. Still, I wound up with an antenna installation that performs well, without creating any neighborhood strife.

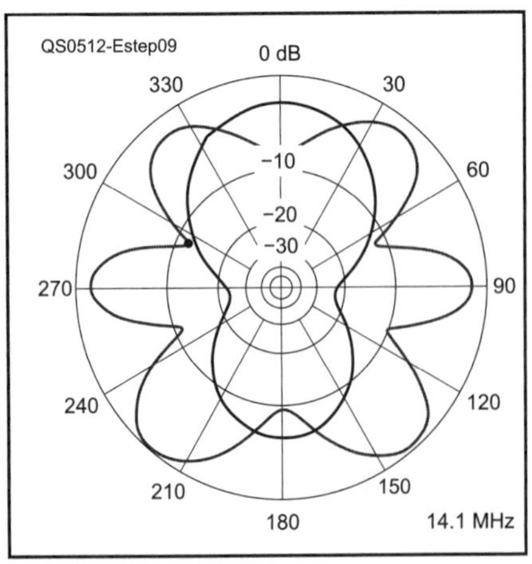

Figure 9—The wire Yagi on the roof works only on 20 meters, and only in one direction (black), but it fills in the null left by the other antennas, in the direction of the DX I need most! For construction, see Figure 10.

St Louis, Missouri in 1957 as KNØLTB at age 17. The sunspots were going crazy at the time, very different from today's conditions. Tony managed to complete his DXCC as a Novice with a Hallicrafters SX-96 receiver and a home-brew transmitter.

He later moved to London, where he held the call G5ANZ. When he returned to St Louis, he got his Amateur Extra class license, set up a shack, and after various tries with wire antennas eventually evolved the antenna scheme described in this article. It has accounted for WAZ, DXCC with QRP, and a lot of fun.

Tony holds a BA degree from Washington University and recently retired from a 43 year career on Wall Street, most recently as Managing Director of Bank of America Capital Management. While Tony is the author of many publications in the field of finance, this is his first in Amateur Radio. Tony can be reached at 541 Hickory Ln, St Louis, MO 63131 or **kt0ny@arrl.net**.

The Compact Quad Multiband HF Antenna

An easy to build loop with a simple matching section offers four band HF coverage.

Andrew Roos, ZS6AAA

When I relocated to Johannesburg I needed a new multiband HF antenna. Since I was staying in a rented house, a tower was out of the question, but fortunately my lovely wife had taken my not-so-subtle hints and given me a Spiderbeam 39 foot telescopic pole for my birthday.[1] The pole is much stiffer than a fishing rod, but still only suitable for a very lightweight antenna. This ruled out trap dipoles such as the W3DZZ, as the traps would have been too heavy. The garden, which measured only 66 × 46 feet, was too small to accommodate a G5RV. The location of the shack on the opposite side of the house from the garden made an open wire fed doublet impractical. I ruled out most of the standard multiband HF antennas and decided to develop my own.

A closed loop seemed to be a good starting point since loops are naturally resonant on all multiples of their fundamental frequency. This matches the harmonic relationships between the most popular amateur bands. They can be fed at the bottom of the loop, so the weight of the feed line does not have to be supported by the higher (and thinner) sections of the mast. Unfortunately, the 39 foot high mast was not tall enough to support a 40 meter 1 λ loop, and a 20 meter loop would only provide coverage of the 20 and 10 meter bands through the harmonic relationships. At this low point in the sunspot cycle, I also wanted to be able to work the 40 and 15 meter bands.

Fortunately a loop which is N + λ/2 long is also resonant (the reactive component of its impedance is zero), although it has a very high impedance — as high as 50 kΩ for the λ/2 loop. This high resistive impedance can be transformed to a low impedance using a λ/4 matching section of the right characteristic impedance. To transform 50 kΩ to 50 Ω would require a matching section with a characteristic impedance (Z_0) of about 1600 Ω, which is not feasible. However a quarter-wave matching section with a Z_0 of 800 Ω (about the maximum practical impedance) would transform 50 kΩ to approximately 13 Ω, resulting in an SWR of 4:1 on 50 Ω coax. This is a very reasonable SWR for a multiband antenna, since it can be easily matched by most wide range antenna tuners and will not cause an unacceptably high loss in the coax. For example, if the antenna is fed by 100 feet of RG-213 at 7 MHz, then the matched loss would be about 0.6 dB, and an SWR of 4:1 would result in an additional loss of only 0.6 dB (15% of the transmitter power); while an SWR of 5:1 would result in an additional loss of 0.7 dB (17% of transmitter power) compared to a perfectly matched transmission line.[2]

The resulting design is a loop that is electrically 1 λ in the 20 meter band, fed by a matching section with a Z_0 of about 800 Ω that is λ/2 long in the 20 meter band.[3] In the 40 meter band the matching section is λ/4 long and transforms the high impedance of the loop to a low impedance that can be fed directly from 50 Ω coax. In the 20 meter band the λ/2 matching section has no effect, simply reflecting the 140 Ω impedance of the one-wavelength loop, for an SWR of about 3:1. In the 15 meter band, the matching section is ¾ λ long, and again transforms the high impedance of the 1½ λ loop to a low impedance, in this case close to 50 Ω. In the 10 meter band, the matching section is 1 λ long, so once again it simply reflects the low impedance of the loop, which is now 2 λ in circumference. The result is an acceptable match to 50 Ω coax with an SWR of 5:1 or less on the entire 40, 20 and 15 meter bands and on the bottom half of the 10 meter band. A 1:1 current (choke) balun should be used to connect the coax to the 800 Ω balanced matching section. The balun should perform

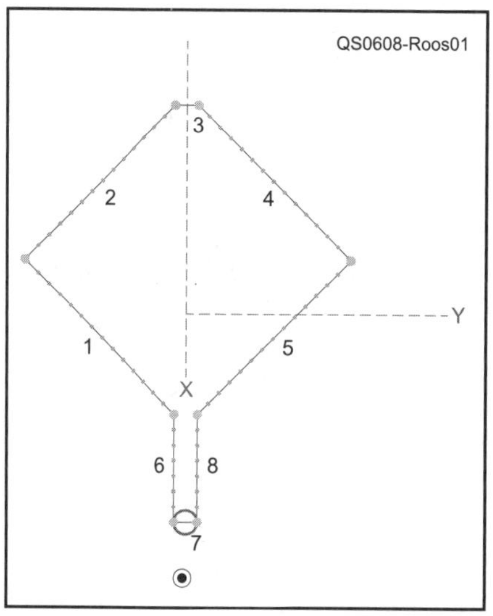

Figure 1 — Layout of compact quad loop.

[1]Notes appear on page 65.

well because the load impedance is low on all bands.

I investigated several different loop shapes using *EZNEC* antenna modeling software.[4] The best predicted performance was achieved with an apex-down delta loop. While this loop shape retains the cleanest pattern at the higher frequencies, it is not mechanically realizable with a single support. The best practical shape is the diamond quad as shown in Figure 1. The matching section runs horizontally along the X axis since, at 36 feet, it is too long to fit vertically below the loop on the 39 foot mast. If your configuration is supported by a 66 foot or taller mast, the matching section can be run down the mast, provided that adequate separation is maintained if the mast is conductive.

Construction

I used 18 gauge enameled copper wire for both the loop and the matching section since bare copper wire is hard to obtain in South Africa, and I didn't want the additional weight of PVC insulation. If you are able to obtain hard-drawn copper wire, it would be a better choice as it is stronger and less likely to stretch.

The antenna and matching section together require about 144 feet of wire — 72 for the loop, and 72 for the 36 foot long matching section. I suggest using a single continuous length of wire for both to avoid the requirement for a joint between the antenna and the matching section. Attach the center point of the wire to the top of the mast using a lightweight plastic insulator (such as those used in electric fences) secured with a couple of cable ties. The corners of the diamond are pulled away from the mast using 0.04 inch diameter monofilament nylon "builder's line" (100 pound test), also attached using lightweight insulators, which should be left free to slide over the antenna wire so they can find the position where the top and bottom halves of the antenna wire are under the same tension. Try to secure the other ends of the builder's line to elevated supports if possible — I used a drain pipe for one side and a telephone pole, that the telephone company had thoughtfully placed in the corner of the garden, for the other.

Cut two spacers measuring 8 × ¾ inches from a sheet of ¼ inch thick Perspex or any other good insulator. Drill 3⁄32 inch diameter holes ½ and ¾ inch from each end of the spacer, centered from side to side. One of the spacers will be used to hold the wires apart at the bottom of the loop. You can attach it to the mast using two heavy-duty cable ties arranged in a criss-cross pattern to hold the spacer horizontal. Feed the ends of the wire coming from the loop through the inner holes and then back through the outer holes, securing with cable ties. Figure 2 shows the assembly of the spacer.

The second Perspex spacer is used at the other end of the matching section to maintain the spacing between the two parallel wires. After being fed through the outer holes, around the back of the spacer and then back through the inner holes, the wires should be connected to the balanced terminals of the balun. I used a W2DU type balun (ferrite beads over a length of coax), but any 1:1 current (choke) balun will work. You can use cable ties to secure the wires to the spacer for strain relief, as shown in Figure 1. You may not need any intermediate spacers if you can tension the matching section sufficiently to keep the two wires parallel. However, if you live in a windy area, I suggest adding a few intermediate spacers made of lightweight plastic to prevent the spacing between the wires from being affected by the wind. The antenna can be fed via any length of 50 Ω coax connected to the unbalanced side of the balun.

The distance of 8 inches between the outer holes of the spacers sets the spacing for the open wire line that forms the matching section. The formula for the characteristic impedance of open wire line is:

$$Z_0 = 276 \times \log(2D/d)$$

where D is the spacing between the wires, 8 inches, and d is the diameter of the wire, 0.0403 inches.[5] The calculated impedance is 717 Ω.

The exact shape of the loop is not critical. In my case, it looks more like an upside down kite than a square diamond as you can see in the lead photo, due to the locations of the available attachment points used to hold the sides of the loop. The important factor is for the loop to have as much internal area as possible — stay away from long thin loops that look more like a folded dipole, as this will compromise the performance.

The sides of the loop double as guys to support the mast. I added another two guy lines (also made of builder's line) at the top of the mast, as well as four guy lines at the bottom of the loop, which is about 13 feet off the ground. I consider this construction to be best suited to temporary use (in my case, for short-term use at a rented property; or perhaps for a field station or DXpedition, since the telescopic mast can be collapsed to less than 4 feet in length). For a permanent installation an aluminum mast, copper-clad steel wire and heavier non-metallic guy wires should give many years of use under all weather conditions.

The lead photo shows the finished antenna. The photograph has been digitally enhanced to show the thin antenna wires, which would be almost invisible otherwise. You can clearly see the inverted kite shape of the loop, and the parallel wires of the matching section, which are attached to the Perspex spacer at the bottom of the loop. The other visible lines are guy wires. The total width of the antenna, from corner to corner, is only 26 feet, making it ideal for use in restricted spaces.

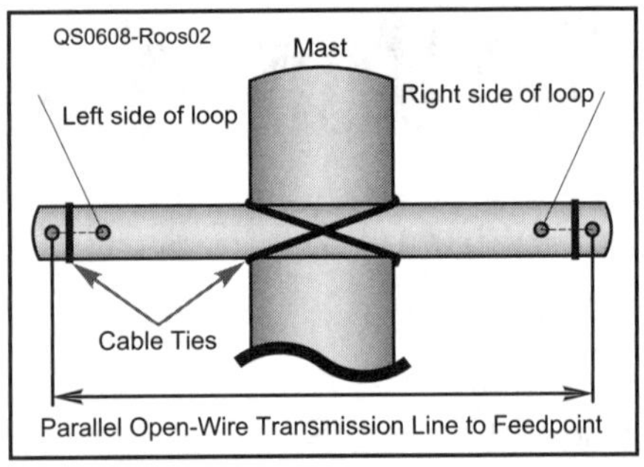

Figure 2 — Details of compact quad loop transmission line attachment.

Tuning the Antenna

You can tune the antenna by gradually shortening the matching section until an acceptable SWR is obtained on all bands. The minimum SWR will not necessarily coincide exactly for all bands, so you should tune it for a good compromise between the SWR on the different bands, rather than simply minimizing the SWR on a single band. Since the antenna is designed to be used with an antenna tuner, an SWR of 5:1 or less is adequate.

The length of the loop, 72 feet in circumference, should not have to be changed unless you are using wire with thick insulation, in which case the length of the loop will have to be reduced by about 2% to compensate for the dielectric effect of the insulation. In this case, disconnect the matching section from the loop, and first trim the loop for minimum SWR at 14.175 MHz. Then connect the matching section, and trim it for acceptable SWR on all bands.

Figure 3 shows the measured SWR of the Compact Quad on the 40, 20, 15 and 10 meter bands. I used a Palstar ZM-30 antenna analyzer with the ZM30-BT balanced transformer adapter to measure the SWR directly at the antenna feed point (the end of the matching section furthest from the loop). The antenna

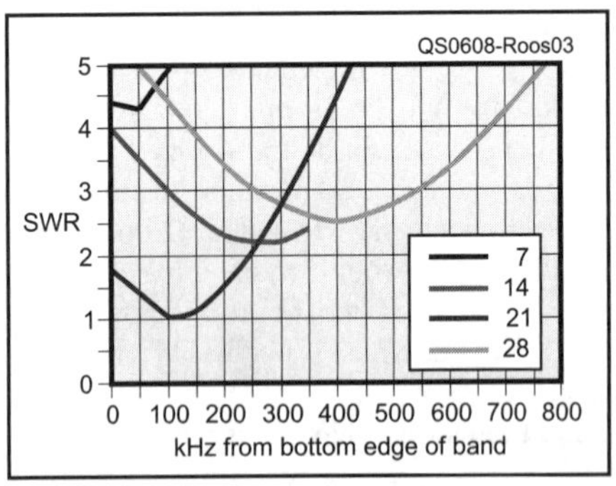

Figure 3 — Measured SWR of compact quad loop at end of matching section. 7 MHz in black, 14 MHz in red, 21 MHz in green, 28 MHz in blue. X axis is kHz from lower band edge.

covers the entire 40, 20 and 15 meter bands and about 700 kHz of the 10 meter band with an SWR of less than 5:1.When the SWR is measured from the shack it is slightly lower on all bands, due to the additional loss of the coax feed line, which is 80 feet of RG213.

I have used two antenna tuners with this antenna, the internal auto-tuner in my Kenwood TS-850S, and a Palstar AT1KM manual tuner, which I use when running with the linear. Both tuners have matched the antenna effortlessly on all frequencies that I have tried. Running up to 400 W CW (the maximum permitted in South Africa) I have not seen any signs of arcing, overheating or other problems in either the antenna or the balun, although if you are planning to use kilowatt power levels then you might consider using larger diameter wire.

Performance

Figure 4 shows the far-field elevation plots for frequencies of 7, 14, 21 and 28 MHz. The plots were calculated using the high accuracy (NEC Sommerfeld) ground model with a conductivity of 0.005 S/m and a dielectric constant of 13. The model included ohmic losses from copper wire with a diameter of 1 mm. The patterns at all frequencies were bidirectional, with only minor departures from bidirectional symmetry (typically less than 0.1 dB) caused by the very slight radiation from the matching section. All the patterns are referenced to the maximum gain plotted, which is 8.0 dBi at 21 MHz.

At 7 MHz the pattern is a squashed sphere that is virtually omnidirectional, with a maximum gain of 3.5 dBi at an elevation of 55°. The pattern at 14 MHz is a classic bean-shaped loop pattern, with a maximum gain of 6.5 dBi at 33°. The best DX performance is found at 21 MHz, where the maximum gain of 8.0 dBi occurs at an elevation angle of only 21°. At 28 MHz the pattern has a major lobe with a gain of 5.0 dBi at 39°, and a minor lobe with a gain of 1.1 dBi at 10°. While this is not perfect, the minor lobe should offer some good DX opportunities considering the relatively low power levels that are often required on the 10 meter band.

Figure 5 shows the azimuth patterns for an elevation angle of 10°, which is typical of the takeoff angle required for long-range communication. The plot scale is referenced to the maximum gain, 4.9 dBi at 21 MHz. As you can see, the pattern for 7 MHz is virtually omnidirectional. The direction of maximum radiation for 14 MHz and 21 MHz is perpendicular to the plane of the loop (along the X axis in Figure 1), while at 28 MHz it has a four lobed pattern with maximums at 42° from the X axis.

I have been very impressed with this little antenna. I managed 377 QSOs spread evenly between the 40, 20 and 15 meter bands while running 100 W in the 2006 ARRL International DX CW Contest. My 109 QSOs on 40 meters were a big improvement over the 41 contacts that I managed on the same band in the previous year's contest. This is especially gratifying, as this antenna is considerably smaller than the V beam that I used in 2005. Over the past few weeks I have managed to start small pileups of Asian and European stations almost every time I've called CQ on 40 meters. This surprised me as I struggled on this band with wire antennas from my previous location. I have also received many good reports on the 20 and 15 meter bands, running between 100 and 400 W, and have even made a couple of contacts on 10 meters despite low levels of solar activity. Of course my rented house feels more like home now that I can communicate with friends around the world.

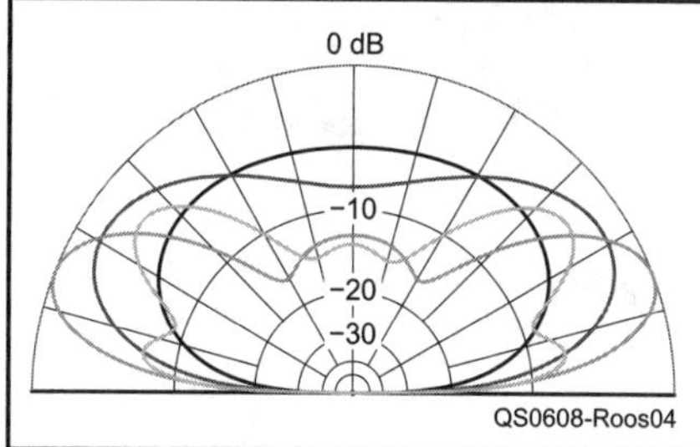

Figure 4 — *EZNEC* elevation pattern of compact quad loop model. 7.15 MHz in black, 14.15 MHz in red, 21.2 MHz in green, 28.2 MHz in blue.

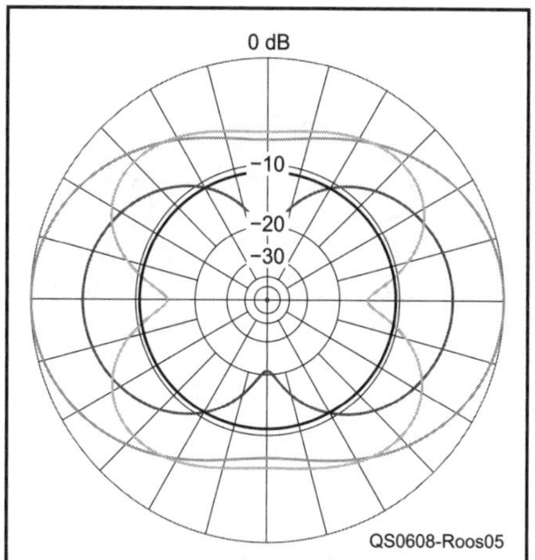

Figure 5 — *EZNEC* azimuth pattern of compact quad loop model at 10° elevation. 7.15 MHz in black, 14.15 MHz in red, 21.2 MHz in green, 28.2 MHz in blue.

If you have limited space, and need a small multiband HF antenna with good performance and a low visual profile, then why not try the Compact Quad?

Notes

[1]**www.spiderbeam.net**.

[2]*The ARRL Antenna Book*, 20th Edition, p 24-10, 2003. Available from your ARRL dealer or the ARRL Bookstore, ARRL order no. 9043. Telephone 860-594-0355, or toll-free in the US 888-277-5289; **www.arrl.org/shop/**; **pubsales@arrl.org**.

[3]I attempted to find references to similar designs on the Internet or in several reference works and antenna compendiums, without success while I was developing this antenna. I have since discovered, however, that Les Moxon, G6XN, mentions this arrangement in his excellent RSGB book *HF Antennas for All Locations*, 2nd edition, p 122. Available from the ARRL Bookstore, ARRL order no. 4300. Telephone 860-594-0355, or toll-free in the US 888-277-5789; **www.arrl.org/shop**; **pubsales@arrl.org**.

[4]*EZNEC* is available from Roy Lewallen, W7EL, at **www.eznec.com**.

[5]J. Kraus, W8JK (SK), *Antennas for All Applications*, 3rd Edition, McGraw Hill, 2002, p 891.

Andrew Roos has been a radio amateur since 2001. He initially was issued the call sign ZS1AN and now is ZS6AAA. He has Bachelor of Arts with Honours degree from Rhodes University. After 15 years in the software development industry, he recently accepted a position as a development engineer at a company that produces specialized antenna systems. Andrew is a past chairman of the Cape Town Amateur Radio Center and regularly teaches classes for the Radio Amateur Examination. You can reach him at **andrew.roos@mweb.co.za** *or PO Box 3294, Pinegowrie 2123, South Africa.*

The Horizontal EWE Antenna

The original EWE shepherd goes in a new direction for low noise reception.

Floyd Koontz, WA2WVL

This antenna was designed to solve a major reception problem at this location by maximizing the received signal to noise ratio (SNR) of 75 meter phone signals coming from Europe.

Background

In 1999 I had up a six element switchable vertical array hung from some of the many 100 foot pine trees in my yard. The array appeared to have good directivity but never received the good signal reports I expected. Eventually I put a dipole between trees at 80 feet elevation as a reference antenna. On nearly every contact the dipole was 10 dB better than the 6 element array. Even though the vertical array wires were 20 feet or more from the trees, the trees seemed to be absorbing the signal. In northern locations, the trees drop their sap to the roots in the winter (DX season on the low HF bands) and many stations have good luck with verticals hung from trees. Here in Florida the sap stays up all year round and the trees are quite conductive (and longer than $\lambda/4$ at 75 meters). Vertical polarization simply doesn't seem to work in my yard.

Recently I decided to look at the EWE antenna[1] in a horizontal plane and discovered that it worked very well. With vertical EWEs, a single element has a deep null off the back. A single horizontal EWE, on the other hand, would have that null at 0° elevation instead. Over real ground, the single horizontal EWE has a front to back ratio (F/B) on 75 meters as shown in Figure 1 for 30° elevation. The low angle gain increases with height above ground, as with most horizontal antennas, but the pattern remains nearly the same at 10 to 30 foot heights. The size can be chosen to fit the available space but four supports (trees, towers, house, etc) are needed to hold up the corners.

Of the different sizes and shapes that can be used, one is optimum. It is a square, $\lambda/8$ on a side (a total of $\lambda/2$ around). For this size (30 × 30 feet for 75 meters), the feed is non-reactive and can be matched with a simple broadband transformer. The calculated feed impedance was 1337 Ω and was matched with a 26:5 two-winding transformer. The F/B was more than 11 dB on 75 meters and 15 dB on 160 meters.

Two Element Design

As noted, a single EWE antenna gives a modified cardioid pattern with a usable, but not dramatic, front to back ratio. The front to side ratio is only about 4 dB so a two element design was established to improve the back and side rejection. Available trees allowed a spacing of 100 feet. Figure 2 shows the layout of this array including the location of the feed points and terminations. Figure 3 is the modeled pattern of the two element array, again at 3.8 MHz and 30° elevation. As is evident, the two element array resulted in a sharpened beamwidth and a significantly improved F/B. While the pattern for 30° elevation is shown, it is similar with lower output at lower elevations.

The antenna was to be 25 feet off the ground. This was determined to be the maximum height that I could safely reach from my 24 foot extension ladder. Vinyl insulated

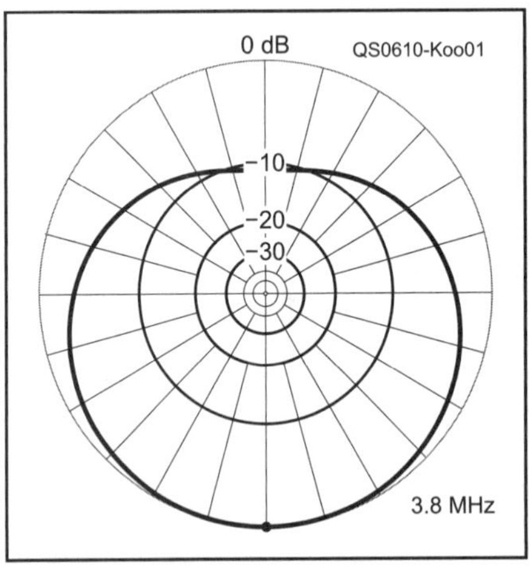

Figure 1 — *EZNEC* modeled pattern of a single element horizontal EWE at 3.8 MHz and 30° elevation.

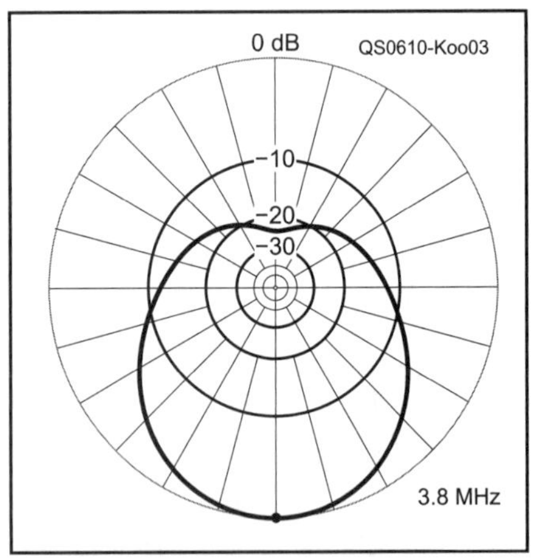

Figure 3 — *EZNEC* modeled pattern of a two element horizontal EWE configured as in Figure 2 at 3.8 MHz and 30° elevation.

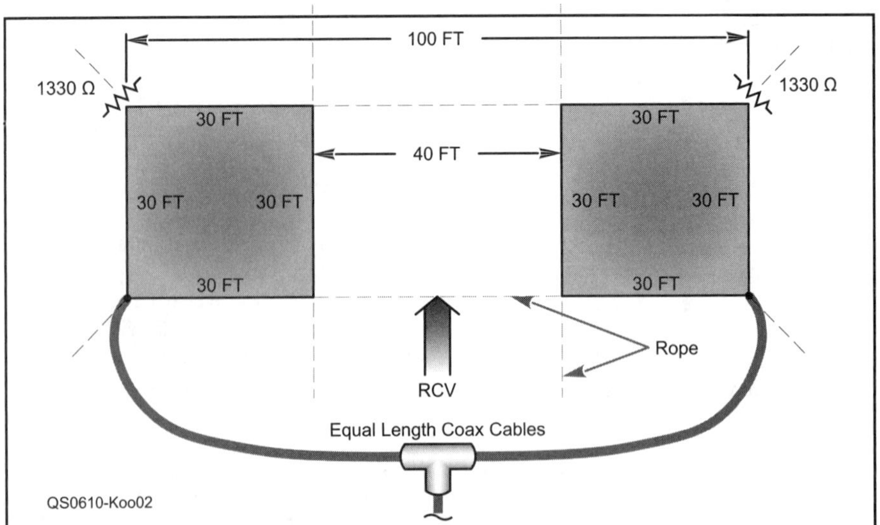

Figure 2 — Orientation of the two element horizontal EWE antenna array.

[1]Notes appear on page 67.

14 gauge "speaker wire" was used, but any wire strong enough to be pulled tight without stretching could be used.[2] The loops were tied at the corners (every 30 feet) by ⅛ inch Dacron ropes. The knots were taped with black electrical tape to prevent the knots from coming untied. The input matching transformer was mounted on a 2 × 4 × ¼ inch piece of UHMW material.[3] The feed coax (RG-58A/U) was woven through three tight holes vertically in line about ⅝ inch apart to support its weight.

The termination was also mounted on a 2 × 4 × ¼ inch piece of UHMW. Both types of boards had lifting holes at the top for attaching the ropes. After all the soldering was completed, the boards were coated with silicone window sealing except for one of the 50 Ω transformer connections. The two loops must be fed in phase to develop the pattern. I had a 50% chance of getting it right. If I guessed wrong, the pattern would have a 30 dB null off the front and receiving performance would be poor. I was successful the first time and did not have to reverse the connections, so I proceeded to coat these connections as well.

During the assembly process the wire is lying on the ground in the space where it will be pulled up. In my case the SE loop enclosed only one tree while the NW loop has three trees inside the loop. A ladder was placed against the trees to be used for supports and a small hole was drilled in each tree. A screw eye was installed as high as I could reach at each support point. A piece of the ⅛ inch Dacron line was fed through each screw eye and pulled back to ground and tied to another screw eye at 4 feet. After doing this on eight trees, I was now ready to lift the antenna in place without further climbing.

Both feed points had equal length coax cables attached. The length was determined by the location of the ninth tree to which both coax cables would go (I used 100 feet for each run length). I ran the cables overhead at about 12 feet from tree to tree. A BNC T was used to parallel the 100 foot coax cables and the third port connected to the coax that went to the shack.

Antenna Output Level Considerations

The calculated gain of the two element horizontal EWE (at 20°) at 3.8 MHz was about –18 dBi (a λ/4 vertical has a gain of about 0 dBi). For a height of only 10 feet the gain drops to –24 dBi. Most modern receivers and transceivers have sufficient sensitivity so that even at this level there is sufficient gain that the signal to noise ratio is limited by external noise.

If the noise level doesn't increase by at least 10 dB when replacing a dummy load with the receive antenna, you will need to have additional gain to get full benefit from this antenna. In order to directly compare the performance of this antenna to your transmitting antenna, you may want to add gain so that the signal levels are similar and the S-meter moves in the same way. A preamplifier with a gain of 18 to 25 dB should be just what you want for 80 meters, more if you use the antenna on 160.

A suitable amplifier for 160 meters was included with a recent antenna article.[4] A dual band amplifier was also published in an earlier issue of *QST*.[5] Gary Nichols, KD9SV, has assembled preamps for either 160 meters or 160 and 80 meters available through Radio-Ware.[6,7]

It also might be that switching your receiver's preamp on provides just the amount of gain required to bring the signal level up to that from your transmitting antenna, and that is a suitable way to equalize the gain. Whatever preamp option you take, make sure that the preamp is protected from pick up of RF from your transmitting antenna. In many cases, the receive antenna will bring it directly into the preamp, if provisions aren't made to the contrary.

Operation with this system has been gratifying with the SNR often being improved by 10 to 20 dB on 75 meters compared to my reference antenna. On 160 meters, many SSB stations can now be copied that were below the noise before.

Notes

[1]F. Koontz, WA2WVL, "Is This EWE For You?" *QST*, Feb 1995, pp 31-33, and "More EWEs For You," *QST*, Jan 1996, pp 32-34. Available on **www.arrl.org**. The antenna gets its name from its resemblance to an inverted U.

[2]"Speaker wire" is available from Home Depot in 250 foot lengths.

[3]Ultra High Molecular Weight (UHMW) Polyethylene is available from plastics suppliers for less than $15 per square foot.

[4]D. Stroud, W9SR, "An Effective Receiving Loop," *QST*, Jun 2006, pp 35-38.

[5]D. DeMaw, W1FB (SK), "A Preamplifier for 80- and 160-Meter Loop and Beverage Antennas," *QST*, Aug 1988, pp 22-24.

[6]G. Nichols, KD9SV, "A Variable Gain 160-Meter Preamp," *Ham Radio*, Oct 1989, pp 46-48.

[7]Click **KD9SV Preamps** at **www.radio-ware.com**.

Floyd Koontz, WA2WVL, is a retired electrical engineer with 40 years' experience designing communications systems, radio transmitters and antennas. He was first licensed in 1955 as WN9JQA and has been WA2WVL since 1961 He has written numerous QST *articles about antennas over the years. An ARRL life member, he can be reached at 8430 W Park Springs Pl, Homosassa, FL 34448, or* **wa2wvl@tampabay.rr.com**.

Two Element Lazy Looziana Loops for 15 or 20 Meters

An easy to make antenna, with good performance and the potential for a bonus (lagniappe) band.

Rick Rogers, K5RCR

My first ham radio station in 1968, as a 15 year old Novice. was set up for 15 meters. It was quite a rig, too: a 1953 vintage 50 W Elmac AF67 "Trans-citer" (transmitter-exciter) with a homebrew power supply, one crystal for 21.120 MHz and a Hammarlund HQ-150 receiver. The antenna, which my dad and I built one summer, was a "One-Element Rotary," a shortened dipole featured in the 1956 *ARRL Antenna Book*.[1] That *Antenna Book*, along with a magnificent pile of old *QST*s, came from a friend's periodic garage clean-out exercise.

My station occupied a none too comfy footprint of about 3 square feet of workbench space in our garage. It was snug, and with the summers in Los Angeles, sweltering. Yet, in the year or so that I kept this station on the air, I managed to work about 30 states and several countries. I have had a sentimental feeling toward 15 meters ever since. By the way, the antenna was still up 30 years later!

That was Then, This is Now

After many moves and many stations, I landed here in south Louisiana on rural property with few restrictions. Even so, the density of tree growth and the reptile population reduced my enthusiasm for wandering around in the woods stringing up large wire antenna arrays. I decided to go with a conventional rotator and tower arrangement. I wanted to try my hand at making a compact but effective 15 meter beam antenna to rekindle my interest in 15 meters. Even with the decline in the sunspot cycle, daytime openings on 15 are relatively common. The antenna had to be light enough to be turned by a modest rotator. However, the design should not be compromised in performance

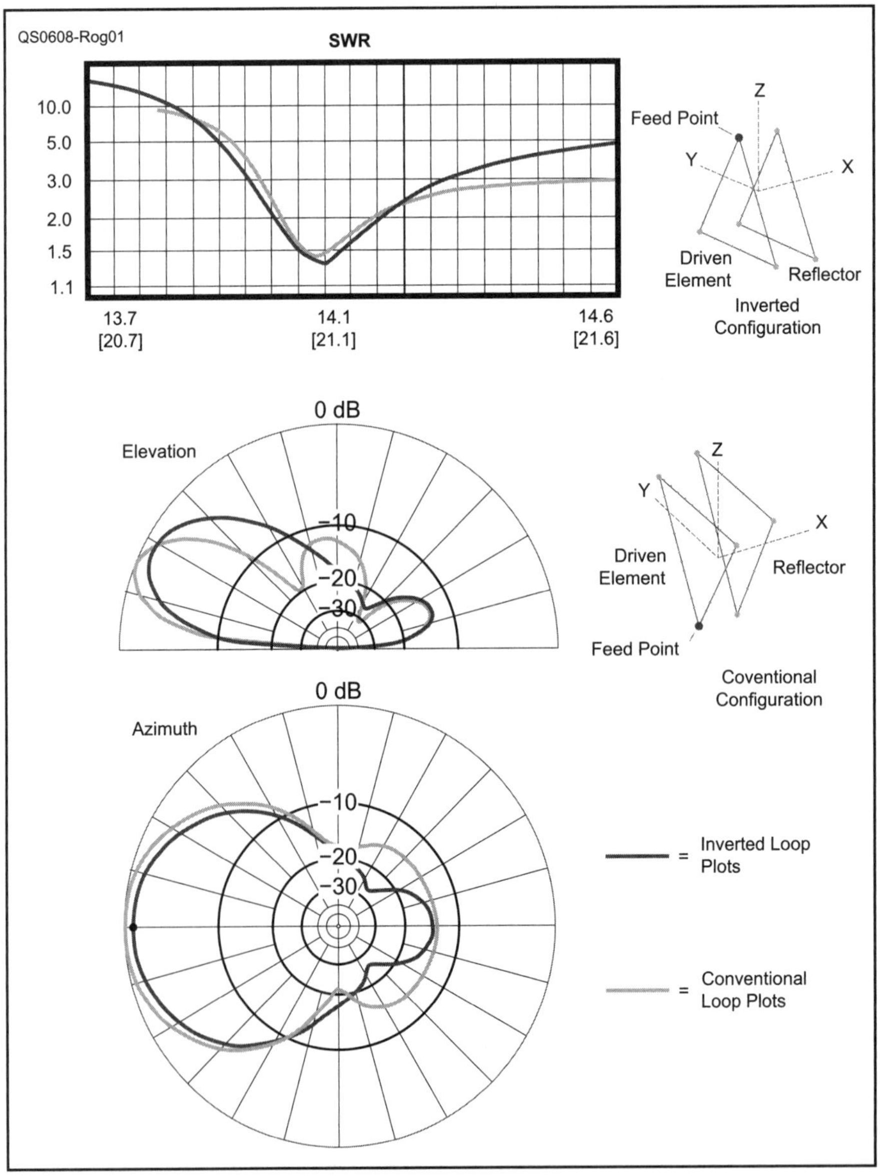

Figure 1 — Orientation, SWR and *EZNEC* model performance of two-element delta loop arrays in the conventional and inverted configuration. Both models are fed at the apex, 45 feet over "real MiniNEC" ground. SWR for either configuration [for 15 or 20 meters] are practically identical. Predicted antenna radiation patterns (20 meter version shown) are also very similar, with expected forward gains of around 10 dBi.

[1]Notes appear on page 70.

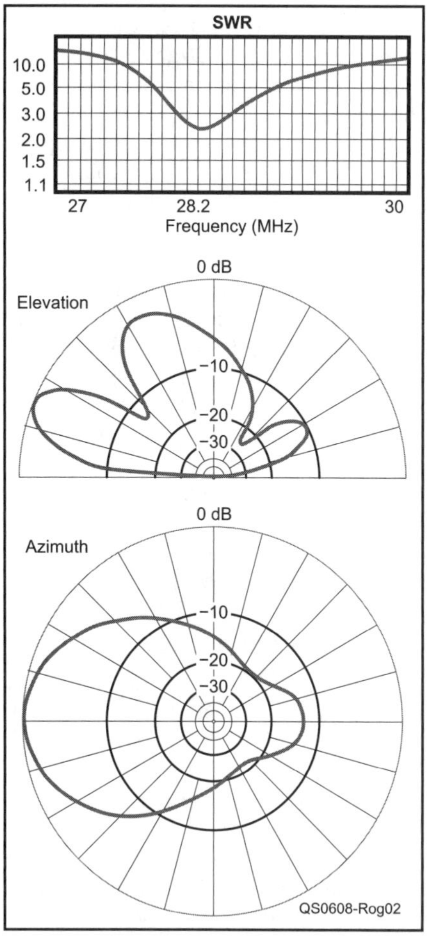

Figure 2 — *EZNEC* prediction of the performance of the 20 meter loop on 10 meters. The model predicts a forward gain of about 9 dBi with the array placed 45 feet over real ground. An antenna tuner easily allows this design to be useful on 10 meters in addition to 20.

Figure 3 — V struts of aluminum tubing fixed to the boom. Buck, the wonder-pony, is indifferent. (At 40 years old, it's a wonder he's still around to be indifferent.)

by small dimensions and the power absorbing, bandwidth narrowing fixes necessary to make them resonate.

Modeling Normal and Droopy-Loops

I settled on a two element delta loop design.[2] This antenna, like the more traditional cubical quad loop with two elements, provides about the same performance as a full size, three element Yagi but with about one-half the boom length and three-quarters the "wingspan."[2,3] It is built with readily available materials. Assuming that everything must be bought new, the total cost will run about $100.

One problem with quads and delta loops in the conventional configuration, with the feed point at the apex and the base of the "delta" pointed up, is that they are awkward beasts to get onto the tower mast. So, I did an *EZNEC* study of two element delta loops in the conventional and the inverted configuration.[4] By inverted, I mean that the delta is still fed at the apex and with the feed point the same height above ground, but the base is now pointed down at the ground as shown in the title photo and in Figure 1.

The model assumed an element spacing of 0.12 λ, 6 feet at 15 meters, a driven loop length of 1.01 λ, 47 feet at 15 meters, a reflector of 1.07 λ, 49 feet at 15 meters, and a feed point 45 feet above "normal" ground. The model results of the conventional and inverted configurations were very similar. The conventional arrangement had about a 1 dB forward gain advantage and a few degrees lower take- off angle. The difference is probably not detectable with the typical ham receiver and ear. Both models presented 2:1 SWR bandwidths of around 250 kHz, as shown in Figure 1. Neither requires any special matching with the specified spacing of the reflector element.[3]

The inverted loop configuration is much easier to assemble and mount on the tower and offers a degree of inherent stability because the bulk of the antenna is *below* the rotator. This revelation convinced me that a full-size 20 meter version of the antenna would work just as well. Additionally, a 20 meter loop will be good to have around when 15 meter openings get chancier during the current sunspot cycle minimum. With a total circumference of around 70 feet, the 20 meter version would be a real challenge to assemble in the conventional way, but easy in the inverted configuration. An additional factor that would make the inverted loop a snap to install at this location is that the 35 foot Rohn tower at K5RCR has a Glen Martin Engineering Hazer trolley system. I can raise and lower experimental antennas at will without climbing the tower. I could essentially pull the inverted loop array up the tower as I built it. Of course, one could accomplish much the same thing by attaching a line to the array's boom and then pulling the antenna up the mast with a pulley attached to the top of the tower or to a ginpole.

In Louisiana, when you get something extra for nothing, the Cajuns call the bonus a *lagniappe* (pronounced *lan-yap*). There is some lagniappe in the 20 meter loop — it is also resonant on 10 meters, with an SWR of less than 3:1 between 28.0 and 28.5. This range is easy for a tuner to handle without much coax loss. According to *EZNEC*, the 20 meter loop should also produce respectable gain and directivity at 10 meters as shown in Figure 2.

Construction

Both the 15 and 20 meter antennas are built on a boom made from a piece of 3 × 10 foot lightweight PVC drain pipe (3000 pound crush rating). This pipe has one end expanded to receive another length of pipe. For either the 15 or 20 meter version, cut off 2 feet of the expanded end and save it. Cut the remaining section of pipe to a 5 foot length for the 15 meter boom, or to 7 feet for the 20 meter boom. Using a square, caliper or protractor, make four marks on the edge of each pipe-boom tube 90° apart. Extend a straight line back from the edge 4 inches from each mark. Make marks 2 and 4 inches back from the end of the boom on both sides of the tube. Drill a ⅞ inch hole for the V antenna struts centered on these marks so that the ⅞ inch aluminum tubing forms a V with the struts at the apex displaced 2 inches from each other as shown in Figure 2.

Insert the four ⅞ inch pieces of 6 foot long tubing into the holes and temporarily fit the two lengths of pipe boom together. Twist the shorter length tube until the V struts on either end of the boom line up with each other. Make some registration marks on the two pieces of pipe, pull them apart and liberally apply plastic pipe cement to the mating surfaces. Now stick the pipe boom sections together to make a big, spindly sawhorse-looking contraption as seen in Figure 3. The boom was assembled in two pieces because I found that it is almost impossible to drill holes at the ends of a 6 or 8 foot long piece of pipe that result in the V struts being parallel with one another. This way, drilling errors can be compensated for.

Drill ⅛ inch holes into the tubing on either side of the boom and put self-tapping screws in the holes to prevent the struts from creeping out of the boom. Now attach a coax feed line to the driven element Vs. There are two ways to attach coax to the array. The first is to flatten the terminal 2 inches of tubing forming the apex of the driven element at the boom. Drill a ⅝ inch hole in the middle of the flattened part and attach a coax bulkhead connector. Connect a 5 inch piece of 14 gauge wire between the center conductor of the bulkhead connector and attach it using a ring terminal and a self-tapping screw to the adjacent V strut as shown in Figure 4. Alternatively, strip 6 inches from one end of a length of RG-8 cable. Separate the shield from the center insulator and solder ring terminals to each conductor. To waterproof the cable end, cover the ring terminals with tape and then spray the assembly liberally with plastic "tool grip" such as Rustoleum "Grip and Guard." Attach a PL-259 to the other end and screw the ring terminals to the ends of one set of struts. Cut a 6 inch piece of 14 gauge wire and connect this between the ⅞ inch tubing sections at the apex of the reflector loop. Either feed line attachment method works well. Cover all connections with silicone sealant to prevent water entry and corrosion. Fit a 3 inch plastic pipe end cap over each end of the boom to give the apex of the antenna added strength as shown in Figure 5. At this point, attach the boom-to-mast plate and attach the boom to the rotator mast on the Hazer or to a cable on the gin pole.

For the 15 meter version, assemble the V struts for the driven element by adding telescoping 6 foot lengths of ¾ and ⅝ inch tubing. Total length of the driven element

Figure 4 — One method of attaching a feed line. Flattened tubing end is drilled for a coax bulkhead connector. Exposed connections are covered with silicone sealant to prevent exposure to water and corrosion.

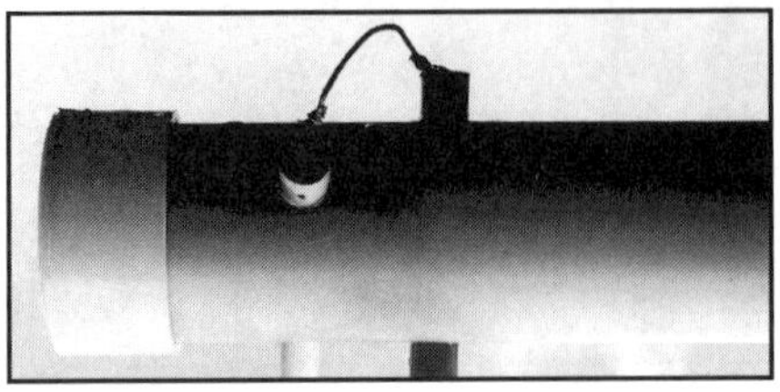

Figure 5 — Pipe cap (for 3 inch pipe) placed over the end of the boom end to stiffen the apex. Note also the tie wire between the reflector V struts.

struts are 16 feet 2 inches. For the reflector struts, the length is 16 feet 8 inches. Drill a ⅛ inch hole through the overlapping tubing sections and secure them with a self-tapping screw. To finish the loop, cut two pieces of 17 gauge aluminum fence wire 16 feet long for the driven element and 17 feet long for the reflector. Attach the wire between the ends of the V struts. The wire will be under modest, but easily manageable, tension. Assembly of the 20 meter version requires 23 foot, 4 inch struts for the driven element and 24 foot struts for the reflector. The base wire is 24 feet long for the driven and 26 feet long for the reflector. Resonant points for either antenna can be adjusted by changing the length of the base wires. It is easy to assemble this antenna by simply hoisting it up on the Hazer or the gin pole as each section is added to the struts.

Results

The 15 meter antenna showed an SWR of 1.2:1 at 21.1 MHz, almost exactly as modeled. The 20 meter version resonated at 14.085 MHz with a SWR = 1.1:1. The antennas appear to be superb performers. The antenna exhibits a front-to-back ratio of about 3 S-units and a front-to-side ratio of 5 S-units as measured on the meter of my Yaesu FT-847. The loops receive considerably better than a very good, 53 foot tall multiband vertical installed over a good ground. I have worked 33 countries in casual operating with the 15 meter antenna in just a few weekend afternoons, including VQ9LA on Chagos Island, about as far from my station as you can get without coming back around. On the first day I had the 20 meter version up, I worked all continents in about three hours. The stations included F6CTL/FO, ZF1PM, YV4GD, LZ2KV and 5Z4IA. Signal reports were uniformly excellent. I found that *lagniappe* was at work in the 20 meter loop as it loaded easily on 10 meters with a tuner and produced good contacts with South America during a brief mid-afternoon opening.

The inverted loop has proven tough in the face of some severe thunderstorms with wind gusts up to 75 miles per hour. The design is stable and easy to assemble. The approach could be extended to lower frequency bands as well. For example, a full-size 40 meter array would need a boom 17 feet long and loops with sides of about 47 feet.

Notes

[1] *The ARRL Antenna Book*, 8th Edition, Chapter 9, "Antennas for 14, 21 and 28 Mc," ARRL, 1956.

[2] R. Haviland, W4MB, *The Quad Antenna*, CQ Publications, 1993.

[3] *The ARRL Antenna Book*, 20th Edition, Chapter 12, "Quad Arrays," 2003. Available from your ARRL dealer or the ARRL Bookstore, ARRL order no. 9043. Telephone 860-594-0355, or toll-free in the US 888-277-5289; **www.arrl.org/shop/**; **pubsales@arrl.org**.

[4] Available from **www.eznec.com**.

Rick Rogers is a Professor of Neuroscience at the Louisiana State University, Pennington Biomedical Research Center, in Baton Rouge. He was first licensed in 1968 as WN6HGY, then WA6EZT, then N9COO, then N7GEF, then KI8GX and now K5RCR. He is well on his way to the not so coveted "licensed in all call areas" award. He can be reached at 9831 Bank St, Clinton, LA 70722 or **rogersrc@pbrc.edu**. QST

A 2 Element Quad for 17 and 12 Meters Using a Combined Feed

L. B. Cebik, W4RNL, and
Robert Cerreto, WA1FXT

For many hams, 17 and 12 meter operation calls for smaller and less complex beams than we use for the other HF bands. A two band, two element quad beam with a common feed point is one good solution to the antenna challenge on these bands — if we can obtain performance equivalent to monoband quads for the same bands. The following project is a joint venture between two old friends, W4RNL and WA1FXT, separated by two state lines and many miles. W4RNL provided the design work and analysis, while WA1FXT constructed and tested the prototype. Despite the distance between collaborators, the project proved to be very successful.

Where to Start

The best place to begin the design of a dual-band quad is with a good monoband design. Several years ago, W4RNL developed a number of monoband quad programs and models based on regression analysis of a detailed series of models. The results are available in several formats, including a *NEC Win Plus* model and spreadsheets.[1]

The program requires you to enter the design frequency and the wire diameter. The calculations produce two element quads with good gain, good front-to-back ratios and the broadest bandwidth that is obtainable for any given element diameter. Figure 1 shows the general outline and the design-frequency free-space E-plane pattern of one sample from the program. With slight changes in gain that result from the selected element diameter, all quads produced by the program yield essentially the same characteristics. The feed point impedance is about 135 Ω resistive.[2]

[1]Notes appear on page 75.

Design Approach

We specifically selected this design because it offered broad bandwidth, as quads go. Generally, quad beams struggle for SWR bandwidth on the wider ham bands. They struggle even harder to obtain a good front-to-back ratio, which tends to peter out more quickly than the SWR bandwidth. A wide operating bandwidth ensures against the inevitable home workshop construction variables that throw a design off its modeled precision. As well, when we combine designs into a multi-band beam,

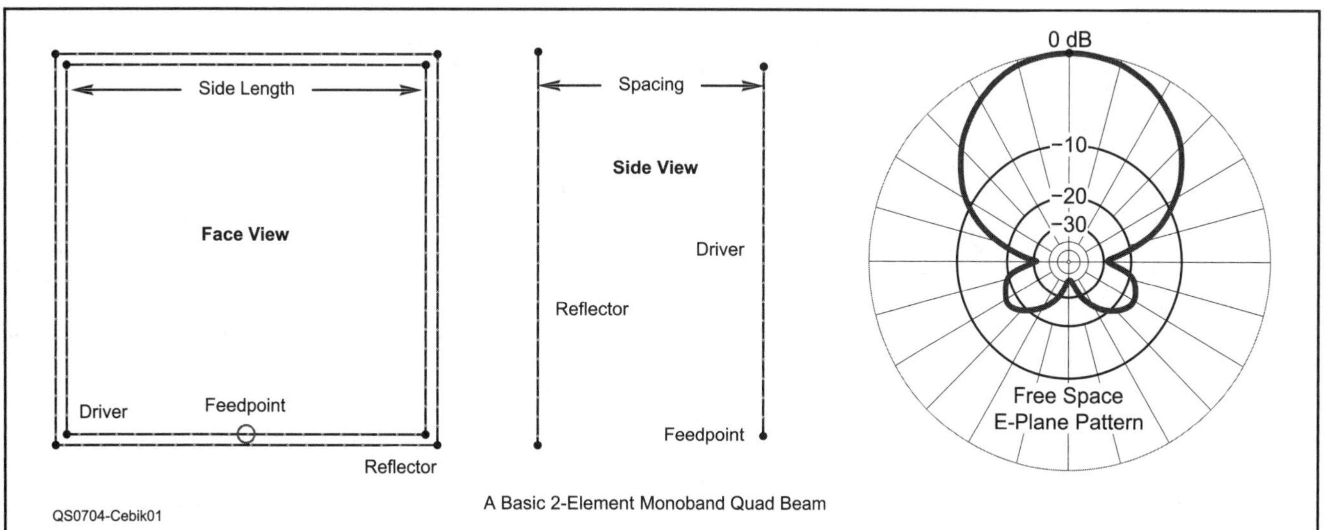

Figure 1 — General layout and free-space E-plane (azimuth) pattern of a basic two element monoband quad beam designed for optimal performance.

the bandwidth on at least one of the bands tends to decrease.

Beginning with a broadband design helps us to obtain good performance across both 17 and 12 meters. Narrow-band designs are usable, but only with precision construction and considerable field adjustment (another name for trial and error tuning). Table 1 lists the dimensions for the 17 and 12 meter monoband quads that form the starting point. Also listed are a few key free space performance figures that we can later use to see how successful our dual-band quad might be. Since the gain values are for free space, you can add about 5 dB more for placement at least ¾ λ above ground. The exact ground-reflection contribution to the gain will vary a bit with antenna height and ground quality.

From Monoband Quads to a Dual-band Quad

Translating a pair of monoband designs into a dual-band quad is not so simple a task as it might seem. Even without a common feed point, element interaction will force us to alter some of the dimensions. Adding a common feed point forces us to bend a horizontal wire of one or both driven elements to meet at a common point. Since we only have two bands, and since the higher frequency elements usually suffer more from misalignment, we bent the 17 meter driven element lower horizontal wire to meet the level 12 meter driver. Table 2 summarizes the dimensions and modeled performance of the resulting two band quad.

The tabular data show that we had to alter three out of the four elements. As well, the dimensions do not show the extra length of 17 meter driver wire needed to reach the combined feed point. The gain level is about a 0.5 dB higher than for a two element Yagi, an amount unlikely to be readily noticed in operation. Although the peak 180° front-to-back ratio suffers a bit relative to the monoband quads, it remains well above what we usually obtain from a two element driver-reflector Yagi. Figure 2 shows the free-space patterns that correspond to the data.

The element lengths of the loops in the multiband quad are different than those of the monobanders because currents are coupled between them. Note that whichever band we choose, the wires for the other band have low level but significant activity. The seemingly unused driver wires show considerable activity, and the 12 meter reflector tries to work as a "wrong-way" director on 17 meters.

The gain tends to change only a little across the narrow 17 and 12 meter bands, but more so than with a monoband quad. On 17, the gain varies by about 0.3 dB across the band. The gain on 12 meters is more stable, with a variation of only about 0.1 dB. The 180° front-to-back ratio remains above 15 dB across both bands.

Figure 3 also shows the structure of the design model. The common feed point is a three-segment wire whose ends form a junction with the two driver loops. The modeling method is adequate in *NEC-4*; however, the small angle created by the driver wires that meet at the center common wire present *NEC-2* with a challenge. As a result, the reported gain values in *NEC-2* are more than 0.8 dB high (7.88 on 17 meters and 7.74 on 12). Monoband quad beams using *NEC-2* produce accurate results since they lack the acute wire angles produced by the dual-band driver elements.

Table 1
Dimensions and Modeled Performance of Monoband Two Element Quads for 17 and 12 Meters
All quad loops use 14 gauge copper wire. Modeling done in *NEC-4*. Both have a feed point impedance of 134 Ω.

18.118 MHz			*Modeled Performance*		
Element	*Side Length*	*Spacing*	*Gain dBi*	*Front-Back Ratio (dB)*	*Worst-Case F-B Ratio (dB)*
Reflector	173.4″				
Driver	164.6″	101.4″	7.01	46.16	17.12
24.94 MHz			*Modeled Performance*		
Element	*Side Length*	*Spacing*	*Gain dBi*	*Front-Back Ratio (dB)*	*Worst-Case F-B Ratio (dB)*
Reflector	126.4″				
Driver	119.6″	74.5″	7.06	41.74	17.36

Table 2
Dimensions and Modeled Performance of a Two Element Quad for 17 And 12 Meters
All quad loops use 14 gauge copper wire. Modeling done in *NEC-4*.

18.118 MHz			*Modeled Performance*		
Element	*Side Length*	*Spacing*	*Gain dBi*	*Front-Back Ratio (dB)*	*Worst-Case F-B Ratio (dB)*
Reflector	171.6"				
Driver	163.0"	101.5"	7.19	29.61	16.68
Feed point	121 – *j*5 Ω				
24.94 MHz			*Modeled Performance*		
Element	*Side Length*	*Spacing*	*Gain dBi*	*Front-Back Ratio (dB)*	*Worst-Case F-B Ratio (dB)*
Reflector	126.4"				
Driver	122.0"	74.5"	7.06	25.99	17.08
Feed point	106 + *j*6				

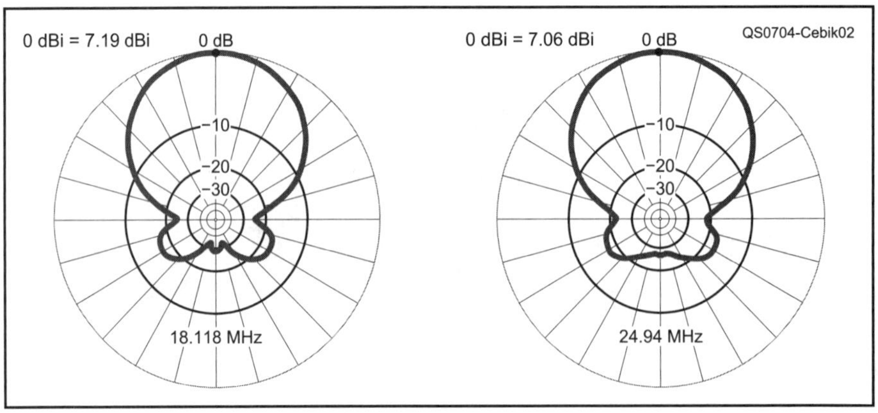

Figure 2 — Free-space E-plane (azimuth) patterns of the two element quad beam for 17 and 12 meters.

Feeding the Dual-band Quad

For neither the monoband nor the dual-band quad is the feed point impedance ready for a 50 Ω coax connection. On 17 meters, with a combined feed point, the feed point impedance is approximately 106 Ω. On 12 meters, the impedance is about 121 Ω. If we were using a monoband quad, we might employ a λ/4 matching section of 70 or 75 Ω cable to transform the impedance down to a value close to 50 Ω. However, λ/4 line sections are frequency specific. Moreover, the simplicity of calculating the impedance transformation of a λ/4 line has frozen our thinking about such lines: we must either use an exact λ/4 line or abandon the use of line transformation systems altogether.

Nothing is further from the truth. Every

length of line transforms the impedance at the antenna end by an amount that is a function of the electrical length of the line. A quarter-wavelength at 18.118 MHz is 163 inches and at 24.94 MHz is 118.3 inches. Perhaps an intermediate line length might yield acceptable terminal impedance values on both bands. A length of 140 inches of 75 Ω cable works best, but most cables today are closer to 70 Ω. A 145 inch length of 70 Ω cable has an electrical length of about 0.22 λ on 17 meters and 0.31 λ on 12 meters. The result is a 17 meter 50 Ω SWR that is below 1.4:1 across the narrow band. On 12 meters, the 50 Ω SWR is under 1.3:1 across the band.

The electrical length of the suggested 70 Ω matching section is 145 inches. The physical length will be 145 inches times the velocity factor of the line actually used. Velocity factors of commonly available 70 Ω line range from about 0.66 to 0.8, depending on the dielectric material used between the braid and the center wire. If you can, measure the velocity factor of the line that you have, since the value may vary from the published or catalog value. However, the precise line length for the matching section is not exceedingly critical. For example, the prototype used a slightly shorter line with completely acceptable results.

The distance between the feed point plate and the hub is about 72 inches. Therefore, the line can go over the hub and partway down the mast. If you use a rotator, there is enough matching section line at the hub to form the stress relief loop that goes around the rotator. Tape the coax to a support rope, since coaxial cables are not designed to bear weight loads. At the junction of the 70 Ω match line and the main 50 Ω cable, install a common-mode current attenuator. You may use a 1:1 balun, a ferrite bead choke, or a coil of coax. All three systems appear in the current edition of *The ARRL Antenna Book.*[3]

Building the 17 and 12 Meter Quad

The dual-band quad uses "spider" quad construction techniques to allow optimum spacing between driver and reflector elements for each band. Optimum spacing helps assure best performance from the antenna. The component parts for this type of quad are the spider hub with spreader support arms, spreader arms to support the wire elements, a mast attachment for the spider hub and spreader arm braces to keep the driver and reflector elements spacing fixed. Table 3 provides a list of materials needed for the project.

Spider Hub

The spider hub consists of a short square boom, eight spreader support arms, a bracket to attach the hub to the mast plate, and a mast plate to attach the hub to a mast. The spider hub can be fabricated with a number of different materials. The choice of materials is dependent upon the builder's preferences, availability, cost, and the tools available to the builder. We chose our materials from what is commonly available at WA1FXT's location — ⅛ inch steel stock for strength, low cost and availability.

Our quad prototype is built very robustly so it will stand up to the severe winter weather from Lake Erie. If you believe that lighter materials will endure the weather at your location, you may use thinner steel stock or aluminum angle stock, along with smaller diameter spreaders. Steel surfaces will require appropriate corrosion prevention techniques. Aluminum, in addition to being lighter, has the added benefit of not requiring any corrosion protection techniques, but aluminum stock is more expensive.

Fiberglass tubes are available from a number of sources and are reasonably priced. So, we decided to purchase our spreaders. We used telescoping fiberglass tubes. Telescoping the fiberglass tubes helps ease some adjustments. The telescoping tubes should be cemented in place after final adjustments. Also, protect all fiberglass surfaces from damaging sunlight with a coat of flat black paint over a suitable primer.

Other than welding of the spider hub, fabrication of the quad does not require the use of any exotic tools. The welding required is simple and straightforward. So, paying a professional welder should not break the budget if you cannot do your own welding.

Boom

The main boom was fabricated by welding two 12 inch angles together to form a

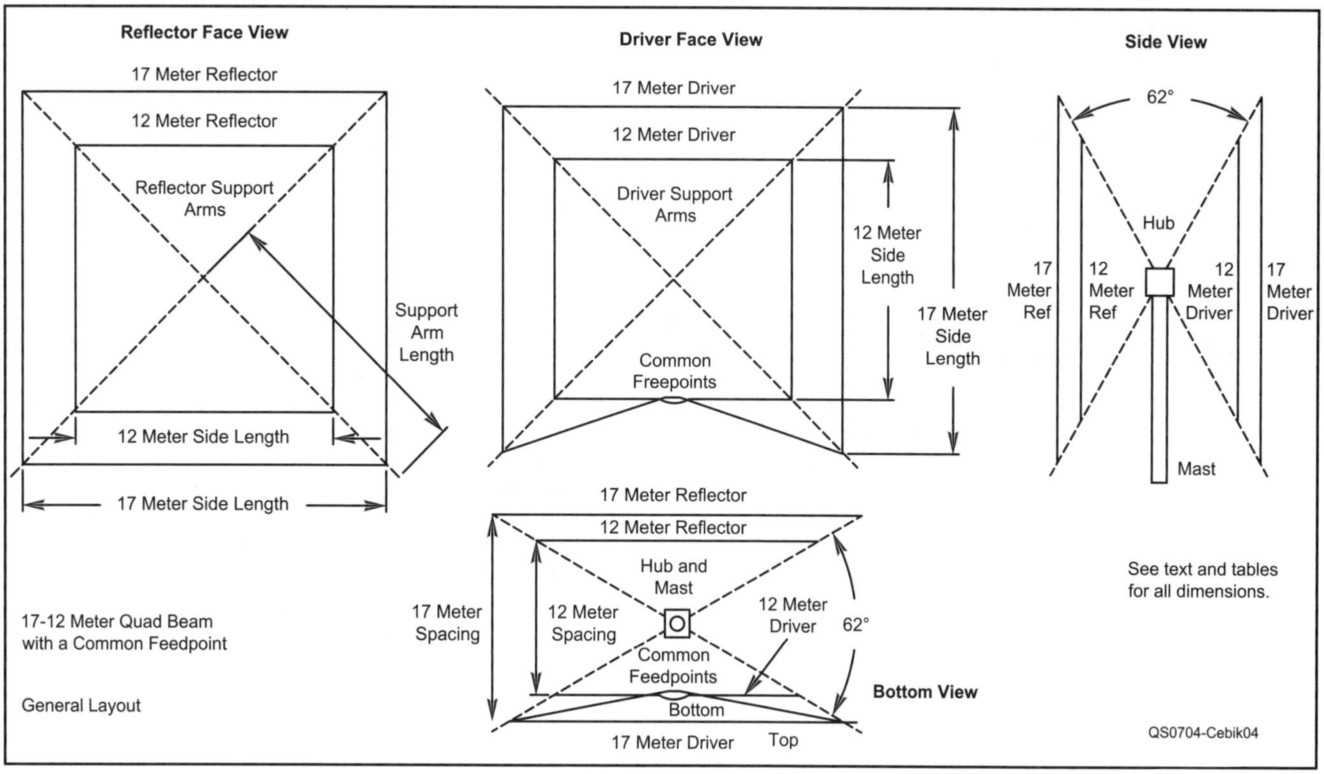

Figure 3 — General structural layout of the 17-12 meter two element quad beam with a common feed point.

Figure 4 — The spider hub immediately after welding.

Table 3
A List of Materials for the Dual-band Quad

Quantity	*Material*
8	Fiberglass tubes for the spreaders, 8′, 1½″
8	Fiberglass tubes for the spreaders, 8′, 1″
12′	Angle stock for spreader support arms, boom, and boom bracket of 1½ × 1½ × ⅛″
1	8 × 12 × ⅛″ flat plate.
16	1¾″ stainless steel hose clamps.
2	¼ × 2″ stainless steel U-bolts.
20′	Fiberglass rod or tube for spreader braces.
1	18 × 18 × ¼″ Polycarbonate sheet for element attachments and feed line attachment plate.
	75 Ω coax for the impedance matching transmission line.
Stranded bare 14 gauge copper wire, as needed for the elements.	

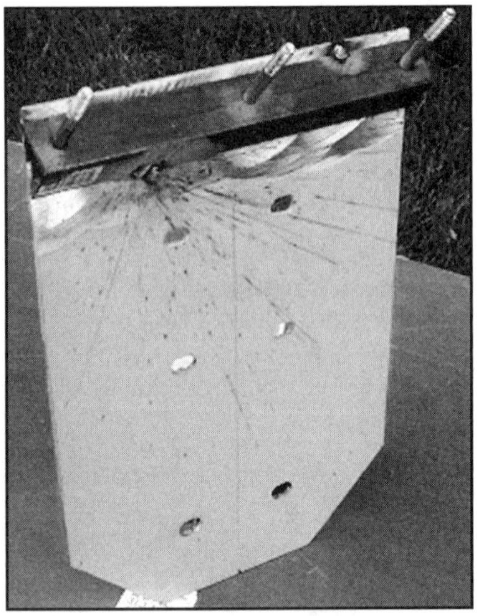

Figure 5 — The boom to mast plate attachment: note the 45° angle at the top of the plate and its bolts. This is used to attach the spider hub to the mast plate.

Figure 6 — The assembled hub and mast-plate, with a coat of protective primer.

perfectly square tube. This square boom sets the 90° angles between the spreader arms. The spreader support arms are 12 inches long with a 21° angle cut into one end. The angled ends of the support arms are welded in place about 9 inches apart on the boom. The spacing between the support arms and the angle at which they meet the main boom roughly set the spacing between the drivers and reflectors. Figure 4 shows the newly welded hub assembly. Fine adjustments to the element spacing can be made later using the flexibility of the spreaders. The element spacing is fixed using braces between the spreader arms.

The boom is attached to the mast plate by bolting it to a 12 inch piece of angle stock welded at a 45° angle to the mast plate. See Figure 5. Before welding the parts, measure the parts twice and cut them once. Be sure that each spreader support arm is at the correct angle, that it is spaced on the boom correctly, and that the assembly is symmetrical. These precautions will assure you that the antenna elements will be placed close to the modeled dimensions. A carefully constructed spider also results in a pretty quad! We found that a simple wooden jig helps hold the support arms in the correct position during welding. Figure 6 shows the assembled hub after it received its protective primer coat.

Spreaders

We chose fiberglass tube diameters of 1¼ inch and 1 inch for the telescoping spreaders. 1 inch and ¾ inch tubes can be used in lighter designs. If you substitute different materials or sizes for the spreaders, keep in mind that the longevity of your antenna will be determined in part by the strength of the spreaders. Fasten the spreaders to the spreader support arms using stainless steel hose clamps.

The spreaders should be braced to keep the spacing between drivers and reflectors constant when the winds become strong. These braces are located at approximately 71 inches from the boom end of the spreaders. We used ½ inch PVC and conduit brackets for our braces because that is what we had available. For lighter structures, ⅜ inch solid rod should be sufficient. These braces can be lashed with cable ties and cemented in place. Figure 7 reveals the type of arm-to-brace connections used in the prototype.

Element Construction

The wire elements are attached to the spreaders using circular polycarbonate plates. Figure 8 displays a sample. The plates are drilled for the spreader diameter and are held in place with cable ties and epoxy cement. We found that stringing the 17 meter elements before stringing the 12 meter elements works best. The approximate attachment points for the elements (as measured from the spider boom) are 17 meter driver — 126 inches, 12 meter driver — 93 inches, 17 meter reflector — 129 inches, 12 meter reflector — 94 inches.

Precise symmetry is not as important as the overall loop circumferences. Try to maintain the loop circumferences as specified by the model plus some extra slack for adjustments and be as symmetrical as you can. Do not permanently fix the loops in place until after final adjustments. Temporary cable ties and electrical tape will hold the elements in place until you finalize the adjustments. Then cement the attachment plates to the support arms permanently.

Feed Point Details

The feed point plate is a polycarbonate or similar plate, as shown on Figure 9. We used a small piece of UV protected PVC that we had in stock. The transmission line matching section attaches to the plate with standard coax fittings and is supported from the mast plate and hub. Transmission line to the shack can be cable tied or taped to the mast, and a current balun should be placed at the junction of the matching line and line to the shack. We used slip on ferrite beads for our balun. Type 43 material worked just fine.

Measurements and Matching

After construction, we lifted the quad about 11 feet above ground for initial testing. We also remodeled the quad at its test height to have some performance numbers with which to compare test measurements. Next, we measured and adjusted each set of elements and measured the feed point impedances using separate feed points.

The 17 meter feed point impedance initially showed a negative reactance, so the element loop lengths were changed until we were close to what the modeling predicted over real ground. The 12 meter feed point impedance was then measured and adjusted in the same manner. We then joined both drivers for a common feed point. Next, we measured the combined feed point impedance and we found that the 17 meter impedance was approximately 130 Ω, while the 12 meter impedance was approximately 132 Ω. These values will decrease at higher positions above ground.

The actual prototype loop lengths for the elements were within 1.5% of the modeled dimensions, even at the low test height. These loop lengths were 17 meter driver — 663 inches, 12 meter driver — 482 inches, 17 meter reflector — 691 inches, and 12 meter reflector — 511 inches. The spacing between the drivers and reflectors was 102 inches for 17 meters and 75 inches for 12 meters. Relative to a tight correlation between final construction dimensions and initial models, the use of closed loops in quads tends to create a larger set of construction variables than we usually find for beams with linear elements. We attributed differences between the modeled data and measured data to such factors as slack in the wires, mechanical errors, proximity to ground, and ground composition assumptions.

The combined feed point impedance is not the 50 Ω our rig likes to see. So, it had to be transformed from about 130 Ω to 50 Ω on each band. The coax that we used has a nominal impedance of 75 Ω and the velocity factor is 0.8. Using these values, the approximate feed point impedance of the quad, and the frequency of the model, a utility program calculated physical line lengths of 11.46 feet (or about 95° electrical length) for 17 meters and 8.33 feet (or about 95° electrical length) for 12 meters. The average physical length for both these lines is about 115 inches (equivalent to about 144 inches electrically). Using the average value of the physical lengths, the predicted SWR for both bands is reasonable. So, we cut our matching line to 115 inches and added it to the quad feed point.

Table 4 summarizes the measured SWR results that we obtained at the transmitter. These measurements were acceptable to us, so we made no further adjustments. You may choose to optimize them further for your needs. Keep in mind that the test readings are at the test height of 11 feet. They will change as the antenna is raised above the test height. Hence, you may have to adjust the matching line section length when you raise the antenna to its final operating position.

We also measured the front-to-back ratio at the test height, even though the height is low compared to a normal operating position. Our measurements were done at a distance of about 200 feet from the quad using a calibrated S-meter and known power source driving the antenna. We measured 14 dB for 17 meters and 20 dB for 12 meters. These values compare favorably with the modeling numbers for this height. They will improve as the antenna is raised to its final height.

Table 4
Measured 17 and 12 Meter 50 Ω SWR Values for the Combined-feed Quad using 115 Inches of 75 Ω Coax with a Velocity Factor of 0.8

17 Meters

Freq MHz	*50 Ω SWR*	*R (Ω)*	*X (Ω)*	*Z (Ω)*
18.074	1.15	51.1	5.5	51.4
18.104	1.17	49.6	7.0	50.2
18.134	1.18	48.4	8.2	49.2
18.144	1.20	47.9	8.8	48.8
18.174	1.22	46.9	8.7	47.7

12 Meters

Freq MHz	*50 Ω SWR*	*R (Ω)*	*X (Ω)*	*Z (Ω)*
24.89	1.55	62.8	20.9	66.3
24.92	1.55	62.2	21.1	65.8
24.95	1.55	61.9	21.4	65.5
24.98	1.55	61.6	21.6	65.4

Figure 7 — The arm-to-brace connectors used in the prototype quad.

Figure 8 — Here is our polycarbonate element attachment plate: do not use a loop of wire at the corners to reinforce the elements. The loops can detune the elements.

Figure 9 — The prototype feed point plate.

Conclusion

The design exercise and the resulting prototype confirm that it is possible to create a very good dual-band two element quad beam for 17 and 12 meters using a common feed point if we adhere to a few key principles. First, we must begin with the best possible monoband designs and combine them with attention to the interactions between active and supposedly inactive elements. An initial broadband design minimizes both design and adjustment time on the version with combined feed points. Second, we must construct the antenna carefully with special attention to the support structure. Third, we should go to the trouble of carefully adjusting the assembled quad to produce the desired performance. A home-brew dual-band quad with a common feed point is not a "kit" project that automatically assures success. Instead, it is a project in which we can achieve success according to the care and patience we put into each stage of the work.

Notes

[1]For a spreadsheet version of the design program, see **www.cebik.com//trans/ant-design.html**.

[2]Models used in the design phases of this project appear at the ARRL Web site at **www.arrl.org/files/qst-binaries/cebik0407.zip**.

[3]R. D. Straw, Editor, *The ARRL Antenna Book*, 20th Edition. Available from your ARRL dealer or the ARRL Bookstore, ARRL order no. 9043. Telephone 860-594-0355, or toll-free in the US 888-277-5289; **www.arrl.org/shop/**; **pubsales@arrl.org**.

Bob Cerreto, WA1FXT, is a Senior Engineer with a major dc power systems supplier. Back in Connecticut, Bob was first licensed in 1965 as WN1FXT. Since then, the ham radio world for Bob has been filled with many homebrew projects. He feels most comfortable in his radio lab and woodworking shop with fellow hams and grandchildren. Since moving to Ohio in 1984, his backyard has become a three-acre antenna breadboard. Bob's wife Judith, KC8BOM, serves as a catalyst for his ham radio ventures and is an accomplished builder, too. Bob can be reached at 49015 Middle Ridge Rd, Amherst, OH 44001 or **tonybob@adelphia.net**.

Since his retirement as a professor at the University of Tennessee, L. B. Cebik, W4RNL, has written extensively for radio amateurs about antennas and related topics in many venues, including his own Web site, **www.cebik.com**. *Licensed since 1954, LB has served ARRL in many capacities, including Educational Advisor and Technical Advisor. He is a contributing editor to* QEX *and is technical editor for* antenneX. *LB can be reached at 1434 High Mesa Dr, Knoxville, TN 37938-4443 or* **cebik@cebik.com**.

A Two Element "Wonder Bar" Beam for 17 Meters

Who says nostalgia isn't what it used to be? As with the original, this antenna provides good performance in a compact, almost invisible package.

Gary A. Cook, W9JSN

Nostalgia, plus a need for a better 17 meter antenna, recently led me to revive and adapt an antenna design popular during the late 1950s. The result was more than a trip down memory lane: I now own a compact, 2 element, 17 meter beam that works very well.

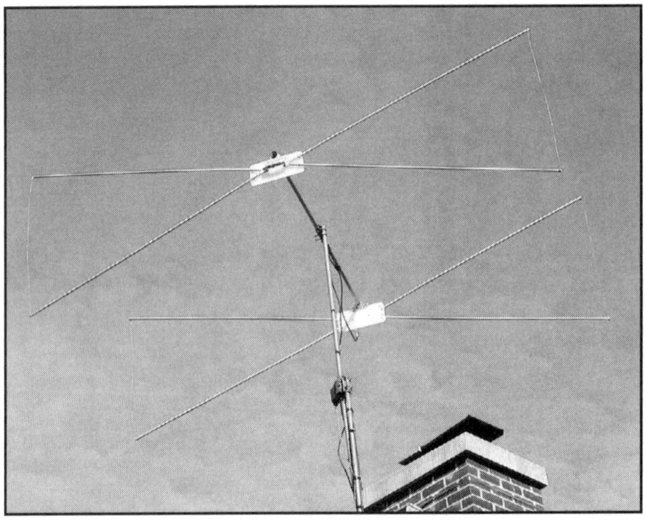

Figure 1 — The finished Wonder Bar beam, ready for action.

So What's a Wonder Bar?

When an article describing the easy construction and outstanding performance of a "Wonder Bar" antenna for the 10 meter band appeared in the November 1956 issue of *QST*, I was a 17 year old high school student, licensed as WØVNX and living in a small Iowa town.[1] Lacking the money to invest in a factory made beam, I jumped at the idea of building an inexpensive antenna that might help me work some DX stations on 10 meters.

I immediately sent off to Allied Radio in Chicago for the B&W Miniductor and standoff insulators I would need for the center loading coil of this shortened, bow tie shaped, rotatable dipole. But, unhappily, that was as far as I went with my antenna project. Somehow other things distracted me and I never managed to round up the remaining materials needed for the antenna, much less actually build it.

The Miniductor and stand-off insulators eventually got used for something else and disappeared from my junk box. Meanwhile I went off to college, followed by graduate school, and then spent nearly 40 years teaching college students. But I never forgot about the Wonder Bar. I held on to the original 1956 article, and even accumulated a file folder of related articles — one about a 20 meter version, and two *Hints & Kinks* items written by hams who had added an element or two to the original 10 meter antenna to make a beam.[2, 3]

That Was Then, This is Now

Last fall I began thinking seriously about building some kind of antenna for 17 meters, and the design of the old Wonder Bar came once again to my mind. I was looking for an antenna that would be directional, rotatable and have some gain over a dipole. I also wanted it to be relatively small and inconspicuous, so as not be the cause of any adverse comments from either my neighbors or my spouse.

I decided that a 17 meter Wonder Bar used as a driven element, along with another Wonder Bar tuned and mounted to function as a reflector, would be just the ticket. A little work with a calculator allowed me to enlarge the dimensions of the original 10 meter version of the antenna for 17 meters. Each of the two elements would be only 12 feet 8 inches across — approximately half the length of the elements on an ordinary full sized beam for this band. A look at several antenna handbooks revealed that I could expect good results with the two elements spaced just 6 feet 7 inches apart. I also concluded that I could wind my own coils and build the whole antenna from materials readily available at local hardware and discount stores. The result was a new and improved version of the project I had envisioned, but never completed, almost 50 years ago. See Figure 1.

For the benefit of readers who might want to try their hand at building a similar antenna, here is a detailed description of the materials and construction involved in my completed 17 meter Wonder Bar beam. I am indebted to the authors of all the articles cited in the notes for suggestions that helped me arrive at the final design of this antenna.

Construction

I cut the two center supports of the beam from a single 15 × 20 × ⅜ inch Poly kitchen cutting board purchased at a local household goods store. (This construction idea occurred to me one morning while I was cleaning up the kitchen!) I used a small power saw to cut across the cutting board 7 inches from each of

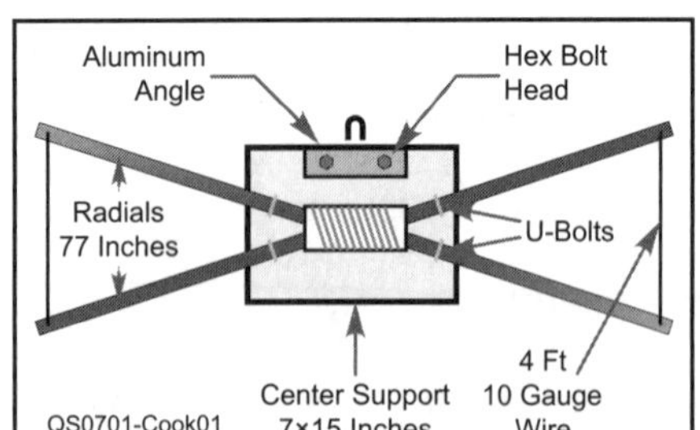

Figure 2 — Construction details of Wonder Bar beam.

[1]Notes appear on page 78

the ends on the long sides to obtain two 7 × 15 inch pieces. I then used an electric sander to smooth the edges and round the corners of the newly cut side of each piece. The eight diagonal radials are each 77 inches in length, and can be made from ¾ inch diameter aluminum tubing, copper tubing or steel electrical conduit. I opted to go first class with aluminum radials (cut from 8 foot lengths obtained at a nearby building supply store) to hold down the overall weight of the beam. Copper tubing might be a good compromise since it is less expensive and not much heavier than the aluminum. The steel conduit would be the least expensive choice but it would add considerable weight to the finished antenna.

One end of each radial must be flattened for a length of about 2 inches. I tried at first to accomplish this by compressing the tubing in a shop vise but found that my vise was not heavy enough to do the job. After a little frustration, I took the radials outside and did it the way I would have done it when I was 17: I pounded one end of each radial flat with a hammer on a cement block. Once this was done, I drilled a ¼ inch diameter hole half an inch from the end of each flattened section.

After I cut the radials to the appropriate length by means of an inexpensive tubing cutter (easier and neater than a hacksaw), I fastened the flattened ends to the center supports using ¼ inch hex bolts (2½ inches long) and ¾ inch U bolts, as shown in Figure 2. I had to spread the legs of the U bolts open a fraction of an inch farther so that they would slip over the radials. It is a good idea to use washers and lock washers under all nuts to hold them in place and also to keep them from cutting into the cutting board center support.

You may also want to use extra washers and nuts as needed to make spacers between the center support and the flattened ends of the radials. I mounted the two center bolts for each element 5 inches apart — that is, 2½ inches on either side of the center of the support board — to allow room later for the center loading coil. You should fan out the radials slightly more than 4 feet at the far ends before they are fastened down with U bolts. This will put a little tension on the wire tie bars once you attach them to the far ends of the radials. Pencil lines drawn vertically and horizontally through the center point of the support board help in lining up the center bolts and U bolts so that the four radials of each element appear visually symmetrical.

For the tie bar at the end of each element, I used a piece of 10 gauge bare copper wire slightly less than 4 feet long. I soldered a ring terminal to each end of this wire, and then used a small screw through the terminal to attach the end of the wire to the radial half an inch from the far end. As mentioned earlier, you will want a slight amount of tension on these tie bars to hold them tight and to keep them looking straight.

Figure 3 — Close-up of the back side of the modified cutting board used as a center support and insulator.

I made the identical loading coils for the centers of the two elements from 7 foot lengths of 12 gauge bare solid copper wire, wound on 6 inch sections of 1 inch (inside diameter) PVC plumbing pipe. The local hardware store sold me a single foot of this pipe very inexpensively. I could obtain only the insulated version of the 12 gauge wire for the coils, so I fastened one end of each wire in a vise and used a wire stripper to carefully remove the insulation without nicking the wire. Each coil occupies the center 4 inches of a PVC form. Drill small holes for the ends of the wire about 1 inch from each end of the form. Insert one end of the wire far enough through the hole to allow several inches for eventual mounting and hookup. Hold the other end of the wire in a vise to keep it tight; then turn the form slowly to wind the coil on the form. I also drilled a ¼ inch diameter hole ½ inch from each end of the coil form so that I could eventually slip the ends of the form over the tips of the two center bolts fastening the radials to the center support board. Solder ring terminals to the wires extending from the ends of each coil, and fasten these to the two center bolts using lock washers and nuts.

I used a smaller coupling coil to link the feed line inductively to the driven element of the antenna. This coil consists of two turns of 12 gauge insulated solid copper wire wound fairly tightly around the loading coil for the driven element. I soldered one end of this coupling coil to the inner conductor, and one end to the outer conductor, of a short length of 52 Ω coax. The other end of this coax was terminated in a standard PL-259 plug, to which I attached a PL-258 coax coupler, so that I could later connect my 52 Ω coax feed line to this adapter by means of another PL-259 plug.

I was lucky enough to have on hand an 8 foot aluminum boom, along with a boom to mast mount, left over from an old citizens band (11 meter) beam that someone had given me. I cut this boom down to a length of 6 feet 8 inches centered on the mast mount. Any suitable length and diameter of aluminum tubing or steel pipe would, of course, work just as well for the boom. And a small sheet of metal or heavy plastic (such as the leftover material from the kitchen cutting board mentioned above), along with two pairs of U bolts (one pair for the boom and one pair for the mast), could be used to fashion a boom to mast mount.

I used a 6 inch length of 1½ inch aluminum angle stock (1/16 inch thick) and a U bolt to mount each of the center supports of my two elements to the boom. (This aluminum angle stock is available in 4 foot lengths at the local hardware store.) I used two hex bolts, ¾ inch in length, mounted 5 inches apart, to fasten the inner side of the angle stock to the side and over the top of the center support. Two holes drilled in the top side of the angle stock to accommodate the legs of an appropriately sized U bolt made it possible to hang the center support board underneath the boom. See Figure 3. Be sure that you center the angle stock and the holes for the U bolt horizontally on the center support board. Use a small carpenter's level during the final installation to make sure that each element is hanging properly.

Although it was probably not necessary, I used a mast clamp with serrated edges (borrowed from an old TV antenna) for the U bolt attaching each element to the boom. It should be easy to slip the U bolt for each element over the end of the boom and to slide the element back and forth on the boom if necessary while assembling and tuning the antenna.

Making it Play

Once you've built the two bow tie elements and have them mounted temporarily on the boom, the only part of this project calling for critical measurements begins. Tune the center loading coils of both elements to the proper frequencies, and then adjust the coupling coil on the driven element. Back in 1956 this part of the project would have been a real chore, but by putting off my antenna project until 2006 I was able to enjoy the fruits of recent technology. For these tasks you will want to own or borrow a good antenna analyzer. I used an MFJ-259 portable antenna analyzer to tune both coils to the appropriate frequencies and to adjust the coupling coil for the lowest possible SWR.

Based on a good deal of trial and error, I would recommend the following procedure for this stage of the project.

- Temporarily mount the antenna on a short mast or on top of a self supporting ladder where it will be well away from all metal objects and yet within reach for easy adjustment of the loading coils.
- Tune the coil on the reflector element, mounted by itself temporarily on the boom. Tune it so that it is resonant at a frequency approximately 5% below the bottom of the 17 meter band (around 17.165 MHz). I removed one prong from a two pronged spade terminal and bent the remaining prong so that it would fit snugly on the wire of the loading coil but at the same time slide easily around the coil for purposes of adjustment. I soldered this spade terminal to an insulated 12 gauge stranded copper wire with an alligator clip at the other end. Once the spade terminal was set in place, I temporarily fastened the alligator clip to the wire at one end of the coil, thereby shorting out the turns between that end of the coil and the spade terminal.
- Begin by shorting out six or seven turns at one end of the coil. Then temporarily link the antenna analyzer inductively to the coil and adjust the frequency of the analyzer until you see a small dip in the meter, indicating the resonant frequency of the element. The inductive link connecting the antenna analyzer to the loading coil of the reflector can be a coupling coil of a two turns wound temporarily around the outside of the antenna loading coil, or a small coil — such as the one that comes with the dip kit for the MFJ 259 — inserted inside the coil form of the loading coil.
- The other end of the inductive link should be connected to the antenna analyzer via a short length of coax and the appropriate plug. Once you determine the initial resonant frequency of the bow tie element you will know whether you need to short out more turns and thereby raise the resonant frequency, or the reverse. Keep adjusting the tap on the loading coil until you find the point at which the element resonates at the desired frequency.
- Use the same procedure to tune the driven element to the center of the 17 meter band (about 18.12 MHz). In this step the two turn coupling coil that will be a permanent part of the antenna can serve as your link between the loading coil and the antenna analyzer.
- Finally, adjust the coupling coil on the driven element, by sliding it to various positions along the length of the loading coil, and also by varying the tightness around the outside of the loading coil, for minimum SWR. This is a crucial step and requires some experimentation. I found that I achieved the lowest SWR by locating the coupling coil near the end of the loading coil farthest from the shorted turns.

Figure 4 — The front side of the center support showing the loading coil, the coupled coil and the element radial mounting. Note the two plastic cable clamps that hold the coax and the coupling coil mechanically solid.

The readings you will get in the three final steps above are somewhat interdependent, so you may have to repeat the steps several times to find the best settings. It is likely that the resonant frequency of the driven element will shift upward somewhat once the antenna is raised to its final position. By proper adjustment (and this definitely requires both persistence and patience), you should be able to obtain an SWR of less than 1.7:1 across the entire 17 meter band.

Once you find the best settings for the loading and coupling coils, solder all the connections with a heavy duty soldering gun. I used two ½ inch plastic cable clamps fastened to the center support of the driven element with small screws to hold the coupling coil, and the short length of coax attached to it, firmly in place. See Figure 4. This is important, since even a slight movement of this coil can affect the SWR of the antenna.

Performance

Prior to the construction of this beam, I had on a number of occasions tried operating 17 meters with an all band Windom antenna that had served me well on 40, 80 and 160 meters. But the results of these attempts were far from impressive. I typically received mediocre signal reports from stateside contacts and seldom made any contact at all with the European stations I called.

In the month or so that I have used the two element Wonder Bar beam, my results with 100 W on 17 meter phone have been most gratifying. With the antenna mounted at a height of only 30 feet, and using a small TV rotator, I have worked stations throughout the US and Canada with many reports of S9+. Better yet, I have been able to work European stations with ease, receiving signal reports ranging from S5 to S9, whenever the band has been open to that part of the world.

Casual tests run with several stateside contacts indicate that the beam has a front to back ratio of approximately 10 to 15 dB, and that both the gain and the directional pattern of the antenna resemble those you would expect from a full sized two element Yagi. Not bad, I would say, for an antenna design that comes fully equipped with warm feelings of days gone by. My only regret now is that I was not smart enough to build a 10 meter version of this beam back in 1956 or '57. I could have used it to work some serious DX when the sunspot activity of Solar Cycle 19 was reaching its historic peak!

Notes

[1]E. Bishop, K6OFM, "The 'Wonder Bar' Antenna," *QST*, Nov 1956, pp 32-34, 138. See also S. Leland, W1JEC, "The Old Timer's Notebook: Remember the Wonder Bar Antenna — A 10 Meter Bow Tie?" *QST*, Apr 1980, pp 59-60. Another version of the original 10 meter Wonder Bar (this version by Jim, K2JXW) can be seen at **www.hamuniverse.com/wonderbar.html**.

[2]R. Rosenbaum, W5ECP, "A 'Wonder' on 20 Meters," *QST*, Jun 1957, pp 44-45.

[3]G. Ryan, " 'Wonder Bar' Beam," *QST*, Feb 1957, pp 43,134; F. Masho, W3CBM, "A Three Element Wonder Bar for 10 Meters," *QST*, May 1981, p 46.

Gary Cook, W9JSN, was first licensed as WNØVNX in 1954. He grew up in a small town in northeast Iowa, and after earning a BA degree from Drake University and a PhD from Yale University, taught philosophy for many years at Beloit College in Wisconsin before retiring in 2004. When not doing research and writing on the history of American philosophy, he enjoys experimenting with antennas and working on vintage ham gear. He traded in his WØ call for his current W9 call when he moved to Wisconsin in 1965. He has held an Extra Class license since 1978. You can reach the author at 737 Milwaukee Rd, Beloit, WI 53511, or at **cookga@beloit.edu**.

A Short Boom, Wideband, Three Element Yagi for 10 Meters

L. B. Cebik, W4RNL (SK)

Although there are many two and three element Yagi designs for 10 meters, most cover much less than the full band. For this Yagi, I set four requirements:

- The beam must show less than 1.5:1 SWR from 28 to 29 MHz.
- The driven element must use a direct 50 Ω feed with no matching network, such as a gamma or beta match, required.
- The antenna should have the shortest feasible boom length and turning radius.[1]
- Finally, the beam may require careful but not finicky construction.

When we combine all of these requirements, some of which are admittedly just builder/operator preferences, we find only a few candidates.

Traditional Yagi designs that are worth considering have been around since the 1980s. If we check the 10 meter performance of these antennas, we have a basis for comparison.

A normal wideband two element reflector-driver Yagi is about 5.5 feet long. The beam manages about 6.0 dBi free-space gain across the band with only 10 to 12 dB front-to-back ratio (F/B).

A wideband three element Yagi with a direct 50 Ω feed point is close to 11.5 feet long. It provides about 7.1 dBi free-space gain and averages about 20 to 21 dB F/B across the band.

Now suppose that we could find a way to achieve about 6.7 dBi gain with about 16 or 17 dB F/B and still keep the beam only 5.5 feet long. That beam would be worth at least a second look. The proposed design consists of a three element beam, but instead of the usual reflector, driven element and director configuration, it uses a close spaced dual driven element and a single director.

The differences among the beams amount to more than numbers. Figure 1 overlays the free-space E-plane (azimuth) patterns for the three antennas. The weaker gain of the two element design shows up almost as clearly as the poorer F/B. The gain deficit of the proposed design is far less evident. As well, even though the worst case F/B of the proposed design is a bit weaker, the average F/B is not too dissimilar from the rear pattern of the standard longer-boom three element Yagi.

[1]Notes appear on page 80.

LB CEBIK, W4RNL

The proposed design shares a possibly significant property with the standard three element Yagi. Both use directors and therefore show a rising gain value as we increase frequency across the band. As shown in Figure 2, this characteristic contrasts sharply with the descending gain curve of the standard driver-reflector two element Yagi.

All three small Yagi designs share a common trait — they cover the first MHz of 10 meters with less than 1.4:1 SWR. They all provide for a direct connection to a 50 Ω coaxial cable with no matching network (although a common-mode current balun is always a good idea). In fact, if we reset the element lengths for a slightly higher frequency (about 28.9 MHz), the beams will cover all of 10 meters (28-29.7 MHz) with less than 2:1 50 Ω SWR.

The Short-Boom Wideband Phase-Driven Three Element Yagi

The performance improvement over a standard two element wideband Yagi with the same boom length comes at a price — an extra element and a phase line. If we were working with a 40 meter Yagi, the third element would add a considerable load on the mast and rotator. At 10 meters, however, elements are much lighter. So a third element does not come close to stressing any part of the support system. The array that we are exploring has two driver elements spaced 25 inches apart with a single director 39 inches forward of the first driver. The total length is 64 inches plus a few inches at the boom ends. A 6 foot boom using 1.25 inch aluminum or similar material would serve very well. The antenna is short enough to allow for a PVC schedule 40 boom without significant sag. Let's examine the structure in small steps.

The Elements

We can build beams to be light or to withstand heavy winds. To give you a choice, I shall provide two sets of dimensions. One set uses heavier elements for wind loads up to about 90 to 100 mi/h. The lighter version might be rated to the 60 to 70 mi/h level.

The dimensions in Table 1 apply only to the two different element diameter taper schedules. If you change the material diameters or the interior lengths of wider tubing, the beam may not perform as advertised. Remember to add about 2 to 3 inches to the lengths of the smaller tubes to ensure sufficient material for a secure overlap.

Both versions provide essentially identical performance across the band. The gain varies from 6.4 dBi at 28 up to 7.1 dBi at 29 MHz. The F/B peaks at almost 18 dB at mid-band. Its lowest value is about 14.5 dB at 29 MHz.

The Overall Design

The general layout of the beam is shown in the lead photo. In this design, I started with a simple narrow band driver director array. All two element driver director Yagis have a very narrow bandwidth. I then changed the driver system to a pair of phased drivers in order to line length. There will be enough spacing between the phase line and the boom to minimize interaction if ¼ inch thick insulation plates and saddle U bolts are used.

The coax connector can sit on a small metallic plate attached to the forward edge of the forward driver plate. Just be sure that the screws you use to secure the coax connector plate do not contact the boom. Of course, all hardware should be stainless steel, which you can obtain from most home centers these days. I use stainless steel nuts to separate copper wire ring connections from the aluminum in the element.

Final Adjustments

Final adjustments should be negligible if you have constructed the phase line carefully. I performed initial tests with the antenna 10 feet above ground on a temporary mast. The usual test for determining correct construction is to sample the SWR across the operating passband (28.0 to 29.0 MHz, in this case). Modeling showed that the 50 Ω SWR curves vary only a little as you increase the height from about 10 to 33 feet (1 λ). Thus, you can be confident that any adjustments you make at a low test height will hold at the ultimate operating height.

For the initial tests, I left the ⅜ inch diameter tip sections unfastened. I fastened all other parts of the antenna in final form. I held the tip sections in place with electrical tape for possible adjustment. However, my prototype needed no changes in the lengths of the elements.

Performance

The antenna performs as modeled. The added gain over a two element driver reflector design is less evident than the improved F/B. The SWR measurements overlay the modeled curves in Figure 4 too closely to need a new chart.

Part of the successful translation of the design models to a physical antenna involves the use of high quality parts throughout. Home center tubing — when available — tends to use thinner walls than 6063-T832 tubing with 0.058 inch wall thickness. The latter nests very well so that sheet metal screws used as element section fasteners do not vibrate loose over the seasons. UV protected polycarbonate boom to element plates endure the sun for many years. Stainless steel hardware is weather impervious. Unless you are building a short-lived trial version of this (or virtually any) antenna, I do not recommend junk box materials.

There are only two HF amateur bands that might call for a design like the short-boom wideband three element phased-driver Yagi: 10 and 40 meters. In fact, this 10 meter version evolved from a 40 meter design exercise. I lack the shop space and the support system needed to build and test a 40 meter prototype. I have included all of the models referenced in these notes for availability on the ARRL Web site.[2] The 40 meter design that covers 7 to 7.3 MHz with under 1.5:1 50 Ω SWR is in the collection. When modeling these antennas in either *NEC-2* or *NEC-4*, be certain to apply the Leeson corrections. Even *NEC-4* will drift off the mark with the large element-diameter taper required by 40 meter elements. The Web material also includes additional figures and design details that should prove helpful to the builder.

The 10 meter version of the antenna is not for everyone. However, it supplies a Yagi design that fills a niche between standard two element and three element wideband Yagis. It improves performance while retaining the short two element boom. The wideband aspect of the design may prove as important in allowing for construction variations as it is in obtaining a low SWR across the band. Hence, it is a beam that the home craftsperson can — with care — successfully build.

Table 2
250 Ω Transmission Line Dimensions

AWG Wire Size	*Wire Diameter*	*Center-to-Center Spacing*
#14	0.0641"	0.262"
#12	0.0808"	0.330"
#10	0.1019"	0.416"
# 8	0.1285"	0.525"

Notes

[1]Turning radius is the distance from the mast to the farthest tip of any element.

[2]**www.arrl.org/files/qst-binaries/.**

LB CEBIK, W4RNL

Figure 3 — Closeup view of the feed point assembly.

The last *QST* article authored by longtime contributor L. B. Cebik, W4RNL (SK)

L. B. Cebik, W4RNL, ARRL Technical Advisor and antenna authority, passed away in April 2008. He had submitted this article prior to his death and we publish it as a memorial to a life that added an immense amount of knowledge to the study of antennas by amateurs and professionals throughout the world.

Cebik was 68. An ARRL Life Member, he was known to many hams for the numerous articles he wrote on antennas and antenna modeling. He had articles published in most of the US ham journals, including *QST, QEX, NCJ, CQ, Communications Quarterly, Ham Radio, 73, QRP Quarterly, Radio-Electronics* and *QRPp.*

Those of us who worked with L. B. along with the entire Amateur Radio community miss him very much. — *Joel R. Hallas, W1ZR, QST Technical Editor*

A 10 Meter Moxon Beam

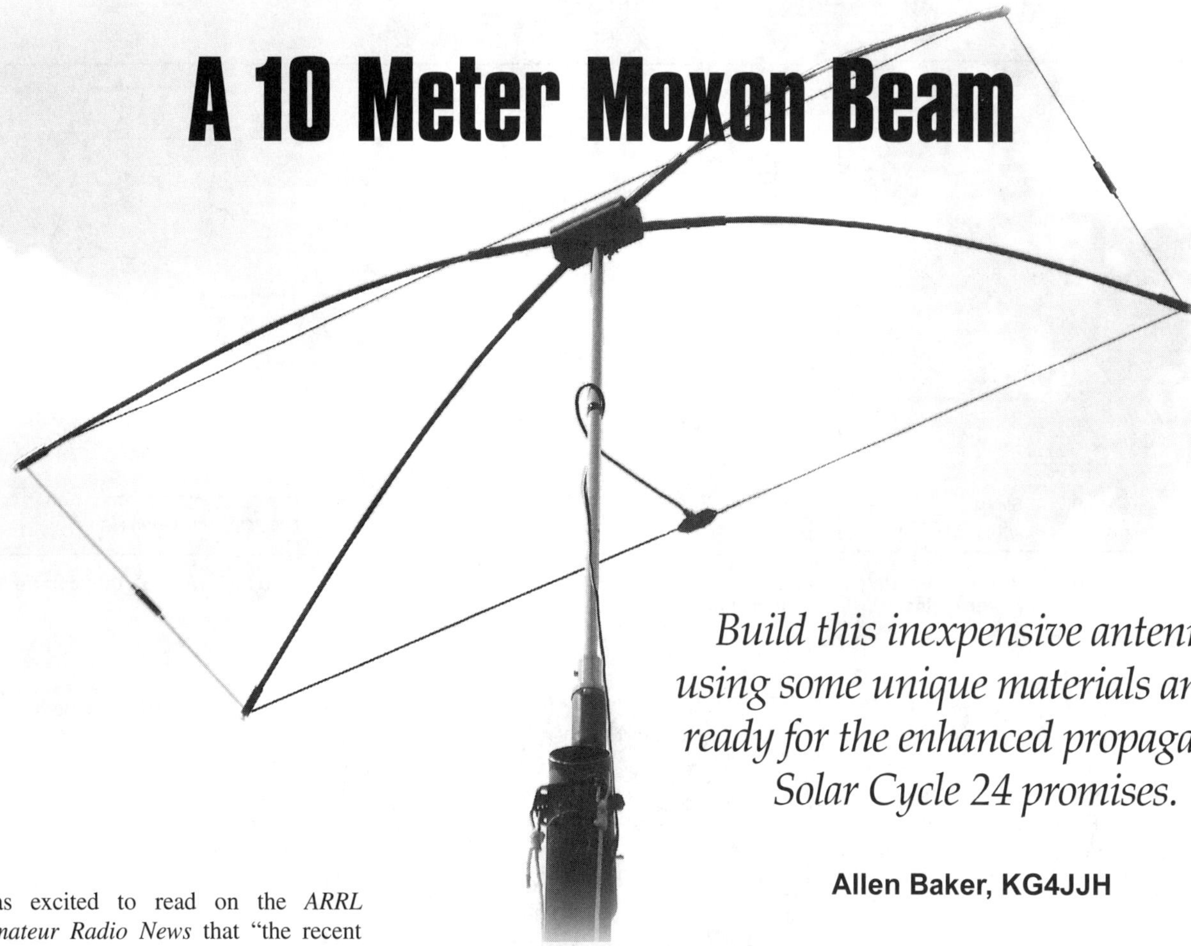

Build this inexpensive antenna using some unique materials and be ready for the enhanced propagation Solar Cycle 24 promises.

Allen Baker, KG4JJH

I was excited to read on the *ARRL Amateur Radio News* that "the recent appearance on the sun of two so-called 'backward sunspots' may mean solar cycle 23 is drawing to a close and Cycle 24 now is under way or soon will be."[1] During the peak years of solar cycle 23, some years back, I consistently made worldwide contacts on 10 meter PSK31 using 5 W to an attic mounted Moxon rectangle.

The Moxon rectangle was developed by Les Moxon, G6XN, as a derivative of the VK2ABQ square, with further refinements to natural beam dimensions by L. B. Cebik, W4RNL (SK).[2] The antenna is essentially a two element Yagi with the element ends folded toward each other. It has about the same gain, a higher front to back (F/B) ratio, direct 50 Ω feed and is approximately 30% smaller than the usual two element Yagi.

For this Moxon Beam or *MoxBeam* project, I describe a portable version of my attic antenna that is built from composite lumber, fiberglass tubing, and Home Depot wire for about $75. Your efforts will be rewarded with a 7 pound antenna that, if mounted 25 feet above typical ground, delivers up to 11.3 dBi gain, and has a 40 dB F/B ratio (see Figure 1), a 19° take-off angle and a turning radius of 6.5 feet. For amateurs interested in other bands, I have prepared a table of construction dimensions and performance details for the 6 through 20 meter amateur bands using the same technique. This information, along with additional drawings and photos is available at the *QST* binaries Web site.[3]

[1]Notes appear on page 84.

Construction

Hub

You may have seen the new composite lumber that is used to build outdoor decks. At least one company offers this material in dimensional sizes such as nominal 2 × 4 inch. It is an ideal material for building the antenna hub.[4] It saws and drills as easily as wood but is impervious to weather extremes. It has low thermal expansion and contraction and is ultraviolet protected. The smallest piece available is 12 feet long and, according to my local lumber yard, it is expensive at $26 each. You only need an 8 inch piece for this design and the same hub dimensions can be used for all bands, or you can make 18 hubs. You may be as lucky as I was and find a free scrap at a construction site. All required materials are listed in Table 1.

Print a full size template of the hub (see Figure 2) and temporarily fasten it to the composite material using a glue stick. Mark a centerline down the length of the material and use it to line up the template. Using a table or miter saw, cut two 20° angles on each end to form a "V" along the centerline (see Figure 3). The total length should

Table 1
10 Meter MoxBeam Materials List

Qty	*Description*	*Source*	*URL*
1	Hub, Trex composite lumber	Home Depot	**homedepot.com**
2	½" OD × 8' fiberglass tubing*	Max-Gain	**mgs4u.com/fiberglass-tube-rod.htm**
4	¾" OD × 8' fiberglass tubing*	Max-Gain	**mgs4u.com/fiberglass-tube-rod.htm**
33'	#14 PVC insulated stranded wire	Home Depot	**homedepot.com**
4	Nylon screw 5⁄16-18 × ½"	Ace Hardware	**acehardware.com**
1	Budwig HQ-1	RF Connection	**therfc.com/anttacc.htm**
1	Thumb screw, stainless, ¼-20 × 1½", #91745A546	McMaster-Carr	**mcmaster.com**
1 box	Ring terminal #34540, 16-14 gauge	Ace Hardware	**acehardware.com**
4	Nylon screw and nut, 6-32 × ½"	Ace Hardware	**acehardware.com**
1 pk	Quick release button, stainless, #92988A510.	McMaster-Carr	**mcmaster.com**

*Max-Gain Systems will cut tubing to specification, which can significantly reduce shipping charges

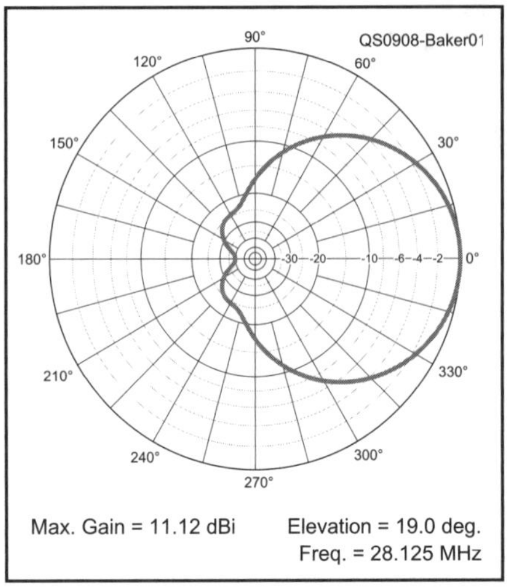

Figure 1 — 10 meter MoxBeam azimuth plot at a height of 25 feet over typical ground.

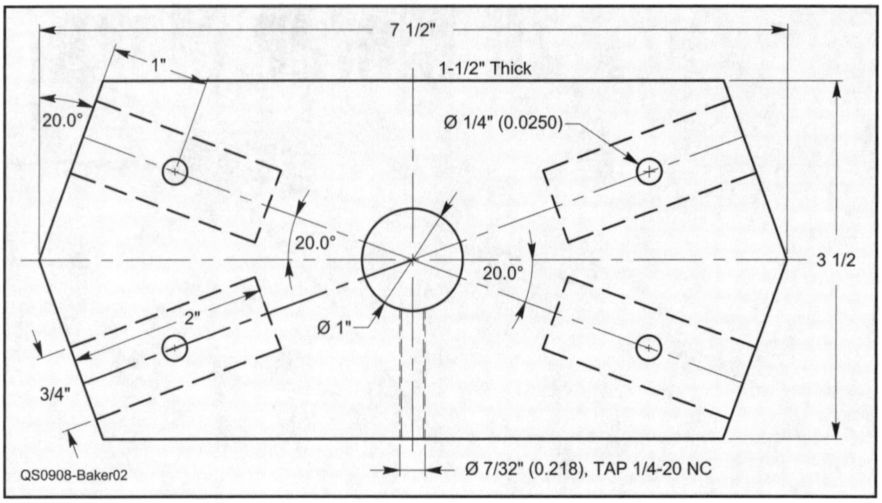

Figure 2 — Template of MoxBeam hub. If printed full size (see Note 3), this can be used as a cutting template.

be 7½ inches from end to end. Extend the spreader hole centerlines down the ends and mark the centers. The angles on the ends make drilling the spreader holes easy. By clamping the piece on its end you automatically get the required 20° spreader angle. I find it helpful to drill an ⅛ inch pilot hole first and then enlarge it to the final diameter.

Using a drill press with a ¾ inch Forstner bit, drill a 2 inch deep hole at each location. Trial fit a piece of ¾ inch OD tubing to ensure that the insertion depth is 2 inches. Next, drill a 1 inch diameter hole in the hub center for the mast. A tapped hole on the side allows the hub to be secured to the mast with a ¼-20 × 1½ inch thumb screw. And last, drill four ¼ inch holes on the top of the hub for the spreader quick release buttons. If desired, remove sharp corners using a file or router with a round over bit. The hub material can be painted, but the manufacturer recommends that you let it weather for 8 to 12 weeks, and then use any paint or stain that adheres well to wood.

Spreaders

The 83 inch long spreaders (see Figure 4) are made from ¾ inch and ½ inch outside diameter (OD) fiberglass tubing (⅛ inch wall). Cut four 61 inch long pieces of ¾ inch OD tubing (Figure 4, dimension S2), and four 28 inch long pieces of ½ inch OD tubing (Figure 4, dimension S3). On one end of the ¾ inch OD tubing, drill a 17/64 inch hole through one wall, 1 inch from the end. Deburr the inside of the hole with a round file and insert a quick release button. Insert the ½ inch tubing into the other end of the ¾ inch tubing with a 6 inch overlap. Fasten the tubes with a 6 × ½ inch thread-cutting screw in the center of the overlap.

Corner Clamps

The four corner clamps (see Figure 5)

Figure 3 — Preparing a hub on a table saw using a printed template.

are made from ¾ inch OD (⅛ inch wall) fiberglass tubing. Cut four pieces, 2½ inches long. Using scraps from the spreaders, cut four 1 inch long pieces of ½ inch OD fiberglass tubing. Roughen the inside of the ¾ inch tubing with a tubing brush. Then, glue the ½ inch OD tubing inside the ¾ inch OD tubing, flush with one end, using fiberglass epoxy. Allow 24 hours for the glue to set and then remove any excess dried glue inside the ¾ inch tubing with a ½ inch Forstner bit. Next, tap the inside of the ½ inch tubing using a 5/16 NC-18 tap. The last step is to drill an ⅛ inch diameter hole all the way through both tubes, ½ inch from the flush end. Use a countersinking bit to smooth the edges and prevent chafing the wire insulation. A 5/16 NC-18 × ½ inch long nylon hex head screw is then screwed into the threads. When a wire element is threaded through the ⅛ inch hole the nylon screw is tightened enough to hold the wire, and the corner clamps become part of the element assembly.

Elements

The elements are cut from regular Home Depot 14 gauge THHN insulated wire that was left over from dipole construction. This *machine tool wire* (MTW) has 0.015 inches of PVC insulation plus a 0.005 inch outer nylon jacket. I initially suspected that the nylon jacket affected the insulation's dielectric constant because the first set of elements I made had a resonance that was considerably lower than the model indicated (*EZNEC4 4.0* includes the effects of wire insulation).[5] Closer inspection, however, revealed that the ½ inch metal screws used to attach the elements to the insulators were to blame. The metal screws were replaced with nylon and the element resonance was on target.

The insulator ends are terminated with ring terminals by crimping and soldering. The feed point ends are soldered to a Budwig HQ-1 center insulator as shown in Figure 6. Mark the wire corners with a permanent marking pen and place marks ⅜ inch on either side of the corner mark. This will allow you to center the corner clamps on the corners. Trim the wire to length leaving ¼ inch of wire insulation stripped off for connecting the ring terminals. This gives 7/16 inch of additional length from the end of the wire to the tip of an Ace Hardware 16 to 14 gauge ring terminal. Check the ring terminals that you intend to use and adjust the wire lengths as needed to meet the dimensions.

First, trim ¾ inch off both sides of the Budwig wires for a total end-to-end length of 5 inches. For the driven element, solder a ring terminal on the end of the wire. Place the ring terminal on a metal measuring tape (cloth tapes can stretch) with the terminal tip on zero and clamp the measuring tape and terminal in a vise. While keeping constant tension on the wire and tape, measure out 1 foot 10 3/16 inches (see dimension details on the binaries Web site) and mark the wire. Continue measuring to 7 feet 10 5/16 inches and mark the wire, recheck the measurement

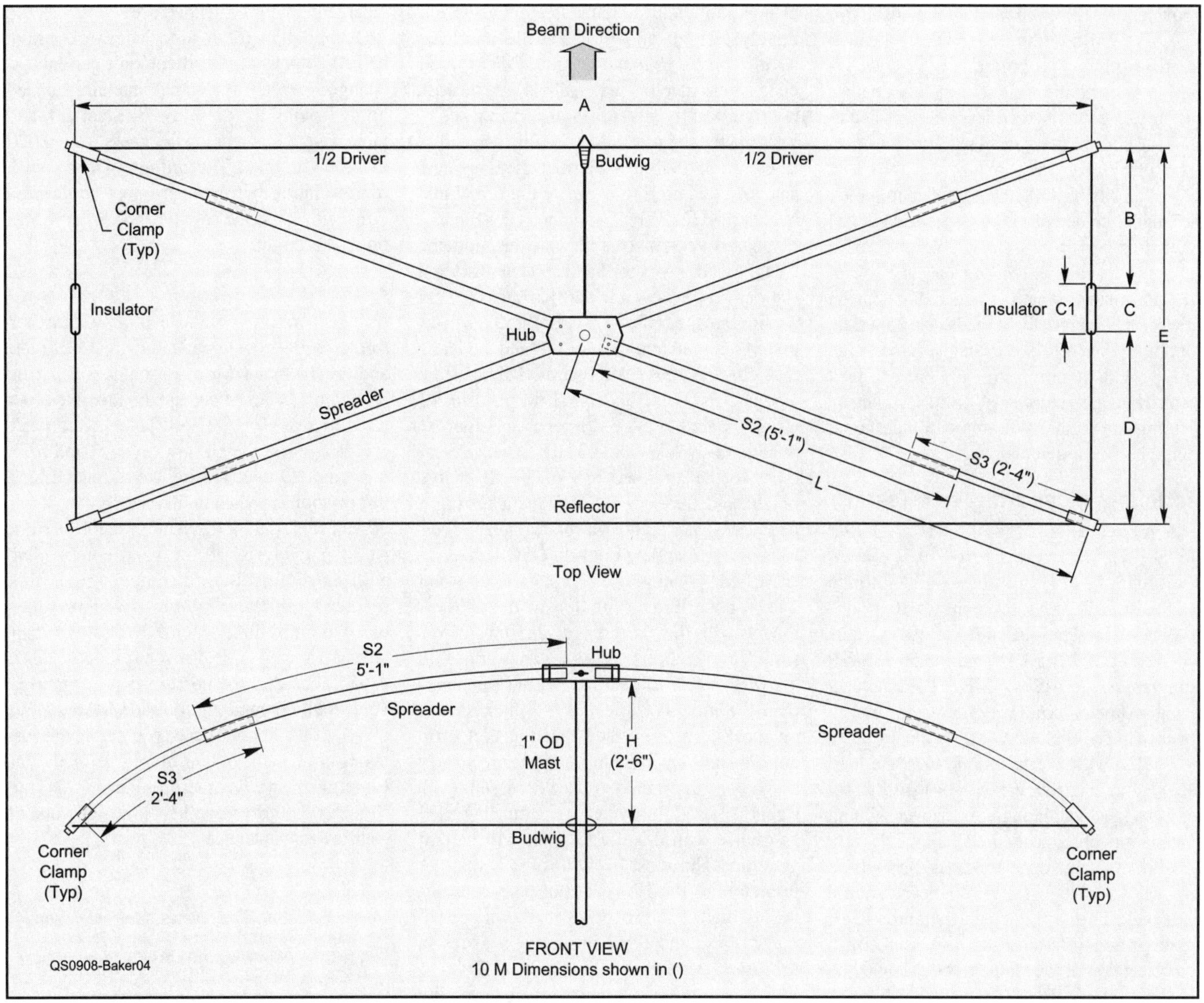

Figure 4 — Dimensioned drawing of the 10 meter MoxBeam.

and cut. Slip on a corner clamp and then a 2 inch length of ¼ inch OD heat shrink. Trim 1 inch of insulation from the wire and wrap the stripped wire around the Budwig wire. Adjust until the element wire insulation just touches the Budwig wire and solder. Slide the heat shrink tubing over the solder joint and shrink. Center the corner clamp between the marks and tighten the nylon screw finger tight, then tighten ½ turn with a wrench. Don't over tighten as it doesn't take much torque to secure the wire. Repeat this procedure for the other half of the driven element.

The reflector is longer than the driver and has no Budwig connector. Solder a ring terminal on the end of the wire. Using the clamping method mentioned above, measure out 2 feet 3⁷⁄₁₆ inches and mark the wire. Continue measuring to 16 feet 9¹¹⁄₁₆ inches and mark the wire, recheck the measurement and cut. Slip on two corner clamps, strip ¼ inch of insulation and solder a ring terminal on the end. Measure 2 feet 3⁷⁄₁₆ inches from the tip of the second ring terminal and mark the location of the second clamp. Center the corner clamps between the marks and snug the nylon screws.

Insulators

The insulators provide the required spacing between the driver and reflector elements. Cut two strips, ½ inch wide × 4¾ inch long, from plastic screen base material (used for screened-in porches). This material is stable, flexible and light weight. The element ring terminals are fastened to the insulators using 6-32 NC × ½ inch nylon screws and nuts. It is important not to use metal screws here as it extends the element lengths and detunes the antenna. Mount the element terminal rings so that the tip-to-tip measurement is 4¹⁄₁₆ inches.

Assembly

Assembly of the MoxBeam requires no tools and can be completed in less than 5 minutes. Place the hub on the ground and plug each spreader into the hub. Rotate the spreaders until the quick release buttons snap into the hub holes. Lay the wire elements out and slip the corner clamps over the spreaders, flexing the spreaders up as you progress.

There is sufficient tension created by the elements to hold the corner clamps on the spreaders and keep the feed point taut. Flip the antenna over, place it on the mast and tighten the thumbscrew to lock it in place. The 6 and 10 meter versions can also be installed with the spreaders flexed upward.

Due to higher lateral forces on the spreaders, the 12 through 20 meter versions must be installed with the elements flexed downward. Connect the coax feed line to the Budwig center connector and secure the coax to the mast. Adjust the coax for minimum pull or sag on the driven element. I often include a 1:1 choke balun, to prevent common mode current on the coax shield, by strapping it to the mast. Disassembly requires a small diameter rod to depress the spreader quick release buttons in order to remove the spreaders from the hub.

For portable use, the antenna can be hung from a convenient tree limb. A 1 inch OD wooden dowel inserted in the hub with an eye hook screwed into the top serves as a convenient tie point. The bottom of the dowel is used to suspend the weight of the coax before it is connected to the feed point. Strings tied to opposite ends of two spreaders allow the beam to be rotated from the ground.

Another method I find useful is to mount the smaller antennas with the spreaders flexed upward on a painter's extendable pole. With the acme threads extending past the top of the hub, a short length of wooden extension pole is screwed into the threads using a plastic coupling. The feed line is then secured to the pole, providing support for the antenna feed point. This setup works well while camping by securing the painter pole mast to the camper using plastic conduit clamps.

Modeling, Measuring, and Testing

A modeled azimuth plot for a 25 foot elevation over average ground is shown in Figure 1. Maximum in-band gain is 11.28 dBi at 28.0 Hz dropping to 10.05 dBi at 29.0 MHz. The peak F/B of 39.66 dB is achieved at 28.125 MHz, and the 50 Ω SWR minimum occurs at 28.25 MHz. The antenna's response favors the lower end of the band to cover the CW, digital, and SSB modes up to 28.7 MHz with a 1.5:1 SWR bandwidth. If you wish to alter the response or use different wire I suggest validating any changes with an antenna modeling program.

Builders who want to check the outline dimensions of their finished antenna should understand that the corner clamps alter these dimensions slightly. By securing a $^3/_4$ inch section of the elements at an angle, the clamps shorten the antenna length by approximately $^3/_4$ inch and increase the width by $^{11}/_{16}$ inch on the 10 meter version. *EZNEC* modeling shows that this dimensional change does not significantly affect overall performance.

With the 10 meter MoxBeam installed at an elevation of 25 feet, an MFJ-259B antenna analyzer was used to compare the actual and modeled 50 Ω SWR. The antenna SWR curve compared very favorably with the modeled data. I tuned across the band and heard several stations on CW and a few PSK signals.

Jim Innis, K5SP, of Gainesville, Texas was coming in loud and clear at S-8 and we had a nice chat on PSK. Afterwards, I rotated the antenna 180° and his signal dropped to S2. Using the traditional 6 dB per S-unit approximation, this works out to 36 dB, which is very close to the modeled F/B of 39 dB at this height and frequency. Subsequent testing a few weeks later yielded several solid PSK contacts during the New England QSO Party.

All MoxBeams use a dielectric constant of 6 for the wire insulation in the *EZNEC* models with the exception of the 6 meter version. This version repeatedly came up with a 1 MHz lower measured resonant frequency than the models indicated. A quick search on the Internet revealed that the dielectric constant can even change with wire insulation color. An e-mail to (since deceased) L. B. Cebik, W4RNL, provided some much needed assistance with this mystery, for which I am grateful: "The most likely answer is that the insulation of the Home Depot wire is changing its dielectric properties in the upper HF range. PVC and other insulating plastics in the same general range often have additives to improve their durability or some other property such as fire retardation, and these materials can alter the dielectric properties, especially in frequency regions that go untested, since they fall outside the wire's intended use."

Figure 5 (right) — Close-up view of one of the corner clamps.

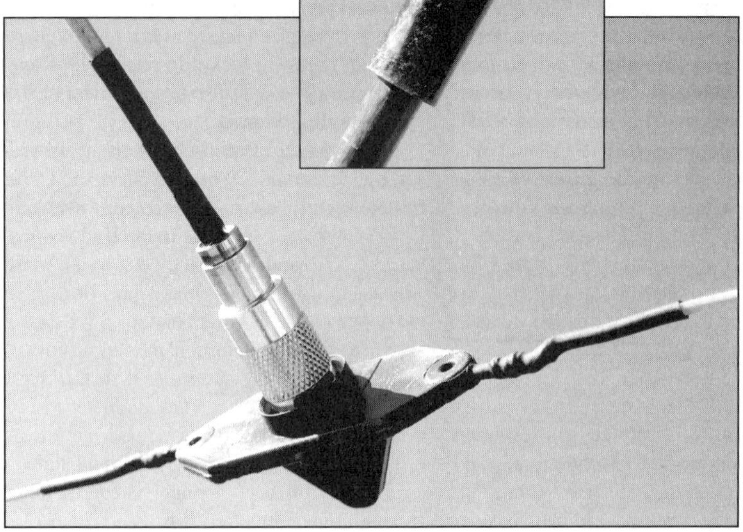

Figure 6 (below) — The center insulator and coax fitting adapted from a Budwig antenna insulator.

Although increasing the dielectric constant helped somewhat, I still couldn't get enough change to match the actual antenna. I ended up by leaving the dielectric constant at 6 and increasing the insulation thickness from 0.020 to 0.031 inches. The antenna (made with red insulation wire) was trimmed to the new dimensions and the resulting SWR curve was finally on target.

Final Notes

The Moxon rectangle is a proven performer with decent gain and a great F/B ratio, and the 10 meter MoxBeam makes a terrific directional beam. If you are anxious and don't want to wait for the next solar cycle peak, you can get on the air now on any band from 6 through 20 meters using the techniques and information supplied in this article.

Ten meters is a fascinating band to work. At solar cycle peaks the band comes alive with extremely long distance signals that refract from the F2 layer. Even in times of solar minimum when F2 is rarely available, 10 meters still has some long-distance capabilities due to sporadic E propagation.[6] This small antenna will provide hours of fun, and, because it's portable, you can use it just about anywhere. If you make use of the latest rugged and weatherproof materials, the 10 meter MoxBeam will soon become one of your favorite antennas.

Notes

[1] *ARRL Amateur Radio News*, "Backward sunspots may herald start of Solar Cycle 24," Aug 30, 2006; **www.arrl.org/?artid=6730**.

[2] L.B. Cebik, W4RNL (SK), "Moxon Rectangles"; **www.cebik.com/moxon/moxpage.html**.

[3] **www.arrl.org/files/qst-binaries**.

[4] "Trex Traditional Railing," Trex Company (available at Home Depot); **www.trex.com**.

[5] Several versions of *EZNEC* antenna modeling software are available from developer Roy Lewallen, W7EL, at **www.eznec.com**.

[6] **en.wikipedia.org/wiki/10_meters**

Photos by the author.

ARRL member and General class licensee Allen Baker, KG4JJH, received his license in 2000, fulfilling a lifelong dream of becoming a ham. He holds a BS in Industrial Engineering from Tennessee Tech and works as an Instrumentation and Controls Engineer for the company that operates the US Department of Energy weapons plant in Oak Ridge, Tennessee. Allen is active on SSB and in digital modes, enjoys the challenge of working low power (QRP) and loves to experiment with antennas and radio gear. He can be reached at 211 Brochardt Blvd, Knoxville, TN 37934 or at **kg4jjh@arrl.net**.

A 20 Meter Moxon Antenna

Moxons work great — and they take up less space than full size two element Yagis.

Larry Banks, W1DYJ

I was first licensed as a novice in 1962 as KN1VFX in southern Connecticut, and put a homebrew 807 CW rig on the air. I fed it to an 80 meter vertical outside of my second story bedroom window — with no counterpoise. (The advertisement looked great — how naïve I was!) I was able to make contact with a few local hams, and my best DX was an ARRL Official Observer (OO) report for transmitting a second harmonic on 40 meters, outside of the Novice band. Thus began my interest in antennas.

I quickly gained my Technician class license and built a six element, 6 meter Yagi that I put on a small roof tower, and a homebrew 6 meter AM transmitter based on designs in ARRL publications now long lost. I fell in love with building antennas. My best DX with this antenna and rig was double-hop E-skip to California — I was hooked on antennas.

Fast Forward to Today

I rediscovered Amateur Radio in 1994 and purchased a new home in 1996 on a 210 foot hill ten miles north of Boston — a home I purchased partially because of my love of VHF communications. I consider myself very lucky in that I was able to purchase a home where my ham interests were one of the major requirements. I put a 6 foot tower on my roof with a newly built seven element 6 meter homebrew Yagi, from *The ARRL Antenna Book*, reliving my youth.[1] Shortly thereafter I added an 11 element, 2 meter FM Yagi (a hand-me-down from my dad, WA1INL, now SK) used for repeaters and foxhunting and another 6 foot tower with a homebrew five element, 10 meter Yagi, again from *The ARRL Antenna Book*.[2] A K1FO SSB 2 meter Yagi soon followed, from Directive Systems. (I was in a hurry, and decided to purchase — not homebrew — this one.)

But after completing VHF/UHF Century Club (VUCC) on 6 meters and Worked All States (WAS) on 10 meters, I needed another challenge. I had dipoles for 80, 40, 20, 17 and 15 meters in my backyard, but they were all too low to be really very effective. Early in 2004 I had started working on 20 meter PSK31 WAS and was not happy with the 20 meter dipole. I needed a better antenna.

[1]Notes appear on page 88.

The Moxon Rectangle

I started thinking of putting a simple rotatable dipole at the top of my HF roof tower. I then looked at the 20 meter Yagis in my *ARRL Antenna Book*, but they had a larger turning radius than I could use, due to the spacing of my two roof towers. I also felt I wanted to try something different. Along came the April 2004 *QST* and the article by Allen Baker, KG4JJH, "A 6 Meter Moxon Antenna."[3] Somehow I had never run into the Moxon, but it seemed to have the characteristics I was looking for — smaller than a two element beam with about the same gain.[4] It also offered the front to back ratio of a three element beam. Allen also included some very nice construction methods using insulated tubing support blocks.

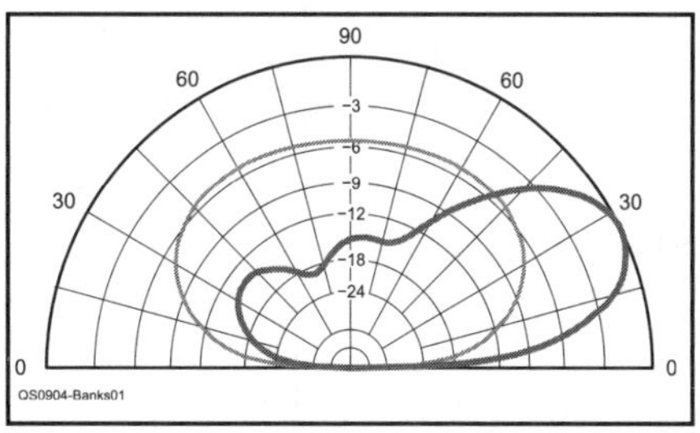

Figure 1 — A comparison of the lower radiation angle and greater gain of the Moxon (red) compared to the dipole (blue).

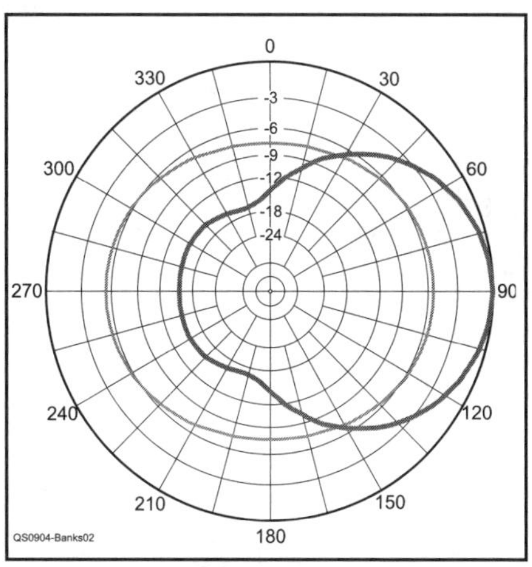

Figure 2 — This shows the superior front-to-back and greater gain of the Moxon (red) compared to the dipole (blue). Both plots are at the Moxon maximum radiation angle of 26° elevation.

Allen's article also referenced the Moxon Web page of L. B. Cebik, W4RNL.[5] I always enjoyed reading LB's antenna columns and articles in *QST*, *QEX*, *10-10 International News* and elsewhere. Somehow I had not run across LB's extensive writings on the Moxon and after spending quite a bit of time perusing the site I was convinced this was the antenna to experiment with.

I had previously used *YA* Yagi analysis software, included in my *ARRL Antenna Book*, to analyze a number of antennas. It was time to upgrade to newer modeling software so this project gave me a reason to purchase *EZNEC* from Roy Lewallen, W7EL, and add that to the fun.[6]

Comparison of Antennas

My dipole was at 18 feet elevation on average, and the Moxon would be at 36 feet at the top of my roof tower's mast. Using *EZNEC* was very informative. I quickly confirmed why my dipole was so poor — it was so low that most of the radiation and reception was directed above 50°. The *good* news, I suppose, was that at the maximum gain (about 5.4 dBi at 60° elevation), the dipole was almost omnidirectional. My dipole could hear high angle noise from all directions!

Figure 1 compares the elevation plots of the two antennas and Figure 2 the azimuth. These plots were very encouraging. The Moxon should do much better, with a lower radiation angle and about 5.3 dB more gain, not to mention a front-to-back ratio of about 15 dB.

Construction

The antenna is built from 6 foot sections of telescoping aluminum tubing. Standard construction techniques were used, with a 3 inch overlap, slotting the end of the larger tube, and using stainless steel pipe clamps to secure the joints.

The mast-to-boom and boom-to-element plates are $\frac{3}{16}$ inch aluminum. Stainless steel saddle clamps and insulated tubing support clamps are used with these plates. I used $1\frac{3}{4}$ inch aluminum (EMT) conduit for the mast and boom. Stainless steel hardware was used throughout. I also decided to use a commercial 1:1 balun at the feed point as it lent itself to a convenient way to connect the coax and transition to a balanced feed.

Table 1

Required Parts and Suggested Suppliers

Part	*Size*	*Quantity*	*Supplier*
Aluminum tubing	3/4" × 6'	4	Texas Towers
	5/8" × 6'	4	
	5/8" × 1'	1	
	1/2" × 4' 9"	2	
	1/2" × 3' 7"	2	
	1/2" × 1' 3"	4	
Aluminum plates			Local surplus aluminum dealer or
Mast-to-boom	3/16" × 8" × 8"	1	home supply store
Boom-to-element	3/16" × 4"×12"	2	
Corner plates, 45°	1/16" × 4" × 4"	4	
Tubing support	1/16" × 1 7/8" × 7/8"	6	
EMT boom	1 3/4" × 10"	1	Local hardware store
Fiberglass rod			McMaster-Carr
Driven element	5/8"× 1'	1	
Sides	3/8" × 1' 4"	2	
Corners	3/8" × 6"	8	
Insulated tubing support	3/4"	6	DX Engineering
Stainless saddle clamp	1 3/4"	12	DX Engineering
1:1 balun		1	Unadilla "W2AU" 1/1
Stainless steel pipe clamps	Various	12	Home supply store
Stainless steel hardware	Various	Various	Home supply/McMaster-Carr

The corners need some explanation. Aluminum tubing bent at 90° would normally be used. I had read of several hams having difficulty bending aluminum tubing, however. Based on another one of W4RNL's articles suggesting the use of "L-stock," I decided to use corner plates to sandwich the tubing (see Figure 6).[7] I added fiberglass rods inside the tubing to take care of any deformation of the tubing by the hardware. Although a more complex design, this was easier for me to fabricate with my modest workshop.

See Figures 4 through 10 for construction details and Table 1 for a list of parts and the suppliers I used.

Results

Results for a home-brew antenna come in two forms. The primary measure is how well the antenna performs electrically, but how long the antenna stays up is also important! The mechanical design of a home-brew antenna is just as important as the electrical design, especially in New England.

Mechanical Performance

Let me first address the mechanical design. I erected this antenna in the fall of 2004. As I write this in the winter of 2007-8 I am in my fourth New England winter since erecting it. This has allowed me to see how it handled wind, snow and ice. Each of the first two winters saw a failure — during the first winter a corner plate came off and the antenna quickly became a complex 1½ element beast with very strange RF behavior. It still worked, sort of. When I investigated this failure I realized that the #6 hardware I originally used was not sufficient, especially because I just used lock washers and nuts. Switching to #10 hardware with lock washers and locking (nylon insert) hex nuts fixed this design deficiency (see Figure 6).

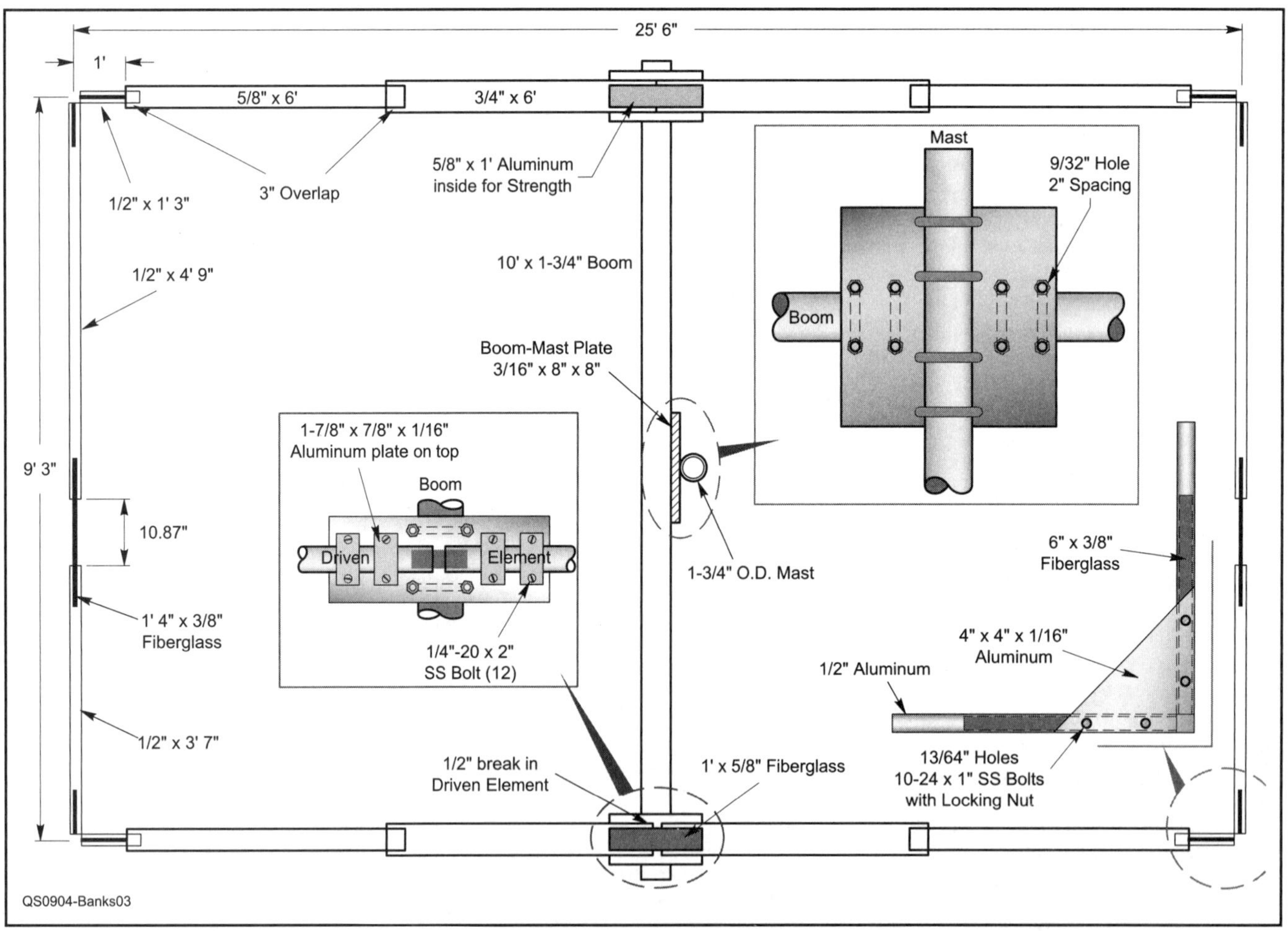

Figure 3 — Details and dimensions of the antenna and various mounting brackets.

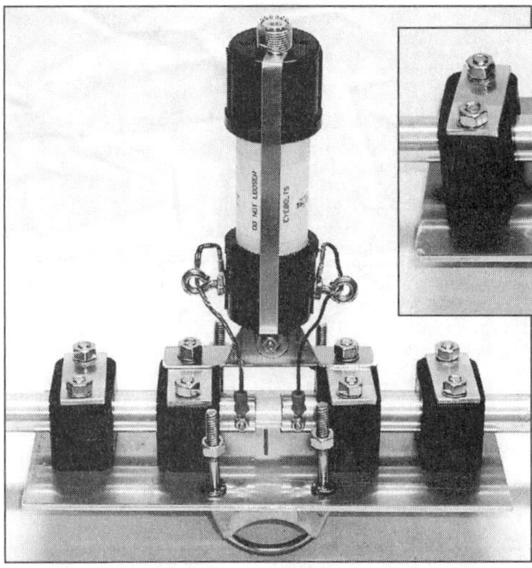

Figure 4 — Driven element boom-to-element detail. Note that the assembly is upside down as shown and the Balun hangs down in use. The fiberglass rod can be seen inside the ½ inch break in the driven element.

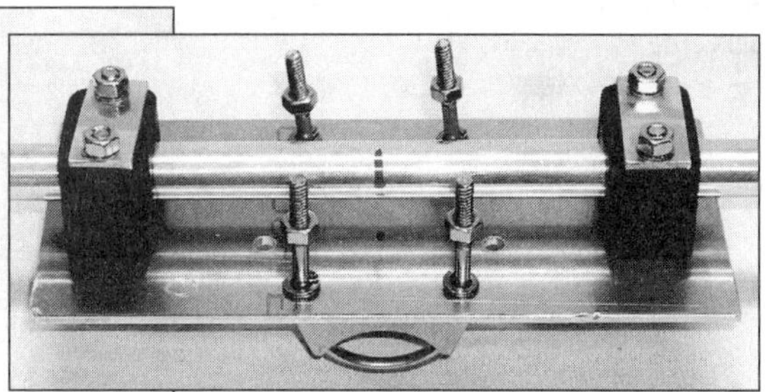

Figure 5 — Reflector boom-to-element detail. The small aluminum plates on top of the insulated pipe clamps were added for strength.

Figure 6 — Detail of the corner construction showing the two triangular plates and locking hex nuts. For an explanation of why the hardware looks new and the aluminum looks old, see the discussion about mechanical performance.

During the second winter the SWR became intermittent. Usually fine, it would go to infinity as soon as I started a QSO with a station in an unconfirmed DXCC entity. Again, I discovered this was due to the #6 hardware that I used to connect the balun to the driven element. I should have taken the antenna down from the tower after the first failure and changed *all* of the hardware as above, but that was a lot of work and I didn't — lesson learned!

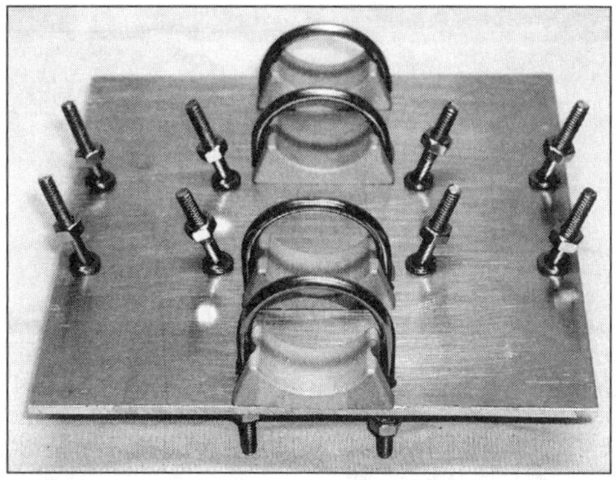

Figure 7 — Detail of the mast-to-boom plate.

SWR Performance

Accepting the initial mechanical failures, which only reduced the amount of time I have been able to use the Moxon, the really important information is how does it work with RF. I have tried to be as quantitative as possible with minimal test equipment. My SWR testing has been with my Autek Research RF-1 SWR analyzer. My receiver testing has been with my Kenwood TS-2000 using its S-meter for gathering data. Since modern transceiver manufacturers usually do not adopt the classically accepted S-unit of 6 dB, I have used 5 dB for the TS-2000. (I have seen postings of from 4 to 6 dB for the TS-2000 S-meter calibration, and as typical transceivers do not have linear S-meters, I took this as a reasonable compromise.)

Figure 8 — The antenna under construction in my backyard.

Naturally, the first thing I checked was the SWR. Figure 12 compares the *EZNEC* prediction with my measured SWR. The test data is with the Moxon on my test stand on my back deck using my Autek analyzer and about 20 feet of LMR-240 coax, and when it was installed on my roof. Note that the Autek resolution is 0.1 units, so the measured data is stepped. I cannot explain the upward shift in frequency while on my roof, other than to assume it is the effect of the real life environment and the 10 meter Yagi 6 feet below

Figure 9 — Moxon pointing straight up on my test stand for initial testing. My HF roof tower with the five element 10 meter beam can be seen waiting for the antenna to be mounted.

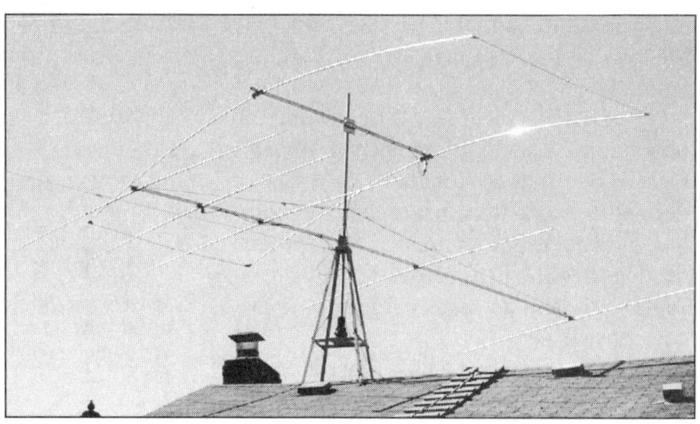

Figure 10 — Moxon mounted on the HF roof tower above the five element 10 meter beam. The *roof ladder* shown helps save the shingles and enhances safety.

Table 2

Comparison of Signal Strength, Field Day 2005

The dipole runs roughly north and south

Station Location			S-units		Station Location			S-units	
State	*Distance (Miles)**	*Azimuth (°)*	*Moxon*	*Dipole*	*State*	*Distance (Miles)**	*Azimuth (°)*	*Moxon*	*Dipole*
ID	750	270	6	5	MS	1200	225	9	8
IL	950	270	7	5	CA	2500	255	9	7
AL	1050	225	5	—	MS	1200	280	9 +20	9
IA	1100	270	8	6	CA	2500	255	8	7
TX	1700	240	7	6	SD	1500	270	7	5
NC	600	240	9	7	AB	2000	300	8	5
AR	1200	240	9 +15	9	CA	2500	270	7	3
AZ	2200	240	9	7	MI	700	270	7	3
CO	1700	270	9 +20	9	NC	600	240	7	5

*Estimated

Figure 11 — The two roof towers, with my VHF antennas to the right: K1FO 2 meter, 2 meter FM and seven element homebrew 6 meter Yagi.

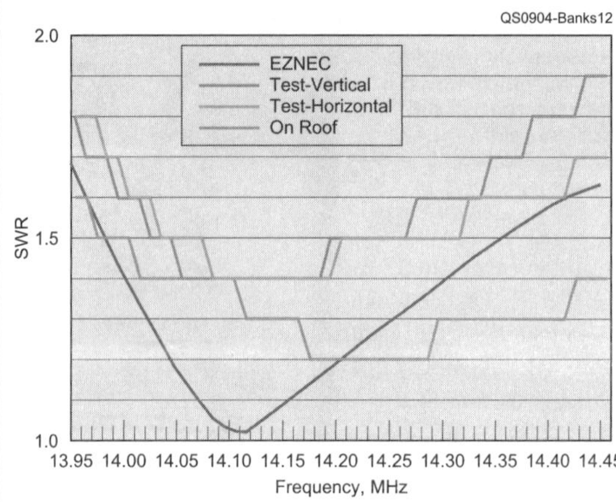

Figure 12 — Comparison of calculated and measured SWR. The test curves are for vertical (pointed straight up) and horizontal orientation.

it. (This would be a good future study for *EZNEC*.) Note that the on-roof measured SWR is measured in the shack and therefore is better than reality due to the attenuation of about 70 feet of LMR240 coax.

On Air Performance

Casual operating showed that the Moxon was one to two S-units better than the dipole in all directions, about what I expected based on the *EZNEC* analysis. Testing with W1PW in Arizona (about 2200 miles from Boston) showed up to 4 S-units improvement over the dipole with different propagation, and reasonable front-to-back ratio.

I needed to be more analytical about the Moxon compared to my dipole if I wanted to really understand how it was working, however. ARRL Field Day 2005 gave me an opportunity. Usually a member of W1HP's Field Day effort, for a number of reasons I decided to work from home. After confirming a contact, I documented the S-meter readings from both the Moxon and my dipole. Table 2 shows the results from this quasi-qualitative experiment.

As can be seen in the table, the difference between the Moxon and the dipole ranged from one S-unit to around 20 dB, if I can believe my TS-2000 S-meter. In all cases the Moxon was better.

I plan on continuing these measurements to obtain a better understanding of the Moxon's beam pattern. I believe the Moxon is a great addition to my RF arsenal.

Acknowledgments

Many people and groups have helped me along with ham radio. I need to mention a few. First is my friend Phil Walker, W1PW. Phil was instrumental in encouraging me to become re-involved in ham radio in the early '90s and without him I would not be on the air. Next, the late L.B. Cebik, W4RNL, for his extensive writing on antennas. Finally and most importantly, my wife, Maren, who puts up with towers on our roof, dipoles in the backyard, sometime TVI, a not always clean workshop and projects for her that are often delayed.

Notes

[1] *The ARRL Antenna Book*, 17th edition, 1994, p 18-18.

[2] See Note 1, p 11-11.

[3] A. Baker, KG4JJH, "A 6-Meter Moxon Antenna," *QST*, Apr 2004, pp 65-67.

[4] See **www.moxonantennaproject.com/backround.htm** for information about the Moxon antenna. It was developed by the late Les Moxon, G6XN, author of *HF Antennas for All Locations*, Available from your ARRL dealer or the ARRL Bookstore, ARRL order no. 4300. Telephone 860-594-0355, or toll-free in the US 888-277-5289; **www.arrl.org/shop; pubsales@arrl.org**.

[5] Moxon Rectangles, **www.cebik.com/moxon/moxpage.html**.

[6] Several versions of *EZNEC* antenna modeling software are available from developer Roy Lewallen, W7EL, at **www.eznec.com**.

[7] "An Aluminum 2-Element Moxon Rectangle," **www.cebik.com/moxon/mox.html**.

An ARRL member, Larry Banks, W1DYJ, was first licensed as a Novice in 1962 as KN1VFX in Stratford, Connecticut and is now an Amateur Extra class licensee. Seldom active during the '70s and '80s due to marriage and kids, he became active again in 1994. Larry holds three degrees in Electrical Engineering from MIT, and worked at Hewlett-Packard/Agilent Technologies for 36 years. He spent 24 years in Medical Instrumentation R&D, first as a design engineer of Electrocardiographs and then as a Project/Section Manager in Cardiac Ultrasound Systems. He moved to HP Corporate Education in 1993 and was the Global Program Manager for Agilent's Enterprise-wide Learning Management System after Agilent split off from HP. "Retiring" in 2005, he now consults with companies who need to implement eLearning and Learning Management Systems. He can be reached at **larryb@alum.mit.edu** *or at* **www.qsl.net/w1dyj** *where you can see other photos of Larry's shack and location north of Boston.*

A High Gain Single Wire Beam

Here's an inexpensive, easy to duplicate, beam antenna that can hang between trees.

Robert Wilson, AL7KK/VE7ZKK

Did you ever want a cheap beam that could be put up between two trees, yet have the gain of a Yagi? Well here it is. I call it a high gain single wire, or HGSW beam. Yes, only one wire is required. A 20 meter version will fit in the same space as a 75 meter dipole.

How it Works

As you may remember, a ⅝ λ antenna will give a sharper pattern and higher broadside gain than a ½ λ dipole. Here I have put three of them in line with certain critical adjustments to allow a 50 Ω feed, and also to allow self-feeding of the end two end sections with shorted 450 Ω phasing stubs. Unlike many antennas, the center point is reactive, but this problem is canceled by using an inductor across the coax feed point. Since the small inductor is only about 5 μH, it is a cheap and easy thing to build. Remarkably, this simple device allows a person to resonate the antenna in any desired portion of the 20 meter band.

EZNEC modeling results indicate that with the antenna at ½ λ (32.8 feet) high, the gain will be about 11.2 dBi with a peak of the elevation lobe at 29°.[1] Azimuth, elevation and SWR plots at ½ λ are shown in Figures 2, 3 and 4. Even more gain is available, and more importantly lower elevation angles of the main lobe, with greater heights. For example at ¾ λ, the peak elevation drops to 20° and to 15° at 1 λ.

The one downside is that rotation is not possible when the HGSW is installed between two trees, or towers. Still, if you want a strong signal in two directions it's tough to beat. Of course, with multiple supports, you could have a number of them for your favorite directions. Your maximum signal will be in a 28° wide beam in each of the two broadside directions.

A second height factor is important. Elevation also tunes the SWR of the antenna. If it is less than a λ/2 high you may need to make center section length adjustments to lower the SWR. Based on my own experience I strongly recommend using a minimum of λ/2 altitude.

[1]Several versions of *EZNEC* antenna modeling software are available from developer Roy Lewallen, W7EL, at **www.eznec.com**.

Building your HGSW

The parts list looks like this: This antenna will require 5 ceramic insulators, 129 feet of 14 gauge antenna wire (I prefer hard drawn copper or copper plated steel wire), 28 feet of 450 Ω window line, 60 inches of 16 gauge enameled copper wire (for the inductor), and a 1⅜ inch diameter by 2 inch long coil form (plastic water pipe or equal). There will be eight insulator connections each with a 6 inch foldover, which should be soldered for strength and conductivity. Last will be a coax connector and a pair of 6 inch 14 gauge wires for the feed point.

Construction details are shown in Figure 1. It is straightforward and will require a total of eight simple insulator connections. Each will have a 6 inch wire foldover twisted and soldered at an insulator. First, make up two sections of wire exactly 46 feet long with insulators at each end. This will give two wires with the correct 45 foot lengths. These will be the two end sections of the antenna. Next, assemble the center section of the antenna by making up two 18½ foot sections of wire and attaching them to a single center insulator, again with a 6 inch overlap each.

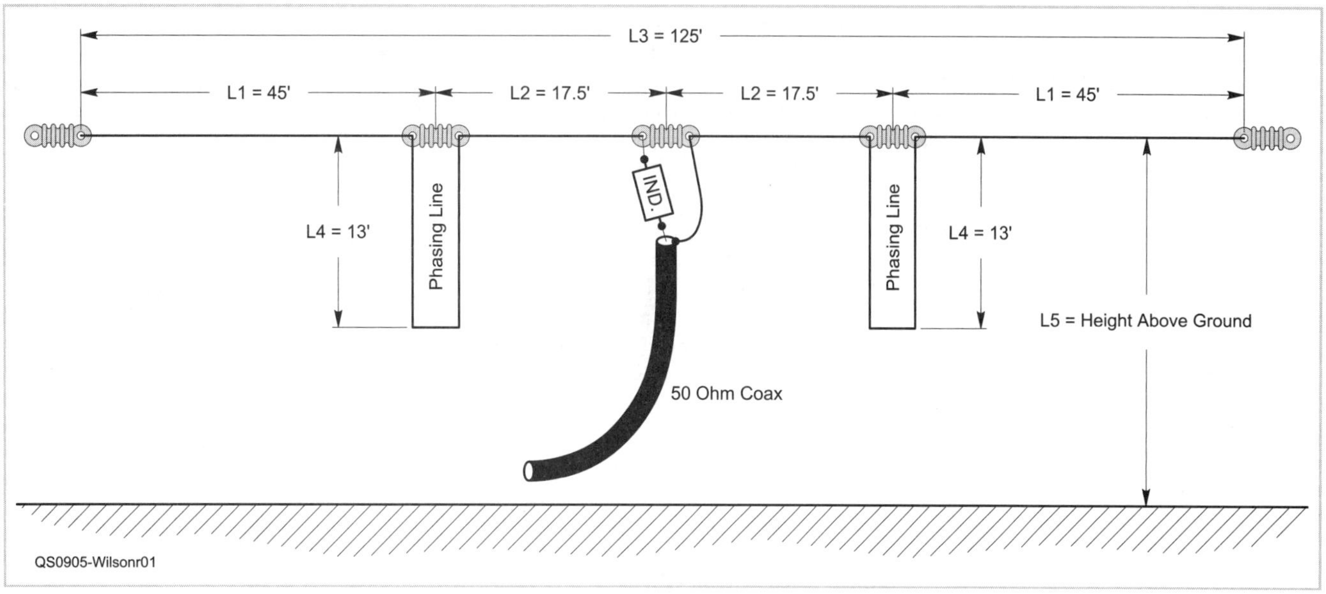

Figure 1 — Drawing of HGSW antenna with dimensions for the 20 meter band.

Table 1

Dimensions of an HGSW Beam for 10 Amateur Bands

Lengths in feet.

Band (meters)	*75*	*40*	*30*	*20*	*17*	*15*	*12*	*10*	*6*	*2*
L1 end (2)	168	89.4	63.2	45	35.3	30	25.6	22.4	12.7	4.37
L2 center (2)	65	34.7	24.6	17.5	13.7	11.7	9.97	8.72	4.95	1.70
L3 total size	467	248	176	125	98	83.3	71.2	62.3	35.4	12.2
L4 stubs	48.6	25.8	18.3	13	10.2	8.66	7.4	6.47	3.68	1.26
L5 height	120	64	45	32	25	21	18	16	10	10
Inductor (μH)	25.9	11.7	7.3	4.9	3.4	2.8	2.2	1.9	0.85	0.13
Gain (dBi)	11.4	11.4	11.3	11.2	11.1	11.0	11.0	11.0	11.4	10.9
Freq (MHz)	3.8	7.15	10.1	14.2	18.12	21.3	24.93	28.5	50.2	146

This center point will be the feed point shown in Figure 1. Here there will be two attachments, first a coax connector, second will be an inductor of 5 μH.

The inductor for the 20 meter HGSW antenna can be made by wrapping 13 turns of 16 gauge enameled wire around a small section of 1⅜ inch diameter PVC water pipe. The two holes should be drilled 1 inch apart for the ends of the wire. Clean the insulation from the coil tail ends, and attach it with solder in series between the center conductor of the coaxial cable and one side of the center insulator. Now take the two unconnected ends of the center wire and connect them to one end of the 45 foot wires, again using a 6 inch overlap. The antenna is nearly complete except for the phasing stubs.

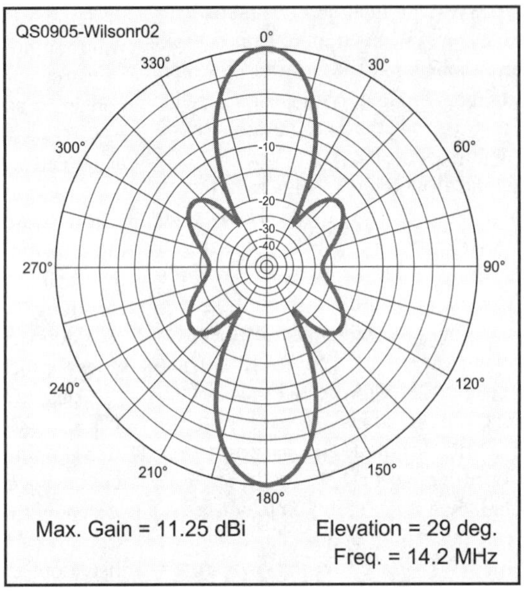

Figure 2 — *EZNEC* azimuth plot of the 20 meter HGSW at a height λ/2.

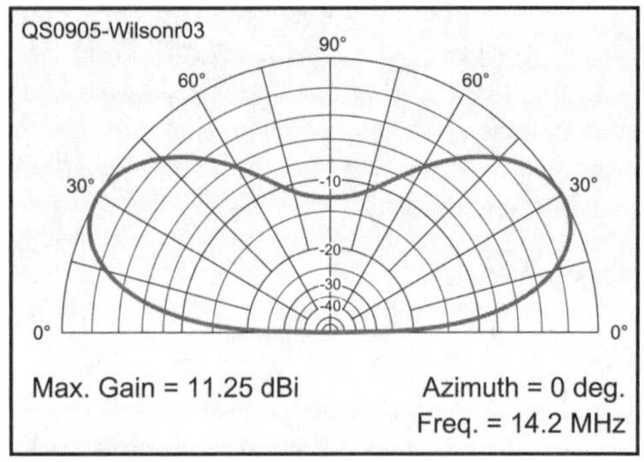

Figure 3 — *EZNEC* elevation plot of the 20 meter HGSW at a height λ/2.

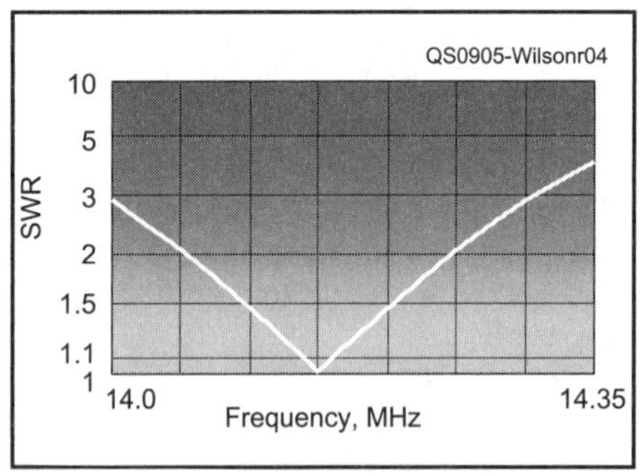

Figure 4 — Predicted SWR of 20 meter *EZNEC* model of HGSW at a height λ/2.

The two phasing stubs should be cut so that their active length will be 13 feet from the antenna wire connection to their shorted ends. Connect the ladder line stubs across the appropriate insulators as shown in Figure 1. Strip and prepare the top ends of the ladder line stubs to wrap and solder to the insulators as shown. The lower ends of the two lines should be stripped and bent over and soldered together. The resultant active line length must be 13 feet. The distance from the center insulator to the ladder line should be 17½ feet.

If you have a lot of wind in your area you might want to tie a 1 once lead fishing sinker to the bottom of each of the phasing lines. Alternately a string can be attached and tied to some secure point below the antenna. I have had no problem with my phasing lines except that they curl slightly, which is not ordinarily a serious difficulty.

Constructing the antenna took me only one morning. Using a sling shot and clothes line rope to hoist it into the local trees took a couple of hours more. Also I wound up about five turns of coax near the feed point into a 6 inch diameter coil to act as a cheap but effective choke balun. This balun was secured with electrical tape, but tie wraps may be even better in the long term.

The HGSW on Other Bands

Table 1 shows the lengths necessary to build a HGSW beam for 10 amateur bands. These were scaled from the 20 meter model that I built and tested. *EZNEC* was used to calculate the gain and inductor values.

The results have been good with my own beam strung very roughly north and south. I have had contacts on my main east and west beam lobes to Europe and Australia, and even a few contacts way off the main beams too. Note that there are some valuable small side lobes that allow contacts with strong stations in almost all directions, but the two main lobes at right angles to the wire will always be best. If you build an HGSW beam let me know how it works. I think you will enjoy its remarkable low cost, simplicity and the good gain.

ARRL member Robert Wilson, AL7KK and VE7ZKK, was first licensed in 1947 while in high school and now holds an Amateur Extra class license. Over the years he obtained eight years of university study in physics and aircraft technology. Early he was also a passenger ship radio officer. Later he worked for the Department of Defense, the Voice of America and the FAA as an engineer. He also built some of the first U.S. satellites at the State University of Iowa. He has written about 60 magazine articles and a book about low frequency antenna construction.

He can be reached at 1350 South Bagley Creek Rd, Port Angeles, WA 98362 or at **al7kk@yahoo.com**.

Building a Five Band G3TXQ Broadband Hexagonal Beam

The hexagonal structure makes for a light and compact Yagi you can build.

Leo Shoemaker, K4KIO

Be ready for the sunspots' return with a home brew five band directional antenna with small footprint but top-notch performance. You can build this beam in six steps with parts from your local hardware store and a few online sites. I bought all my parts for under $350.

Why a Hexagonal Beam?

The G3TXQ broadband hexagonal beam is essentially a two element Yagi compressed to about ½ full size. It looks like nothing so much as an inverted umbrella. Its compact shape and symmetrical construction mean low weight and low wind loading so the beam can fit snugly into a garden home backyard. It can be mounted on a push-up mast and turned with a light duty rotator. It blends in well with trees providing a measure of stealth. The umbrella shape allows nesting of the wires of the five bands and use of a single feed line for them all. No traps or other tuning devices are needed and the antenna presents a 50 Ω SWR of under 2:1 over nearly all of the five bands 10 through 20 meters.

This hexagonal beam offers a number of features as follows:

- Forward gain and front/back nearly comparable to a two element Yagi.
- Five band operation with low SWR without traps or tuning devices.
- Broadband characteristics.
- Low noise performance.
- Low weight and low wind load.
- Construction from general hardware components and ease of adjustment.

The beam consists of two elements for each band. The driven element is in the shape of an M while the reflector element is wrapped around the four spreaders to the rear of the driver wires as shown in Figure 2. The elements are made of wire instead of the tubes used by most Yagi antennas. That means you need the supporting structure made up of six flexible fiberglass tubes attached to a base plate. Viewed from above, it is in the shape of a hexagon, giving the antenna its name. The antenna shape is held together by the base/rod/center post structure, the wires and Kevlar/Dacron cords.

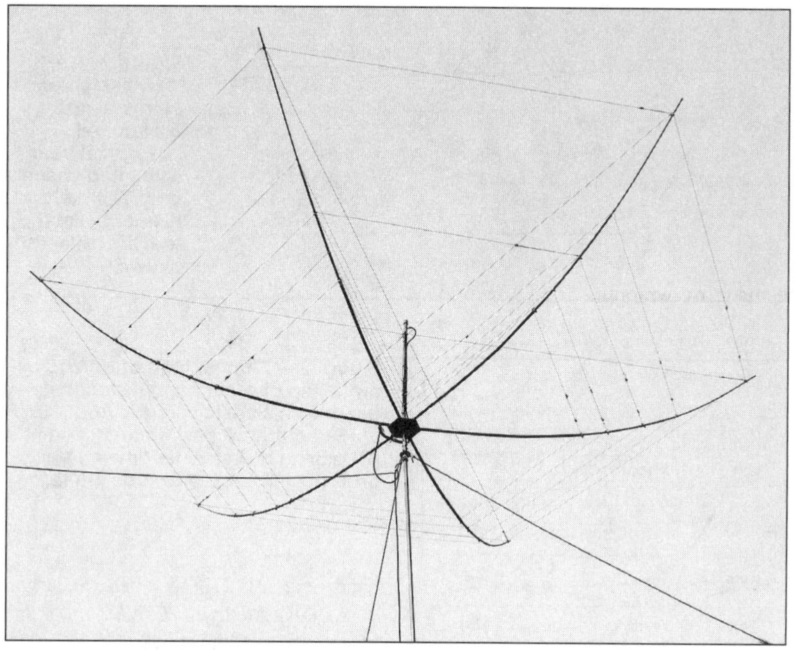

Figure 1 — G3TXQ broadband hexagonal beam built by K4KIO.

The antenna is just under 22 feet in diameter, is constructed of six fiberglass tubes and 16 gauge stranded copper wire, and weighs about 22 pounds. The center post is a 5 foot piece of fiberglass or PVC. The single 50 Ω coax snakes up the side of the center post where it feeds all the driver wires of the hexagonal beam from the top.

G3TXQ Broadband Hexagonal Beam

This beam resembles in structure the Hex-Beam® described by M. I. Traffie, N1HXA, in *Communications Quarterly*.[1] Many homebrew versions of this antenna, referred to here as the *classic* version, have been built in the past decades. They are built on the same frame as the G3TXQ broadband hexagonal beam described above but are slightly smaller with a diameter of about 19 feet because the reflector wire is bent into a W configuration while the driver wire remains in the shape of an M.

While the classic version is a great antenna, its size compression results in a relatively narrow SWR bandwidth as shown by *EZNEC* modeling outputs with the red trace in Figures 3A and 3B. This means you have to choose which portion of the band, CW or phone, you want to operate in if you want maximal performance. Another unfortunate fact is that the optimal performance "sweet spot" of the classic version is on the downward slope of the SWR curve in the range of 2:1 rather than at the minimal SWR point. This just means you would need to use a tuner to correct the SWR if you are operating away from the optimal design frequency.

The classic version was given a significant performance boost with a design improvement by Steve Hunt, G3TXQ.[2] Steve experimented with and modeled various changes in the reflector seeking to overcome the narrowbanded performance of the classic version. The result was discovery of a change in reflector shape that gives the hexagonal beam a lower Q and big boost in bandwidth while actually improving both gain and SWR performance. The only penalty for this improved performance is a slight increase in size. The classic version of the five band hexagonal beam is 19 feet in diameter; this new broad-

[1]Notes appear on page 94.

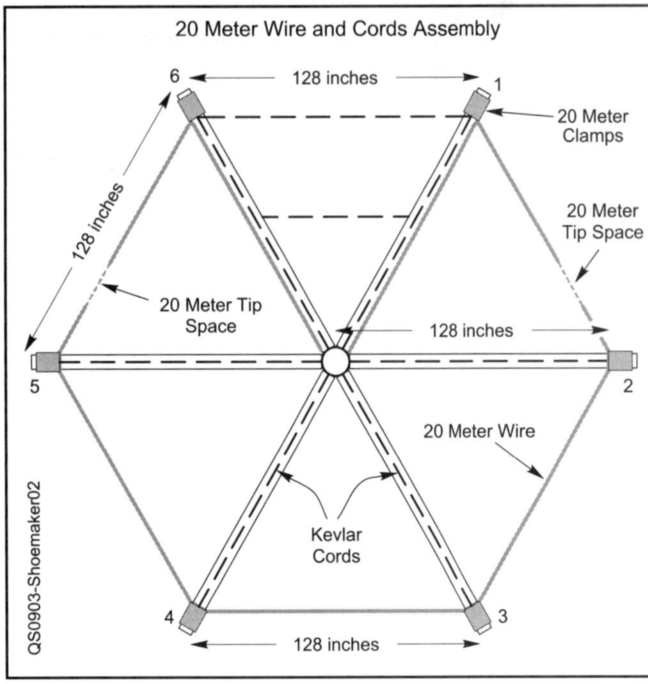

Figure 2 — Top view sketch of single band broadband hexagonal beam, wires and spacers only. The driven element is at the top, the reflector below.

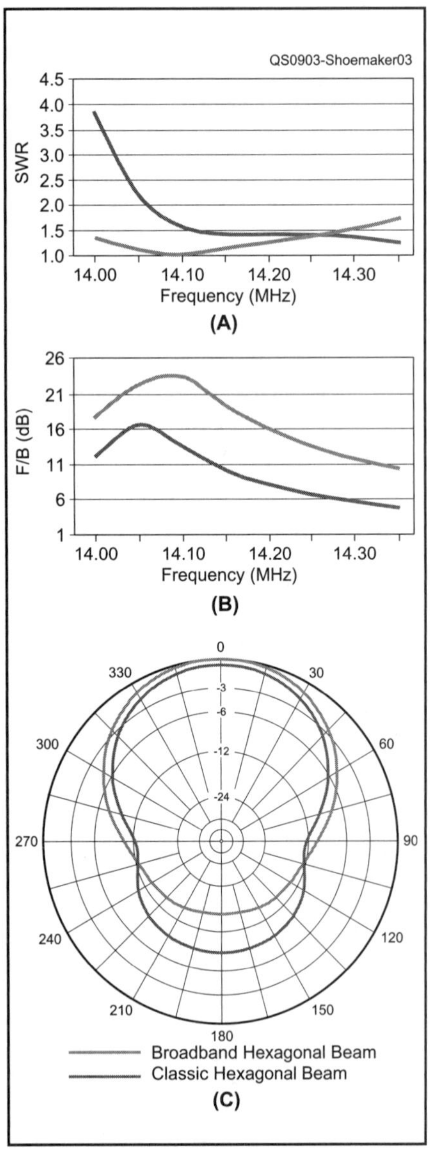

Figure 3 — Comparison of *EZNEC* modeling of broadband hexagonal beam with classic version. At (A) SWR, at (B) front/back performance and at (C) gain in dBi at 30 feet over average ground at optimal elevation angle.

band version is 21.5 feet, or 13%, larger.

As illustrated with *EZNEC* modeling, the blue trace in Figures 3A and 3B show this new broadband design has higher front to back performance and better SWR response over a significantly wider frequency range than the older version of the hexagonal beam. The front to back performance (F/B) for the G3TXQ broadband hexagonal beam is in excess of 10 dB over the entire 20 meter band, whereas for the classic version F/B exceeds 10 dB only in a range of 150 kHz. In addition to the improvements in SWR and F/B, the broadband model has a little more forward gain as shown in the *EZNEC* azimuth radiation pattern in Figure 3C. Finally, the *sweet spot* for design is not only broader but is centered on the lowest SWR point making it easier to operate without a tuner. Performance characteristics and comparisons for the other four bands are similar to those for 20 meters.

Building a Broadband Hexagonal Beam

The broad band hexagonal beam has now been built by a sizable number of hams who enthusiastically report on its on-air performance even in the pits of the present sunspot cycle. You can join this group by following these six steps for constructing a G3TXQ broad band hexagonal beam using mainly hardware store components with some parts from online suppliers.

Step One — The Base plate

The base plate is made of a 12 inch square type 6061 T6 aluminum plate that is 3⁄16 inch thick. This particular type of aluminum is harder than pure aluminum and less likely to bend while being resistant to weather. Sketch the hexagon using the approach shown in the drawing on the *QST* Binaries Web site.[3] You can cut the plate with a hacksaw and drill the center hole using a 1¼ inch metal hole saw. U bolts are used to secure the spreaders to the base plate. A square base floor flange normally used for handrails is ideal for mounting the center post to the base plate. The black arrow shown in Figure 4 helps keep you oriented as you install the wires later in step six. Stainless steel hardware should be used to minimize corrosion and lock washers will help keep the fixtures from loosening with movement of the hexagonal beam by wind and rotation.

Step Two — The Spreaders

Fiberglass tubes are recommended for the spreaders. Three thicknesses are needed: 1, ¾ and ½ inch outside diameter. Max Gain Systems offers fiberglass tubes that are ideal for this.[4] The smaller sizes slip perfectly into the next larger sizes. In fact, due to popular demand, Max Gain recently began packaging the specific sizes for this broadband hexagonal beam as a special kit. Fiberglass has the strength plus the spring needed to hold the hexagonal beam in shape. I know of many who have tried to use PVC for spreaders but none reported good results. PVC is just too heavy and limber for use in this application, although it will work fine for the center post.

The spreader sections can be glued together with adhesive, bolted together or attached together with hose clamps for easy disassembly for DXpedition or ARRL Field Day use. In order to get the proper curvature

Table 1

Table of Wire Lengths and Tip Spacer Lengths

Measurements in inches

Band (Meters)	*Design Frequency (MHz)*	*½ Driver Wire Length*	*Full Reflector Wire Length*	*Tip Space*	*Height of Wires Above Baseplate*
20	14.150	218.0	412.0	24.0	42
17	18.100	169.5	321.0	18.5	20
15	21.250	144.5	274.4	16.0	14
12	24.950	121.7	232.0	13.5	10
10	28.400	106.8	204.4	12.0	6

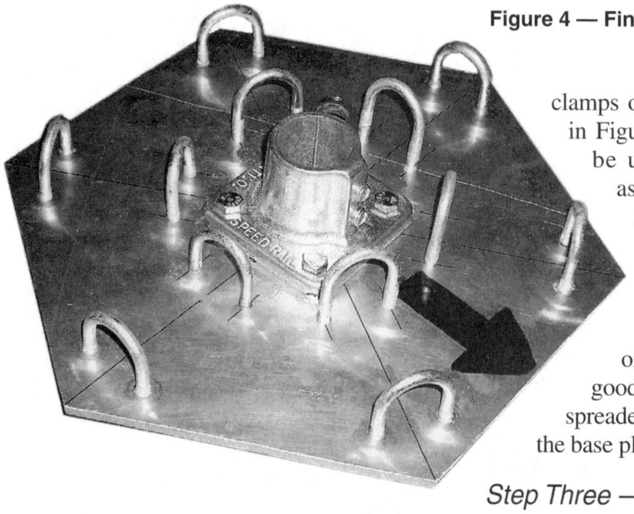

Figure 4 — Finished base plate.

of the spreaders the construction should follow the guidance in Figure 5A. If you depart from this approach, your curvature might be different and it might, in turn, require different vertical center post positions for the wire terminals. You will need 139 inches in length for each spreader. Slip five stainless hose clamps on the spreaders as shown in Figure 5B. These clamps will be used to anchor the wire assemblies to the spreaders and the exact locations will be adjusted later in Step Six. Painting the spreaders with a primer and your choice of colors for UV protection is a good idea. Figure 5B shows the spreaders ready for installation on the base plate.

Step Three — Center Post

The center post provides the anchor for the driver wires and the support cords, as well as a feeding arrangement for the driver wires. It is 1¼ inch outside diameter fiberglass tube or 1 inch inside diameter PVC 5 feet long with European style double terminals for connections of the driver wires and coax feed line. These terminals are sold in 12 piece assemblies at RadioShack and can be easily cut into doublets. Holes for #6-32 by 2 inch brass round head bolts should be drilled at the vertical locations as shown in Figure 6A for mounting the terminals.

The transmission line itself is 50 Ω coax and should be connected at the top of the post to the terminals for 20 meters. Modeling as well as actual testing has shown that top feeding is slightly preferable to connecting the feed line at the bottom. Each set of driver wires is connected via small pieces of coax as shown in Figure 6B. Use RG-58 coax if you plan to keep power levels under 500 W; otherwise RG-213 is recommended. After completing the installation of the feed line components, it would be a good idea to do an ohmmeter check to make sure there are no short circuits of coax braid to the inside wire. Later, after the wires have been connected, smearing liquid tape over the exposed coax braid will keep moisture from wicking in and prevent rusting of the terminals.

Drill a hole for a stainless steel bolt through a PVC cap as shown in Figure 6B for the center post. Install a bolt to be used as an anchor point for the radial support cords in step four. The center post is slid through the floor flange on the base plate and the flange set screws are used to hold it in position. It should be positioned about 4 feet above the base and 1 foot below for insertion into the mast.

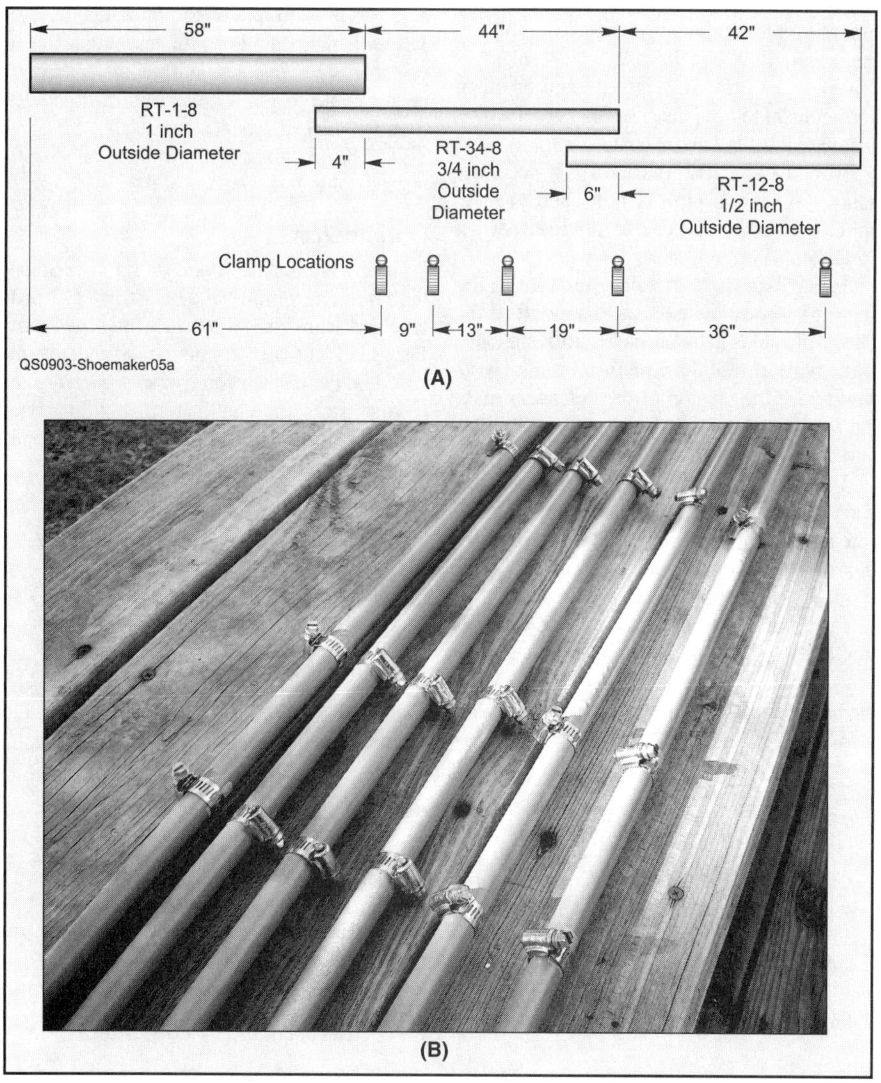

Figure 5 — Spreader details. At (A) spreader components and clamp locations, at (B) spreaders with clamps in place for assembly.

Step Four — Making the Wire/Tip Spacer Assemblies and Support Cords

The driven elements consist of two pieces of wire for each band. The reflectors are a single piece of wire for each band. Each driven element piece will be shaped by the hex beam structure into an inverted V and when arranged next to the other piece, will form an M. Each reflector piece will be wrapped around the four spreaders and attached to non-conducting cords called tip spacers, which in turn are attached to the driver wires.

Lengths of the driver and reflector wires for 16 gauge stranded wire and tip spacers are shown in Table 1. Be precise in cutting the wires as an inch can make a difference in the resonant frequency. A ring terminal should be soldered on one end of each driver wire and both ends of the reflector wire to tie the tip spacers between the driver wires and the reflector. About ½ inch of the other end of the driver wires should be bent into an L shape for insertion into the terminals on the center post. Add this L length to the specifications in Table 1.

The tip spacers are made up of very strong Kevlar/Dacron cord. It does not stretch and resists UV radiation. Add a few inches for knotting the tip spacers to the specifications in Table 1.

The two half driver wires, the reflector wire

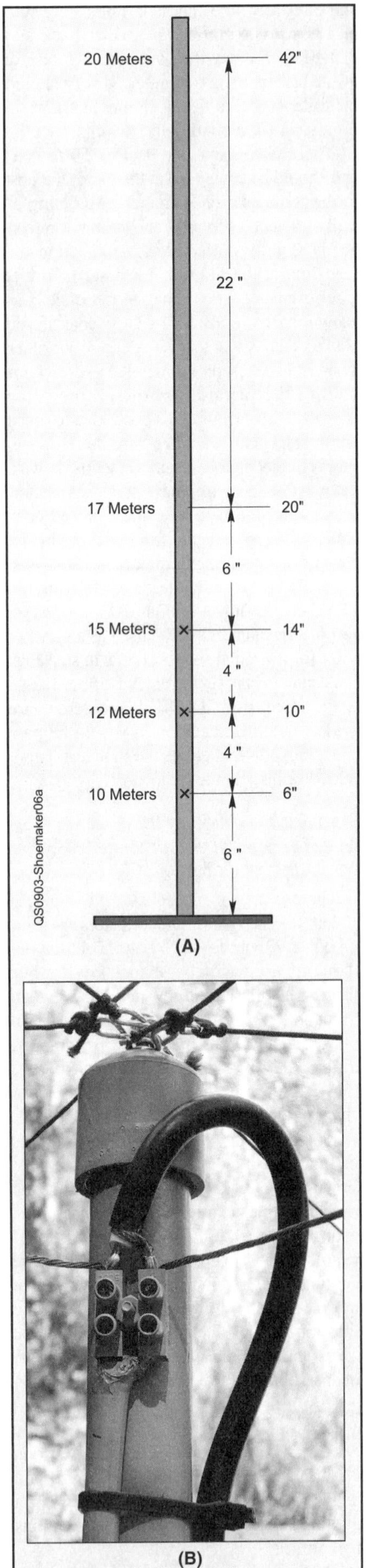

Figure 6 — Center post details. At (A) vertical locations of terminals on center post. At (B) center post with terminals and wires installed. Note the coax feed line up the side of the post.

and the two tip spacers should be measured and assembled as single end to end sets before installation on the spreaders using the approach below for each band. It is easy to get the five bands wire/tip spacer assemblies mixed up so careful labeling of them is a good idea.

Twelve support cords are also needed as shown in Figure 2. These cords should be made of Kevlar/Dacron cord 129 inches long, not including the little S hooks on the ends. Six cords are used for radial support from the ends of the spreaders to the top of the center post. The other six cords are used for perimeter support around the ends of the spreaders to keep them in position. Together these cords establish and maintain the upside down umbrella shape that is needed to mount wire/tip spacer assemblies for the five bands.

Step Five — Assembling the Frame

You will need a temporary test stand for mounting the hex beam during assembly of the wires. This can be a 2 inch steel pipe driven in the ground, a table, stand or other arrangement to hold the assembly erect while wires are installed and adjusted.

Attach the spreaders to the base plate using the ½ inch pipe U bolts and ¼ inch split lock washers. Be careful not to crush the spreaders with the U bolts.

First, establish the inverted umbrella shape of the hex beam using 12 support cords. This will make it much easier to install the wires with a minimum of fuss because you won't have the tension of the spreaders making it difficult to pull the clamps and wires into the right position on the spreaders.

Hook one of the support cords to a spreader end and then to the center post cap, pulling the spreader up into the bowed position. Repeat this for the other five spreaders. Now, hook support cords from the ends of each of the spreaders, as shown in detail on the *QST* binaries Web site, to the adjacent spreader until you have 12 cords installed and the shape of the hex beam is established in preparation for installation of the wires.

Step Six — Installing the Wire/Tip Space Assemblies

Attach the wire/tip spacer assembly to the spreader hose clamps closest to the center post using cable ties starting with the 10 meter band. Leave the cable ties open so the wires can move freely. Insert the ends of the driver wires into the bottom center post terminal and tighten the set screws. If the wire is too short to reach the terminal, then loosen one of the hose clamps and let it slide inward creating more slack. After connecting to the terminals, adjust the position of the hose clamps to take most of the slack out of the wire/tip spacer assembly. But do not make it too taut since this could distort the overall shape of the hexagonal beam. Loose wires do not harm performance, within limits of course. Repeat this process for 12 meters and continue on until all five bands' wire/tip spacer assemblies have been installed and adjusted.

Adjust to make the wire sections level, and parallel with one another by only making small changes in clamp positions. Look along the vertical alignment of the bent spreaders to make sure none are twisted out of shape. This is more art than science, but if you are patient, you will obtain a well balanced, symmetrical hexagonal beam.

The beam is now ready for mounting on a mast or tower. You should run SWR tests first to ensure that the wires are tuned properly. At near ground level, the SWR will not be nearly as good as what you should expect when the beam is 20 to 40 feet high, but you should be able to see a dip in SWR to the range of 2 or 3:1 while the beam is on the test stand. Use a portable SWR test set and a short coax connector cable to get a better dip on this first round of tests at ground level.

The beam can now be mounted on a push-up mast or tower.

Conclusion

Many hams have longed for a directional high frequency multiband beam that is affordable and more compact than a full scale Yagi. This G3TXQ broadband hexagonal beam could be the answer for you as it was for me. If you rise to the challenge of building one while awaiting the return of decent propagation you will be ready for the sunspots.

Notes

[1]Registered Trademark of Traffie Technology. M. Traffie, N1HXA, "Miniature Antennas," *Communications Quarterly*, Spring 1996, pp 99-103.

[2]S. Hunt, G3TXQ, "A Broadband Hexagonal Beam," *AntenneX* Online Issue No. 128, Dec 2007.

[3]Detailed parts list and drawings may be found at **www.arrl.org/files/qst-binaries**.

[4]**www.mgs4u.com**.

Leo Shoemaker, K4KIO, was first licensed as a Novice in 1956. He holds an Amateur Extra class license, a degree in electrical engineering from the University of Louisville and he spent his career with BellSouth. He enjoys working DX on both CW and SSB and experimenting with wire beam antennas. Leo is an ARRL member and can be reached at 2610 Acton Rd, Birmingham, AL 35243 or through his Web site at **www.leoshoemaker.com/k4kio.html**.

Construct a Rugged, Easy to Build, 2 Element Yagi

Compact performance in a lightweight beam.

Chuck Hutchinson, K8CH

In June of 2000, Sylvia, K8SYL, and I moved from Connecticut to rural Michigan. For us the new setting was ideal, but after the move our budget wasn't. In short order I had a couple of wire antennas in the air. One of them was an experimental antenna for 17 meters that I had quickly installed. It worked — sort of. My first contact was with my friend Dave Patton, NN1N. Thanks to Dave's skill, we were able to have that contact. He gave me a signal report of "puny weak." Clearly, I needed something better.

In April of 2001, I retired after 20 years with ARRL. Shortly after, I started work with Dean Straw, N6BV, on *Simple and Fun Antennas for Hams*.[1] One of the first projects designed especially for that book was a 2 element Yagi for 17 meters. It would work just fine with the 30 feet of TV type tower and rotator that I had. And it did and does!

I live on a small hill that is mostly surrounded by open fields, which means that the winds blow unimpeded across my antennas. For that reason, I want an antenna that can handle at least 90 mph winds. A rating of 80 mph might be good enough, but I'd already seen wind gusts close to that speed. As you can see, for me *rugged* is a necessity.

From the beginning, my plan was to share this design so that you could build one too. I know that some of you have limited resources available, and that others have no desire to be a machinist. For that reason this antenna requires no special tools or assemblies. You can save some money by fabricating your own element-to-boom and mast-to-boom plates.

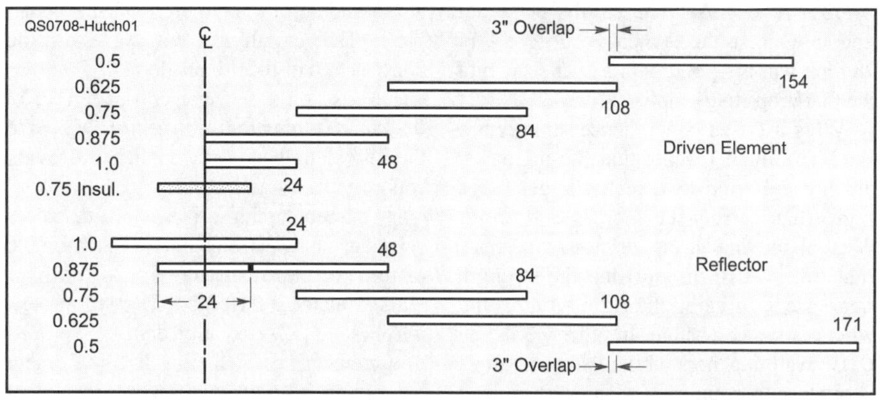

Figure 1 — Dimensions of the reinforced elements showing the tubing overlap of one side of the 17 meter Yagi. On the left are element diameters. All dimensions are in inches. The center line represents the boom location. See text for additional details.

The Parts

Perhaps the most obvious component of a Yagi is aluminum tubing. I've based the elements and boom on 6 foot lengths of aluminum tubing, which are available from Texas Towers and can be shipped via UPS.[2] (They also sell 12 foot lengths, but those have to be shipped by freight.) Elements are made from 6063-T832 drawn aluminum tubing with a 0.058 inch wall thickness. These tubes telescope inside one another very nicely. For the boom, I chose 2 inch OD 6061-T6 extruded aluminum tubing with a 0.120 inch wall thickness.

Stainless steel U-bolts connect elements, boom and mast to each other. DX Engineering sells saddle clamps with all necessary hardware included.[3] The saddles provide a solid grip on the parts being joined. DX Engineering also sells pre-drilled plates for mounting elements to boom and boom to mast.

The center insulator for the driven element is a length of 0.75 inch fiberglass rod. This is available in 8 foot lengths from MaxGain Systems.[4]

Driven element insulators are made from short lengths of 1 inch PVC conduit. This is available from building supply dealers and is gray and marked as "sunlight resistant."

[1]Notes appear on page 98.

Table 1
Major Parts for the 2-Element 17 Meter Yagi

Qty	*Item*
2	Drawn 6063-T832 aluminum tubing, 1" OD, 0.058" wall, 6' length.
3	Drawn 6063-T832 aluminum tubing, 0.875" OD, 0.058" wall, 6' length.
4	Drawn 6063-T832 aluminum tubing, 0.75" OD, 0.058" wall, 6' length.
4	Drawn 6063-T832 aluminum tubing, 0.625" OD, 0.058" wall, 6' length.
4	Drawn 6063-T832 aluminum tubing, 0.5" OD, 0.058" wall, 6' length.
1	Drawn 6061-T6 aluminum tubing, 2.0" OD, 0.12" wall, 6' length.
2	Element to boom plate, 8×3.5×0.19" (6061-T651 aluminum — DX Engineering ETB190).
1	Boom-to-mast plate, 8×10×0.25" (6061-T6 aluminum — DX Engineering BTM810).
8	Saddle clamp, 2" (DX Engineering SAD200).
2	Saddle clamp, 1.25" (DX Engineering SAD125)
1	Saddle clamp, 1" (DX Engineering SAD100).
2	PVC conduit, 3" lengths of 1".
49	Aluminum pop rivets, ⅛".

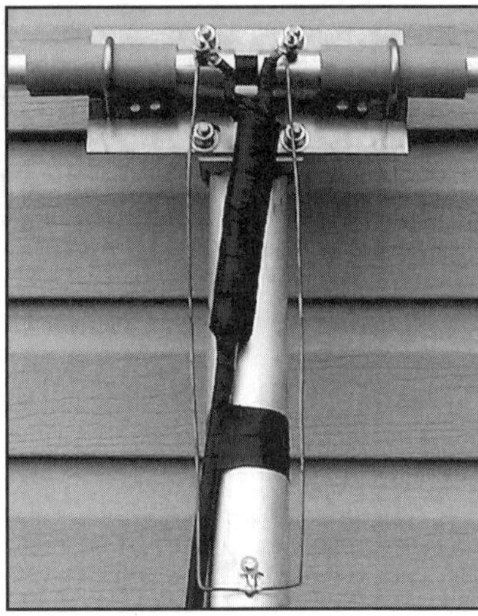

Figure 2 — Detailed view of the "hairpin" matching network.

I used 49 aluminum pop rivets to assemble the elements. The feed point connections are made with #8 hardware. You'll find a list of the major parts in Table 1.

Element Design

The elements are spaced 5 feet 8 inches from each other. That's about as much as you can get on a 6 foot boom, and it's adequate.

For a short Yagi like this one, element strength is the key factor in determining the antenna's ability to handle winds. I thought about using an un-reinforced element. Would it be strong enough? I calculated the strength using equations from the ARRL book by Dave Leeson, *Physical Design of Yagi Antennas*.[5] The results were not encouraging, as the element is only good for 78 mph winds — and with a quarter inch of ice that drops to 61 mph.

What if I used larger, hence stronger tubing? The math reveals that the increased weight and wind load of the larger tubes more than offsets the increased strength. What about smaller, lighter, less wind resistant tubes? It turns out that the strength decrease is greater than the weight and wind resistance change. In other words the 0.058 wall thickness tubing I started with is very close to optimum for an un-reinforced 17 meter Yagi element. It all meant that I needed to reinforce the elements.

In consultation with Dean Straw, N6BV, we came up with the element design and dimensions shown in Figure 1. These elements will handle 112 mph winds — and even with a quarter inch of ice they will still handle 81 mph. Fortunately, high winds and ice loads rarely happen simultaneously, so this design should be adequate for most needs.

Construction

Start by cutting the elements as shown in Figure 1. When you begin element assembly, use fine sandpaper to clean that portion of each element section that will telescope inside another (next larger size) section.

After sanding, apply a thin coat of Penetrox (or equivalent) to the sanded portions. Push the sections together, twist and move back and forth a little bit. This will lightly coat the inside of the larger tube. This prevents oxidation of the aluminum and small metallic particles in the Penetrox help to maintain electrical conductivity through the joints. It also has the excellent side effect of making it much easier to disassemble the elements, should you ever want to do that.

Use three ⅛ inch pop rivets at each tubing joint. In the 24 inch length of 0.875 inch OD tubing that spans the center of the reflector, use a single rivet to hold the piece in place at the middle of the larger (1.0 inch) center piece. That length was left over from the construction of the driven element. It allows you to use only a single 6 foot length of 0.875 inch tubing for the reflector. Of course the 6 foot length has to be cut in half, leaving you a pair of 3 foot lengths.

At the center of the driven element, leave a ¾ inch gap between elements. Wrap a couple of turns of electrical tape around the fiberglass insulator at the center. This will protect it from solar (UV) radiation damage. For feed line connections, drill the driven element halves and insulator 1¹⁄₁₆ inch each side of the center (holes 2⅛ inch apart). These holes should be ⁵⁄₃₂ inch diameter for #8 hardware.

Cut two 3 inch lengths of 1 inch PVC conduit. (These will be used to insulate the driven element.) Be sure to use sunlight resistant PVC — it is marked as such on the conduit. With a hacksaw, cut a slot down the length of each PVC insulator. Make a second cut parallel to and ³⁄₁₆ inch from the first cut. This will allow the insulator to firmly close against the driven element.

Use 1 inch U-bolts and saddles to fasten the reflector to the mounting plate. Be sure to put some oil on the U-bolts *before* you install the nuts. Stainless steel is prone to seize up if you don't apply some lubricant to the threads *every time* you move the nuts. Trust me on this; there is nothing more frustrating than to have the nut freeze on a U-bolt as you are installing your Yagi at the top of your tower — even if it's only 30 feet above the ground.

Fasten the plate to the boom with 2 inch U-bolts and saddles. Don't forget to lubricate the threads.

The driven element mounting scheme and connections are shown in Figure 2. You'll need 1.25 inch U-bolts and saddles to fasten the driven element and its insulators to the mounting plate. Add the boom-to-mast plate and its four 2 inch U-bolts and saddles and the basic beam is finished.

Feeding and Matching Your Yagi

For a 50 Ω feed-point impedance, you'll need 0.50 µH of inductance across the feed point. An easy way to obtain that is to use a *hairpin match*, called that because it does look like a hairpin. The hairpin also allows you to ground the center of the driven element. Details are shown in Figure 2. I used #12 solid copper wire to make my hairpin. A lug mounted at the hairpin center is used for making a connection to the boom. This brings the element to ground potential and thus provides some measure of lightning protection. This connection can be made with a sheet metal screw. Because you will be joining dissimilar metals, be sure to clean the contact areas and apply a thin coat of Penetrox.

Look again at Figure 2 to see how the hairpin and the feed line attach to the driven element. I used six ferrite beads to make a choke balun. These are available from Palomar Engineers as their BA-8 kit (**www.palomar-engineers.com/Balun_Kits/balun_kits.html**). Amidon also sells beads (**www.bytemark.com/amidon/**). Alternatively, you can wind 8 feet of the feed line into 7 turns to make your choke balun. Use electrical tape to

Table 2
Tubing Dimensions (inches) for 2 Element Yagis using Outer Diameter Tubing Shown

Band (Meters)	*1.0"*	*0.875"*	*0.75"*	*0.625"*	*0.5" (Reflector)*	*0.5" (Driven Element)*	*Hairpin (Inches)*	*Survival (mph) clear/0.25" ice*
10	n/a	n/a	n/a	36	70	61	23	96/68
12	n/a	n/a	36	33	53.5	40.5	16.57	100/73
15	n/a	18	36	57*	69	55	13.83	92/65
17	n/a	36	72**	28	69	53	12.625	89/64
20****	36	36	62***	49	69	51	11.2	83/60
20*****	36	36	62***	49	69	50.5	16.53	83/60

Notes
*The 0.625" extends 18" into the 0.75" which extends 18" into the 0.875" tubing.
**The 0.75" extends 36" into the 0.875" tubing.
***The 0.75" extends 36" into the 0.875" which extends 36" into the 1.0" tubing.
****Element spacing 68".
****Element spacing 92".
These are the sections from the center to each tip. There will be the same arrangement on the other side.

hold it in place.

I use white PlastiDip and crimped connector lugs to prepare the coax for connecting to an antenna. PlastiDip, and other similar products, are advertised for coating tool handles. I use it because it is flexible and keeps water out of the coax. To be safe, I also solder the crimped lugs. The solder keeps water out of those joints. I apply layers of black electrical tape over the PlastiDip coating to protect against UV (ultraviolet) radiation from the sun.

Installation

With help from my wife, I installed the Yagi at the top of my 30 foot tower, which is bracketed to the garage as shown in the lead photo. The center of K8SYL's 75 meter dipole was mounted at the top of the mast above the beam on the tower. My 40, 20 and 15 meter dipole was a couple of feet below the Yagi.

I ran the feed line to my radio to check out the beam. As I tuned the band, I found a DX station. To my joy Yassar, TA3D, came back to my first call, and we exchanged 59 signal reports. To my disappointment, however, the SWR was nearly 3:1. Further, my transceiver was putting out less than 50 W. The internal antenna tuner could handle that, but the SWR should be very close to 1:1. I had to determine the reason and correct the problem.

I lowered the Yagi and sat it a few feet above the ground pointing straight up. In that position, the SWR looked good — just as it had before. You've probably already guessed what was wrong when the beam was on the tower. If you thought it was interaction with those dipoles, you're right.

I took down the 40, 20 and 15 meter dipole and lowered K8SYL's 75 meter dipole to a position about 3 feet below the Yagi. *Voila!* The SWR was now less than 1.2:1 anywhere in the band.

17 Meter Results

Even at 30 feet high, this Yagi is an

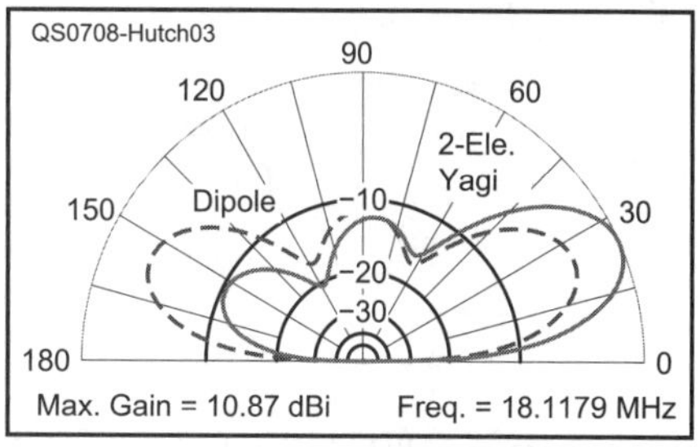

Figure 3 — Computed elevation pattern of the 2 element 17 meter beam (solid line) at 30 feet compared to a dipole (dashed line) at the same height.

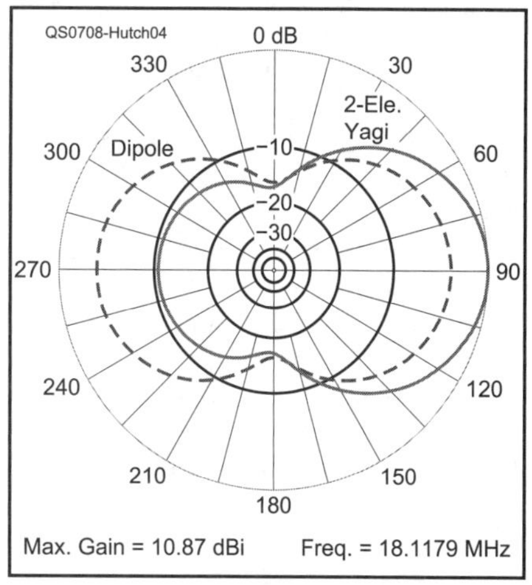

Figure 4 — Computed azimuth pattern at 25° elevation of the 2 element 17 meter beam (solid line) at 30 feet compared to a dipole (dashed line) at the same height.

excellent performer. You can see in the antenna modeling results of Figures 3 and Figure 4 how this Yagi compares to a dipole. I've enjoyed QSOs with many DX and domestic stations. This antenna has brought me a lot of fun.

Conversion to 15 Meters

After about a year and a half of 17 meter fun, I found myself thinking about the upcoming contest season. I love 15 meters and wanted to use something better than the 40 meter dipole that I had been using. It was time to modify the 17 meter Yagi for operation on 15 meters.

I opened the *EZNEC* model of the beam on my computer. I shortened the reflector and driven element tips to see how much I'd have to change the hairpin match for 15 meter operation. To my surprise and delight, I only needed to change the element tip lengths. The new dimensions are the same as in Figure 1, except the 0.5 inch section of the driven element is telescoped for a total half length of 130.5 inches, while the reflector is shortened in the same manner to 145.5 inches.

I made a new set of four element tips so there would only be one set of rivet holes in each. Alternately, at the outer ends of the 0.625 inch sections, make a pair of "X" shaped cuts about an inch deep with a hack saw and use a stainless steel worm-screw hose clamp to secure each tip. Just extend or retract the 0.5 inch tips inside the 0.625 inch tubes to shift between bands.

15 Meter Results

With the 2 element Yagi now operational on 15 meters, I found that it worked even better than it did on 17. That's probably because the antenna is higher and the element spacing is wider, both in terms of wavelength, on 15 meters. There were plenty of stations that were easy to work — both domestic and DX. In October of 2003 I exchanged 59 reports with XZ7A in Myanmar (Burma). Being on a hill is an advantage when you're using a low antenna.

I used Dean Straw's program *HFTA* to analyze the path toward XZ, BY and JA.[6] I discovered that I have an advantage on 15 meters in that direction. The terrain makes my 30 foot high antenna perform similar to an antenna at over twice that height on flat land.

Short-Boom 2 Element HF Yagis

Perhaps you'd prefer a less rugged version of the 17 meter beam, or maybe you'd like to have a Yagi for another band. If so, the

Table 3
Half-Element Dimensions for 2 Element Wire Yagis

Band (Meters)	*Reflector (Inches)*	*Driven Element (Inches)*	*Spacing (Inches)*	*Matching Inductance (μH)*
10	105.0	98.75	66	Not Needed
12	118.25	111.0	70	0.54
15	140.4	131.5	68	0.57
17	164.6	154.5	69	0.54
20	210	199	92	0.73
40	420	400	234	Not Needed

Note: The reflector and driven element dimensions are from the center to a tip. These dimensions will have to be doubled to determine tip-to-tip dimensions.

following family of small 2 element beams should be interesting for you. Make the elements from 6063-T832 drawn aluminum tubing with 0.058 inch wall thickness. You'll notice that these beams use nothing longer than 6 foot lengths of aluminum tubing. Again, that's because those lengths will ship via UPS with no special handling required. Make an exception for 20 meters, where you'll want to have an 8 foot boom. The good news is that UPS will ship up to 9 foot long packages. The good folks at Texas Towers tell me that they will cut a 12 foot length into unequal lengths for you. Shipping will cost you a little bit more, however. On the subject of booms, I like to use 2 inch OD 6061-T6 aluminum tube with a 0.12 inch wall. It's very strong — no doubt stronger than I need.

Here are a couple of guidelines for assembling the beam elements. When joining two pieces of tubing, use 3 inches of overlap (2 inches for the 10 meter reflector). When doubling the thickness of a section, extend the inner piece of tubing about an inch beyond the next joint toward the element center.

Since only one half of each element is shown, the first section nearest the boom will actually be made up of a continuous length of tubing for the reflector. For example, the first section of the 10 meter version is made of 0.625 inch diameter tubing that is 72 inches long, which is our standard 6 foot length, centered on the boom.

Table 4
Coil Dimensions of 2 Element Yagi Matching Inductances

Inductance (μH)	*Diameter (Inches)*	*Length (Inches)*	*Turns*
0.54	2	2.05	4
0.57	2	1.875	4
0.73	2	2.3	4

I've calculated hairpin matching networks for each of the beams. I prefer to use a hairpin match because it's easy to ground the center of the hairpin to the boom using a solder lug. Each hairpin is made from 12 gauge copper wire 2.5 inches wide and the length given in Table 2. The element spacing is 68 inches for all the bands except 10 meters (66 inches) and 20 meters, which has two choices as noted in the table.

While compact, the 20 meter antenna with the 69 inch boom will only cover the upper (phone portion) of the band. The SWR at 14.0 MHz is over 4:1. For that reason I do not recommend that you use this boom length on 20 meters. The alternate spacing, 92 inches (an 8 foot boom), allows coverage of the entire band with an SWR of less than 2:1.

Wire Yagis

Of course, you don't need to use aluminum tubing to build a Yagi. Wire will also work. Oh yes, you'll have to provide supports, but some of you may want to try a wire Yagi.

N6BV designed a series of wire Yagis just for you. These were meant to use straight elements — *not* inverted V elements. The dimensions in Table 3 are based on #14 wire. The values in the table are half element values — in other words, from the center to the element tip. For tip-to-tip (overall) lengths, you'll have to multiply the table values by two. For these versions, lumped inductances are easier to work with in shunt with the feed. The required inductance is shown for each band in Table 3. Coil dimensions for matching inductances are given in Table 4.

Notes

[1]C. Hutchinson and D. Straw, *Simple and Fun Antennas for Hams.* Available from your ARRL dealer or the ARRL Bookstore, ARRL order no. 8624. Telephone 860-594-0355, or toll-free in the US 888-277-5289; **www.arrl.org/shop/**; **pubsales@arrl.org**.

[2]Texas Towers: **www.texastowers.com/**.

[3]DX Engineering: **www.dxengineering.com**.

[4]Max-Gain Systems: **www.mgs4u.com**.

[5]D. Leeson, W6NL (formerly W6QHS), *Physical Design of Yagi Antennas,* published by ARRL. Now out of print.

[6]The program *High Frequency Terrain Analysis, HFTA*, comes on the CD included with *The ARRL Antenna Book*, R. D. Straw, Editor, 21st Edition. Available from your ARRL dealer or the ARRL Bookstore, ARRL order no. 9876. Telephone 860-594-0355, or toll-free in the US 888-277-5289; **www.arrl.org/shop/**; **pubsales@arrl.org**.

Chuck Hutchinson, K8CH, was first licensed as KN8UDJ in 1960 and has been an ARRL member ever since. He enjoys contesting, DX chasing and experimenting with antennas. Chuck and his wife, Sylvia, K8SYL, are graduates of Michigan State University. He retired from ARRL HQ in 2001 after 20 years on staff. Today, Chuck and Sylvia live on 11 rural acres in west Michigan. You can reach him at 9145 Bliss Rd, Lake Odessa, MI 48849 or at **chuck@k8ch.net**.

A 20 and 40 Meter Vertical on 'Autopilot'

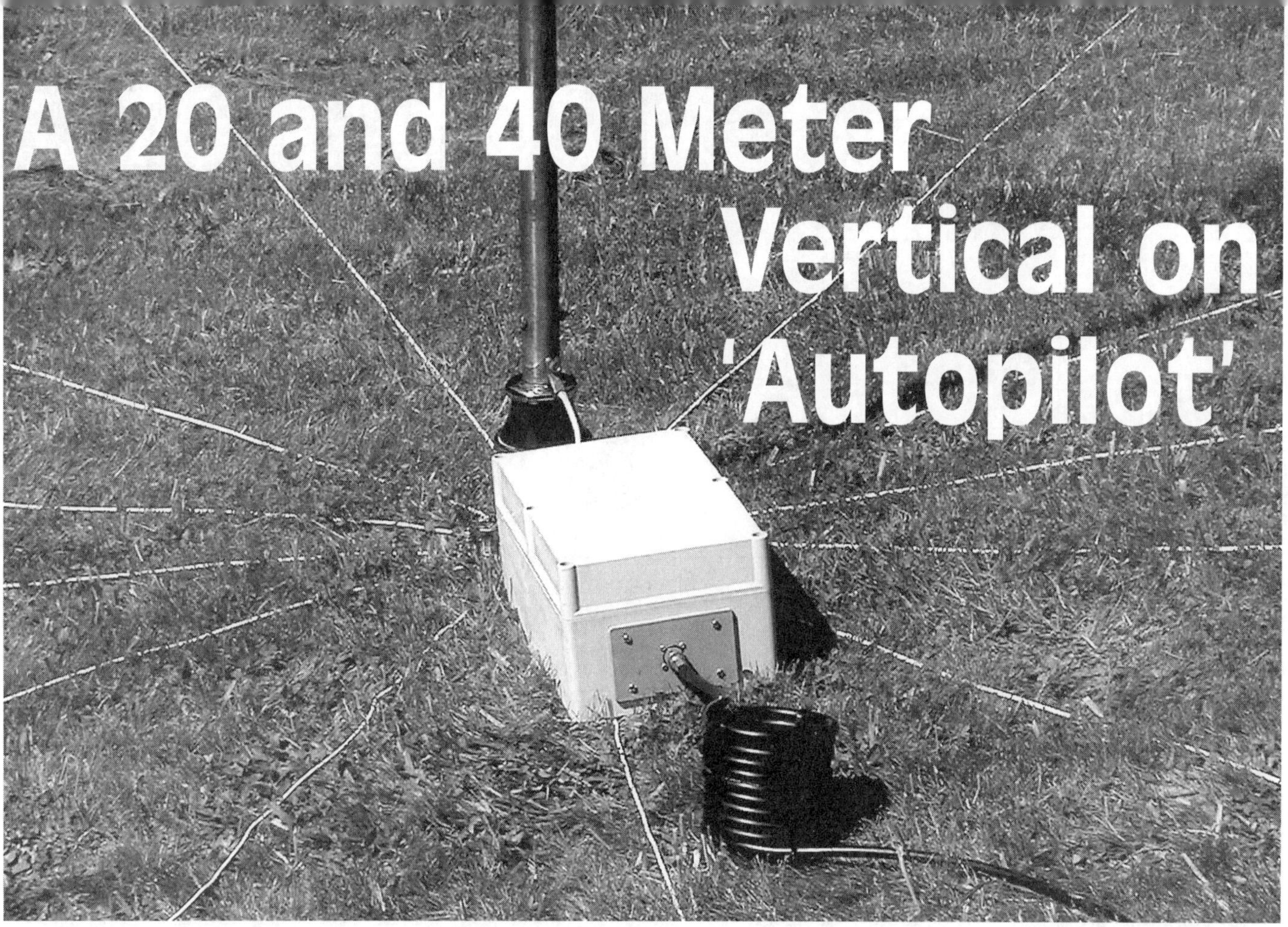

An old "Hints & Kinks" item results in an excellent Field Day antenna.

Dan Richardson, K6MHE

With its size, ease of installation, and reliable performance, the 40 meter quarter wavelength vertical antenna is a very popular HF antenna. What's more, with proper matching, it also performs very well on the 20 meter band. Being a half-wavelength on that band, the antenna provides an efficient low-angle radiation pattern, with roughly a dB of gain over the shorter standard quarter-wave vertical.

This article describes a dual-band vertical antenna system for 20 and 40 meters that I built for portable Field Day use. It uses a somewhat innovative tuning method that, after the initial setup adjustments are made, permits operation on ether band with no further tuning required—and it works great!

I discovered this scheme while perusing some past issues of *QST*. I had a very nice telescoping 35 foot marine vertical whip that I wanted to use for an upcoming Field Day

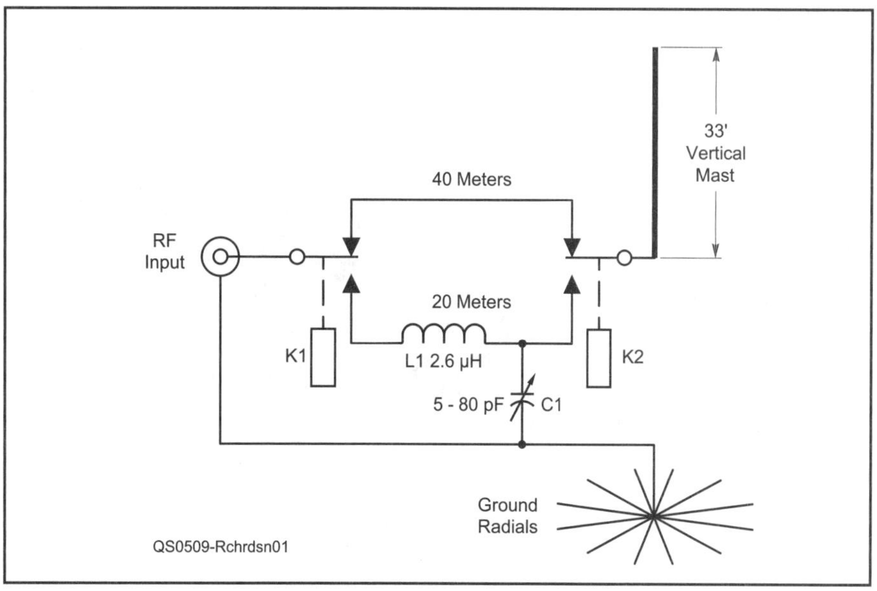

Figure 1—Using remote relays (K1, K2) for dual-band operation. One DPDT relay could be substituted for the two relays.

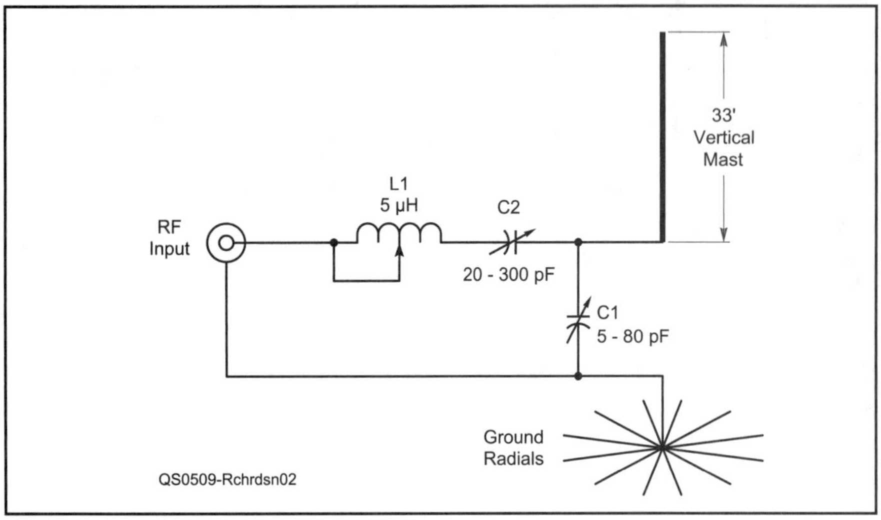

Figure 2—The modified L-network for dual-band operation—without relays.

and, while looking for some ideas, I came across this novel method for dual band operation. It was described by Wes Hayward, W7ZOI, in an old *Hints & Kinks* column.[1]

Typically, a quarter-wave vertical is fed by directly connecting the coaxial feed line to the base of the antenna. As the base impedance is low (typically 40-70 Ω, depending upon the ground/radial system utilized) this technique provides an acceptable SWR and works well. However, the base impedance of a half-wave (20 meters, in this case) vertical is much higher (around 1100 Ω); therefore a matching network is required to successfully feed the antenna with coaxial transmission line. Prior to discovering this method, I had only seen this type of dual-band operation performed using remote relays (K1, K2) in conjunction with an L network (L1, C1), as shown in Figure 1.

This innovative approach involves placing an additional capacitor in series with the inductor of an L network so that, when properly adjusted, a conventional 40 meter, 1/4 wavelength, vertical will operate on both the 40 and 20 meter bands with an acceptable SWR with no need for switching schemes or any additional adjustments.

How it Works

W7ZOI's modified network, shown in Figure 2, is easily understood when compared side-by-side with the simple L network of Figure 1. L1 and C2 form a series resonant circuit that behaves—on 40 meters—as a short circuit. The result is the same as having the relay contacts in the 40 meter position, as shown in Figure 1. The shunt capacitor (C1) reactance on 40 meters is roughly ten times that of the antenna's base (feed-point) impedance and therefore has a negligible effect on matching. The net result is the antenna's low base impedance essentially remains unchanged to the coaxial transmission line on 40 meters.

Twenty meter operation is accomplished by adjusting L1 and C2 so that the net inductive reactance is the same as it would be in the basic L network. The value of C1 remains the same in both networks.

So, we end up with a matching system that does double duty. The modified L network is adjusted so that L1 and C2 provide the proper amount of inductive reactance such that, when used with C1, matches the high base (feed point) impedance of the antenna on 20 meters. And, simultaneously, it maintains the series resonance that is required for matching the antenna's low base impedance on 40 meters. A pretty neat idea!

Construction

I built the entire network solely from junk-box parts. As is usually the case when building a project like this, my treasure trove did not contain capacitors of the exact values recommended by W7ZOI. I substituted a 100 pF capacitor for C1 and a 500 pF capacitor for C2. Originally, I was concerned that, by using larger values than those recommended, particularly for C2, the minimum capacitance values might be too large for proper tuning; however, this turned out not to be the case. The inductor was made from a section of coil stock (2 inch diameter; 14 gauge wire; 10 t at 8 tpi). All components were mounted on a 1/8 inch thick plastic panel, which, in turn, was installed in a watertight housing that had seen previous life as part of a shipboard antenna system. The completed network is shown in Figures 3 and 4.

If you choose to build this project take care to use capacitors that have a sufficiently high voltage rating. Mine are rated at 4.5 kV and should easily handle 500 W or more. [Vacuum capacitors would be ideal for this application, but they are expensive.—*Ed.*]

Tuning

As I stated earlier, my antenna's height is about 35 feet; therefore, its resonant frequency falls below the 40 meter band. This caused me no problems when tuning; however, I did have problems trying to adjust the network with an antenna height that was slightly less than 1/4 wavelength. I will discuss this in a moment.

Tuning was done empirically. Using an MFJ-259B antenna analyzer, the tap on the inductor was set near maximum inductance and capacitor C2 was adjusted for the best indicated SWR on 40 meters. Next, the analyzer's frequency was changed to the 20 meter band and C1 was adjusted for the best SWR, but the match obtained was not good. I next moved the tap on the coil in the direction of less inductance and C1 and C2 were adjusted again for their respec-

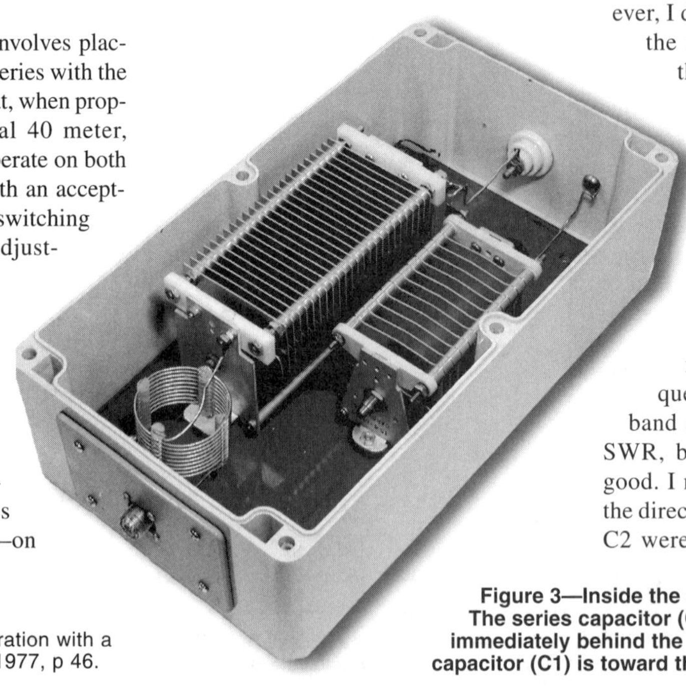

Figure 3—Inside the dual-band modified L-network. The series capacitor (C2) is at the top left, immediately behind the inductor (L1). The shunt capacitor (C1) is toward the right.

[1]W. Hayward, "Dual-Band Operation with a 33-Foot Vertical," *QST*, Jun 1977, p 46.

Figure 4—Tuning the network with an SWR analyzer. See text for tuning suggestions.

tive bands. This process was repeated until satisfactory SWR readings were obtained. (My requirement was to have the SWR fall below 2:1 on both bands. Many solid-state rigs start reducing their power or, worse yet, shut down at an SWR of 2:1.) The resulting SWR curves are shown in Figures 5(A) and (B). I should point out that this system was tuned using a marginal radial system (only twelve radials) and the SWR curves are therefore rather broad. Those curves would be more pronounced had the antenna been adjusted utilizing a more efficient ground system.

My telescoping mast comprises six sections that, when extended, reaches about 35 feet. As a matter of interest, I tried tuning the antenna with one of the sections retracted, resulting in an overall antenna height of about 30 feet, which is, of course, self-resonant above the 40 meter band (about 7.8 MHz). Using the shorter length antenna, I could achieve a good match on one band or the other but not on both simultaneously. It turned out that the additional inductance required to obtain a good match on 40 meters with the shorter antenna height resulted in having too much net inductive reactance in the L1/C2 series combination for a satisfactory match on 20 meters. Conversely, reducing the amount of inductance that provided a good match on 20 meters did not afford enough inductance for an acceptable match on the lower frequency. Hence, for this system to work properly, the antenna height should be no shorter than $^1/_4$ wavelength for the lowest frequency of operation. I found no detrimental consequences, however, when it was slightly longer.

Conclusion

After the initial adjustments, the network performed flawlessly. Alternating back and forth between 20 and 40 meter operation required nothing more than changing the frequency/band selection on the radio, with no concern whatsoever about the antenna system.

I am surprised I haven't seen this technique in wider use. Given that more than 25 years has passed since it was first published, I felt it worthwhile to present it again for those who may not be aware of this practical and versatile design.

I wish to thank to Wes Hayward, W7ZOI, for his review and suggestions in the preparation of this article.

First licensed in 1955, Dan Richardson, K6MHE, enjoys experimenting with antennas. Dan was an electronic technician in the US Coast Guard during the 1950s and '60s; he then joined the family foundry business, retiring about 10 years ago. After he had moved to rural Northern California, Dan's devotion to antennas flourished, and he's published several articles in leading amateur journals. Dan can be reached at PO Box 2644, Fort Bragg, CA 95437 or **k6mhe@arrl.net**.

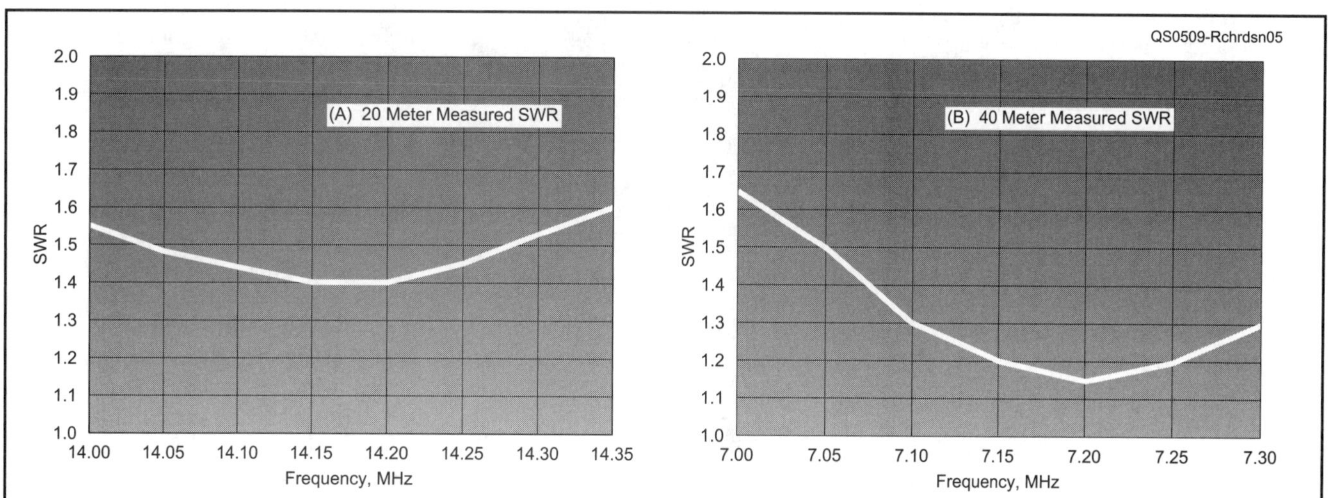

Figure 5—The SWR at 20 meters (A) and 40 meters (B).

A Rose Garden Array

An empirical study of elevated, sloping radials in a phased vertical array.

Jack Falker, W8KR

Every November, after our roses are put to bed for the Minnesota winter, my garden sprouts a 20 meter 4-square phased vertical array. I've been experimenting with elevated ground planes ever since my wife talked me into selling my 80 foot tower and Yagi several years ago, on the grounds that I was getting too old for climbing.

Horrors, you say! How could a DXer of nearly 50 years give up his tower and beam? Well, I have to admit to thinking long and hard about that. But I recognized that my summers were increasingly spent on my roses (more than 100 at the moment) and less on DXing. So the idea of giving up the tower after so many years, and doing something different with vertical antenna technology in the winter months, was kind of exciting. I also recognized from my early years in the hobby that the old saying, "A vertical antenna is one that radiates equally poorly in all directions," is just not correct for vertical antennas with extensive radial systems. This is especially true if you provide some gain and front-to-back ratio with multiple elements and appropriate phasing. [For optimum low-angle results, the ground conductivity must be relatively high in the area around the antenna for up to a few hundred feet—not a condition found everywhere.[1]—*Ed.*]

I started with a simple 20 meter $^1/_4$ λ ground-plane, elevated about 6 feet, with 12 sloping quarter wave radials. That turned out to be a better antenna than I expected. It was pretty competitive in most CW pileups, particularly with the linear on line. Unfortunately, it clearly didn't have the "ears" for the really long haul stuff, such as long path to VK or the Middle East.

I started reading about phased arrays and the radial systems necessary to create a low-loss ground. Since this array was going to be in my garden and, by definition, go up in the fall and down in the spring, there was no way it could be a ground-mounted system with the several hundred radials required at ground level. In reading the literature, especially the excellent book, *ON4UN's Low Band DXing*,[2] I became increasingly fascinated by the fact that a vertical antenna elevated $^1/_4$ λ needed four or even fewer radials. Now, $^1/_4$ λ of elevation on the low bands is a very big deal, but on 20 meters and above, it's not that hard to imagine. I talked to some low-bander friends who said they were getting good results on 80 meters at elevations of 10 feet (0.038 λ) with only four radials. I also learned that WWV uses nine radials elevated $^1/_4$ λ for each of their vertical antennas.

Figure 1—The ComTek Systems hybrid quadrature phasing coupler at the center of the array.

Figure 2—A ferrite-bead 1:1 current balun is at the base of each element.

The Past is Prologue

My memory flipped back to my earliest days of DXing in Detroit. In 1955, Karl, W8GB (SK), virtually ruled the low end of 40 CW every night with his homebrew 807s running 150 W, and his three-wire vertical with four radials elevated about 8 feet (0.061 λ) off the corner of his garage on a tiny city lot. (The three wire, 7 MHz vertical is featured in my very dog-eared copy of the 1955 *ARRL Antenna Book.*) Karl worked a lot of DX we kids couldn't

[1]Notes appear on page 105.

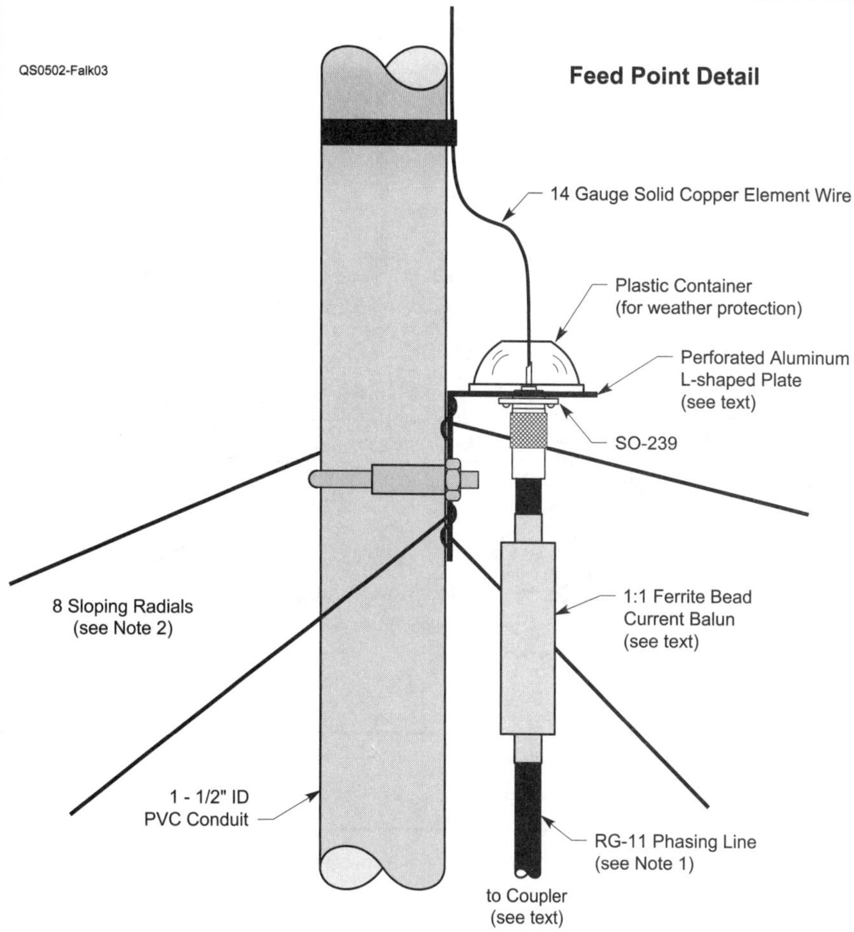

Notes:

1. Phasing lines are RG-11, with .78 velocity factor.
   Length in feet = .78 × (234/f)
   Each phasing line has 1:1 current balun at feed point
   (see text)

2. Each element uses 8 elevated, sloping radials.
   L = length in feet = 234/f
   Radials are connected together, where they cross.

3. Elements are 14 gauge solid copper.
   L = length in feet = 234/f
   Wire is taped to fiberglass poles and PVC tubing.

4. S = element spacing in feet = 246/f
   D = diagonal spacing = $\sqrt{S^2 + S^2}$

Figure 3—Details of each element feed point.

even hear until I emulated his antenna on the corner of my dad's garage. The difference was that my three-wire vertical was for 20 meters, which made my four radials, elevated 8 feet (0.122 λ), twice as high, relatively. I earned DXCC with that antenna and my Hallicrafters SX-71 and Heathkit DX-100 combo, before heading off to college in 1958.

Design to Fit

With all that empirical evidence and some encouragement from current literature, I set out during the winter of 2001-2002 to design an elevated phased array for the fall of 2002. I had visions of big rolls of coax, toroids and relays, until Bob, WØBV, showed me a little box containing a hybrid quadrature phasing coupler made by ComTek Systems.[3] This device, shown in Figure 1, is based on a Collins-defined phasing method. The hybrid quadrature coupler is discussed in some detail, including schematic and formulas, in *ON4UN's Low Band DXing*.[4] ON4UN also devotes significant discussion to the performance of the ComTek coupler on these pages.

While I'm game for just about any antenna construction project, my lack of test equipment and experience in building this kind of gear, plus my desire to get something on the air that fall, convinced me that using a phasing "module" was the right decision for my project. Just building the four antennas with their interconnected radial system offered enough construction fun!

I got in touch with Jim, K4SQR, at ComTek and ordered an ACB-4 Hybrid Coupler, designed for the middle of the 20 meter CW sub-band. In addition to the coupler and in-shack control box, Jim supplied all of the custom-cut RG-11 coax cable phasing lines, complete with a ferrite-bead 1:1 current balun for each line. The baluns (Figure 2) are necessary at the feed points (detailed in Figure 3) of elevated vertical arrays to reduce RF currents on the coax shield. For those who might plan to make their own phasing lines, note that the baluns can affect the electrical length of the lines, requiring additional measurements and/or experimentation. Balun kits are available both from ComTek and The Wireman.[5]

Putting it Together

With the *guts* of the system in hand, I set out to build four antennas from my junk box/pile. I wanted something that would be attractive and very modular for annual assembly, disassembly and storage. Those nice blue fiberglass quad arms I had been saving for more than 30 years came to mind, but they were only 13 feet long, so I needed something to lengthen them. PVC $1^1/_2$ inch ID conduit filled the bill. With a little duct tape as a shim around the base of each quad arm, they slip nicely into the conduit, which also serves as a mounting base for each antenna. I taped $^1/_4\ \lambda$ of #14 solid copper wire to each quad arm to form the vertical elements. An L-shaped piece of perforated aluminum (left over from an old linear amplifier project) makes up a little mounting platform for an SO-239 coax connector, leaving lots of holes for stainless-steel screws to connect $^1/_4\ \lambda$ radials made of the same #14 wire. This is all U-bolted to the PVC at the right spot to accommodate the full $^1/_4\ \lambda$ of radiating wire. Each SO-239 is covered with a little plastic baby food container (from the grandkids) to keep the snow off.

Each antenna is mounted on an 8 foot steel fence post (which freezes solid in the ground around here) and is elevated to about 8.5 feet (0.13 λ). Eight radials are then attached to each vertical, with several doubling as guy wires. As you can imagine, 32 sloping radials create a bit of a maze within the square, but that's important. Broadcast engineering literature tells us that these radials should be

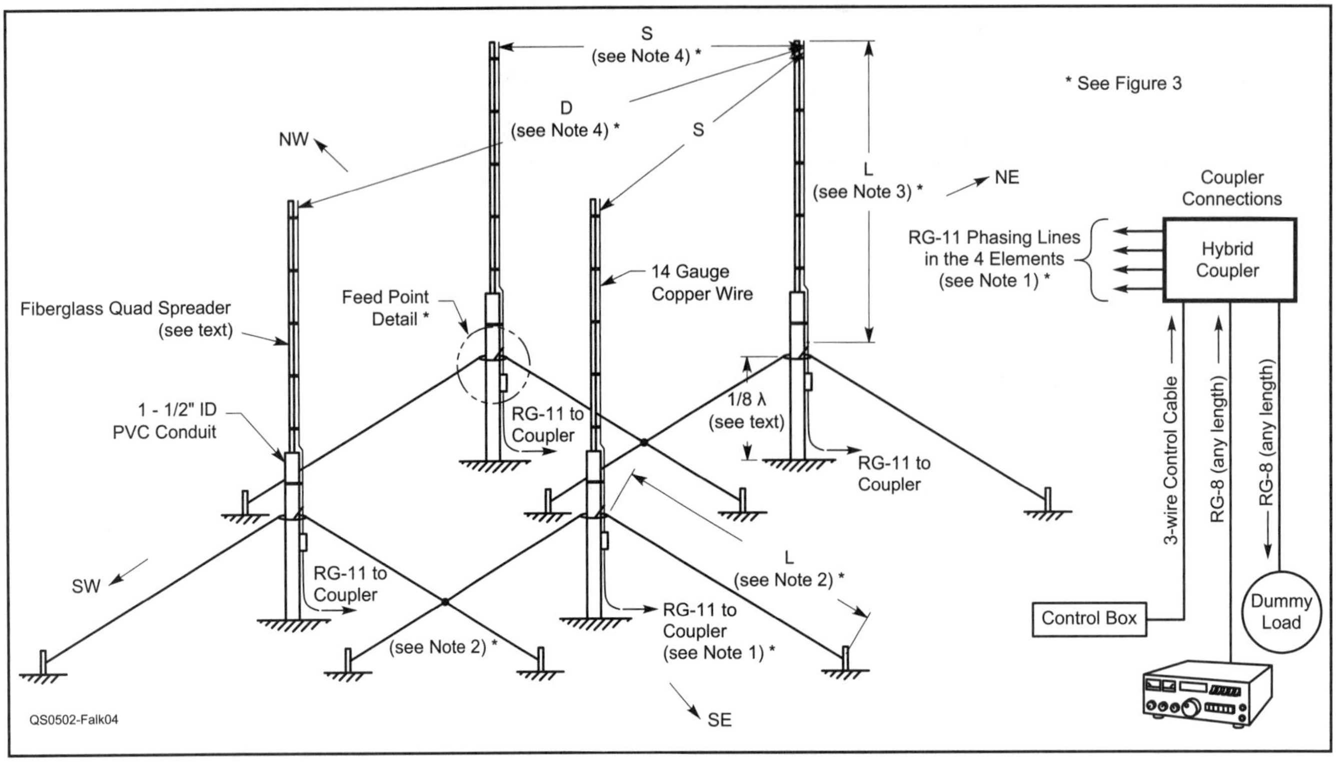

Figure 4—Drawing of the entire array.

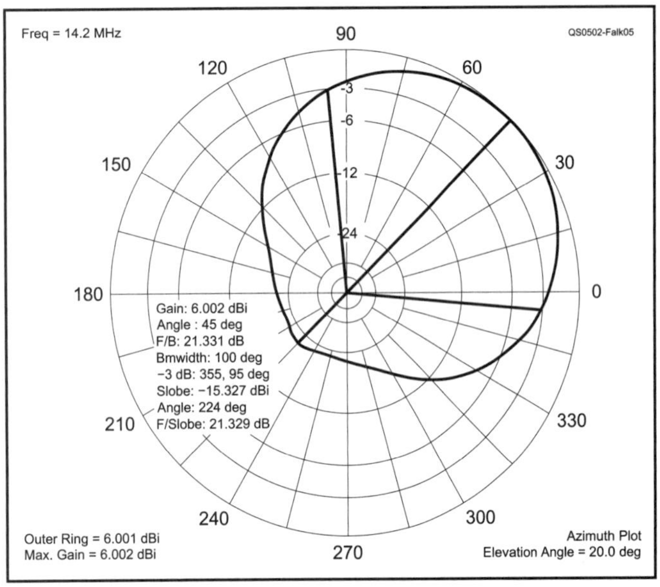

Figure 5—*ELNEC* plot of antenna azimuth pattern.

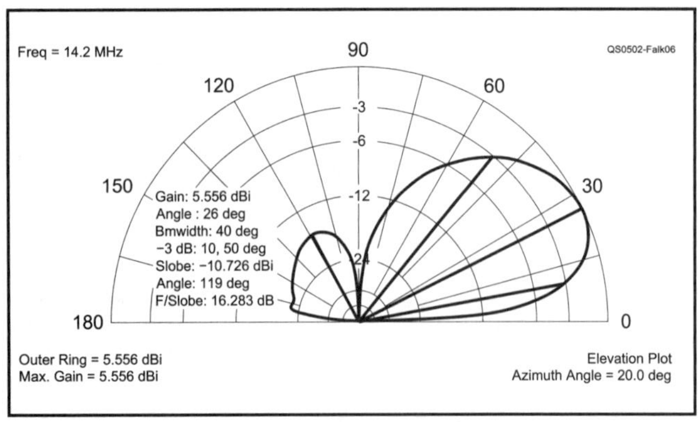

Figure 6—*ELNEC* plot of antenna elevation pattern.

interconnected wherever they cross within the square. This results in each antenna being connected to all the others, thereby creating a very uniform ground screen. In my mind, interconnection compounds the effectiveness of the radials. I'm tempted to try 12 radials per antenna, to make 48 in the maze, but I have a hunch (and *ON4UN's Low Band DXing* confirms) it won't make a difference, or perhaps even be a negative.

The hybrid coupler module is mounted in the center of the square with the four phasing lines running to it. The coupler also has to be elevated because if it is much lower than the antennas the phasing lines won't reach it. In addition to the phasing lines, three other cables attach to the coupler: the 52 Ω transmission line from the shack (any length), the control cable to switch antenna directions and another 52 Ω line running to a dummy load, either at the antenna or in the shack. The array is shown in Figure 4.

The purpose of the dummy load is to dissipate *dumped* power at port four of the coupler, which occurs if all four port impedances are not the same. In other words, if something happens to one of the antennas, causing it to not be resonant at the same frequency, its impedance will differ from the other three elements. As a consequence, power will be dissipated in the dummy load resistor.[6]

In that regard, the final step in building the array is to measure power to the

Figure 7—W8KR in the midst of his rose garden array—it's winter in Minneapolis.

dummy load with a wattmeter. I found it easiest to run a piece of coax back to a dummy load in the shack where I can measure power to the load a couple of times a year, just to make sure nothing has changed. ComTek indicates that dumped power of 5% or less of transmitter power is considered normal. Note that if you were to build a separate enclosure to keep the coupler and the dummy load out of the weather, the extra run of coax could be eliminated. ON4UN shows just such an arrangement in his book.

How Does it Play?

My experience with big towers and beams tells me that this dog hunts on 20 meters. I measure this by what I hear and how many times I need to turn on my linear to crack a pileup, which is not very often. This antenna has great *ears* and performance on the longest hauls because its radiation takeoff angle is low. The longer the haul the better the front to back: 20 to 25 dB (by my S-meter observations), dropping to around 15 dB on short-haul, ie, higher-angle signals. The *ELNEC*[7] plots, shown in Figures 5 and 6, indicate a forward gain of 5 to 6 dB over a single vertical, front-to-back of 21 dB; and a takeoff angle of 26°. That's about how it feels on the air. It isn't a Yagi at 70 feet, but you don't have to climb it, it's inexpensive, and it goes up and down with minimal effort. Figures 7 and 8 provide the two-season view of the multifunction backyard.

One final note: We feed lots of birds in the rose garden during the winter and deer wander around the feeders at night. I've noticed that the birds are smart enough to fly around the maze of radials in the square. However, the other night I watched as a deer wandered into the radial maze. Had she been startled it could have been a mess but, as I held my breath, she backed her way out slowly. Now, if I could just keep them out of there in the summer!

Special thanks to Bob Garwood, WØBV, for his help in drawing the schematic of the array, and to my son, John Falker, for taking the high resolution digital pictures for this article.

Figure 8—Summer in Minneapolis and it's back to a rose garden!

Notes

[1]*The ARRL Antenna Book, 20th edition*, Chapter 3, pp 11-15. Available from the ARRL Bookstore for $39.95 plus $8 shipping in the US, $10 elsewhere. Order number 9043. Telephone toll-free in the US 888-277-5289, or 860-594-0355; **www.arrl.org/shop/**; **pubsales@arrl.org**. Compare your ground conductivity in Figure 19 to that of Minneapolis to get an idea of relative performance.

[2]J. DeVoldere, *ON4UN's Low Band DXing*, available from the ARRL Bookstore for $19.95 plus shipping. Order number 7040.

[3]ComTek Systems, tel 704-542-4808; **www.comteksystems.com**.

[4]*ON4UN's Low Band DXing*, Chapter 11, pp 14-15.

[5]The Wireman, **www.thewireman.com**.

[6]*ON4UN's Low Band DXing*, Chapter 11, p 14.

[7]*ELNEC* has been superseded by *EZNEC*, available from Roy Lewallen, W7EL, at **www.eznec.com**.

Jack Falker, W8KR, was first licensed in 1954 in Detroit, Michigan as W8SRK. He began DXing and building antennas in 1955. Jack spent more than 40 years in the practice and teaching of corporate finance and now owns his own investment practice in Minneapolis, Minnesota, where he and his family have lived since 1977. He is a graduate of the University of Michigan and the University of Detroit, with BA and MBA degrees. He is a life member of ARRL. Jack can be reached by e-mail at **jack@falkerinvestments.com**. QST

SHORT Dipoles for 160 Meters

Get on the "top band" with limited real estate—build one of these shortened dipoles. Apply these principles to 80 and 40 meters, too.

Floyd Koontz, WA2WVL

Many hams do not operate 160 meters due to the size of a resonant antenna required for this band. Using the accepted formula of 468/f, the length of a dipole cut for 1.84 MHz would be 254 feet, 4 inches. What we shall endeavor to do here is to design a simple structure that approaches the performance of a full-length dipole, but is less than half that size. For simplicity, most of the antennas (in 7 examples) shown here are 120 feet long and at a height of 60 feet. Shorter and lower dipoles can be designed, but they will have degraded performance (impedance, gain and bandwidth all decreasing).

Arguably, the three most important antenna electrical parameters are impedance, gain, and the SWR bandwidth. If the bandwidth can be wide, then a tuner will not be needed, since today's broadband transceivers will operate directly into 50 Ω. To operate over the full band, however, a tuner will be needed with any of these antennas.

The Full-Length Dipole

The full-length dipole is the standard to compare against. By using *AO* software for the initial modeling and then *NEC Wires* software to take into account the ground effects, the dipole was computed to be 257 feet, 5 inches long.[1] For average soil at 60 feet:

Impedance = 45.8 + $j2$ Ω at 1840 kHz.
Gain = +1.27 dBi at 60 feet and +3.15 dBi at 100 feet (at 30° elevation).
2:1 SWR Bandwidth = 65 kHz at 60 feet and 95 kHz at 100 feet.

As expected, higher dipoles have more gain and greater bandwidth. In all of the figures that follow, the numbers (parentheses) represent that antenna point's height above ground. The non-parenthetical numbers represent length.

Example #1

The first approach (Figure 1) is to add coils in both sides of the dipole to reduce its physical length. The design of this type of dipole was covered in a recent *QST* article.[2] The coils were computed to have an inductance of 129.85 μH. The performance was calculated using a coil Q of 500. For average soil at 60 feet:

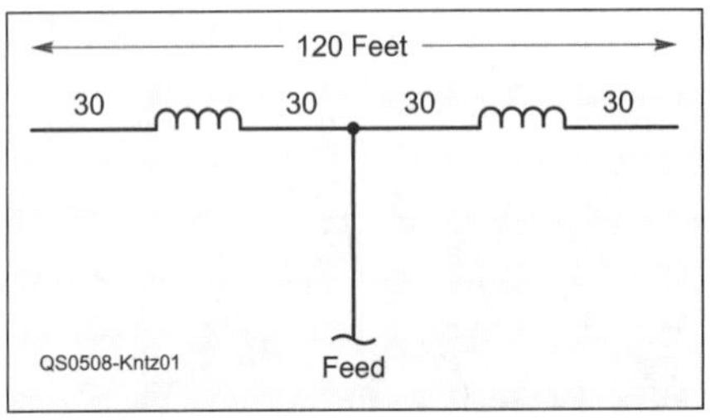

Figure 1—A 160 meter dipole shortened with loading coils. All single-line feeds represent coaxial cable with an insulator at the antenna/feed line junction.

Impedance = 21.2 + $j0$ Ω at 1840 kHz.
Gain = –0.51 dBi at 30° elevation.
2:1 SWR Bandwidth = 10 kHz.

No evaluation was done at 100 feet.

Example #2

Figure 2 shows one of the simplest dipoles of reduced length. The ends of the wire hang straight down and approach (6 feet) the ground. The ends of the wires should be tied down to keep them from flopping around in the wind. The total length of the wire is the same as a full-length dipole.

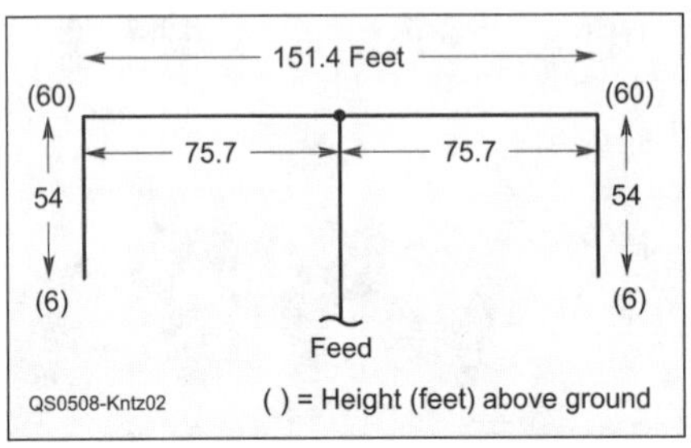

Figure 2—A "hanging ends" shortened dipole. The ends are perpendicular to the antenna main radiator.

Impedance = 42.8 + $j3$ Ω at 1840 kHz.
Gain = –0.36 dBi at 30° elevation.
2:1 SWR Bandwidth = 65 kHz (same as a full-length dipole).

Example #3

Another way to gain wire length while keeping the width of the dipole at 120 feet is to bring the drooping wires back toward the center of the antenna and tie them to the ground with small ropes. The impedance can be raised by using "off-center feed," such as feeding at the corner or in the center of one of the sloping wires. Figure 3 shows this example.

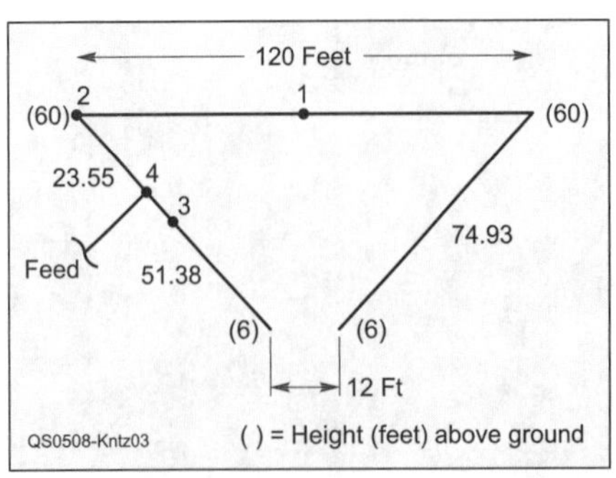

Figure 3—Bringing the wire ends close to the antenna center can increase the antenna's effective length.

Center feed impedance at point (1) = 13 – $j0$ Ω at 60 feet.
Corner feed impedance at point (2) = 23 – $j0$ Ω at 60 feet.
Sloper center feed impedance at point (3) = 45 + $j0$ Ω at 60 feet.
Optimum feed impedance at point (4) = 50.1 + $j4$ Ω at 60 feet.

[1]Notes appear on page 107.

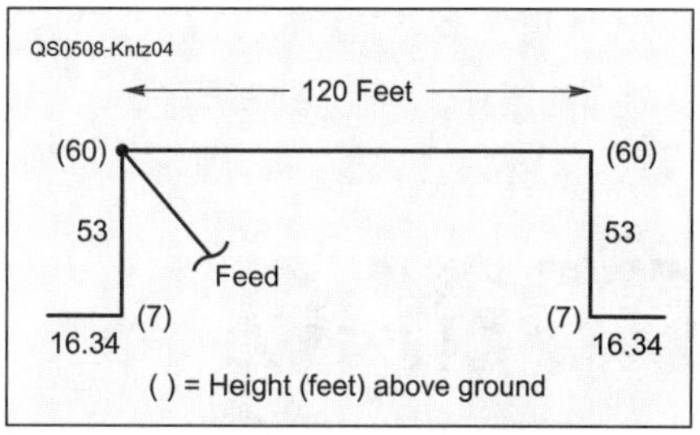

Figure 4—The main radiator is still 120 feet long, but the effective length has been increased with additional horizontal members. The feed point is offset to produce the required match.

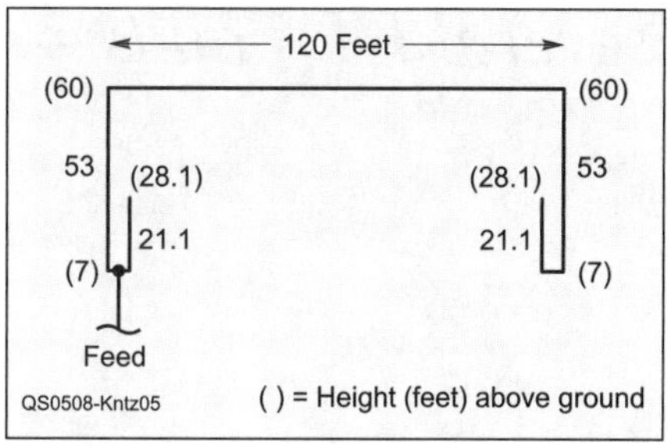

Figure 5—The feed can be closer to the ground with this arrangement.

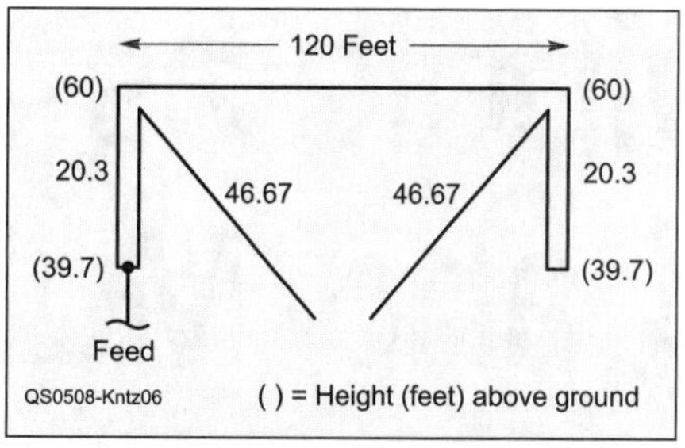

Figure 6—Linear loading is used for lengthening the antenna, while the offset feed provides a good match in this example.

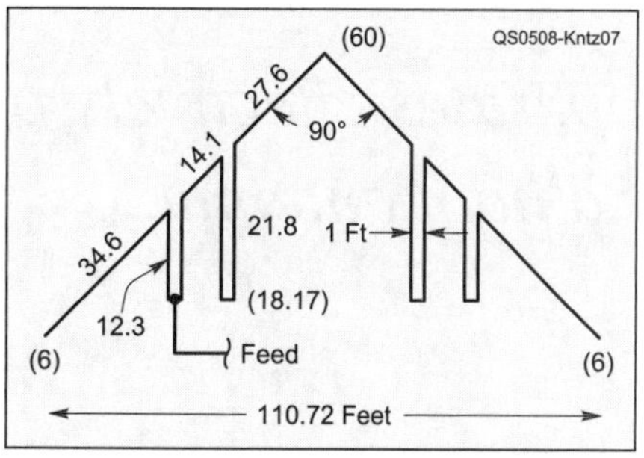

Figure 7—The popular inverted-V can use linear loading to increase its length and an offset feed to provide a match.

The antenna has the same performance at each feed point and its parameters are:

Gain = –1.04 dBi at 30° elevation.
2:1 SWR bandwidth = 24 kHz (because of "out of phase" currents in the sloping wires).

Using "off-center feed" gives the same antenna pattern as center feed and can simplify the support of the coax weight and may reduce the length of the coax run. With any of the dipoles presented here, a balun is recommended at the feed point.

Example #4

This dipole is similar to Example #2 except the ends of the wires are bent to run horizontally at the 7-foot height to attain the full wire length required (Figure 4). The feed is in the corner to get a good 50 Ω match. The wire ends could also run toward the center, if desired.

Impedance = 58.9 +j0 Ω at 60 feet.
Gain = –1.56 dBi at 30° elevation.
2:1 SWR Bandwidth = 53 kHz.

This design has good bandwidth and allows for "tweaking" of the tuned frequency from the ground. With clips, the wire ends can be extended to optimize tuning for the CW end of the band.

Example #5

This is similar to the previous antenna except the wire ends are bent back up and run parallel to the vertical wire. Spreaders (using ½ inch PVC) are required to hold the wires apart and the ends have to be supported from above by a small rope. Commercial open wire line can be used with some reduction of gain and bandwidth. This will change the length of the wire by a few feet. Figure 5 shows the example.

Impedance = 54.8 –*j*4 Ω at 60 feet.
Gain = –1.16 dBi at 30° elevation.
2:1 SWR Bandwidth = 49 kHz.

Example #6

This time, "linear loading" and "off-center feed" are used to provide a perfect 50 Ω match. The feed point is at the bottom of one of the loading stubs, as shown in Figure 6. The length of the stubs sets the impedance, while the sloping wires set the frequency.

Impedance = 49.5 – *j*1.6 Ω at 60 feet.
Gain = +0.71 dBi at 30° elevation.
2:1 SWR Bandwidth = 45 kHz.

Example #7

The inverted V is a favorite of many hams, so a short V was designed for 160 meters. This time, two linear loading stubs are needed on each side, to get down to 1840 kHz. Again, a perfect 50 Ω match can be obtained by adjusting the stub lengths. The center frequency is adjusted at the end of the wire. The antenna is shown in Figure 7.

Impedance = 47.8 – *j*2.0 Ω at 60 feet (45° slope; each wire).
Gain = –4.05 dBi at 30° elevation.
2:1 SWR Bandwidth = 23 kHz.

The low gain and bandwidth are due to the low average height of the inverted V (33 feet).

Conclusion

A word of caution about using "off-center feed": This technique works perfectly on the computer, but in real life the effective height of the antenna is a guess and the ground characteristics are a "wild" guess. These parameters will have major effects on the feed impedance. A better approach to getting a good match in the shack is to use a small L network at the transmitter end to set the match (center feed). By switching between two networks, a separate CW and PHONE match is possible. This would be especially useful on an 80 meter antenna.

The design procedure for the L network is to estimate (or measure) the impedance at the transmitter end of the coax. Then compute the values needed (I used software written by N6BV).[3] If you are going to estimate the impedance, it is helpful make the coax length multiples of a ¼ wavelength.

I hope this presentation gives you some ideas as to how best fit a 160 meter dipole on your property to enjoy our "top band." You can put these techniques to work with 80 and 40 meter antennas, as well.

Notes

[1]*AO* and *NEC Wires* are antenna programs developed and marketed by Brian Beezley, K6STI. These are no longer available. *EZNEC* by R. Lewallen, W7EL, can now be used to perform this modeling (**www.eznec.com**).

[2]L. Lopes, CT1EOJ, "Designing a Shortened Antenna," *QST*, Oct 2003, pp 28-32.

[3]*Transmission Line for Windows (TLW)* by N6BV as well as *EZNEC ARRL* and other modeling software is included with *The ARRL Antenna Book*, available from your dealer or the ARRL Bookstore. Order no. 9043.

Floyd Koontz, WA2WVL, is a retired electrical engineer with 40 years' experience designing radio transmitters, antennas and communications systems. He was first licensed in 1955 as WN9JQA, later as KØLTV, and since 1961 as WA2WVL. Floyd was an operator at the XE5L (Sucoro) DXpedition in 1965, and has written numerous antenna articles for QST *and other ARRL publications about wire beams, the EWE, the Diamondback and other antennas. An ARRL life member, he can be reached at 8430 W Park Springs Pl, Homosassa, FL 34448, or* **wa2wvl@gowebco.com**.

The Fan Dipole as a Wideband and Multiband Antenna Element

It's broadbanded and operates effectively on several HF bands—a nice combination for a ham antenna.

Joel R. Hallas, W1ZR

Multiband HF antennas are certainly an attractive concept, especially as new bands are added to our authorized set of frequencies. Now with 9 bands between 3 and 30 MHz, it is difficult to have separate antennas for each band on a reasonably sized piece of property without significant interaction. Over the years there have been a number of common approaches to multiband antennas:

- *Antennas that take advantage of multiple resonances in a single element structure.* This category includes 40 meter dipoles operated on 15 meters and G5RV dipoles. Although these can be effective, the feed impedances are generally different on different bands, often resulting in potentially significant transmission line losses if coax is used.
- *Antennas using resonant circuits as traps to isolate sections.* These can often be effective as well. The traps may restrict the frequency range on each band and will contribute to system losses, however.
- *Antennas using parallel dipoles resonant on each band.* These have been used successfully by some, although I have not been very successful with this approach. Unless elements are orthogonal (two antennas at right angles), the coupling can restrict the frequency range and make adjustment difficult. Mutually orthogonal dipoles for more than three bands are hard for me to visualize.
- *Antennas using tuned feeders and low-loss line.* The "double zepp" approach has gained in popularity with the common use of "no tune" restricted matching range transmitters and the availability of wide range antenna tuners. The typical antenna is a half wave at the lowest band, center-fed with open wire or ladder line. The impedance on any band is considered irrelevant since the tuner can deal with the resultant impedance at the bottom of the low loss line and transform it to the 50 Ω unbalanced impedance needed by the transmitter. This can be a very effective system. The fact that the pattern changes from broadside to a complex multi-lobe pattern above a full wavelength can provide additional geographic coverage from the single antenna. In addition, the whole antenna is always used, providing gain in some directions at the higher frequencies. The typical configuration is shown in Figure 1.

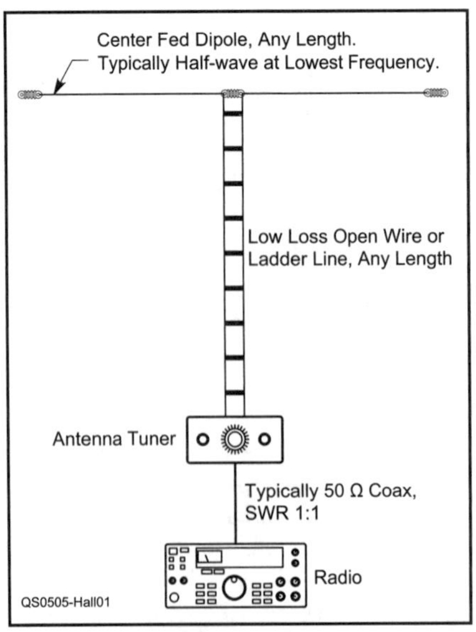

Figure 1—"Double Zepp"—typical configuration.

The loss can be low if connections are made carefully. As shown below, however, the SWR can be above 10:1 on the transmission line on some bands. *The ARRL Antenna Book*[1] indicates that a matched 100 foot length of clean, dry (people used to wax their twin lead!) ladder line has a loss of 0.15 dB or less at HF *if matched*. An SWR of 10:1 adds about 1 dB of loss,[2] still probably less than typical trap antennas.

Another potential area of concern is the impact of the SWR on the voltage and current within the tuner. A high SWR can result in maximum currents and voltages on the line equal to the matched value times the square root of the SWR. These will occur at different spots along the line depending on load, line length and operating frequency. With all the different frequencies involved, it is likely that the maximum of each will be near the tuner on one band or another, however. For a 100 W output and a 10:1 SWR, this results in a maximum of 630 V or 1.6 A. For 1.5 kW it increases to 2.5 kV_{rms}, 3.5 kV_{peak} or 6.1 A, however, which can explain the arcing or smoking observed in some tuners!

Enter the Fat Cylindrical or Biconical Dipole

John Kraus, W8JK, shows the way to solve this problem and simultaneously reduce the frequency sensitivity of such antennas in his classic text *Antennas*.[3] Figure 9-9 indicates that while a cylindrical di-

[1]Notes appear on page 110.

Figure 2—Looking up the tower at the center spreaders of the Cage antenna at W1AW.

Figure 5—W1ZR 20 meter biconical array, hidden in the trees.

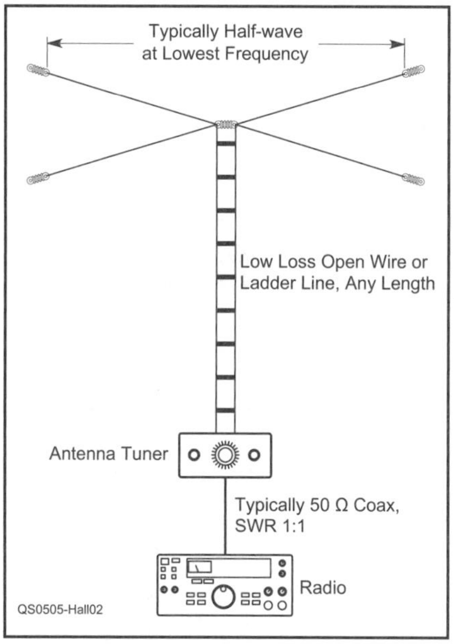

Figure 3—Biconical (fan) dipole configuration.

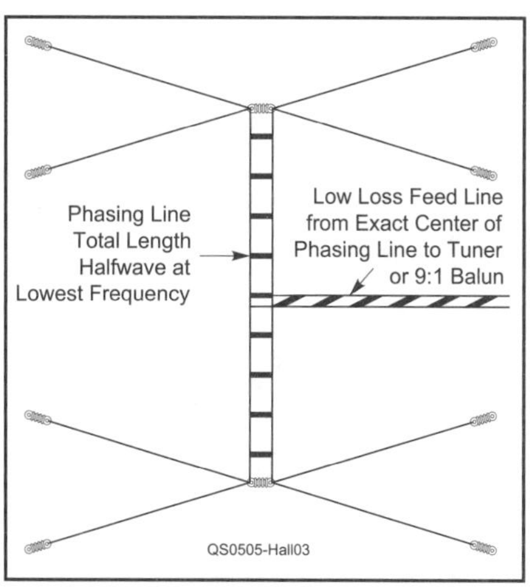

Figure 4—Biconical lazy-H antenna configuration.

pole with a length to diameter ratio of 2000 (about #4 AWG wire at 20 meters) will have an impedance variation between its half wave and full wave resonances of 70 Ω to about 3300 Ω, with corresponding variation in reactance, the 2λ resonance is at about 2300 Ω and the average is 873 Ω. In contrast, the figure shows that for a length to diameter ratio of 60 (a 6.6 inch cylinder at 20 meters) the half wave impedance is about the same while the full wave is reduced to around 900 Ω, with the average a convenient 454 Ω.

This approach has been used frequently to broaden the bandwidth of a half wave antenna, for example, to allow better coverage of both 80 and 75 meters with a single dipole, but has even more impact on harmonic operation where the concept has not often been employed in amateur service.

The straightforward approach to this is to use larger diameter wire or tubing for the elements. This can work for VHF, but supporting an 80 meter dipole made with 2 foot diameter tubing is hard to imagine! Fortunately, the same effect can be accomplished with a skeleton structure, and a "cage" antenna has been popular in various applications (it was used for the first transatlantic ham QSO and one is now in use at W1AW; see Figure 2). The cage typically uses a crossed spreader and four wires for the elements (six were used for the T/A tests). W1AW Station Manager Joe Carcia, NJ1Q, designed the one at the ARRL HQ station to operate well on all W1AW 80 and 75 meter frequencies. Joe used 3 foot PVC spreaders (reinforced with dowels and sealed using PVC caps and tees) at right angles with #14 stranded wire to construct this highly effective antenna.

An alternate is to use a skeleton biconical dipole as shown in Figure 3. This has the advantages that only two wires are used per side, and that the spreader structure is much simpler. I use a single 6 foot spacer on each end of the elements of a 20 meter array as described below. An 80 meter antenna might require some intermediate spacers to avoid tangles during raising. Note that while the upper wire must support the antenna weight, the lower wire needs only to provide conductivity. Although the amount of wire required is comparable to a parallel dipole antenna, the performance is better.

Fortunately, Kraus provides the capability to transform between the cylinder structure, the thin rectangle (equivalent radius is $^1/_4$ of the width) and the biconical. The impedance of a biconical, or fan, with an end radius of 2.8 times the cylinder radius (in this range) has the same impedance characteristics, yielding a half angle of 2.7°. Extrapolating to the flat biconical (triangle), we would expect to need a triangle base of $4 \times 2.8 = 10.2$ times the cylinder diameter. For a 20 meter fundamental antenna this is a manageable 5 foot 7 inches. I also considered a "mini-cage" made of 400 Ω ladder line (typically sold as "450 Ω" ladder line, but really measuring closer to 400 Ω) with the wires connected together at the feed point.

Why Bother?

What does it buy us? I have run *EZNEC*[4] simulations of 80 meter fundamental and 20 meter fundamental skeleton biconical dipoles and compared the results

Table 1

Comparison of SWR of 80/75 Meter Dipoles of Various Cross Section

SWR at 50 Ω, *EZNEC* prediction. Single band coax fed.

Antenna Type	*3.5*	*3.6*	*3.7*	*3.8*	*3.9*	*4.0 MHz*
Thin, #14 wire dipole	5.9	2.8	1.6	2.1	3.6	5.8
Ladder line "cage"	3.8	1.9	1.2	1.7	2.7	3.9
3 foot fan	3.2	1.9	1.1	1.5	2.3	3.2
6 foot fan	3.0	1.8	1.1	1.5	2.3	3.1
22 foot fan	3.3	1.9	1.3	1.7	2.4	3.4
NJ1Q cage	2.5	1.6	1.1	1.4	1.9	2.5

SWR at 400 Ω, *EZNEC* prediction. Multiband ladder line fed.

Antenna Type	*80*	*60*	*40*	*30*	*20*	*17*	*15*	*12*	*10 Meters*
Thin, #14 wire	7.2	6.5	12.9	14.7	8.3	9.0	5.8	9.2	5.0
Ladder line "cage"	8.5	4.1	7.0	10.7	4.7	7.0	3.6	5.2	3.5
6 foot fan	8.2	3.6	4.8	8.0	3.8	4.5	2.8	4.0	2.7
NJ1Q cage	9.1	2.4	2.1	6.6	2.3	2.6	3.6	1.3	4.1
Max gain over $1/2\ \lambda$ dipole	0.85	1.2	2.2	4.9	2.5	3.5	1.7	2.2	4.7
Azimuth of main lobe (number of significant lobes)	90 (2)	90 (2)	90 (2)	90 (2)	129 (4)	118 (4)	152 (10)	138 (6)	149 (10)

with that of a single wire dipole. To make comparison easy, for each case, I "trimmed" the length to provide best SWR from band edge to band edge. At 80 meters, using Kraus' factors, the predicted end spacing is about 22 feet, so I also ran it at a more reasonable 6 feet to find the difference.

I was glad I did, since *EZNEC* predicts that beyond a few feet the additional spacing doesn't matter. The results are interesting at the fundamental as well as on the harmonics (see Table 1). At 80/75 meters, while a thin wire dipole's SWR across the band may be outside of the range of many in-rig auto tuners, even the easiest to make multiwire configurations are within the usual 4:1 range of such devices on this tough band.

Note that with ladder line feed as a multiband antenna, on the higher bands where losses are highest, the SWR variation is reduced significantly. We still have the SWR at the first resonance to deal with and it is very similar to that with the single wire, typically around 8:1 at resonance. Note that the reduction in impedance variation will generally also mean less retuning as frequency is changed across any band. Tables 1 and 2 tell the story.

Table 2

Comparison of SWR of 20 Meter Dipoles of Various Cross Section

SWR at 50 Ω, *EZNEC* prediction. Single band coax fed.

Antenna Type	*14.0*	*14.1*	*14.2*	*14.3*	*14.35 MHz*
Thin, #14 wire	1.7	1.6	1.6	1.65	1.7
Ladder line "cage"	1.3	1.25	1.3	1.3	1.36
6 foot "fan"	1.2	1.1	1.03	1.16	1.22

SWR at 400 Ω, *EZNEC* prediction. Multiband ladder line fed.

Antenna Type	*30*	*20*	*17*	*15*	*12*	*10 Meters*
Thin, #14 wire	40	5.1	6.0	6.7	10.5	10.6
Ladder line "cage"	32	6.5	5.3	4.5	5.9	5.4
6 foot "fan"	26	7.8	4.6	4.4	4.2	4.2
Max gain over #14 dipole (dB)	0.8	1.1	1.4	1.8	2.3	

All at 90/180°

Table 3

Performance of Biconical Lazy-H at 73 Ft Elevation

Gain and SWR	*30*	*20*	*17*	*15*	*12*	*10 Meters*
Gain dBi	8.5	10.3	12.0	12.7	11.4	12.4
Gain over dipole	1.1	2.5	4.7	5.1	3.9	4.4
Gain over 20 m fan	0.6	2.2	3.8	4.0	2.5	2.8
EZNEC SWR	31	4.2	10.6	13	4.9	2.6
Measured SWR	4.6	1.3	1.4	2.7	2.5	1.7

The Biconical Lazy-H

At my station, I have solved the problem of high SWR at the fundamental (half-wave) on the 20 meter fundamental antenna by making a "stacked biconical" (a popular TV antenna in the '50s) or "biconical lazy-H," if you prefer. This consists of two 20 meter biconical dipole elements, the second a half wavelength below the first and fed in phase with 400 Ω ladder line. By transforming each element's low impedance at 20 meters through the resulting quarter-wave matching section to the center of the phasing line, and feeding them in parallel, I end up with about a 4:1 SWR across 20 meters.

This configuration, shown in a diagram as Figure 4 and a hard-to-see photo in Figure 5, provides broadside gain on 20 meters. It also acts like a well matched 4 element combination broadside and collinear array on 10 meters. The downside of this is that I now have higher SWR on some of the intermediate bands. On the plus side, broadside gain is provided on all bands 14 through 30 MHz. It will tune and operate on 30 meters as well, but watch the SWR!

At W1ZR, this antenna is fed with about 150 feet of ladder line and then a 9:1 balun. The SWR readings were taken with a Bird 43 wattmeter at the balun. I expect that the difference between predicted and measured results is due to some combination of loss in transmission line and balun and measurement error. The antenna does seem to perform well.

Notes

[1]R. D. Straw, N6BV, Ed., *The ARRL Antenna Book*, 20th edition, p 24-20.

[2]*The ARRL Antenna Book*, 20th edition, p 24-10.

[3]J. Kraus, W8JK, *Antennas*, New York: McGraw-Hill Book Co Inc, 1959.

[4]*EZNEC* is a registered trademark of Roy W. Lewallen (**www.eznec.com**).

Joel Hallas, W1ZR, is QST *Assistant Technical Editor and Product Review editor. You can reach him at* **jhallas@arrl.org**.

An Effective 160 Meter Receiving Loop

Use this antenna to hear those weak top-band signals by nulling out the noise and interference.

Dick Stroud, W9SR

If you would like to work 160 meter DX but don't have the space for a large receiving antenna, this small loop may be the answer. I was inspired to experiment with single turn receiving loops after attending a recent seminar on low band antennas by Ted Tahmisian, W9WT. A few such loop designs have been described over the years in various articles. A single turn receiving loop developed by Doug DeMaw, W1FB (SK), is shown in several ARRL publications and has been widely used.[1] A loop system made of solid shielded coax with an amplifier was later described by Barry Boothe, W9UCW.[2]

Design Considerations

The performance of several loop designs was evaluated using an 1830 kHz remote transistorized signal source and an Empire Devices NF-105 field intensity meter. I found that the null is typically 4 to 5 dB deeper with loops made of solid shielded coax compared to those made of standard RG-type cable.

Shielding coverage of standard cable can be as low as 85%, which will allow strong signals to pass through. Coax with a solid outer conductor provides almost complete shielding and thus can produce a deeper null. The larger single turn loops appear to be more sensitive up to the point at which mechanical constraints become the limiting factor. Shape does not appear to make any perceptible difference.

The accepted maximum perimeter of a directional single turn loop is 0.1 λ. Those made of standard coaxial cable, however, have been much smaller. With the high cable capacitance, the overall size must be kept small enough to allow tuning of the loop to resonance using an external capacitor.

With this type of loop the impedance of the cable is not critical and, with the loop described in Figure 1, the impedance of the home constructed coax is much higher than standard cable resulting in lower distributed capacitance. This allows a larger antenna to be used thus providing a stronger signal. The loop shown has a perimeter of 0.04 λ, which is a good electrical and mechanical compromise. The loop has an apparent inductance of 10 μH that can be tuned to resonance by 740 pF. The fact that the capacitance per unit length varies slightly within the assembled coax does not affect performance.

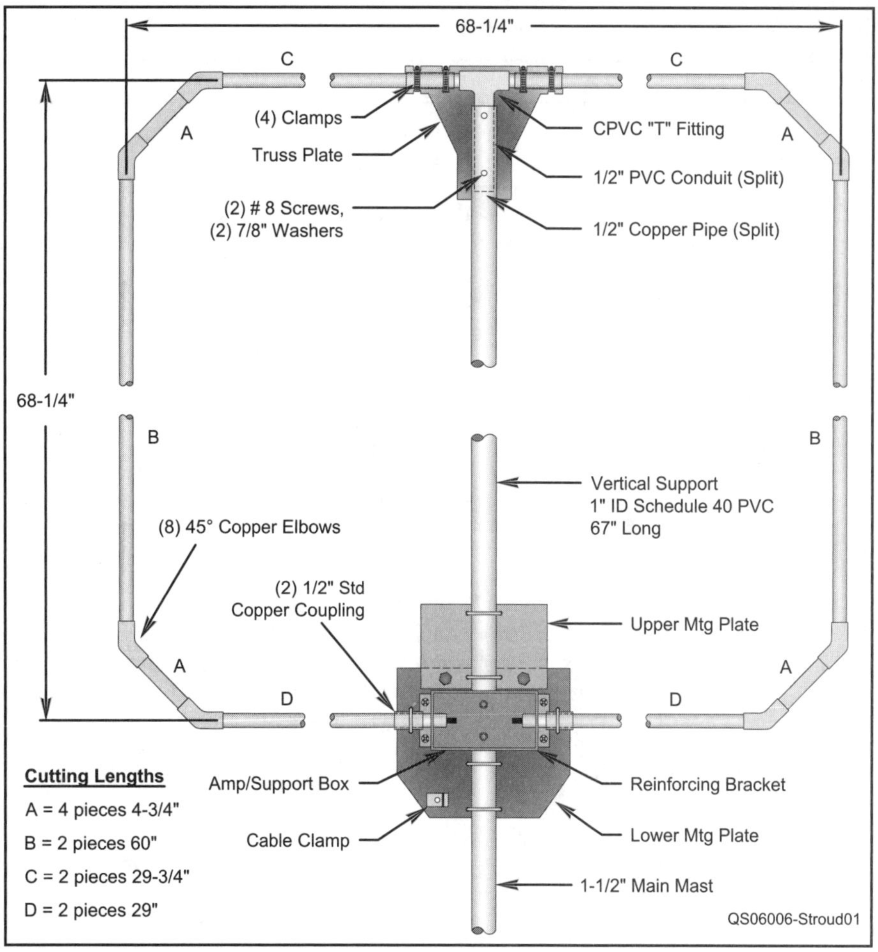

Figure 1 — Loop antenna configuration.

[1]Notes appear on page 114.

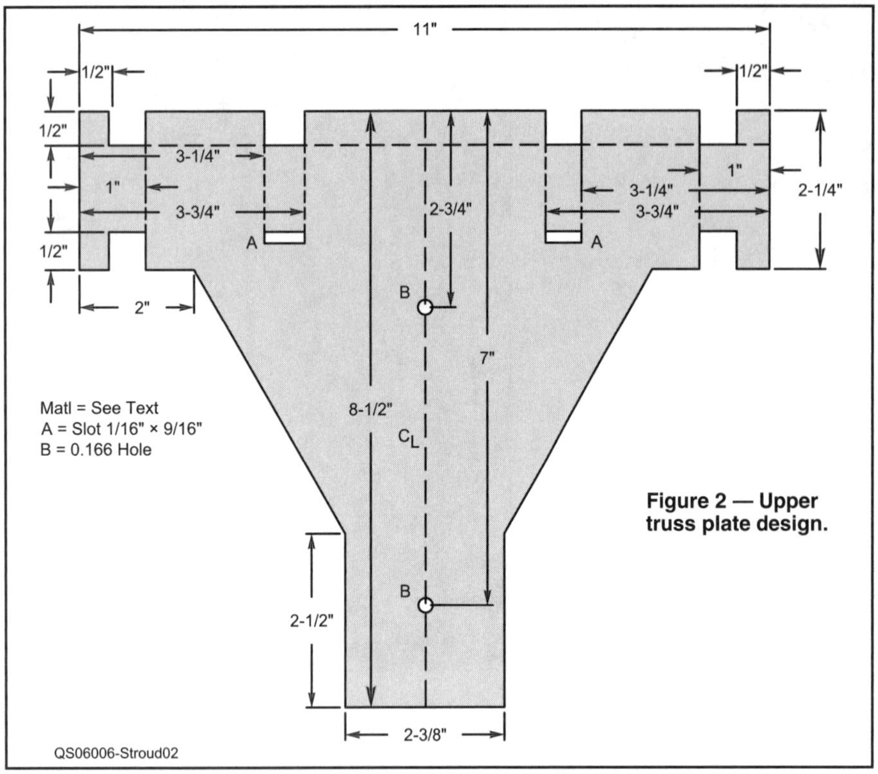

Figure 2 — Upper truss plate design.

Figure 3 — Upper truss plate details showing T connector and element mounting.

Putting it All Together

This loop is very sturdy and can be built at reasonable cost. Most mechanical parts are readily available from home improvement outlets. All hardware is stainless steel. The main element is built of ½ inch copper water pipe and copper plumbing fittings. The center conductor is a single length (about 23 feet) of RG-11 foam dielectric cable with the outer jacket and shield braid removed. The two halves of the copper element can be soldered together on a flat surface, such as a garage floor, and later assembled into the "T" fitting at the top and to the aluminum box at the bottom. The design is octagon shaped, allowing the coax core to be easily threaded through the assembled pipe. A leader wire is soldered to the core conductor to aid in pulling it through the copper pipe.

The center vertical support for the antenna is a 67 inch length of 1 inch ID schedule 40 PVC pressure pipe. The upper truss plate is made from a Farberware 8½ × 11 inch, non-absorbing resin cutting board available at low cost from department stores (see Figures 2 and 3). This material is sturdy and has good RF characteristics at this frequency. The plate is notched and slotted, as shown, to accept the two stainless steel element clamps on each side. Two 4 inch lengths of standard ½ inch PVC electrical conduit are split and placed around the element under the clamps.

The center CPVC T fitting accepts the ½ inch copper pipe and has built-in stops so that when the element is inserted it will leave about a 1 inch electrical gap in the copper tubing. This is necessary for the loop to operate properly. Strips of fiberglass PC material (less copper) are used to shim the element to be in line with the T fitting.

To attach the PVC vertical support to the truss plate and T fitting another 6 inch length

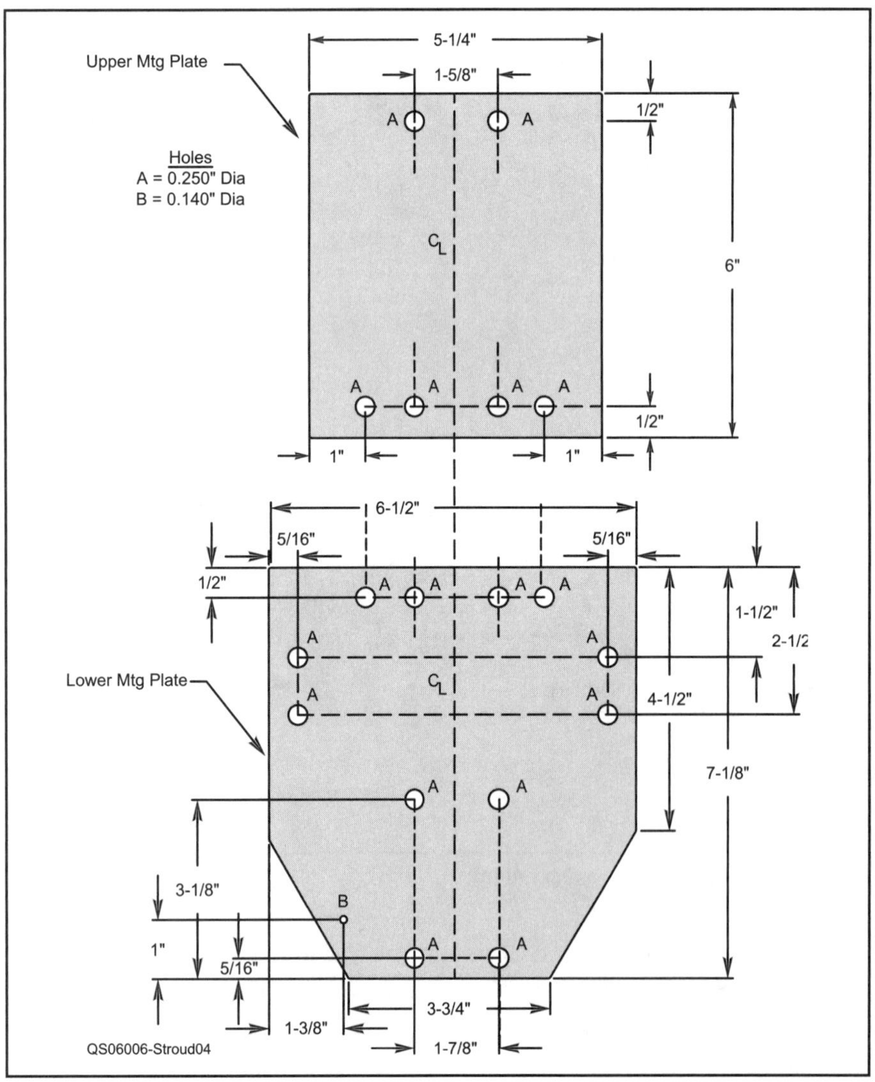

Figure 4 — Mounting plate construction. Use 0.125 inch aluminum.

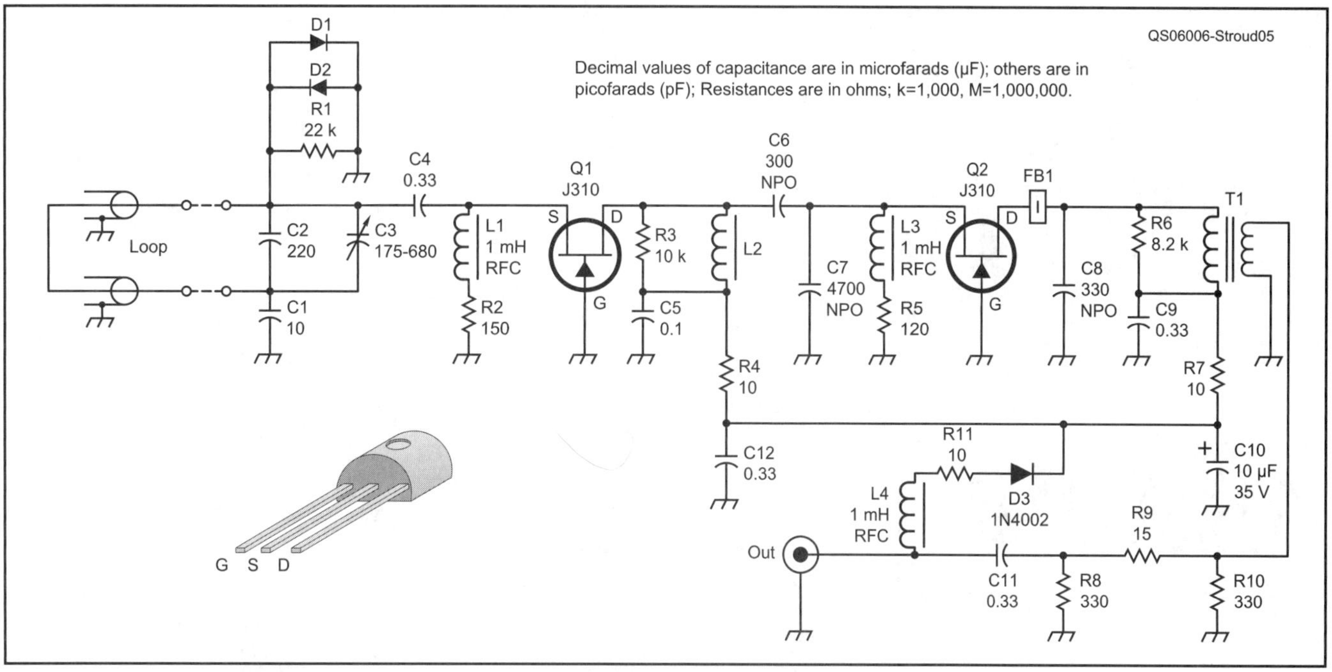

Figure 5 — Amplifier schematic.

C1 — 10 pF, 100 V molded mica capacitor, CM05.
C2 — 220 pF, 100 V molded mica capacitor, CM05.
C3 — 175-680 pF variable capacitor (Arco 468).
C4, C9, C11, C12 — 0.33 µF, 50 V ceramic capacitor.
C5 — 0.1 µF, 50 V ceramic capacitor.
C6 — 300 pF, 100 V molded mica capacitor, CM05.
C7 — 4700 pF, 100 V ceramic NPO capacitor.
C8 — 330 pF, 100 V molded mica capacitor.
C10 — 10 µF, 35 V electrolytic capacitor.
D1, D2 — Silicon small signal diodes, 1N4454, 1N914, or equivalent.
D3 — Silicon diode, 1N4002.
L1, L3, L4 — 1 mH miniature RF choke (J W. Miller 8230-92).
L2 — 75 turns #28 enameled wire on T37-2 form (see text).
Q1, Q2 — J310 low noise field effect transistor.
R1 — 22 kΩ, ¼ W resistor.
R2 — 150 Ω, ¼ W resistor.
R3 — 10 kΩ, ¼ W resistor.
R4, R7 — 10 Ω, ¼ W resistor.
R5 — 120 kΩ, ¼ W resistor.
R6 — 8.2 kΩ, ¼ W resistor.
R8, R10 — 330 Ω, ¼ W resistor.
R9 — 15 Ω, ¼ W resistor.
R11 — 10 Ω, ¼ W resistor.
T1 — Primary, 73 turns; secondary 7 turns, #28 enameled wire on T37-2 form; see text.

of conduit is split on one side and placed inside the upper end of the PVC center support. A 6 inch length of copper pipe is split for a length of about 5¼ inches and placed into the conduit. This leaves about ¾ inch of uncut copper extending out that will be later epoxied into the T fitting. Two 8-32 × 2 inch screws attach the truss plate to the vertical support and also go through the short length of conduit and split pipe. Fender washers, ⅞ inch in diameter, are used under the screw heads.

The aluminum center support box is a Pomona model #2906, available from **www.sales@xcess.com**. The box acts as a lower tie point for the copper element sections and also houses a two stage amplifier. The box measures 4¼ × 2⅝ × 1¾ inches with a ½ inch mounting foot on each side. Three 1⁄16 inch drip holes are drilled in the lower edge of the box before it is bolted to the main mounting plate. A U shaped reinforcing bracket is made of 0.063 inch thick aluminum and formed to fit inside against the ends and back of the Pomona box. The box and bracket are bolted by the same screws to the lower mounting plate. The box mounting feet are also bolted to the plate.

Holes are drilled on each side through the box and bracket to just clear the copper element. A standard straight copper coupling is placed over the element on each side to prevent crimping by the U bolts and the element is clamped to the base plate on each side. A 1 inch long section of ¾ × ¾ × ⅛ inch aluminum U channel is trimmed and placed under the element sections aligning the element with the holes drilled in the box. When the antenna is otherwise complete and square the element is cemented into the center T fitting with 5 minute epoxy and the clamps and U bolts are tightened.

The 0.125 thick aluminum mounting plate, shown in Figure 4, is made of two sections. The upper section is offset forward above the aluminum box and bolted to the lower plate section with ¼ inch bolts. This aligns the vertical center support with the centerline of the copper element. U bolts in the upper plate section clamp the PVC center support while U bolts in the lower section clamp to the main support mast.

The loop can be cleaned with steel wool and painted after assembly is complete, if desired. If you are anxious to try it you can operate without the amplifier by connecting C4 directly to the output connector and omitting C1. Capacitors C2 and C3 must be in place and C3 properly tuned. The full null depth will not be realized in this condition.

Amplifier Design

The amplifier, shown schematically in Figure 5, is built on a 2⅜ × 3¾ inch double sided PC board that is cut out to clear the small ribbed areas of the aluminum box as shown in Figure 6. It is also drilled for mounting on two ½ × ¼ inch diameter metal spacers fastened to the bottom of the box. The board on each end is notched to clear the ½ inch copper element sections that extend about ⅞ inch into the box. The center conductors should be approximately in line with the PC board surface. The output connector is mounted on the lower side of the

Figure 6 — Amplifier construction details.

Figure 7 — Details of PC board construction.

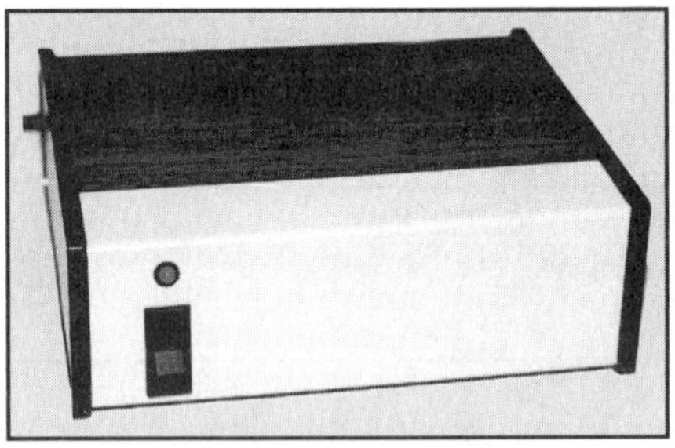

Figure 9 — Loop amplifier power supply.

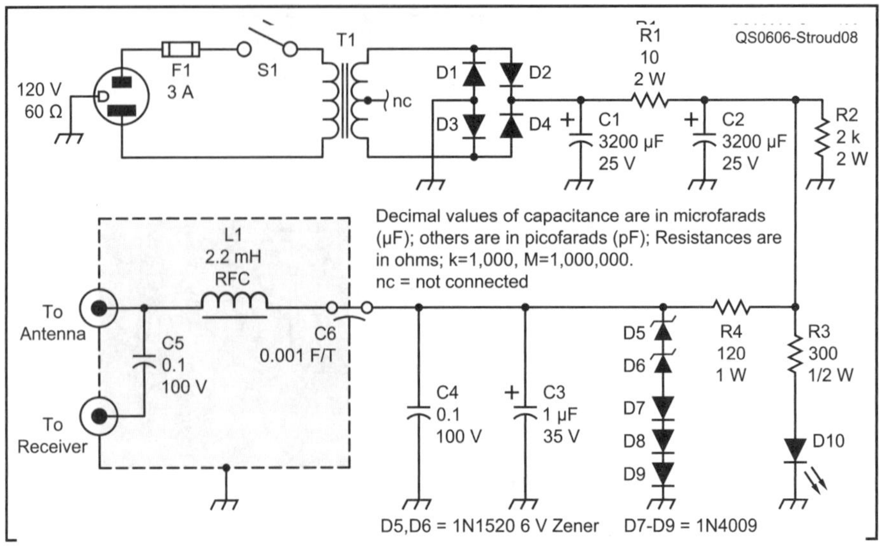

Figure 8 — Power supply schematic.

C1, C2 — 3200 µF, 25 V electrolytic capacitor.
C3 — 1 µF, 35 V electrolytic capacitor.
C4, C5 — 0.1 µF, 100 V ceramic capacitor.
C6 — 0.001 µF, 100 V feedthrough capacitor
D1-D4 — Silicon diodes; 1N4009, or 1 A, 50 V bridge rectifier.
D5, D6 — 1N1520, 6 V Zener diodes (see text).
D7-D9 — Silicon diode; 1N4009 (see text).
D10 — Light emitting diode (LED).
F1 — Fast blow fuse; 3 A, 120 V ac.
L1 — 2.2 mH miniature RF choke (J. W. Miller 9250-225).
R1 — 10 Ω, 2 W resistor.
R2 — 2 kΩ, 2 W resistor.
R3 — 300 Ω, ½ W resistor.
R4 — 120 Ω, 1 W resistor.
S1 — SPST toggle switch.
T1 — 120 V to 12 Vct, 450 mA transformer (RadioShack 273-1365 or equivalent).

box near one end to allow space against the box for the short 1½ inch main support mast. A close-up of the amplifier PC board is shown in Figure 7.

The amplifier consists of two cascaded, low noise, grounded gate stages using J-310 field effect transistors (FETs). C1 balances the sides of the loop against the amplifier input capacitance. This value was first determined with a variable capacitor and then a fixed unit was put in its place. C1 affects the null depth. Back to back signal diodes at the amplifier input protect the FETs from high signal levels generated by your local transmitter.

Inductor L2 is wound by winding one layer of 55 turns, then overwinding back in the same direction from that point to complete the required turns. T1 is wound in the same manner with the secondary wound over the cold end of the primary in the same direction. The pad in the output provides 2.2 dB of fixed attenuation with the resultant overall gain being 19 dB. The noise figure is less than 2 dB. With the values shown, the amplifier center frequency is 1.830 MHz and the –3 dB points are 1.764 and 1.896 MHz. The center frequency of the amplifier can be moved by changing the values of C6 and C8. Current drain at 12 V is 13 mA. The circuit board is available from FAR Circuits.[3]

All amplifier board parts except C3 are available from Dan's Small Parts and Kits. An on-line catalog can be viewed at **www.danssmallpartsandkits.net**. C3 is available from Communication Concepts at **www.communication-concepts.com**.

Power for the amplifier is routed through the output cable from a small power supply at the operating position.

There is no instability and no indication of overload or intermodulation products noted from the amplifier. The unit still performs well at 0°F in the Indiana winter!

The loop is not de-tuned by nearby objects and once it is tuned by C3 to 1830 kHz (or your favorite part of the band) in the shop, it will still be tuned when in the final position. Be sure your station control system will not allow you to transmit into the loop as the result would be catastrophic!

The power supply shown schematically in Figure 8 was built into a surplus Ten-Tec JW-7 cabinet and supplies 12 V dc through the coax to the amplifier. Construction is not critical and any stable 12 V at 13 mA supply can be used. The BNC coaxial connectors located on the power supply chassis should be shielded inside to prevent stray noise pickup. The 6 V Zener diodes were used because they were plentiful in the junk box; however, a 78L12 three terminal regulator would be a good substitute replacing the Zeners as well as the three diodes in series with them. The completed supply is shown in Figure 9.

Results

The null is through the center of the loop or into the page of Figure 1. The maximum signal is to the sides. It is important that the element and center box be waterproof. RTV can be used to seal any questionable area and a thin layer of RTV can be applied under the edge of the amplifier box lid as a gasket before final installation.

The receiving loop is mounted on a steel tower section, with a rotator, about 7 feet above ground. Results have been very gratifying. The peak to null varies greatly with the type of signal received and was measured at 38 dB using the test setup described earlier. The antenna is very effective in nulling out local power line noise, distant thunderstorms, broadcast band harmonics, computer hash and local interference. The null is not as effective on tumbling skywave signals but the loop still does a great job receiving them.

It is great for receiving low angle DX and often outperforms my larger antennas. The difference in gain is well compensated for by the on-board amplifier. I can instantly switch between receiving antennas from the operating position and am constantly amazed by the performance of the small loop. I am working many new European and Asian DX stations that I didn't know were around while I was listening on the transmitting antenna.

Notes

[1]D. DeMaw, "A Receiving Loop for 160 Meters," *QST*, Mar 1974, pp 38-41.

[2]B. Boothe, W9UCW, "Weak-Signal Reception on 160—Some Antenna Notes" *QST*, Jun 1977, pp 35-37.

[3]Send $9 plus shipping to FAR Circuits, 18N640 Field Ct, Dundee, IL 60118-9269.

Photos by the author.

Dick Stroud, W9SR, was first licensed in 1939 and received his Amateur Extra class license in 1952. Dick is a retired electrical engineer who spent over 30 years designing military electronics equipment. You can contact him at PO Box 73, Liberty Center, IN 46766 or at **dikw9sr@citznet.com**.

The Extended Double Zepp Revisited

A vintage favorite still makes a great single or multiband antenna.

Jerry Haigwood, W5JH

The extended double Zepp (EDZ) antenna has been around for a lot of years. Yet, many amateurs still do not know much about it. The EDZ is an excellent, easy to build, wire antenna that produces real usable gain with a broadside pattern. It is also reasonably easy to match to 50 Ω using a ladder line matching section. The popular 88 and 44 foot all band antennas are actually EDZs on 20 and 10 meters respectively.[1] Below, I discuss antenna characteristics, design criteria and matching criteria for the 40 through 10 meter HF bands and give dimensions and construction information for several antennas.

Antenna Characteristics

The EDZ is a 1.25 λ long center fed doublet as shown in Figure 1. It is the longest doublet that will yield a pattern with the major lobe broadside to the antenna. The azimuth plot is shown in Figure 2 and the elevation in Figure 3. Longer lengths will start to produce the classic "butterfly wing" pattern. Even longer lengths will produce multiple lobes broadside with the largest lobes approaching the ends of the antenna.

The EDZ also produces some very usable gain when compared to a ½ λ dipole. The typical gain is on the order of 2.5 to 3.0 dBd as shown in Figure 4. The EDZ derives its gain from the main lobe being compressed making the EDZ a bidirectional gain antenna. The EDZ azimuth plot shown in Figure 2 has a half power horizontal beamwidth of 34.8°. The elevation beam width (Figure 3) is 28.7°. These beamwidths will change depending the height of the antenna above ground. This antenna was analyzed at a height of 30 feet.[2]

The impedance of the EDZ will also vary depending on the height above ground, the ground conductivity and other proximity factors. Typical values at a height of ½ λ over average ground are 150 –*j*850 Ω. As you can see, the feed point is quite reactive. However, a balanced feed line of the appropriate length and impedance can cancel the reactance and move the resistance value closer to 50 Ω. To realize the full benefits of the EDZ, the antenna needs to be installed at a minimum height of ½ λ. Lower heights will result in a distorted pattern. This antenna needs to be installed as a flat top and not as an inverted V. Installation as an inverted V also distorts the pattern and results in reduced gain in the main lobe.

Length

The formula for calculating the length of an EDZ is:

Length (ft) = 1240/frequency (MHz).

Table 1 shows some typical antenna lengths.

Matching

The EDZ can be closely matched to 50 Ω by using a 0.155 λ (56°) matching section of 600 Ω balanced line.[3] A 1:1 balun or choke connects to 50 Ω coax of any length for the run to the station. The specified length of balanced line will cancel out the capacitive reactance of the antenna and transform the resistance value to about 50 Ω. Table 1 lists the actual length of the matching sections (Lf) needed to match the antennas shown.

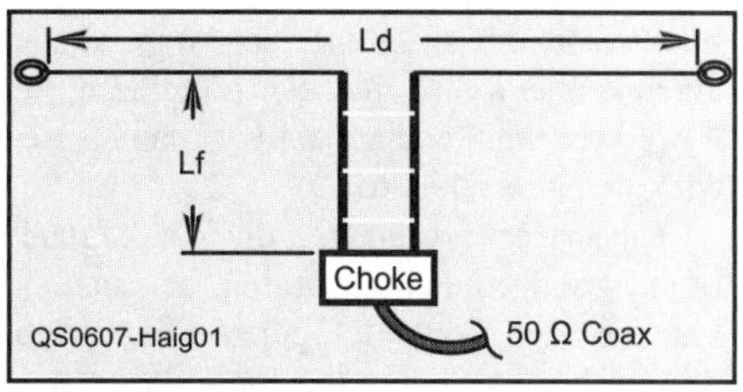

Figure 1 — The extended double Zepp.

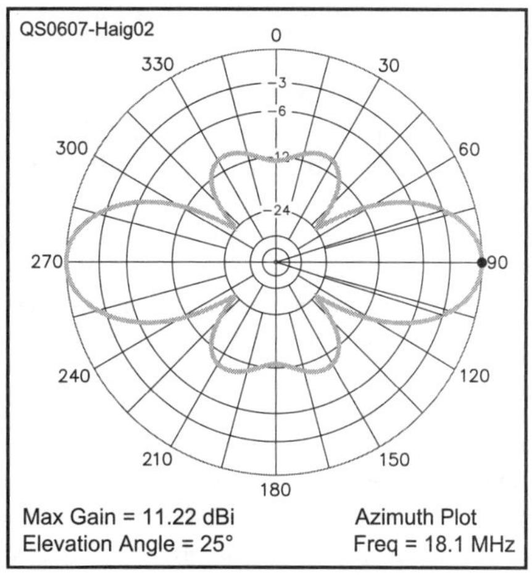

Figure 2 — Azimuth pattern of the 17 meter EDZ antenna shown in Table 1.

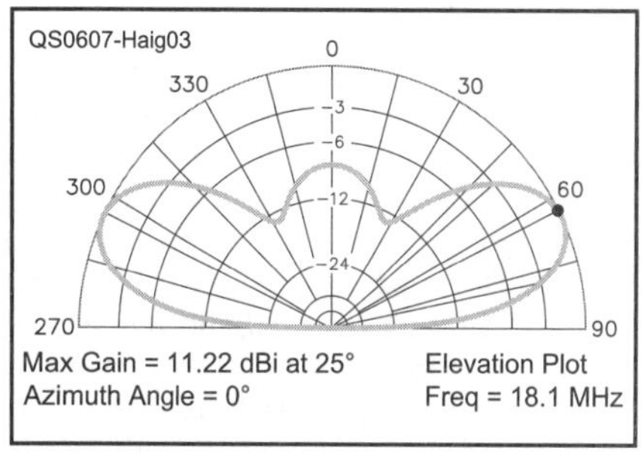

Figure 3 — Elevation pattern of the 17 meter EDZ antenna shown in Table 1.

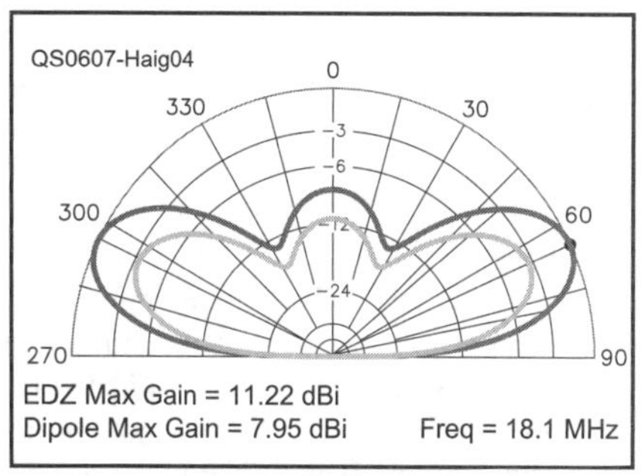

Figure 4 — Comparison of the 17 meter EDZ antenna shown in Table 1 (red) versus a ½ wave dipole at the same height (blue).

Table 1
EDZ Antenna Dimensions for Different Bands

Freq (MHz)	*Ld (ft)*	*Min Height (ft)*	*Antenna Z*	*Feed-point Z*	*Lf (ft)*	*SWR*
7.075	175.2	66	170.5 –*j*976.1	47.10 +*j*0.25	21.65	1.062:1
10.11	122.6	47	163.0 –*j*934.1	47.80 –*j*0.07	14.85	1.046:1
14.175	87.5	34	155.3 –*j*889.6	48.62 –*j*0.36	10.35	1.029:1
18.1	68.5	30	133.4 –*j*848.9	44.63 +*j*0.38	7.92	1.121:1
21.2	58.5	30	132.1 –*j*799.9	47.65 –*j*0.28	6.56	1.050:1
24.9	49.8	30	156.7 –*j*772.3	58.60 –*j*0.10	5.51	1.172:1
28.2	44	30	169.8 –*j*772.4	63.24 +*j*0.26	4.88	1.265:1

[1]Notes appear on page 116.

Bandwidth

The 2:1 SWR bandwidth of the EDZ is approximately 2.2% and varies with height above ground. This fractional bandwidth will cover the CW portion of 40 meters, all of 30 through 12 meters (separate antennas for each band) and about 600 kHz of the 10 meter band.

Multiband Operation

As designed, the EDZ is a single band antenna. The same pattern can be achieved by feeding the EDZ length dipole with low loss feed line all the way to an antenna tuner, either at the station or remotely operated in the antenna field. The antenna will load and work well on all bands at which it is ½ λ or longer, and be only slightly less efficient at the next lower band. The pattern will continue to be bidirectional for frequencies lower than the design frequency. At higher frequencies, as noted earlier, it will start to have multiple lobes, but still operate well into the covered areas.

Building the EDZ

The EDZ is built much like any other doublet. Choose a strong conductive wire that is not prone to stretching. I typically use 7/22 gauge hard drawn copper wire for home antennas and 22 gauge silver plated, Teflon covered, stranded copper wire for portable use. The wire is measured allowing a little extra for the ends. Cut the doublet wire into two equal pieces and attach insulators. The wire is pushed through the hole in the insulator and wrapped back around itself. The wrap should then be soldered. Unsoldered connections will lead to noise and may become intermittent. The center insulator should be about the same width as the balanced line. End insulators can be any size that will handle the power you plan to use.

Homebrewing 600 Ω Balanced Line

Balanced line can be homebrewed in several ways. First you need to determine the size of wire you want to use and then determine the distance apart the wires must be using the formula below. The characteristic impedance (Z_0) of a balanced line with air dielectric can be found from:

$$Z_0 = 276 \text{ Log } (2S/d)$$

where S is the center to center spacing between conductors, and d is the diameter of the wire used in the same units as the spacing.

Therefore, two 14 gauge (0.06408 inch diameter) wires spaced 4.8 inches apart will have a characteristic impedance of approximately 600 Ω. Larger diameter wire could be used for higher power and would have less loss but the spacing may be too great for the 10 meter band. The spacing should be held to less than 0.01 λ for the highest frequency used. Larger spacing may cause the feed line to start to radiate and look more like an antenna than a feed line. For portable use or low power, smaller diameter wire can be used without appreciable loss.

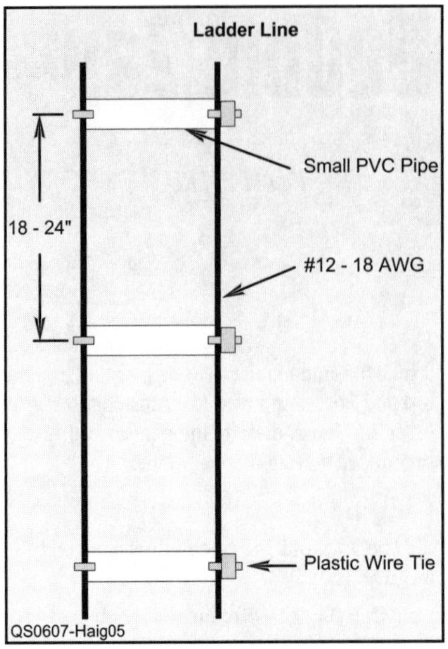

Figure 5 — Diagram of type 1 ladder line.

The spacers for the feed line can be made of any good insulating material. One method I have used is shown in Figure 5. The trick to this method is finding the long plastic wire ties. Another method uses plastic clothes hangers cut to the appropriate length and used as spacers. The spacers are drilled near the ends, and plastic wire ties are inserted in these holes and used to secure the wire to the end of the spacers. Alexander Krist, KR1ST, has documented this method of building ladder line on his Web page.[4]

Installation

Before installing this antenna you will need to decide what areas you want to cover and orient the antenna accordingly. For example, if you are in the Midwest and you are interested in working Europe or VK/ZL DX, the antenna should be oriented so that the ends point toward the NW/SE. Since this antenna has a bidirectional broadside pattern, both areas would be covered. Keep the balanced line away from metallic objects and off the ground.

Tuning

The matching section will need to be cut to correct length, or tuned, to achieve the desired match to the coax. The tuning process will require a 1:1 balun or choke, and preferably an antenna analyzer such as an MFJ-259B.[5] The balun can be a 1:1 voltage or current type or a choke. For 40 through 10 meters, I use eight turns of twisted pair wire wound through an FT140-61 ferrite core. This choke will isolate the balanced line from the coax shield and force equal currents in the balanced line conductors. Since the velocity factor can vary on homebrewed line by a few percent, the balanced feed line should be cut to be about 5% longer than the calculated value. The balanced line is connected to the center of the antenna, the choke is connected to the other end of the same line and the antenna analyzer should be connected as close as possible to the choke (see Figure 1).

Set the antenna analyzer to the frequency that the antenna was designed for. Measure the resistance and reactance. If the feed line reactance is positive (inductive), shorten the balanced feed line by 1 to 2 inches and measure again. If the feed line reactance is negative (capacitive), you will need to lengthen the balanced feed line. Continue adjusting the length of the balanced line to eliminate all reactance at the test frequency. Once the reactance has been eliminated, the SWR should be low and close to the values shown in Table 1. If you don't have access to an antenna analyzer, you can use a low power transmitter and SWR indicator to measure SWR. Lower the SWR by adjusting the length of the balanced feed line. When shortening the feed line, remove small amounts at a time to make sure you do not overshoot the correct length.

Performance

The EDZ is truly a high performance wire antenna. It has a clean predictable pattern with enough gain to help seek out weak signals. The bandwidth is comparable to a ½ λ dipole. It is easily matched using 600 Ω ladder line and makes a great DX antenna. What more would you want?

Notes

1. See LB Cebik's Web site. For the 88 foot model **www.cebik.com/88.html**; for the 44 foot model **www.cebik.com/aledz.html**.
2. The antennas shown here were analyzed with *EZNEC* version 3.0. *EZNEC* is available from Roy Lewallen, W7EL, at **www.eznec.com**.
3. The match was determined using *TLDetails*. *TLDetails* can be obtained from AC6LA, **www.qsl.net/ac6la/tldetails.html**. [*TLW*, Transmission Lines for *Windows*, provided with *The ARRL Antenna Book*, can also be used.—*Ed.*]
4. **www.kr1st.com/hfcoath.htm**.
5. MFJ Enterprises, Inc, PO Box 494, Mississippi State, MS 39762, 800-647-1800, **www.mfjenterprises.com**.

Jerry Haigwood was first licensed in 1962 as WN9NZA. He has held several calls including WN9BIU, WN9LOM, WB9LOM, WB7VIO, KY4Z and currently W5JH. He has had many interests in Amateur Radio including EME, DX, FM and repeaters, and satellites. His current interests include QRP, portable operation and paddle construction. He is a Life Member of the ARRL and a member of the Arizona ScQRPions QRP club. Jerry can be reached at 11402 N 98th Dr, Sun City, AZ 85351 or via **jerry@w5jh.net**.

The Horizontal Loop — An Effective Multipurpose Antenna

The horizontal loop need not be resonant and can work well in a number of ways

Scott M. Harwood Sr, K4VWK

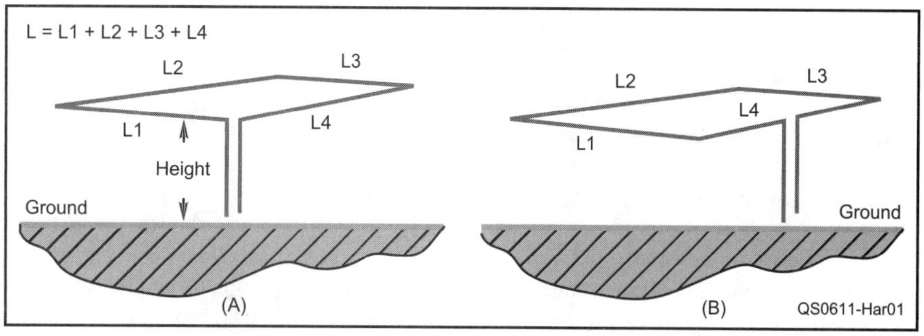

Figure 1 — At A, corner of angle-fed (AF) configuration showing key parameters. At B, center-fed (CF) configuration.

It's been well documented that a large horizontal loop will perform well as an amateur radio antenna. It may also be one of the most misunderstood of antennas. Many hams believe a loop must be resonant on the lowest operating frequency to work well at the design and higher frequencies. The fact is, as I will show later, a loop need not be resonant at all to perform well.

One purpose of this article is to demonstrate how to use computer modeling to perfect a loop for one's needs and location. This paper is not an antenna modeling tutorial. Programs such as *EZNEC* and *NEC Win-Plus* are relatively inexpensive and readily available.[1,2] Thus, I assume readers have a sufficient working knowledge of their respective modeling program to allow modeling the antenna and ground conditions and will be able to interpret the program outputs. Those wishing to learn more should look at the excellent ARRL Antenna Modeling Course, authored by L.B. Cebik, W4RNL, (L.B.) or at least visit his Web site.[3,4]

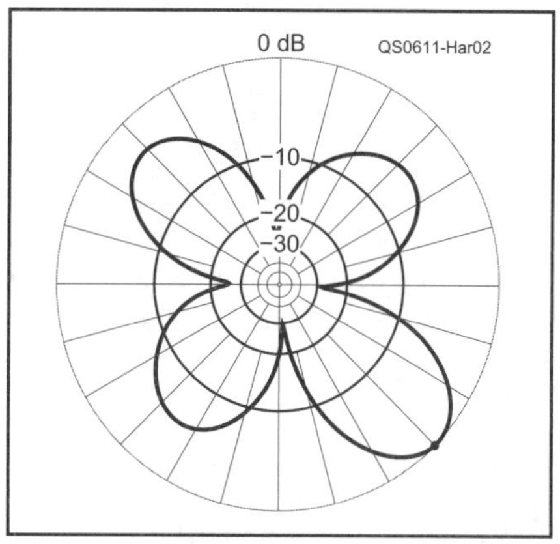

Figure 2 — Azimuth pattern of 300 foot loop on 14.2 MHz at elevation of 25°. Peak gain is 12.5 dBi.

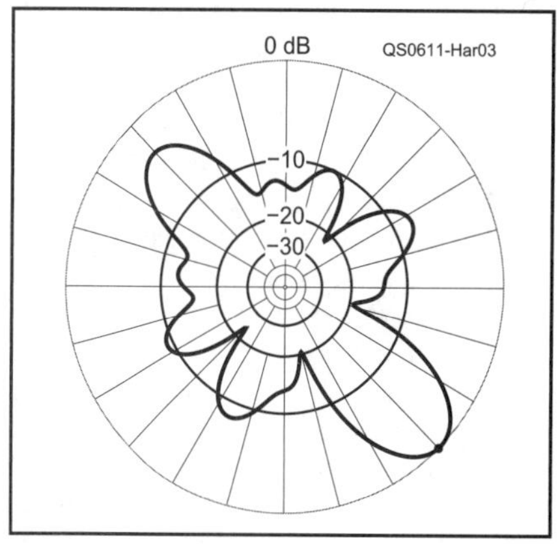

Figure 3 — Azimuth pattern of 320 foot loop on 14.2 MHz at elevation of 20°. Peak gain is 13.3 dBi.

The Misunderstood Loop

L.B. has written several excellent papers on loops that can also be found on his Web site.[5] These should be required reading for anyone contemplating building one. He points out two general misconceptions; the longer the loop the more the gain, and that the loop gives an omnidirectional pattern on all HF bands. The truth is that low angle gain is proportional to height — the higher the antenna, the higher the gain at low angles. Also, at higher than design frequencies, a loop is not omnidirectional. The loop radiation patterns and performance are affected by its shape and feed point location. The key parameters are shown in Figure 1.

[1]Notes appear on page 119.

Table 1
Loop Performance by Shape and Feed Point

	Circular Loop		*Square Loop AF*		*Triangle Loop AF*		*Square Loop CF*		*Triangle Loop CF*	
	Gain (dBi)	*Elev*	*Gain (dBi)*	*Elev*	*Gain (dBi)*	*Elev*	*Gain (dBi)*	*Elev*	*Gain (dBi)*	*Elev*
1.9 MHz	3.93	90°	3.46	90°	2.88	90°	3.70	90°	2.67	90°
3.9 MHz	8.47	90°	8.13	90°	7.76	90°	8.23	90°	8.04	90°
7.2 MHz	7.76	50°	7.35	50°	7.98	45°	7.64	45°	7.30	45°
10.1 MHz	8.34	35°	10.68	35°	7.35	30°	8.17	35°	7.24	30°
14.2 MHz	10.44	25°	12.50	25°	11.27	25°	10.25	25°	8.50	25°
18.1 MHz	11.18	20°	14.03	20°	12.32	20°	11.40	20°	8.84	20°
21.2 MHz	10.16	15°	14.55	15°	12.42	15°	11.28	15°	8.29	15°
24.9 MHz	10.77	15°	13.69	15°	14.09	15°	10.58	15°	10.10	15°
28.5 MHz	11.39	15°	12.85	10°	13.84	10°	11.19	10°	12.66	10°

Is This Antenna For You?

The first step in our planning is to determine if a loop is the right antenna for your needs. If you are interested in 160 and 80 meter DX and can only erect a horizontal loop 30 to 50 feet off the ground, you may be better off with other choices. At those frequencies, and heights below λ/4 to λ/2, most radiation will go skyward, making the loop ideal for near vertical incidence skywave (NVIS) propagation out to distances of a few hundred miles. This provides reliable coverage for emergency and other medium range communications, but not for DX. They do perform well as DX antennas on the higher frequencies. The loop makes a fine antenna for the ham who has space for only one antenna. In my case, I already had very good antennas for 160 and 80 meters, but I wanted something that would perform as a backup for these bands and give me good DX performance on the higher bands. I was particularly interested in 20 and 17 meters.

A good rule of thumb is to start with a loop about 5 λ at the highest desired DX operating frequency. Therefore, using the usual formula for a 1 λ loop, 1005/freq (MHz), a 1 λ loop for 18.1 MHz would have a circumference of approximately 56 feet, so our model loop should be about 280 feet in circumference. I then modeled a square loop 40 feet above real/high accuracy ground with dimensions of 70 feet per side.[6] After some experimentation, I found a leg length of 75 feet with the feed source in a corner of the loop would produce the clean cloverleaf pattern shown in Figure 2 on 20 meters, which was what I desired. Any longer leg length would give more gain, but would make the antenna more bi-directional as shown in Figure 3. Such a pattern may meet your needs. What about other shapes and feed points? Table 1 gives a comparison of these, showing circular, square and triangle loops, with the square and triangle loops being fed in the corners (AF) and midway along a leg (CF).

The circular and triangle loops developed a more omnidirectional pattern on the lower bands, and if that is what one desires, then more experimentation with these types would be beneficial. For my purposes, the apex or corner fed square loop was the best. It gives nice performance on 80 through 30 meters and really shines on 20 and 17 meters. Looking at the figures for 10 meters, one might think it to be an excellent omnidirectional DX antenna. A close look at the actual azimuth pattern (Figure 4) shows that it has just about as many sharp nulls as gain lobes. Also, these lobes are relatively narrow. A windy day or varying atmospheric conditions can cause signals to fade in and out of the nulls. We want these nodes to be as broad as possible with the fewest nulls, so a frequency range of around 5:1 is the best range for a loop. For the DX enthusiast, a loop with a resonant frequency just below 7 MHz should perform well for 40 through 10 meters.

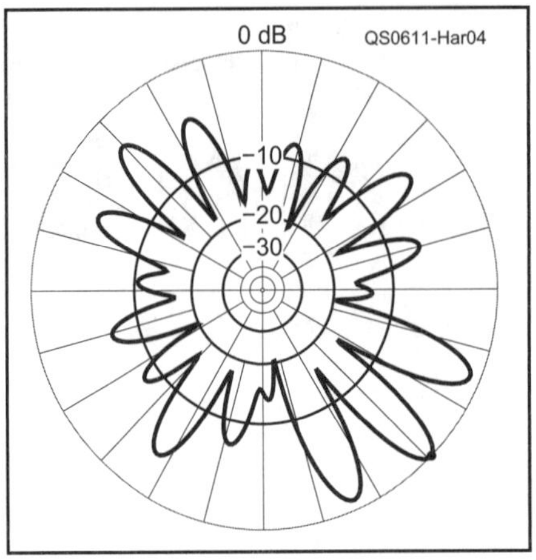

Figure 4 — Azimuth pattern of loop on 29 MHz at elevation of 10°. Peak gain is 12.6 dBi.

Optimizing the Antenna for Your Location

Now that we have a model of the kind of loop we want, the second step is to determine what size and shape loop we can erect. This means conducting a survey of one's antenna space. Take a long tape measure to the site and record distances to trees, towers, fences, other buildings, etc. Take this data and draw out a diagram of the maximum loop antenna possible. Once this is done, you can then attempt to fit your model into your actual location. If you have the supports in the right places, you are indeed fortunate. If not, don't be discouraged. The idea here is to utilize the modeling program, changing source point, shape, leg lengths, height (within limits of the overall yard dimensions), etc, until we have the best antenna for your location that will perform as well as possible on the bands you desire to operate.

Also play with adding loads (inductance and capacitance) at various points. For instance, adding an inductance may enhance performance on 160 meters, while adding a small capacitance may make it perform better on 80 meters. These components can also be utilized to "tame" the antenna on certain bands where matching is difficult. Nothing is sacred here, and playing with models certainly is easier than raising and lowering the loop many times!

One might ask why not erect a vertical loop, since this type of antenna only requires two supports and would seem to also generate good patterns on all bands? Well, as pointed out by W4RNL, loops have a strange behavior. At frequencies of twice design frequency and above, radiation is increasingly off the side of the antenna. Thus, on higher frequencies a vertical loop antenna would tend to radiate straight up and down! This is also why horizontal loops have lower radiation angle at higher frequencies.

Resonant vs Non-Resonant Loops

Now, let's deal with the issue of resonance. Table 2 compares performance of my loop against resonant 160 and 80 meter loops, as well as an inverted V. The non-resonant loop seems to be the better overall performer. One may ask, "What about SWR! Have you considered that? How are you going to feed this thing?" Well, first some basics. In his book, *Reflections*,[7] Maxwell states that all power fed into the transmission line (minus line loss) is absorbed by the load, regardless of the mismatch. Secondly, with open-wire tuned feed lines, we can ignore this mismatch at the junction of the feed line and the antenna, and all matching can be done at the transmitter itself.

One might think, "Isn't this bringing a high SWR (and problems) into the shack?" Well, Maxwell also tells us reflected power by itself is unimportant in determining how efficiently power is being delivered to an antenna. Put another way, if our antenna tuner can properly match the impedance of the input of the feed line, using open wire line, we can transfer just about all power to the antenna. Therefore, with a little planning, our loop can work on all amateur bands utilizing open line feeders and the proper tuner.

Table 2
Loop Performance vs Resonant Antennas

	160 Meter Loop		*My Loop*		*80 Meter Loop*		*Inverted V*	
	Gain (dBi)	*Elev*	*Gain (dBi)*	*Elev*	*Gain (dBi)*	*Elev*	*Gain (dBi)*	*Elev*
1.9 MHz	6.72	90°	3.46	90°	1.64	90°	2.09	90°
3.9 MHz	5.69	55°	8.13	90°	7.88	90°	5.03	90°
7.2 MHz	9.65	40°	7.35	50°	6.54	50°	4.23	45°
10.1 MHz	12.09	25°	10.68	35°	9.51	40°	5.18	25°
14.2 MHz	12.05	20°	12.50	25°	10.96	25°	6.30	20°
18.1 MHz	12.98	20°	14.03	20°	13.89	20°	7.07	15°
21.2 MHz	14.43	15°	14.55	15°	14.11	15°	6.97	35°
24.9 MHz	13.48	10°	13.69	15°	14.61	15°	7.18	10°
28.5 MHz	14.55	10°	12.85	10°	14.25	10°	7.07	10°

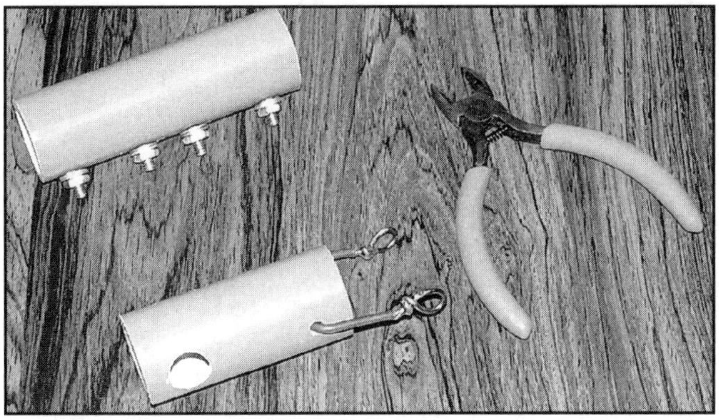

Figure 5 — PVC feed line and corner insulators.

Table 3
Antenna Z_0

	At Antenna	*At End 65 Ft. 450 Ω Line*
1.9 MHz	127.9 –*j*4282 Ω	2.029 –*j*310.1 Ω
3.9 MHz	230 +*j*691.1 Ω	71.72 –*j*160.5 Ω
7.2 MHz	188.3 +*j*475.3 Ω	286.2 +*j*619 Ω
10.1 MHz	313.3 +*j*0.03295 Ω	638.3 –*j*7.773 Ω
14.2 MHz	700.1 +*j*804.4 Ω	1620 +*j*424.2 Ω
18.1 MHz	2452 –*j*578.7 Ω	99.57 +*j*168.9 Ω
21.2 MHz	2280 +*j*609.2 Ω	737.4 –*j*1124 Ω
24.9 MHz	1579 –*j*1257 Ω	107.3 +*j*187.6 Ω
28.5 MHz	815.9 –*j*954.3 Ω	158.9 –*j*460.3 Ω

Tuner Considerations

We must consider the resistive and reactive components presented at the input of the feed line. Our tuner must be able to match these components of the line impedance (Z_0) to an unbalanced 50 Ω resistive load. Table 3 shows the impedances calculated by our modeling program for our loop at the antenna and at the end of 65 feet of 450 Ω transmission line. Here again, one can vary the line type, length, etc. to yield the best combinations for their location and needs. Using a 4:1 balun to a typical unbalanced tuner may work in some cases, but leaves a lot to be desired and will create a lot of problems. One needs a true balanced line tuner to match the antenna to the load properly.

It is not within the scope of this article to go into detail on the design and construction of such a tuner. *The ARRL Antenna Book*, *The ARRL Handbook*, and many other antenna journals have information on such a tuner. I personally prefer the tuner designed by AG6K using two variable inductors and a variable capacitor with a balun on the input side.[8] For those not willing to take the time and effort to design and build a tuner, there are several commercial tuners available.[9] The Palstar BT1500A is a commercial version of the A6GK tuner.[10] It is pricey but very well built, has an excellent metering system and is rated at 1500 W. MFJ has a balanced line tuner for a more attractive price that is rated at 300 W.[11] A surplus Johnson Matchbox should also work well, but likely won't provide a match on 30 meters.

Loop Construction and Erection

Once we have decided on the final design, the next step is construction and erection of our loop. Use your favorite method of putting up rope at the support points, but there should be some method of strain relief at each support such as pulley-counterweights or springs. Loops take a lot of stress from wind and swaying supports. Many years' experience have taught me that in most areas, large loops need to be put up one leg at a time. Therefore, I recommend carefully measuring and cutting each leg of the loop.

Feed line and corner insulators can be fabricated from short lengths of PVC pipe, as shown in Figure 5. These should be spray painted to protect them from the sun. My feed line insulator has short internal jumpers running from the center terminals to the outside terminals. The corner insulators should be connected to the antenna legs by brass or stainless steel hardware.

Lay out the legs and insulators on the ground in the manner they will be part of the antenna. Connect the open wire line and the two extending legs to the feed line insulator and raise this up part of the way. Connect the next leg to one corner insulator. Keep working around the loop until all legs are up and connected and up in the air.

Measure out the proper length of open wire line, and route it along its path to the tuner. Open wire feed line is affected by nearby objects, especially metal objects such as gutters and other wires, so make every attempt to keep it free and clear. It should also be secured as much as possible to eliminate flopping and swaying around. I found 2 inch screw-in porcelain electric fence insulators very handy for making runs along the wall of the house.

There are many ways to run open wire line into the shack. Replacing a window pane with Plexiglas and drilling holes in it is one method. Another way is to run two short pieces of RG-8 coax with the center conductor of each coax connected to one of the open wire feeders. By connecting the shields together a shielded balanced line of twice the coax Z_0 is formed. Note that this transmission line will have the loss associated with the mismatched coax, so make the length of such a section as short as possible.

Final Thoughts

So there you have it. All the information needed to create that new station antenna that will suit your needs and location. Please remember modeling software is a great aid in discovering new high performance antennas, but it is not the absolute gospel. It can point us in the right direction and save a great deal of time in our quest for the "perfect" antenna, but its results need to be tested and verified.

One last thing: L.B. Cebik has stated, "The advantage of the [horizontal loop] will not show itself in any one contact or in a short period. Satisfaction with the antenna grows with time and changes in the propagation paths, a successful communication with almost everywhere shows up in the log." I couldn't agree more. I have enjoyed my loop for some time, and have worked many countries and received excellent signal reports.

Notes

[1] *EZNEC* is available from **www.eznec.com**.
[2] *NEC-Win Plus* is available from **www.nittany-scientific.com/plus/index.htm**.
[3] Information on ARRL modeling course available at **www.arrl.org/cce/courses.html**.
[4] **www.cebik.com**.
[5] For starters read "Horizontally Oriented, Horizontally Polarized Large Wire Loop Antennas" and "Horizontal Loops: How Big? How High? What Shape?"
[6] ASCII files of antenna models used in this article are available from the author.
[7] M. Walter Maxwell, W2DU, *Reflections: Transmission Lines and Antennas*, (Out of print). The entire text of *Reflections II* is available at **www.w2du.com**.
[8] R. Measures, AG6K, "A Balanced Balanced Line Tuner," *QST*, Feb 1990 (updated at **www.somis.org/bbat.html**).
[9] J. Hallas, W1ZR, "Product Review: A New Generation of Balanced Antenna Tuners," *QST*, Sep 2004, pp 60-66.
[10] Available from **www.palstar.com**.
[11] MFJ Enterprises **www.mfjenterprises.com**.

Scott M. Harwood, K4VWK, has been interested in radio since childhood. In the seventh grade he built a two tube regenerative receiver using #30 tubes as a science project. He obtained his Novice license in 1958 and has retained the same call for over 40 years. He now holds an Amateur Extra class license. After college and a tour in the USAF, Scott returned to Virginia where he now resides. An avid antenna experimenter, his main area of interest has been small portable antennas for 160 and 80 meters. He has given talks at local radio clubs on antennas, and has written articles for CQ *and* AntenneX *magazines. You can reach Scott at PO Box 523, Farmville, VA 23901 or at* **scotth@hsc.edu**.

A Tunable Vertical for 75 and 80 Meters

An easy to build vertical for 80/75, with no strings (or radials) attached.

Rich Haendel, W3ACO

Many of us want to get on 75 and 80 meters, especially now that the sunspot cycle is near its minimum. It's time to work DX on the lower bands! Many of us do not have the room to put up a full size horizontal dipole λ/2 high or put up a λ/4 vertical with all its radials.

It takes a height of 130 feet to get to λ/2 and that is what it takes to start getting energy at the low angles best for working DX. So as many do, instead we build a vertical to get that low angle of radiation. That usually means a ground-mounted λ/4 wave vertical, or a shorter antenna with a loading via a coil or a capacitive *top hat*. The hitch is that to minimize losses many ground radials are usually required. How many? Perhaps I am lazy; I don't want to lay down 1000 feet of ground wire and mess up the yard. My wife and my lawn mower would probably become upset!

Are There Other Choices?

If we were to build such a vertical with 16 λ/4 radials on the ground, we get about +0.14 dBi gain and a 25° peak takeoff angle. Others have suggested that we should use raised verticals. That's great, so we build a λ/4 vertical with a top hat and put in 8 raised radials about 8 feet off the ground so we don't get strangled or have RF burns. For all our work we still get about –0.2 dBi peak gain at a 25° elevation angle. I don't know about you, but all the posts and overhead wires will also upset my wife. In addition, my backyard doesn't have the dimensions to place these radials in straight lines, either on the ground or raised up 8 feet. There must be another approach that gives good performance and that leaves the backyard mostly intact.

Design Evolution

Well, here is something a little bit different. Let's start with the concept of a center fed λ/2 vertical 128 feet long with the bottom 10 feet off the ground using bare #12 wire. The feed point impedance at the resonant frequency of 3.75 MHz will be about 90 Ω.[1]

[1]Notes appear on page 122.

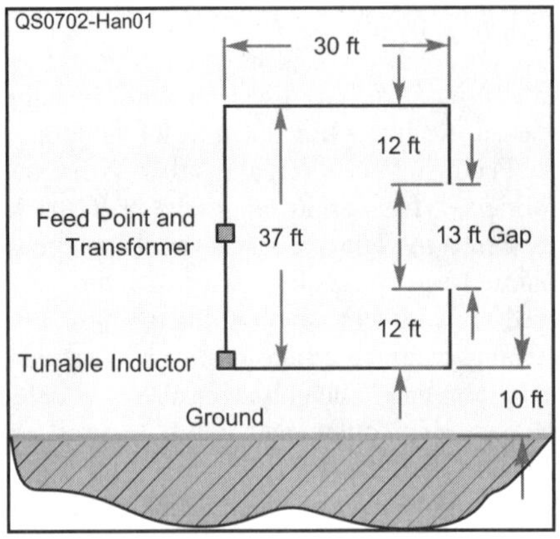

Figure 1 — Diagram of final vertical dipole configuration.

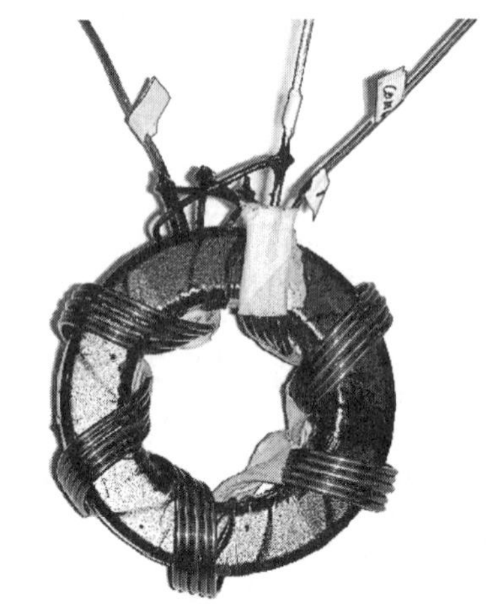

Figure 2 — Detailed view of the 32:50 Ω matching transformer.

Figure 3 — Close-up view of antenna wire attached to brass grommet.

This antenna will have a predicted peak gain of +0.23 dBi and a 16° takeoff angle.

Now let's shorten it quite a bit so that the vertical portion is 37 feet long with the bottom still 10 feet off the ground. I used a top height of 47 feet because that was the highest I could string a wire between my tower and a nearby tree. Now the impedance is highly capacitive because it's much shorter than λ/2 long and the resistance is lower. Next I extended both sides of the antenna with parallel horizontal segments so the antenna forms the letter C. I then adjusted the lengths of these "ends" to tune the antenna to resonance. The resulting horizontal parts are 46.5 feet long to tune the antenna to resonance at 3.75 MHz.

The impedance at resonance is now about 38 Ω and the 50 Ω, 2:1 SWR bandwidth is 125 kHz. For my operating style, especially with the newly expanded phone bands, that's too small a bandwidth to use without a tuner. The antenna provides a gain of –0.6 dBi at 28° elevation. Not bad except for the narrow bandwidth. We could stop here and have an antenna that is quite useful if we used an antenna tuner in the shack.

I wanted an antenna that could work anywhere between 3.5 and 4.0 MHz without using a separate antenna tuner and that had an SWR lower than 1.5:1 over the entire band. Here is what I came up with using a variable inductor and a matching transformer.

I first tried the configuration mentioned above that looks like the letter C, still using bare #12 wire, the vertical portion still 37 feet long with each of the horizontal wires now shortened to 41 feet long and no "vertical end tips." I placed the inductor at the lower end of the vertical segment. The antenna resistance at resonance varied from about 33.5 to 39.5 Ω when tuned from 3.5 to 4.0 MHz using my variable inductor that has a maximum inductance of 14 µH.

I changed the wire from 12 gauge bare copper to 13 gauge plastic covered, which is the wire I had on hand, and repeated the modeling.[2] Polyethylene covered 13 gauge stranded Copperweld wire is available from Davis or Wireman.[3,4] It's good stuff and doesn't stretch. With this configuration the antenna resistance varied from 33.7 to 39.7 Ω when tuned from 3.5 to 4.0 MHz. The value of inductance needed to resonate the antenna changed slightly because the dielectric coating electrically shortened the wire. The bandwidth for a 2:1 SWR averaged about 80 kHz. If you are willing to accept this resistance variation, or your internal tuner is up to the task, the configuration without "end tips" will work okay and you can call it done.

I decided that a 6 Ω resistance variation was a bit too much to match effectively. An impedance of 36 Ω also required that I build a transformer using many windings and I wanted to provide a better match, so I added two vertical "tips" 12 feet long and reduced the horizontal portion to 30 feet as shown in Figure 1. This reduced the predicted resistance variation and lowered the impedance further so the matching transformer would be easier to build.

I selected from a set of ratios determined by (N1/N2).[2] In the table below, N1 and N2 vary between 2 and 7. I didn't want to wind transformers with too many turns, especially when you wind the turns in parallel. Table 1 shows what the various ratios provided, N1 being at the antenna side of the transformer to match to 50 Ω.

I selected the 32 Ω impedance as the design point. The transformer design came from the excellent book by Jerry Sevick, W2FMI, *Transmission Line Transformers*.[5] I wound the transformer using six quintufilar turns with 14 and 16 gauge wire, as shown in Figures 7-8 (B) and 7-11 (lower left) in the book. The toroid I used was an FT-240-61, the same as in the W2FMI book. See Figure 2. W2FMI's measured data shows a mid frequency loss of 0.04 dB. I hope my loss is similar.

The antenna in Figure 1 has a modeled resistance that varies from 27.9 Ω at 3.5 MHz to 33.0 Ω at 4 MHz. Notice that the resistance variation is a bit lower than the earlier design. The matching transformer is at the center of the vertical element. I also used a 1:1 balun between the matching transformer and the feed line. The modeled results are shown in Table 2.

I measured the assembled antenna using my Palstar ZM-30. The ZM-30 has very stable frequency setting capability and step resolution down to 1 kHz which makes measurements easy. Correlation between modeling and measurement was excellent. The modeled 2:1 VSWR bandwidth and the measured values for resonance vs. inductance were very close.

The measurements shown in Table 3 were taken through the 32:50 Ω un-un, a 1:1 balun and 65 feet of RG-8X coaxial cable.

I do not have the capability to measure actual gain. *EZNEC* indicates that the peak gain at 3.5 MHz is –0.54 dBi and at 4.0 MHz –0.52 dBi, both with a takeoff angle at 27°. Note that unlike a true λ/4 wavelength vertical, there is no null at zenith. Gain at zenith varies from –8 dBi to –6 dBi. Not having a null at zenith is not necessarily bad. It improves QSOs at shorter distances. Peak gain is within 0.5 dB of an elevated full quarter wave vertical with 8 raised radials. Not bad! I ran an azimuth pattern at 27° elevation to check for circularity. Gain variation is less than 0.5 dB going around the antenna, which seems reasonable.

Figure 4 — Interior details of remotely tuned loading inductor.

Construction

I used ¼ inch brass tent grommets at every corner and at the ends of the wires. Brass grommet kits are available at hardware and outdoor stores. Try not to buy brass plated steel grommets, they rust rather quickly as I found in earlier experiments. For each corner and end, I stripped the insulation for about 2 inches, made either a 270 or 360° turn around a ¼ inch drill bit, then crimped the grommets and soldered them to the wires. I covered the grommet and exposed solder joint with liquid tape. I have found that an uncovered solder joint tends to corrode over time. The grommets make for an easy connection to the tuning unit. The rounded edges of the grommet also prevent chafing of the Dacron cord. See Figure 3.

The 13 foot gap between the two end "tips" shown in Figure 1 is kept constant by using

Table 1
Transformer Turns Combinations, Impedance Ratios and Associated Antenna Design Impedance

N1	*N2*	*Ratio*	*Antenna Z_o (Ω)*
2	3	0.66	22.18
3	4	0.75	28.13
4	5	0.80	32.00
5	6	0.833	34.72
6	7	0.857	36.73

Table 3
Measured Data

Center Frequency (F_o)	*2:1 SWR Bandwidth (kHz)*	*VSWR at F_o*
3.5	52	1.2
3.6	52	1.2
3.7	55	1.2
3.8	59	1.2
3.9	72	1.1
4.0	86	1.1

Table 2
Modeled Data with 32 Ω Matching Transformer

Center Frequency (F_o)	*Inductance (μH)*	*2:1 SWR Bandwidth (kHz)*	*SWR at F_o*
3.5	11.5	70	1.10
3.6	9.1	70	1.06
3.7	7.0	80	1.11
3.8	4.8	90	1.11
3.9	2.5	90	1.13
4.0	0.5	100	1.16

heavy duty monofilament fishing line (I used 100 pound test line). If you are not familiar with tying knots in monofilament line, look up suitable knots in a fishing book.

I placed the variable rotary inductor at the lower end of the vertical portion. The rotary inductor I used is about 2 inches in diameter and 5 inches long, wound on a ceramic form. It has a silver plated roller wheel that contacts the inductor. Inductor wire size appears to be 12 gauge. The wire and the roller are silver plated. The inductor has a range from 0.5 to 14 μH and is turned by a low speed reversible geared down 120 V ac motor running at 30 rev/m (Herbach and Rademan TM 95MTR2803). A dc motor with similar gearing could also be used. The inductor and motor are mounted to a 3⁄16 inch thick piece of Plexiglas. The inductor is connected to the motor by an O-ring belt and two 2 inch plastic pulleys with metal hubs. The O-ring belt is a replacement part for an under-the-counter kitchen water filter. It is available at most hardware stores and home centers. The O-ring and plastic pulleys conveniently isolate the motor from the RF on the inductor. See Figure 4.

The motor and inductor are housed in a weatherproof plastic electrical box with dimensions about 8 × 8 × 4 inches. These are available at most home centers and electrical supply places for less than $20. I have the housing mounted on a 4 × 4 inch post 10 feet high. The motor control wires run down the pole and into the shack. I can tune the entire range and tweak for minimum SWR in less than 90 seconds. See Figure 5.

The antenna wires are strung between a tree and my tower using Dacron cord. The lower end that is opposite the tuning box is held in place by another Dacron cord tied to the far tree. The transformer and balun are housed in a 4 × 4 × 3 inch weatherproof plastic box. I ran the feed line horizontally from the feed point for about 40 feet before dropping it to ground level to run to the shack.

Figure 5 — Tuning unit mounted on post and ready for operation.

So How's it Work?

The antenna tunes nicely from end to end. My end points are 3.4 MHz and 4.06 MHz. I can hit the minimum SWR point easily. First I set the forward power of my transceiver to about 50 W in tune mode and tune the antenna for minimum reverse power with my in-line SWR bridge. Then I raise the power and change modes as needed. I can do CW DX near 3500 kHz and also some ragchewing at 3900 kHz. I don't need a separate antenna tuner in the shack, I don't have a mess of wires in the backyard and my wife is happy! Maybe you can use this antenna in your backyard.

Many thanks to Rudy Severns, N6LF, for his review and comments.

Notes

[1]J. Devoldere, ON4UN, *Low-Band DXing*, 4th Edition, Chapter 8, Figure 8-28. Available from your local ARRL dealer, or from the ARRL Bookstore, ARRL order no. 9140. Telephone 860-594-0355, or toll-free in the US 888-277-5289, fax 860-594-0303; **www.arrl.org/shop/**; **pubsales@arrl.org**.

[2]All modeling was done using *EZNEC 4.0*, by Roy Lewallen, W7EL, **www.eznec,com**. For all my modeling I used ground parameters of 0.005 S/m conductivity, a dielectric constant of 13 and I included wire loss for copper. For 13 gauge polyethylene covered copper, I used a dielectric constant of 2.25 and thickness of 0.02. I did not include any loss due to the variable inductor.

[3]**www.davisrf.com/**.

[4]**thewireman.com/index.shtml**.

[5]J. Sevick, W2FMI, *Transmission Line Transformers*, 4th Edition, Chapter 7. Available from your local ARRL dealer, or from the ARRL Bookstore, ARRL order no. TLT4. Telephone 860-594-0355, or toll-free in the US 888-277-5289, fax 860-594-0303; **www.arrl.org/shop/**; **pubsales@arrl.org**.

Rich Haendel, W3ACO, holds BEE and MSEE degrees. He has been involved in the development of advanced and anti-jam military radio systems for many years while working at Rockwell Collins. He also designed antennas for military aircraft. Rich served as Senior Technologist in the Communications Section of the Advanced Technology Center and holds nine patents in the field of communications. He was an instructor at Rockwell Collins for more than 20 years, teaching classes in RF engineering.

Rich first got involved in Amateur Radio by building battery operated tube receivers at the age of nine. He has been licensed on and off since 1957 when he was K2YHL. He received his current call in 1963. Now that he is retired, he still visits Rockwell Collins as a consultant. Rich enjoys working DX and building antennas. You may contact Rich at 402 McLean St, Iowa City, IA 52246 or at **rhaendel1030@inabc.net**.

An All-Band Attic Antenna

Nothing beats an "aluminum cloud" on a tall tower, but when choices are limited, an indoor antenna can be amazingly effective.

Kai Siwiak, KE4PT

If zoning rules or aesthetic considerations make outside antennas prohibitive, an indoor antenna might just provide enough performance for casual operation on all bands ranging from 80 to 6 meters. With only 100 W, I've managed contacts with hams in over 130 DXCC countries and all 50 states, including 25 states and 13 countries confirmed on 6 meters. I've even made contacts on the 160 meter band, but the efficiency there is poor, and I don't recommend operating on 160 meters with this antenna.

This isn't a design article for a specific indoor antenna, but rather a description of the performance of one indoor antenna, and some guidelines that might help you understand the limitations, performance and RF safety aspects of antennas of this type.

The KE4PT Attic Antenna

I've been operating with an indoor inverted-L antenna in my attic for several years. I drive the antenna with an ICOM IC-706MKIIG running 100 W through a current balun and an ICOM AH-4 automatic antenna tuner located at the antenna feed point.[1] The tuner is what makes this antenna capable of operation on all of these bands. The basic idea was to place as much of the wire as possible into the clearest space of a cluttered attic in my one floor home in south Florida, as seen in Figure 1.

Design Approach

My basic design approach was to provide a fat radiating conductor using two parallel conductors of fairly thick wire spaced nearly 3 feet apart. This tends to smooth out impedance variations versus frequency and lets the automatic tuner do its work smoothly. I chose aluminum wire for its availability and cost. These wires were placed as far as physically possible from other conducting objects in the attic.

[1]Notes appear on page 127.

I wanted an antenna that could be operated over the widest frequency range possible; so the L length was chosen as long as possible to obtain reasonable efficiency at the lowest frequencies, but with tolerable antenna pattern ripples at the highest frequencies. The actual length was constrained by the available attic space. Next, I used an antenna ground post to act as a counterpoise element. The whole system was match tuned at the feed point with an automatic tuner.

It All Came Together

The horizontal part of the L element comprises two parallel lengths of 9 gauge aluminum wire shorted at the far end, and spaced about 38 inches, as shown in Figure 2. The horizontal portion is about 48 feet long, and a bit more than 14 feet above the ground, under the roof of the house. The horizontal length is approximately a wavelength at 21 MHz so the antenna pattern is very nearly omnidirectional from the 20 down to 80 meters.

The parallel wires are brought together and emerge from the ceiling on a far wall of the house in a storage closet, as shown in Figure 3. Both of the parallel wires are joined together and connected to the antenna post of the AH-4 tuner. A copper ground wire runs from the tuner ground connection to an outside 8 foot deep ground rod. The antenna shares this ground rod with a conductive mast supporting a 2 meter J-pole that tops out at 21.7 feet. This mast also functions as part of the HF radiating system. A length of 50 Ω coaxial cable connects the tuner through an eight turn 5 inch diameter choke balun to the transceiver at the operating position in the ham shack on the other side of the wall of the storage closet.

We're Not Alone

The antenna isn't alone in the attic. There are air conditioning ducts as well as the ac mains power distribution for the house, which are marked by the heavy dashed line in Figure 1. This conduit can also be seen in Figure 4. I modeled the inverted L, the ground post with the attached J-pole mast, and the ac mains including its own ground post by using *4nec2* antenna modeling software.[2] The pro-

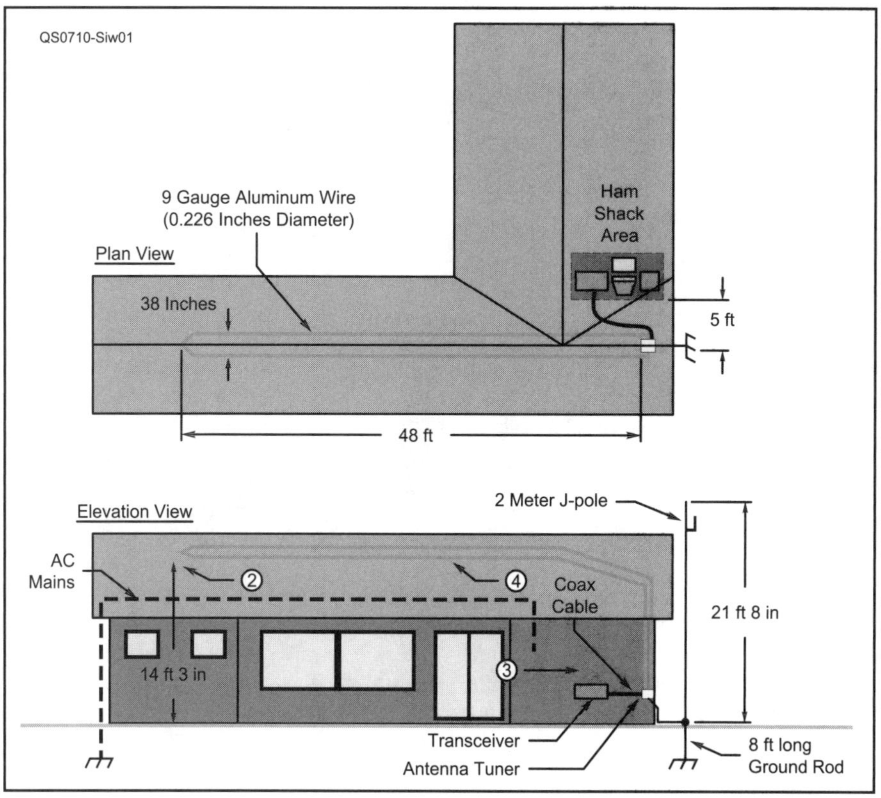

Figure 1 — Plan and elevation views of the attic inverted L antenna. The numbers are keys to the photographic views.

Figure 2 — The inverted L element is supported by egg insulators suspended from the rafters in the attic.

Figure 4 — The ac conduit parallels the antenna elements.

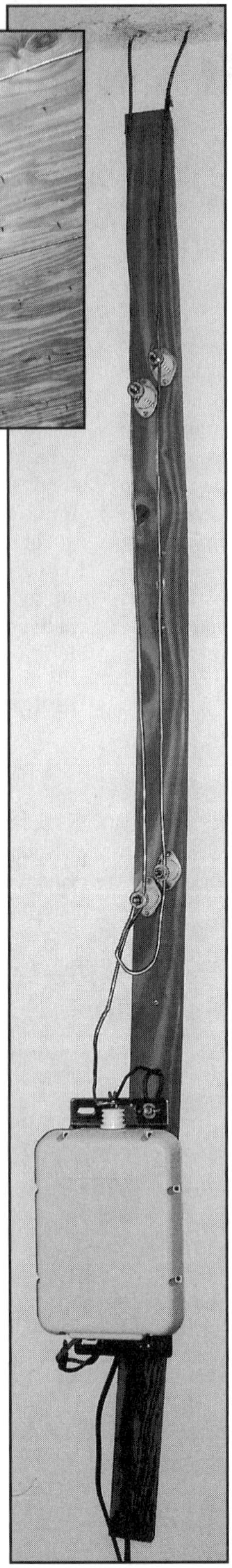

Figure 3 — The ham shack and transceiver are 6 feet to the left of the AH-4 tuner shown in the lower part of the photo.

The Advantage of CW

With a maximum of 100 W from my transmitter, my CW average power is 40 W, but with SSB it is an average of only 20 W (3 dB advantage for CW). At the receiver end of the propagation link the CW receiver noise bandwidth is typically 300 Hz compared with 2700 Hz for SSB (9.5 dB more for CW). Finally, the CW operators appear to listen a bit more intently to my CW, especially many of the DXpedition operators, perhaps tolerating 6 dB SNR whereas comfortable SSB listening needs 10 dB SNR (another 4 dB advantage for CW). The net advantage of CW over SSB might be as much as 16.5 dB or the equivalent of about 3 S-units!

gram uses the *NEC2* (Numerical Electromagnetic Code) calculation engine, and is available for free. I started the *NEC* based analysis to make sure that I could adequately assess the RF exposure that results from this unusual antenna. As a by-product of that effort I was able to learn about the antenna patterns and radiation efficiency of this antenna.

The *NEC* modeling revealed that substantial RF currents exist on the J-pole mast, and they contribute significantly to the overall radiation from the antenna. The J-pole mast contributes a vertical polarization component to the overall radiation pattern that helps keep the far field patterns relatively omnidirectional, especially at the higher frequencies. This was an unplanned benefit, with the interesting lesson that all conductors radiate, even when connected across an earth ground connection!

Indoor Antenna Performance

What can be better that an extended on-the-air test? I kept track of many of my contacts (QSOs) by plotting them using *DxAtlas* by Alex Shovkoplyas, VE3NEA.[3] The results can be seen by the color-coded points on the maps in Figures 5, 6 and 7. The QSOs are between my south Florida location and the mapped points. Figure 5 shows QSOs in the 160 meter band (a few dark red marks), and in the 80, 60, 40 and 30 meter bands. The very close distances are covered well, although there were contacts as far away as Australia, South Africa and India and into Europe on 40 and 30 meters. The green points shown in Figure 6 show 20 meter coverage. A distinct skip zone occurs around my location. Coverage beyond that is worldwide. Figure 7 shows dark blue points for 17 and 15 meter band QSOs, light blue points for 12 and 10 meter QSOs and distinctive gold points marking the 6 meter "magic band" QSOs with 14 states, Puerto Rico and Spain. The 17 through 10 meter coverage has a prominent skip zone of about 1000 km around my location.

Propagation predictions using *HAM-CAP* freeware basically confirm the actual performance of the antenna over the long term, including the skip zones and the tendency for the 17 through 10 meter bands to dominate coverage into South America.[5]

I get my fair share of the rare and distant ones (VU7, 9Q1, VK9N, 1AØ, VQ9 and KH8/S on all five bands), not because of awesome RF power — I use just 100 W — or a high "aluminum cloud" antenna, but by listening and operating carefully, although those rare DX catches are as much a testament to the operating skills of the DXpedition operators! I also take advantage of the extra 2 to 3 S-units of signal-to-noise enhancement that operating on CW provides.

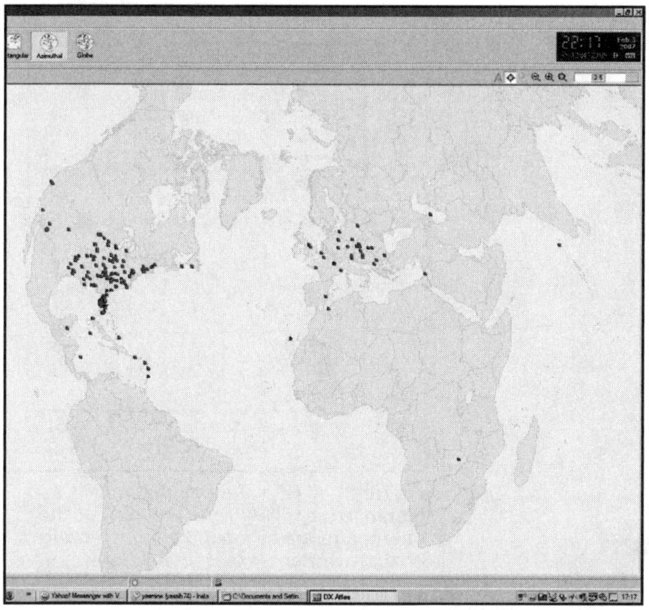

Figure 5 — Some contacts from KE4PT in South Florida. Dark red 1.8 and 3.6 MHz, red 5.4, 7 and 10.1 MHz.

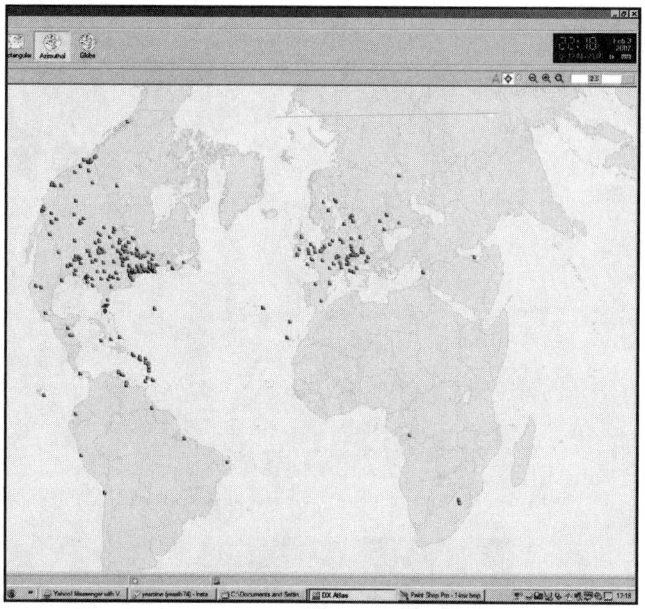

Figure 6 — Contacts from KE4PT in the 14 MHz band.

What *NEC* says

The proof is in the antenna currents. I modeled the antenna and ground system, along with the ac mains and the J-pole mast using *4nec2,* which employs the *NEC2* calculation engine. The basic radiation pattern in the lower bands is a flattened cardioid pattern pointing upward as seen in the 10.1 MHz pattern shown in Figure 8. The antenna pattern in the figure is centered on the feed point of the antenna. To a receiver anywhere on the horizon, the pattern at 10° elevation varies less than an S-unit for 20 meters and longer wavelengths.

As frequency is increased the antenna pattern on the horizon develops more ripples, and the 2 meter J-pole mast contributes more vertically polarized energy to fill in horizontally polarized pattern dips. At 51 MHz, shown in Figure 9, multiple antenna pattern lobes are evident, but coverage again is still roughly omnidirectional in azimuth — effectively within 2 S-units. The total antenna system polarization is randomly elliptical, having both vertical and horizontal components. This can be seen in Figure 10 in which the polarization axial ratio is shown in color on the 51 MHz pattern. Blue indicates a linear polarization, and a trip through the color spectrum shows elliptical polarization culminating in circular polarization in the directions corresponding to the purple colored pattern areas. Informal S-meter tests by a local ham within 20 miles of my location verified that there was substantial energy in both the horizontal and vertical polarization components during an impromptu 6 meter band test.

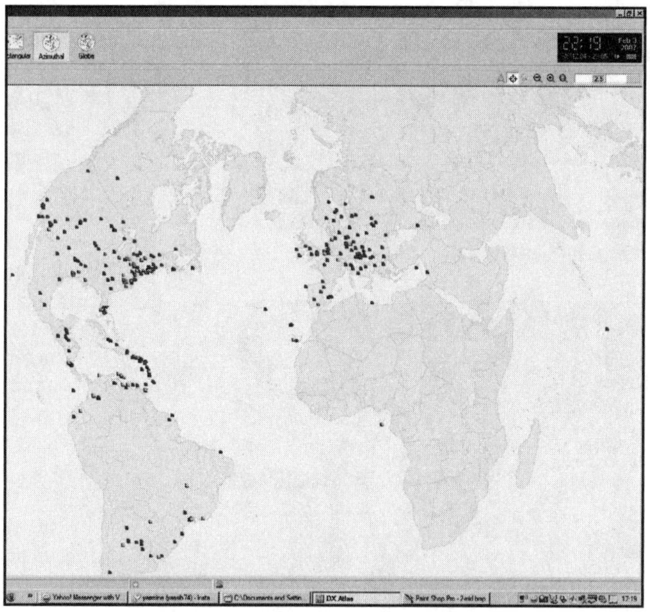

Figure 7 — Higher frequency contacts. Dark blue 18 and 21 MHz, light blue 24 and 28 MHz, gold 50.1 MHz.

The antenna radiation efficiency can be defined as the total power radiated into space (that is, above the ground) divided by the transmitter power. Efficiency, predicted by *NEC2,* is between about –7 dB and –1 dB across 3.5 to 54 MHz. Efficiency dips to a dismal –20 dB in the 160 meter band, so the antenna is not too useful there, although I've made a few contacts in several states on that band. I tried modifications of the design by modeling various ground radials attached to the ground post. Efficiency was not significantly improved, however. At my location, the ground radials could be physically placed only perpendicular to the horizontal element, but that configuration produced undesirable deep nulls in the azimuth pattern.

No attempt was made to predict the antenna input impedances because (1) I always intended to use an automatic tuner, and (2) there were simply too many non-modeled coupling effects in the attic, including a substantial barrel-tile roof which sits inches above the antenna wires. As expected, the *NEC* analysis revealed that the currents on the antenna wires, the ac mains wires, and the J-pole structure are indeed standing waves starting with a null at the open end of each wire. This motivated a relatively simple RF exposure analysis described next.

Some Words about RF Safety and RFI

Indoor antennas should be very carefully considered from the RF safety point of view, especially for those within the dwelling. In addition, there is always potential for RF interference within the home. This applies as much to wire antennas as it does to small loops. Two possible hazards exist: the potentially high RF voltage that can exist on the antenna conductors and exposure to electric (E) and magnetic (H) fields. Both potential hazards are avoided by keeping one's distance!

But How Close is Safe?

I initially evaluated this antenna by real-

FIGURES 8, 9 AND 10 GENERATED BY *4NEC2* (HOME.ICT.NL/~ARIVOORS/)

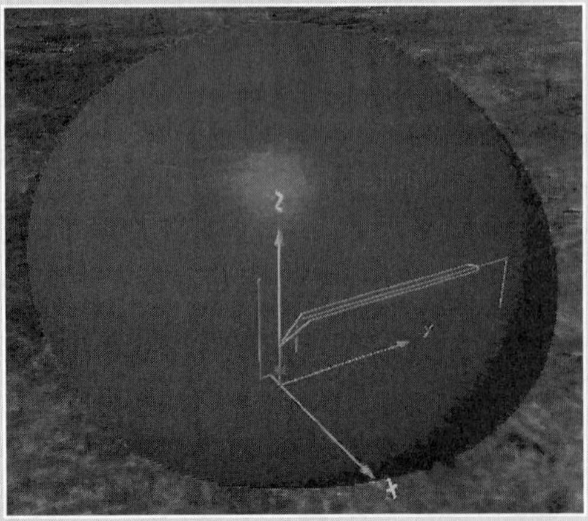

Figure 8 — The antenna and its 10.1 MHz pattern.

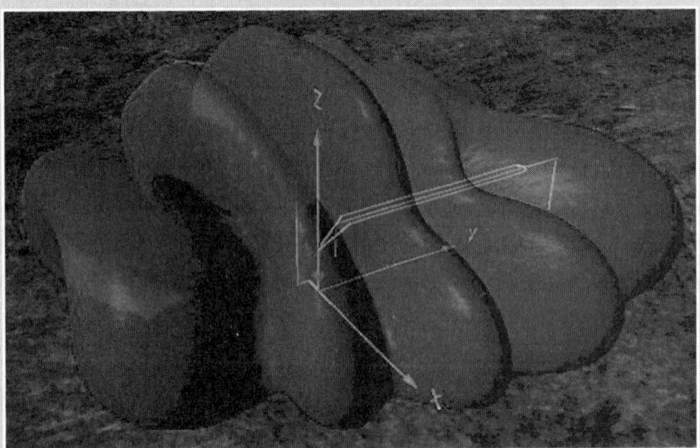

Figure 9 — The antenna and its pattern at 51 MHz.

Figure 10 — Polarization axial ratio at 51 MHz.

izing that at any particular frequency the currents and voltages along the conductors would be standing waves — just as on a dipole. As verified by *NEC* modeling, the electric charge accumulations are "out of phase" with the currents on the wire. Thus, where the standing wave current in the wire (and hence the H fields around the wire) goes through a null the voltages (and hence the E fields) peak, and vice versa.

To estimate the RF exposure, I then assumed that at any point on the antenna wire the currents would resemble those of a resonant dipole with the full power applied at that point. Then I tried the very simple-to-use University of Texas online calculator to determine a compliance distance.[6] I used the "occupational/controlled environment" since I have full control of access to the RF exposed areas.

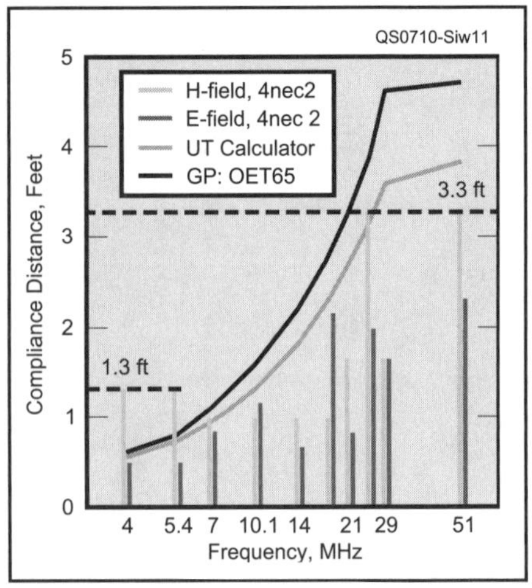

Figure 11 — RF safety compliance distance (at 100 W) from any part of the antenna ranges from 3.3 feet on 6 meters down to under 1.4 feet on 80 meters.

Crunching the Numbers

The calculator needs very few inputs — 100 W RF power, an assumed 2.2 dBi gain of a dipole and a choice of whether there is a ground contribution (I selected no ground contribution). The largest compliance distance of 3.8 feet from any part of the antenna occurs in the 6 meter band, and that was initially the number I used for the compliance distance *at all frequencies*.

As an added check, I also used the "General Purpose Tables" in *RF Exposure and You*, with 100 W, 0 dBi and a controlled environment.[7] Finally, I calculated the near E and H fields using *4nec2* with a "real ground" and 100 W. The composite results are shown in Figure 11 in the form of compliance distance versus frequency. The University of Texas calculator and the general purpose tables give similar results. They appear to be adequate above 7 MHz, but they underestimate the exposure compliance distance at lower frequencies for this particular antenna.

A Second Opinion

The *4nec2* calculations predict compliance distances that are up to a factor of two larger than those obtained with the other approximations for frequencies below 7 MHz. The magnetic fields near the ground on the vertical portion of the antenna are the source of this discrepancy.

The *NEC2* engine in *4nec2* does allow connections of wires to ground, but does not model wires underground such as the ground post that's part of this antenna. It also does not correctly handle the charge distributions at the wire-ground interface, except for the case of a perfect ground. Bill Guy, W7PO, kindly helped me by using *NEC3*, which correctly accounts for buried wires, to check my antenna including the actual buried ground posts.[4] *NEC3* is still under export restrictions and is not generally available.

And the Answer Is —

A spot check of near fields of my antenna both near the vertical and near the horizontal parts of the wires shows that away from the ground connection *4nec2* (using the *NEC2* engine) and *NEC3* predict the same relative field strengths, typically within 10%, for a given radiated power level. Near the ground post connection, however, ground-level field values are similar only if a perfect ground is selected in *4nec2*. For practical RF exposure evaluations, especially for unusual antennas such as this one, modeling should be tried with both a real ground and with a perfect ground, then the most conservative compliance distance should be used.

For this antenna, the 6 meter band *4nec2* result of 3.3 feet gives sufficient compliance distance safety margin on all lower frequency bands. *Lesson: Evaluate unusual antennas very carefully, especially if a ground or ground post is part of the system!*

RFI Rears its Head

Because this indoor antenna is extremely close to wiring in the house, RF interference within the home is a strong possibility. I've noticed coupling RF energy to the ADSL, computer connection, side of my phone line, but only during 160 meter operation. That RFI potential and the generally poor antenna efficiency keep me off the 160 meter band. There is also noticeable coupling to my TV and audio systems, which is remedied by restricting operating during prime family TV viewing times.

Conclusions

An indoor antenna such as this one is not the contester's dream antenna, nor is it a DX hunter's "special," but it can be a useful and effective "stealth antenna" that will get you on the air on all ham frequencies between 3.5 and 54 MHz. Careful operating practices and the use of narrow-band modes, such as CW and digital modes, can yield delightful results. The use of an antenna automatic matching tuner at the feed point allows great flexibility in positioning attic wires, and in my case, allowed for an effective all-band design. Modeling with *NEC* provides great insight into the performance. Finally, great care must be taken in the RF safety analysis.

I'd like to acknowledge and thank Bill Guy, W7PO, for his help with the *NEC3* modeling, and Bob McGraw, K4TAX; Bob Walker, N4CU; Tom Kneisel, K4GFG, and Diana Siwiak, KE4QXL, for their helpful reviews and suggestions.

Notes

[1]R. D. Straw, Editor, *The ARRL Antenna Book,* 21st Edition, pp 26-21. Available from your ARRL dealer or the ARRL Bookstore, ARRL order no. 9876. Telephone 860-594-0355, or toll-free in the US 888-277-5289; **www.arrl.org/shop/**; **pubsales@arrl.org**.

[2]A. Voors, "NEC Based Antenna Modeler and Optimizer," **home.ict.nl/~arivoors/**.

[3]*DX Atlas*, **www.dxatlas.com/**.

[4]K. Siwiak and W. Guy are both members of the ARRL RF Safety Committee, **www.arrl.org/rfsafety/**.

[5]*HAM-CAP*, **www.dxatlas.com/HamCap/**.

[6]Online RF exposure evaluation, **n5xu.ece.utexas.edu/rfsafety/**.

[7]E. Hare, *RF Exposure and You*, ARRL, Newington, CT, 1998.

Kai Siwiak, PhD, KE4PT, was first licensed in 1964. He is a consulting engineer specializing in antennas, propagation, communications systems and ultra-wideband (UWB) wireless technology. Kai has authored several text books on those subjects and also wrote the electromagnetic theory chapter in ARRL's RF Exposure and You. *He holds more than 30 US patents, has been a frequent contributor to IEEE 802 standards, and was an advisor to the US delegation to the ITU-R on UWB technology. He is a member of ARRL, SAREX (Space Amateur Radio Experiment), a life member of AMSAT and a member of the ARRL RF Safety Committee. Kai prefers CW, usually on 40 through 6 meters, depending on those elusive sunspots. He can be reached at* **ke4pt@amsat.org** *for any questions or comments.*

Balloon-Lifted Full-Wave Loop Antennas

Nothing beats a full sized antenna, way up in the air, for launching your signal. Here's how to build and deploy a balloon-lifted, high performance full-wave loop antenna for your next Field Day or contest operation.

Jim DeLoach, WUØI

FRED MARTIN, AE6CW

Figure 1 — Marc, W6ZZZ (left) and author Jim, WUØI (right) inflating a 6 foot balloon during Field Day 2006 in the Rancho San Antonio Open Space Preserve near Los Altos Hills, California.

Temporary operating events such as ARRL Field Day are a great opportunity to experiment with really interesting antennas that just wouldn't fit at home. My club has been flying balloon-lifted antennas since the mid 1990s for Field Day and other events.[1] Last Field Day we lifted full wave loops for 20 and 80 meters and both performed superbly.

In this article I'll show you how to build your own balloon-lifted full-wave loop antenna. I'll explain why full-wave loops work so well with balloons, discuss the unique operational details of flying balloon-lifted antennas, share performance impressions, cover safety procedures, review FAA regulations and provide links to additional information.

Why Full-Wave Loops?

Hams have been using balloons to lift vertical monopole antennas poised against a ground plane for years.[2, 3] There are some key advantages to using a full-wave loop, however, instead of a ground plane vertical.

- Loops are balanced and don't require a ground plane. For a temporary deployment this is a truly big deal, since ground radials are a pain to deploy and are easy to trip over.
- Loops can be horizontally polarized so they have less noise — a key advantage on the lower HF bands.
- Loops perform well even when relatively close to the ground.
- Loops are broadband and are easy to tune. They are far more likely to stay in tune as the antenna blows around in the wind.
- A loop's radiation pattern is quite broad. Thus the radiation pattern remains consistent as the antenna shape, altitude and orientation shifts in the wind.
- Loops fed with ladder line can be tuned for other bands and typically perform well on these bands, too.

So loops are great, but isn't it difficult to keep a diamond shape when lifting a full-wave loop with a balloon? It turns out that it's not that difficult to do.

Construction

The Loop

A full-wave loop is just one long piece of wire, shaped in some form of convex polygon, fed where the two ends come together. While theory says that a circular shape is best, loops still perform well with rectangular or elongated shapes. The balloon-lifted loop has a roughly diamond shape, as shown in Figure 2. One large balloon lifts the apex tip. The loop is fed at the bottom for horizontal polarization, and other smaller balloons spread and lift the two mid points. The size of a full-wave loop antenna is given in *The ARRL Antenna Book* as:

Circumference (feet) = 1005/Frequency (MHz).[4] [Eq 1]

You should use light wire to keep the loop's weight down. The sidebar "Materials" shows several choices for suitable wire. Cut the wire to the circumference length for the desired frequency. If you use stranded copper wire, strip and tin about an inch from each end.

The fishing line tethers serve as insulators for this antenna. These tethers are simply cable-tied to the antenna wire during the lifting phase. To strain relieve the loop

[1]Notes appear on page 131.

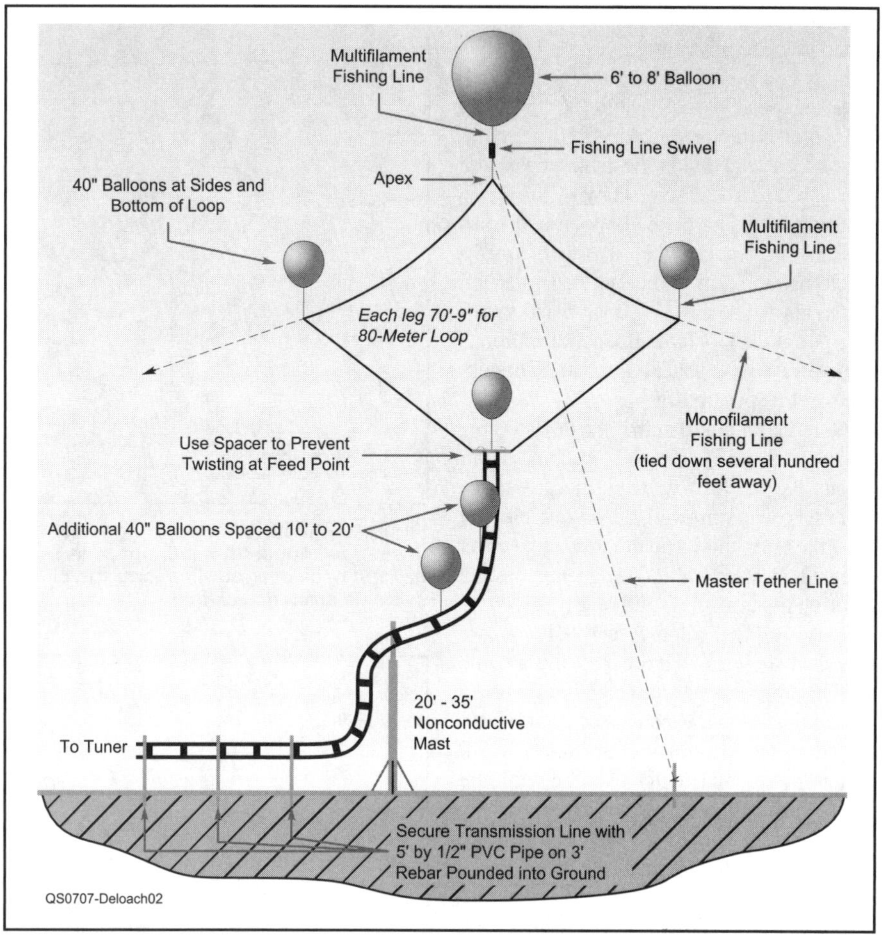

Figure 2 — Overview of balloon-lifted full-wave loop antenna.

wire at the tether points, preposition one or two approximately 1 inch sections of shrink wrap at the top and side corners of the diamond, as shown in Figures 4 and 5.

Feed one or two pieces of shrink tubing down the loop wire for each corner. Gently fold the wire in half to find the apex corner position. Be careful not to kink or bend the wire. Shrink the tubing at this halfway point (this is the apex corner.). Now, gently fold the antenna into quarters to identify the side corner positions. Install shrink tubing at these two positions as well.

Spread the wire out again halved, with the bottom two feed points side by side. Roll up the loop wire starting at these two feed ends. This will put the apex corner on the outside of the roll, which helps when deploying the antenna. Be careful not to kink, bend or knot the wire when rolling it up.

Preparing Balloons

I have used a single 8 foot balloon to lift an entire 80 meter antenna assembly, but my preference is to use one 6 foot balloon, along with several smaller balloons at the mid points. This makes compliance with FAA regulations easier (see the sidebar "FAA Balloon Regulations"), and helps keep the diamond shape better. A 6 foot balloon is also cheaper. Figure 2 shows the location and size of the balloons, and the sidebar provides balloon sources.

Materials

Antenna wire: I use surplus stranded #22 wire from Fair Radio Sales. Part number SEB-400X3, 306 foot spool for $12.95. **www.fairradio.com**. Other authors suggest using "fine aluminum welding wire" from welding supply shops.

Helium: Helium can be purchased from local party balloon suppliers (I use **www.peoplegreeters.com**), industrial gas or welding supply shops. Cost is typically about $0.25/cubic foot. Tank sizes vary widely. I use 176 ft^3 tanks that weigh 80 pounds. I have never needed a regulator given the types of tanks I use. Check with your supplier though. Balloon volume is given by $(4/3)*\pi*(\text{diameter}/2)^3$, but I typically only inflate to 80% of balloon capacity. Thus a 40 inch balloon needs 15.5 ft^3 and a 6 foot balloon needs 90.5 ft^3. If one 6 foot and five 40 inch balloons are desired, and you want to have double the helium just in case, then you need 336 ft^3. I typically get two 176 ft^3 tanks.

Balloons: I use latex or chloroprene "car dealer" balloons from my local helium source or the Web (eg, **www.balloonlovers.com;** also try eBay); 40 inch balloons are $2.50; 6 foot balloons are $18. Note that balloons last longest when stored in a cool, dark place. If you have several hundred dollars to spend and want to get serious, a heavy-duty vinyl advertising balloon will last for years, and you can put your club logo on it! Note that you will need to use much larger tether rope than discussed in this article. Place a drop cloth under vinyl balloons when inflating.

I have successfully used Edmond Scientific and surplus weather balloons over the years. But since these balloons are meant to float free, they tend to be fragile and pop easily when moored. See the University of Hawaii site for additional balloon information.

Nonconductive mast: The MFJ-1910 fiberglass mast (**www.mfjenterprises.com)** works well for this purpose. Cost: $80. High voltage "hot sticks" also work. I have used the tripod from an AS-2236 log-periodic antenna (**www.fairradio.com**) to hold the mast.

Wire connectors: Screw compression type wire connectors taken from small mechanical lugs work well for me (**www.doityourself.com/invt/8017337**). Alternatively, use small "cap" style wire connectors or crimp butt splitters. Source: any hardware store. Cost: a few dollars.

Tether line and swivels: 100-pound-test tether lines are sufficient for balloons up to 6 feet in diameter. Any fishing store can sell monofilament line, multifilament line and swivels. Total cost: about $20.

Tuners: Use only a tuner made for balanced lines!* Others do not work as well. The MFJ-974HB costs $190 (**www.mfjenterprises.com**) and old Johnson Matchbox tuners work well.

Static bleed-off resistors: 1 MΩ or more, non-inductive power resistors rated for over 2 W should do for up to 100 W radiated power (Digikey PPC1.0MW-3JCT-ND).

Use only helium gas to inflate balloons! Hydrogen is *explosive* and not worth messing with. Helium is safe to transport and to use if a few simple rules are followed (see the sidebar "Safety Precautions").

Balloon lift is a function of size, weight, material type, altitude, pressure and temperature. Lift predictions have been published, but in my experience lift never quite matches up to predicted values.[5, 6] At an altitude of 600 feet I've found that 40 inch balloons give about a pound of lift. A 6 foot balloon achieves 3 to 5 pounds of lift, and an 8 foot balloon manages 10 to 15 pounds of lift. Your experience may vary, particularly at different altitudes.

Before inflating the balloons, prepare all the tether lines. Cut pieces of 100 pound test multifilament fishing line to the lengths shown in Figures 4 and 5. Make a bowline knot with a loop on one end at least as big as your fist.[7] Tie a simple slipknot on the other end to prevent unraveling.

Now you are ready to inflate the balloons. Have one person tightly hold the lip of the balloon over the tank nozzle while a second person carefully holds the tank steady and gently opens the valve. Have a third person judge the size of the balloon and warn the inflating person when it is time to stop. Inflate balloons only to about 80% of rated size to allow for some expansion and to prevent popping. Do remember that balloons pop! Always have spare balloons and helium to cover this possibility.

Note that a crack in the nozzle assembly can cause the helium tank to become a dangerous projectile. Any time the protective cover has been removed, be very careful not to let the tank fall. Keep the protective cover screwed on when not inflating, and consider securing the tank to a solid object.

Once inflated, secure each balloon to a tether, as shown in Figure 3, using a cable tie. This approach is simple and has worked flawlessly for me. Anchor each balloon to something solid until you are ready to lift the entire antenna. After use, the cable tie can often be snipped off without damaging the balloon, but note that used balloons are far more prone to popping.

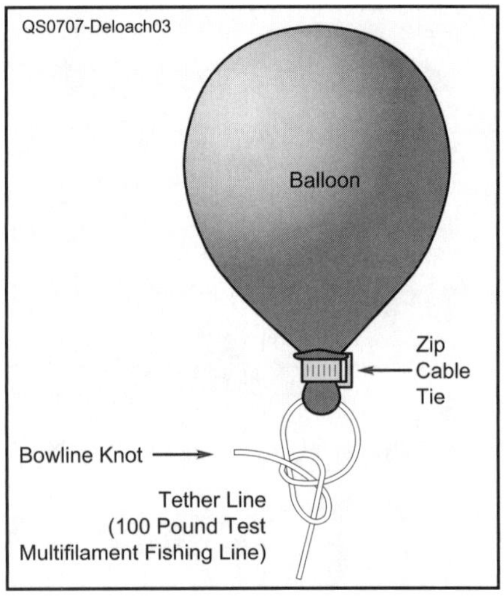

Figure 3 — Method for securing tether lines to balloons. After inflation, fold balloon neck over bowline loop and cinch cable tie around neck fold.

Safety Precautions

Here are some simple rules to follow to fly your antennas safely:

- Make sure you have plenty of space! Never fly a balloon antenna where it could possibly come down on a power line, any other wire, any structure, roads or rail lines.
- Fly balloon antennas only in light to moderate winds. Pull the antenna down when winds rise and wait for the winds to subside.
- Use tether line sizes appropriate for the possible wind load. Do not use frayed or damaged tether lines.
- Wear gloves when handling tethers or antenna wire.
- Never fly a balloon antenna in a thunderstorm.
- Inflate with helium only.
- Always tether the balloon to the ground with a master line separate from the antenna support structure. Tether lines should be stronger than balloons. Use good knots like the bowline.
- Be careful with helium tanks. When transporting, secure tanks so that they cannot roll around in a vehicle. Tanks are heavy, so always have enough people around when lifting. Always keep the valve closed and the nozzle cover secured when not actually inflating balloons. Never let tanks fall — particularly when the nozzle is exposed.
- Secure tether lines to solid objects on the ground. Tie colorful flag tape to the lines so that pedestrians and cars see them and do not become entangled.

Raising the Balloon — the Tethering and Spreading System

Once all balloons are inflated and tethered, you are ready to launch your antenna. Connect the large balloon to the antenna apex, as shown in Figure 4. Also tie the master tether line to the bottom of the swivel. Use bowline knots throughout. The master tether line is for safety. It holds the apex balloon independent of the antenna, so the master tether should be kept clear of the antenna wiring and tether lines. Typically this means that it is secured a little upwind of the antenna.

Let the apex balloon slowly rise upward as you unwind the antenna wire and the master tether. Be careful not to allow the two antenna wires to kink or knot, or to tangle with the master tether. Continue to unwind the antenna wire and master tether until you reach the corner tether points.

Secure the spreading lines and additional balloons to each corner using the technique shown in Figure 5. Spreading lines pull the diamond apart, and they work best if they are light (25 pound test monofilament fishing line works well). Optionally, tie colorful flag tape to the bowline loops to improve corner visibility once the antenna is lifted.

Continue to unwind the antenna wire, master tether and the two spreading lines while walking the spreading lines apart. Continue unwinding until you reach the ends of the loop antenna wires.

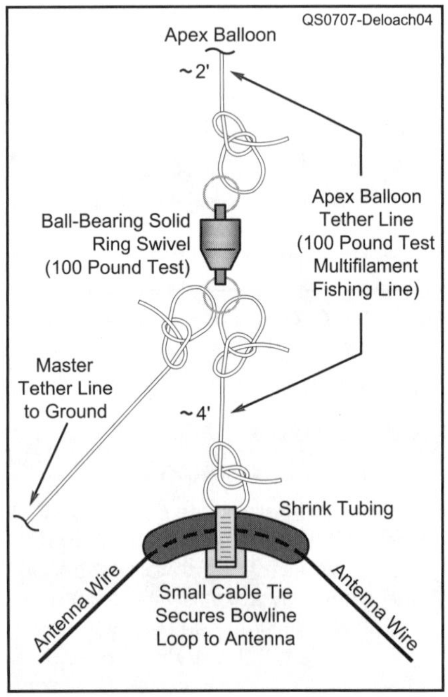

Figure 4 — Securing the apex balloon tether and the master safety tether.

Secure Feed Line

An insulating spacer is essential to prevent the feed wires from twisting and potentially shorting out. Use any convenient light, nonconductive material about 12 inches long for this spacer. Cable-tie the ladder line to the spacer and the balloon tether, as shown in Figure 6. Connect transmission line to antenna wires using screw compression type wire connectors or some other form of wire connectors (see the sidebar). Strain-relieve the antenna wires to the spacer using electrical tape.

Reel out feed line, tying on a 40 inch balloon every 10 to 20 feet, as shown in Figure 2. Once the antenna is in position,

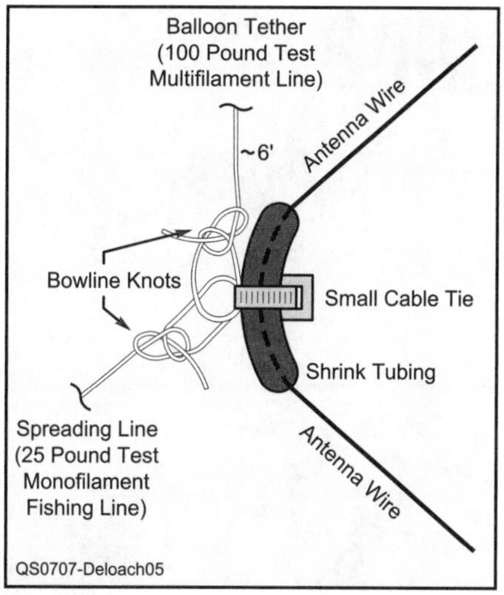

Figure 5 — Securing the corner balloons to tether and spreading lines. Note that the spreading line is looped inside the balloon tether line for safety in case the cable tie should break.

FAA Balloon Regulations

The United States Federal Aviation Administration (FAA) defines regulations governing the operation of moored balloons.* A moored balloon is exempt from all regulations as long as it does not create a hazard to persons or property, it is not more than 6 feet in diameter and its gas capacity is not greater than 115 cubic feet. If not exempt, it must be more than 5 miles from any airport, be operated in good weather and have an "automatic deflation mechanism." If operated above 150 feet, additional advanced notification, lighting and visual warning requirements kick in. Operation above 500 feet is banned outright.

The FAA regulations change from time to time, so read the regulations for yourself before you fly your balloon lifted antenna.

*Federal Aviation Administration regulations concerning "moored balloons" can be found in the *Electronic Code of Federal Regulations* (e-FCR), Part 101 — "Moored Balloons, Kites, Unmanned Rockets and Unmanned Free Balloons," at **ecfr.gpoaccess.gov/**. Look under Title 14, then Part 101.

experiment with tether and spreading line lengths to get the loop to take on as close to a diamond shape as possible.

Transmission Line

Balanced ladder transmission line (often called window line) is far more efficient than coax, tunes much easier and is a great choice for temporary operations. But it needs to be kept away from metal objects and you can't run it on the ground. To prevent the feed line from touching the ground as the balloons bob up and down in the wind, tape the feed line to the end of a nonconductive mast, as shown in Figure 2.

Jim, K6EI, came up with a great way to further distribute the transmission line, if needed. Pound 3 foot sections of steel reinforcing bar (rebar) in the ground, and then slip 5 foot sections of ½ inch PVC pipe over the rebar. Tape the transmission line to the top of the PVC. Space these pipes every 10 or 15 feet all the way to the balanced tuner. See the sidebar for sources for masts, ladder line and balanced tuners.

Static Buildup

Wind-induced static buildup has been reported for large balloon-lifted monopole antennas. Since loops present a short at dc, a differential mode voltage buildup will not occur. Common-mode build up is possible, however. Although we have never experienced this phenomenon, perhaps because we are in a low static area, we still connect two 1 MΩ, high wattage, noninductive resistors between the tuner's balanced antenna terminals and ground to bleed off any possible static charge.

Performance

How well do these antennas work? The ESL Radio Club has had numerous adventures using these balloon loops as part of their annual Field Day low power (QRP) operation. Jim, K6EI, recalls creating a pile-up after calling CQ, holding a frequency through the contest and even working Japan — all on 80 meters while running 5 W. Greg, N6GD, reports, "On 80 CW, there wasn't anybody I heard that I couldn't work with the loop. Pretty amazing considering we were running only 5 W!" These comments are typical. Our Field Day experience has always been that if the balloons stay up, the antenna rocks, even if we're running QRP. So join the fun and try a loop antenna and see what it's like to have the big signal!

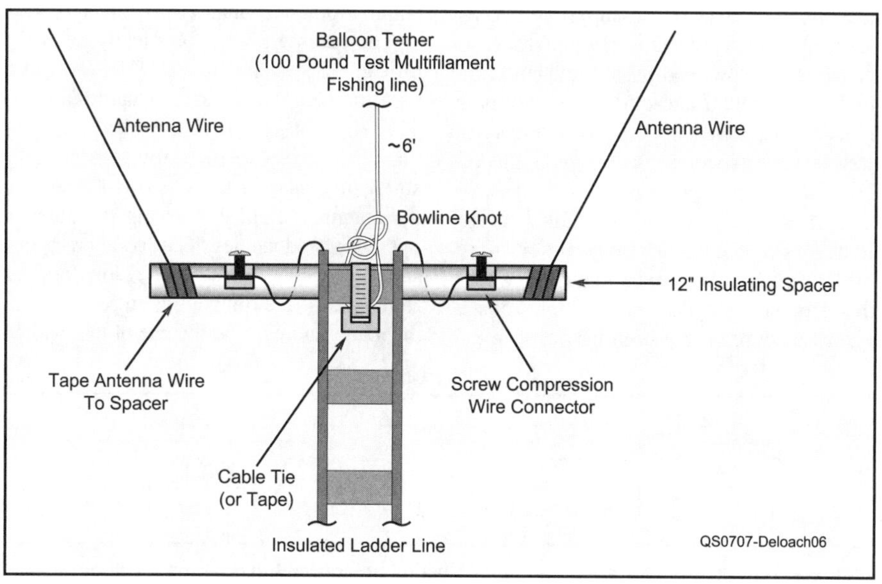

Figure 6 — Feed line assembly detail.

Notes

[1]The ESL Amateur Radio Club of Sunnyvale California, along with its partner club, the West Valley Amateur Radio Association of San Jose.

[2]S. Gibilisco, W1GV, "Balloons as Antenna Supports," *The ARRL Antenna Compendium, Vol 2.*

[3]The Kite and Balloon Antenna Site, YB5AQB (G4VGO), **www.qsl.net/g4vgo/**.

[4]R. D. Straw, Editor, *The ARRL Antenna Book*, 21st Edition. Available from your ARRL dealer or the ARRL Bookstore, ARRL order no. 9876. Telephone 860-594-0355, or toll-free in the US 888-277-5289; **www.arrl.org/shop/**; **pubsales@arrl.org**.

[5]The University of Hawaii has an excellent Web site discussing balloon buoyancy, balloon materials and antenna construction materials at: **www.chem.hawaii.edu/uham/lift.html**.

[6]Another easy to read discussion of helium buoyancy can be found at: **www.howstuffworks.com/helium1.htm**.

[7]See **www.tollesburysc.co.uk/Knots/Bowline.htm** for an illustrated example of the bowline knot. Also see *QST*, Jun 2006, p 57.

Jim DeLoach, WUØI, was first licensed in 1976. He is a Principal Engineer with Qualcomm Inc, working on GPS and other location technologies for wireless devices. He loves to build high performance temporary contest stations, but prefers to get out of the way and let the real operators take over as soon as the station is up and running. He has been building and flying balloon-lifted antennas for Field Day since 1994. Jim lives in Los Altos, California, and can be reached at **jim@deloach.net.** *The author's balloon Web site is found at* **www.deloach.net/balloons**.

Six Band Loaded Dipole Antenna

W8NX's unique design technique makes trap look-alikes do double duty. Here's a wire antenna that covers 160/80/40/30/17/12 meters!

Al Buxton, W8NX

Introduction

This article presents a new loaded wire dipole antenna. It covers the classic 160, 80 and 40 meter bands, plus 30, 17 and 12 meters. I call it the *W8NX Special.* Any amateur who installs this antenna and who has a triband beam for 20, 15 and 10 meters has a very good antenna system for working all the amateur high frequency bands from 160 through 10 meters. I installed my W8NX Special as an inverted V, using the tower holding up my triband Yagi as the center support. See Figure 1.

This antenna is based on the highly efficient *dominant element principle*, requiring only two pairs of load elements to give six bands of operation.[1] The radiation patterns have a single pair of broadside lobes on the classic 160, 80 and 40 meters bands but are similar to those of long wire antennas on the 30, 17 and 12 meter bands.

Radiation takes place along the entire length of the antenna on all bands, providing small but useful antenna gains. Good bandwidth is provided on all bands when used in conjunction with an antenna tuner. With the exception of the 160 meter band, full band coverage is provided on all bands. On 160 meters the effective working bandwidth is typically limited by the size of the capacitors in the antenna tuner. The built-in antenna tuner in my FT-1000MP Mark V transceiver can cover 55 kHz on 160 meters using this antenna.

The antenna length is 134 feet, suitable for installation on most city lots. Mine is installed as a "droopy" inverted V dipole, with the apex at 47 feet on the beam tower and drooping to a height of 20 feet at each end. There is little mutual coupling between the triband beam and the six band dipole, since the working frequencies of the two antennas are sufficiently separated to prevent interaction. Some bending and folding at the ends of the dipole antenna is permissible to accommodate installation on a short city lot.

Figure 1 — W8NX Special antenna mounted at 47 feet on tower used to hold triband Yagi. This is an efficient antenna system that covers 160/80/40/30/17/12 meters with the dipole and 20/15/10 meters with the triband Yagi.

Antenna Performance

Figure 2 shows the schematic diagram of the antenna. The schematic looks the same as that of a standard three band trap dipole. However, the loads do not use the truncating capability of tuned parallel resonant traps. This new type of load acts as either a pair of inductors or capacitors to supply the necessary reactance to bring the antenna into resonance with a low feed-point impedance, on both fundamental and odd-order harmonic modes. This makes the antenna suitable for feeding via a 1:1 current balun with either 50 or 75 Ω coaxial cable.

I advise that you use 75 Ω cable because it makes a typical antenna tuner more effective, especially on 160 meters where the size and cost of the large high voltage tuner capacitors is the limiting factor in the effectiveness of a tuner. The innermost pairs of loads create fundamental resonance on both 160 meters and 80 meters. The outermost pairs create fundamental resonance on 40 meters and third harmonic resonance on 30 meters. The overall antenna gives fifth harmonic resonance on 17 meters and seventh harmonic resonance on 12 meters.

The loads are large physically, with significant stray capacitance. They exhibit a parasitic series resonance at approximately 45 MHz (not shown in the Figure 2 schematic). These parasitic stray effects make small increases in the electrical length of the antenna, slightly lowering the antenna operating frequencies. The loads are necessarily large to achieve high Q, low loss performance. Wide air gaps between turns of the load windings and the use of thin walled

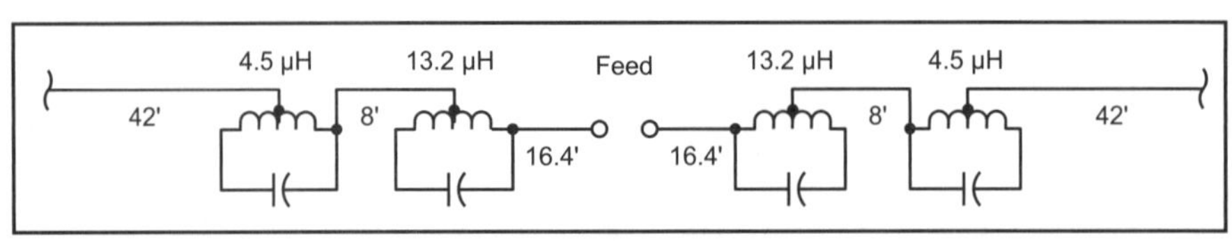

Figure 2 — Schematic for six band W8NX Special using dominant element principle dipole.

[1] A. Buxton, "Dominant-Element-Principle Loaded Dipoles," *QEX*, Mar/Apr 2004, pp 20-30.

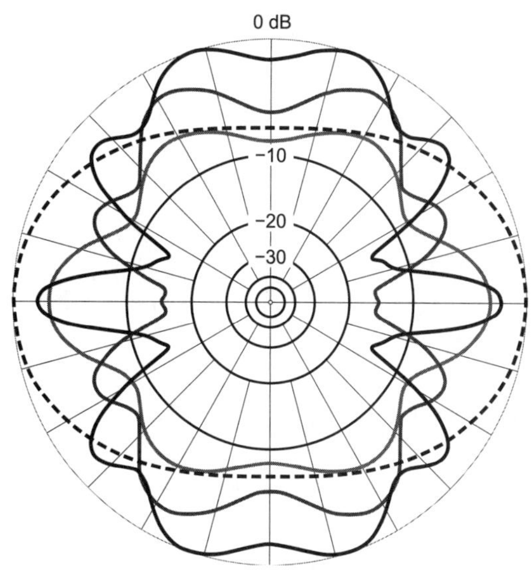

Figure 3 — Azimuthal responses for 40/30/17/12 meters at elevations yielding peak gain for each band.

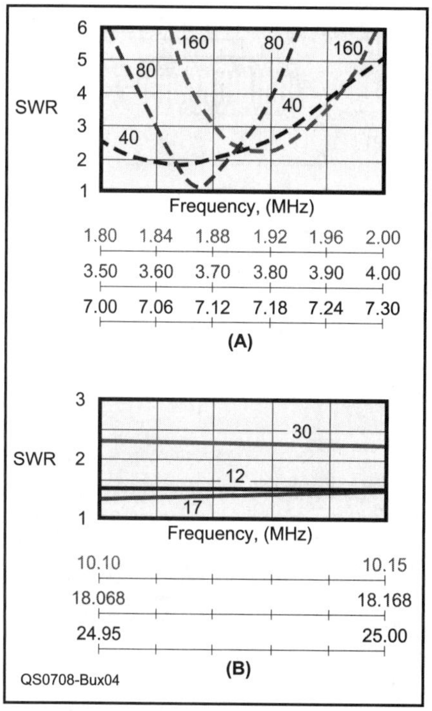

Figure 4 — At A, SWR curves for 160/80/40 meters. At B, SWR curves for 10/17/12 meters.

PVC coil forms minimize dielectric losses in the load elements. The use of RG-8U coax cable with large diameter stranded wire center conductors minimizes skin effect I^2R losses.

The Q of each 160/80 meter load is 260, and the Q of the 40/30 meter loads is 325. Load losses on 80 through 12 meters are less than 0.5 dB, but on 160 meters the loss approaches 3 dB. On 160 meters the radiation resistance of the antenna is low because of the relatively short length of the antenna, reducing the overall radiation efficiency to about 50%.

The radiation patterns have a single pair of broadside lobes on 160, 80 and 40 meters. Figure 3 compares the azimuthal patterns for 40 through 12 meters, at the peak elevation angles for each band. The patterns on the higher frequencies display numerous lobes, characteristic of long wire types of antennas. The peak gain on 40 meters is 1.5 dB above an ordinary dipole. As is the case with an ordinary dipole this has only two lobes. The gain on 12 meters is about the same as on 40 meters but the pattern has 10 lobes.

The measured SWR curves for the 160, 80 and 40 meter bands are shown in Figure 4A; those for the 30, 17 and 12 meter bands are shown in Figure 4B. The SWR curves are those measured at the rig end of an 80 foot long, 75 Ω RG-59 feed line. The curves pretty much speak for themselves.

Those of you interested in getting as much effective working bandwidth as possible on 160 meters can employ the trick of extending the feed line length when operating on 160 to that of a quarter-wave impedance inverter. The length of the extension must bring the total length of the RG-59 feed line to about 100 feet. This maximizes the effectiveness of the antenna tuner, reducing the required size of the tuner capacitors. The tuner now has an easier matching job of keeping your rig or linear amplifier happy. However, you have increased your feed line losses and even though your rig is happy over a broader bandwidth you have somewhat degraded the radiation efficiency of your antenna system. If you carry this trick to extreme measures under high power linear amplifier operation, you could conceivably incur current or voltage breakdown in your antenna feed line. The safe upper limit on 160 meters for SWR for RG-59 feed line is about 6:1 at maximum legal power operation corresponding to a maximum usable bandwidth of 130 kHz.

Remember your rig or linear amplifier never sees this SWR — the antenna tuner shields it from this level of mismatch. Your

Figure 5 — At left, construction techniques for 160/80 meter load element. At right, construction techniques for 40/30 meter load element.

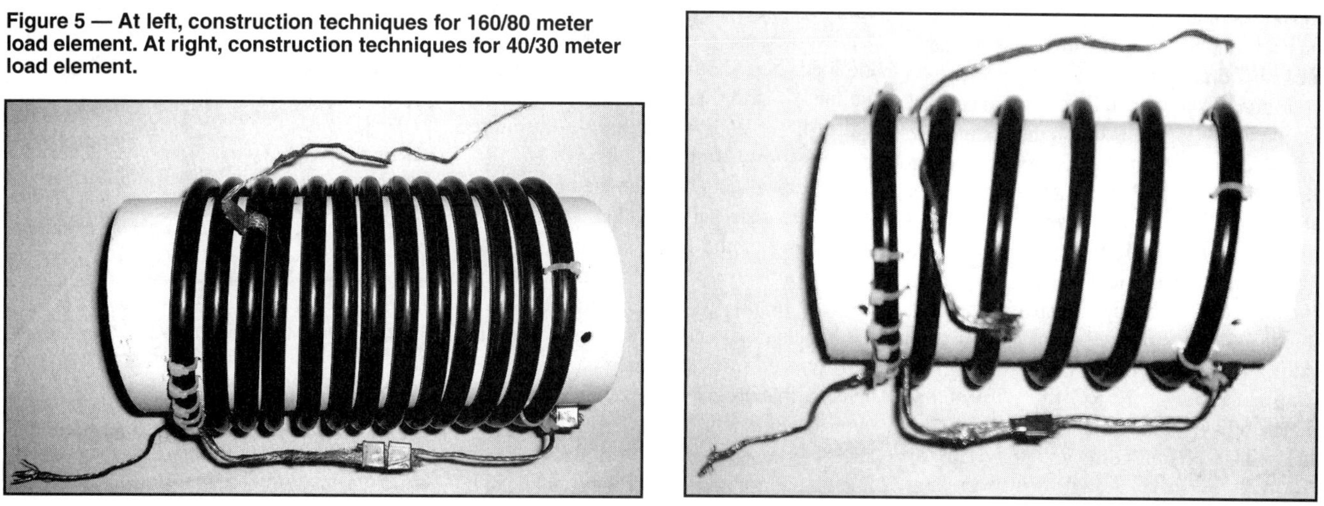

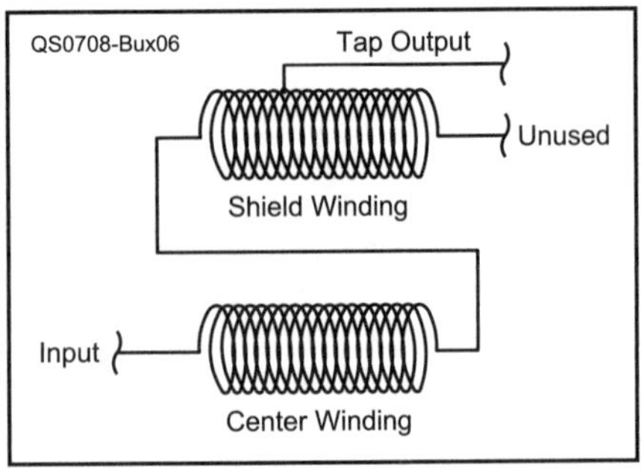

Figure 6 — RG-8U load element schematic.

QS0708-Bux07

Load	Form Dia Inches OD	Form Length Inches	F0 MHz	Dim A Feet	Dim B Feet	Number Turns	Spacing Between Turns Inches
160/80	4.188	9.125	2.38	2.78	12.88	13	0.14
40/30	4.188	7.75	6.62	2.57	4.61	6	0.65

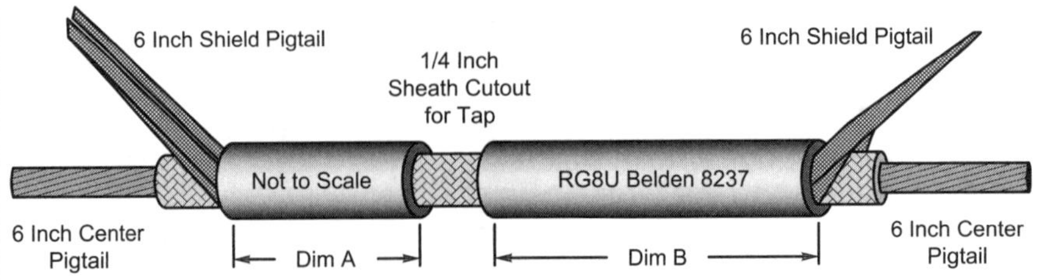

Figure 7 — Details of the load element coax.

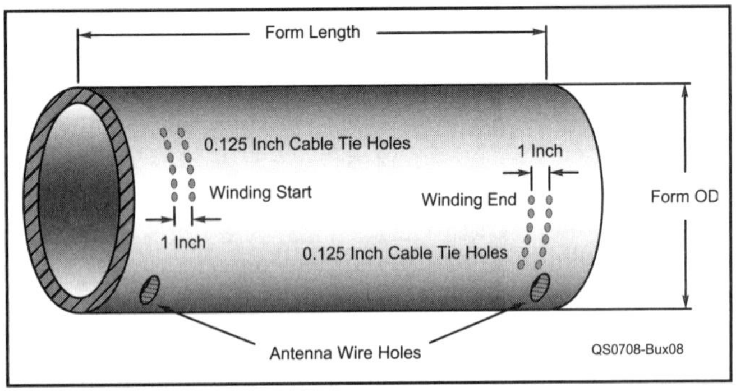

Figure 8 — Details of coil form.

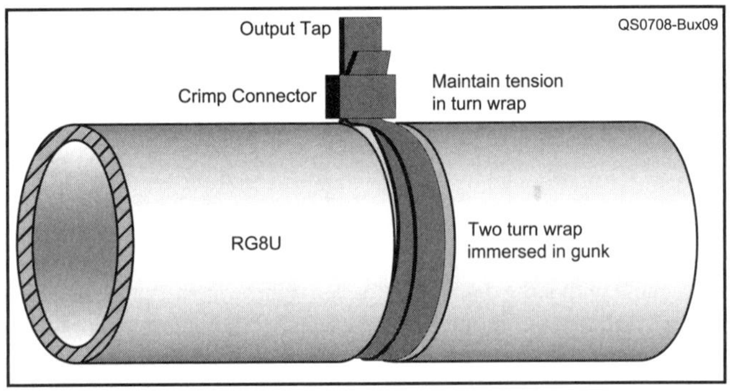

Figure 9 — Output tap detail.

antenna tuner does not reduce the SWR on the feed line. While some amateurs frown upon using this trick, it does give considerably more effective working bandwidth for 160 meter operation.

Construction

The toughest part of constructing this antenna is making the load elements. Figure 5A shows the 160/80 and 40/30 meter load elements. The load element schematic is in Figure 6. Note how the pigtail at the output end of the center winding is fed backward to the pigtail of the input end of the shield winding. The loads are made of RG-8U coaxial cable (Belden 8237) wound on a form made of 4.188 inch outside diameter PVC drain pipe.

Figure 7 shows details of the coax used for the loads. Figure 8 shows the load forms, the critical ones being the lengths and diameters of the forms and the 1 inch edge margins of the windings on the forms. Dimensions A and B fix the output tap location, with the RG-8U laid out flat and straight on a table. The tap is a 15 inch length of silvered braided shield wire cannibalized from RG-58 coax cable. The tap requires two turns tightly wrapped around the RG-8U wire at the ¼ inch break cut in the PVC sheath. Cover this break with anti corrosion gunk (Burndy Products, Penetrox A will do) to prevent resistance developing at the tap.

Care is required in making the output tap to hold the two turn wrap around the RG-8U permanently in tension using a crimp connector. Figure 9 shows the details of the 15 inch output tap of the loads. Unfortunately, soldering at the tap would weaken the electrical properties of the RG-8U coax cable so a mechanical-only connection is necessary.

The input terminal of any load is the near end (nearest the feed line) of the center conductor winding of the coax cable. The far end of the center conductor is fed back to the near end of the shield winding. The output of the load is taken at the tap on the outer shield winding. The output tap acts as an auto transformer, giving the needed L/C ratio for the load. You should fine tune the loads to within 1% of the specified frequency. I used a dip meter and an accurately calibrated receiver for fine tuning the loads.

Air gap spacing between the turns of each load reduces dielectric load losses and permits fine tuning of the load resonant frequency. Expanding the air gap increases the load resonant frequency; reducing the gap lowers the frequency. Do not hesitate to increase the gap between turns as much as necessary to achieve the resonant frequency of the load, even though you may distort the appearance of the load. After completion of the fine tuning, the location of the turns must be stabilized by cable ties, as shown in Figure 5. More cable ties are actually required than are shown, especially around the first and last turns of the load winding. Stabilizing the interior turns is not as critical, as they have less effect on the load's resonant frequency than the outermost first and last turns.

Although making the loads for this antenna may seem like a challenging task, your efforts will be well rewarded. There is nothing as satisfying in ham radio as the successful completion of a good, hands-on homebrew project.

I wish to thank my friend and colleague, Mel Vye, W8MV, whose help and constructive comments are greatly appreciated. I also wish to thank Jeremy (KB8QVF) and Angie Holland for their help with the photography of the antenna and the load elements.

Al Buxton, W8NX, has been a radio amateur since he was first licensed as W7GLC in 1937. A registered professional engineer, Al holds BSEE and MSEE degrees from Tulane University. His career spans industry and academia: 26 years in the defense industry with Goodyear Aerospace, six years with Tulane University and 11 years with the University of Akron. He's an Associate Professor Emeritus of Akron University. In industry, Al worked on the development of computers, automatic controls, radar, aircraft guidance, and navigation and space antennas. You can reach Al at 2225 Woodpark Rd, Akron, OH 44333 or **buxtonw8nx@aol.com**.

A Simple Broadband 80 Meter Dipole

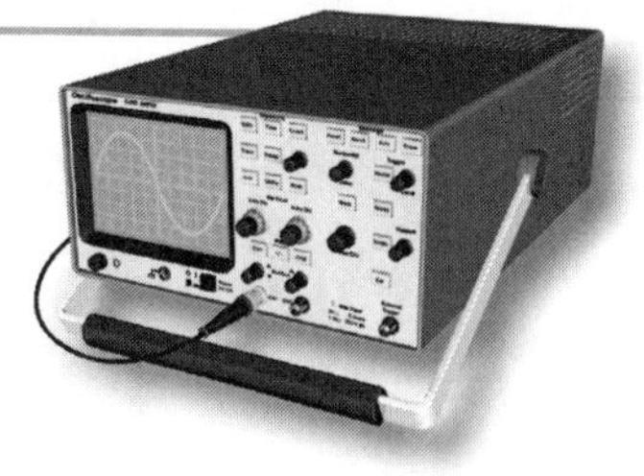

This easy to make antenna can keep you from getting stuck in one corner of this popular band.

George Prince, N6DNA

For most of my Amateur Radio life I have avoided the 80 meter band. The 80 meter coil on top of my vertical gave such little bandwidth and a dipole was just too long for my location. When solid state transceivers came on the market there was the added concern that moderate antenna standing wave ratio (SWR) would cause the final transistors to cut back power. I was electronically raised with the type 6146 final amplifier tube, but was happy to switch to solid state when the products were offered. Nevertheless, an 80 meter dipole was a dream that recently had caused me to do some investigation and antenna construction.

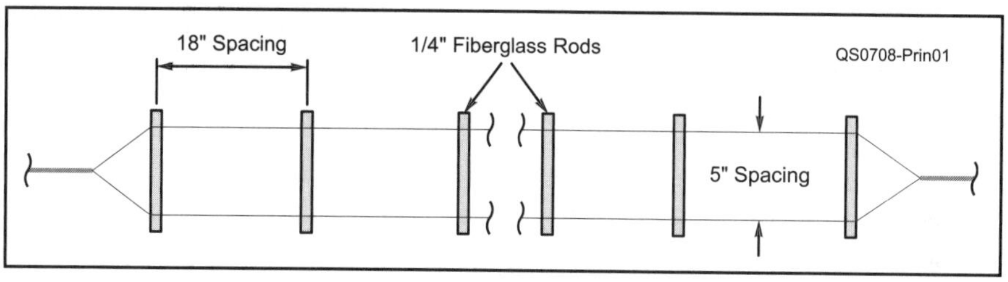

Figure 1 — Broadband 80 meter dipole, final configuration.

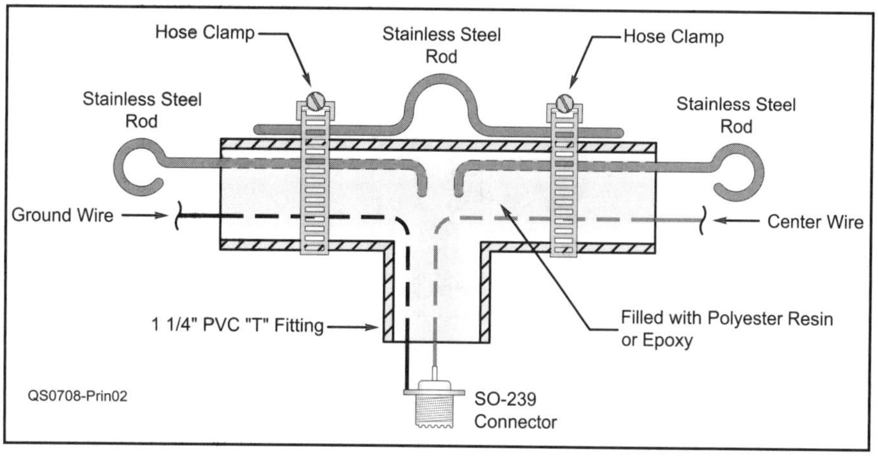

Figure 2 — Detailed view of the author's homebrew center insulator.

Can Problems be Resolved?

The biggest problem I had was the narrow bandwidth associated with the usual single wire dipole. I decided that if I did put up a wire dipole I would just cut it for the phone band to be able to work several of my ham friends who had moved to the Midwest and East Coast. Fortunately in the April 2006 issue of *QST* there was an article giving the lowdown about internal antenna tuners.[1] The article took the time to describe the problems associated with 80 meter antennas, giving wonderful graphs of frequency, power loss and SWR with a dipole antenna. Therefore I will not go into the electronic antenna details but will focus on the product and the results.

One of the bits of information deep in my old electronic brain archives was the fact that if you increase the diameter of an antenna radiator you will increase the bandwidth, or as the big boys say, decrease the sharpness of the radiator Q. I was sure there were all sorts of formulas to give me all the large radiator electronic metrics. I began to ask knowledgeable hams about the use of multiwire dipoles. I hoped two wide spaced wires would act as large diameter radiator. The answer I got was yes, but also it was explained to me that such designs were not new. Nonetheless such designs have been found to be effective and more broadband than the single wire dipole.

Making it Come Together

The Antenna Proper

I decided to use two 14 gauge enameled copper wires on each leg. I would try separating the wires, picture a ladder line, by 5 inches and see what would happen. Certainly more separation than 5 inches could be tried. I ordered enough ¼ inch diameter fiberglass solid rod to separate the wires by 5 inches with a spacer on the wire every 18 inches. I used the center frequency of the 80 meter band for the standard one wire formula to get a starting length. The result was 125 feet, or 62.5 feet for each leg. After rolling out the spool of the antenna wire, with the help of my bride, I cut all the wires and ended up with a horrible rats' nest of wire. If you decide to make this antenna,

[1]Notes appear on page 137.

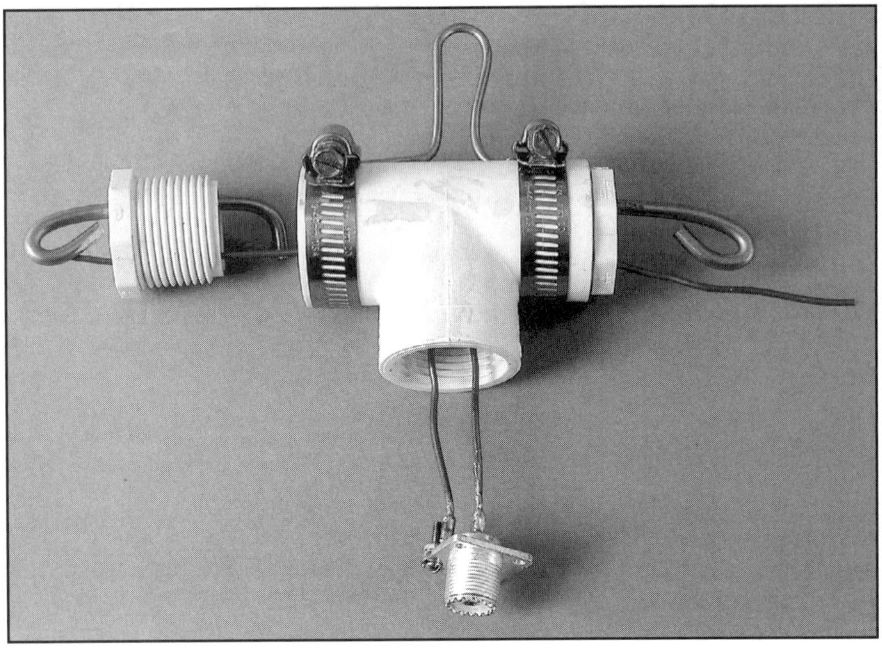

Figure 3 — The pieces of the center insulator before final assembly and sealing.

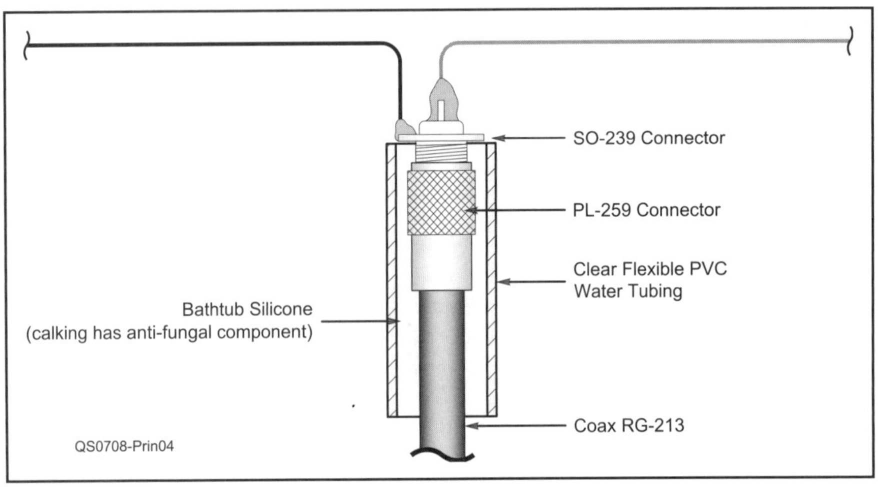

Figure 4 — Waterproof coax cable connection method.

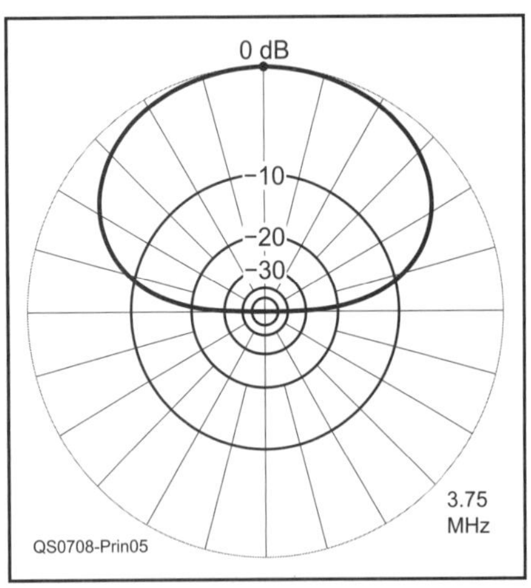

Figure 5 — *EZNEC* plot of broadside antenna elevation pattern.

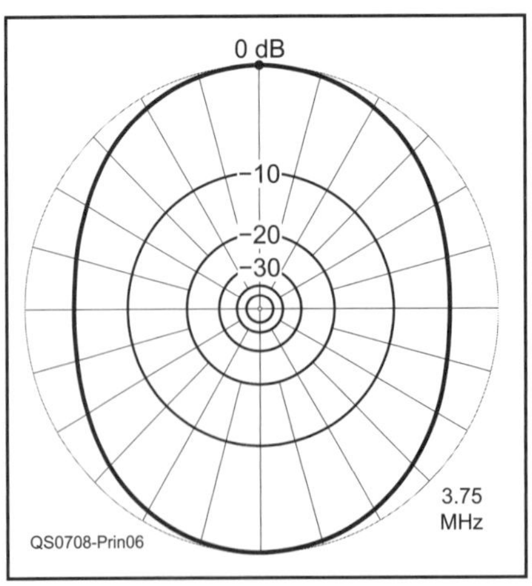

Figure 6 — *EZNEC* plot of antenna azimuth pattern at 30° elevation.

separate the long wires — don't keep them all in a cut bundle!

Spreading the Wires

I drilled each fiberglass rod in a drill press, with a holder to keep the holes aligned; then I threaded each drilled rod onto both wires of the elements. With the wires stretched I then slid the rods to the 18 inch separation and dropped epoxy on each point where the rods touched the wire. This repetitive nonsense almost made me junk the whole project, but I had hopes it would be worth the piecework. The resulting configuration is shown in Figure 1.

Center Insulator

Many varieties of antenna insulators to transition to coax are available on the market, but I chose to make mine from a 1¼ inch schedule 40 white PVC T coupling. The SO-239 female coax receptacle can be pushed into the bottom after the hot and ground wires are soldered onto the connector. I ran the electrical copper wires along the bottom of the fitting and two 10 gauge stainless steel wires along the top of the fitting, with loops on the ends that hold the antenna loop attachment of the ladder line antenna element. Details are shown in Figure 2. The plastic casting held ends of the stainless 10 gauge wire are bent to prevent coming out of the T under strain. Attaching the ladder line to the stainless 10 gauge wire loop will keep the strain off the bottom electrical lines. I then sealed the ends of the T fitting with quality duct tape around the 1 inch PVC plugs with the wire holes and carefully filled the unit with catalyzed polyester resin (epoxy can also be used).

The top center T hanger loop for the fitting was held on top of the T PVC fitting with stainless hose clamps. Sounds complicated but it's very simple and worth the effort. All electrical connections were well soldered. The ladder line ends were terminated in a short piece of stronger stranded 10 gauge copper and soldered together. The pieces ready for final assembly are shown in Figure 3.

Balun Details

My balun consisted of 10 turns of RG-213 coax, 9 inches in diameter hanging below the center T fitting. The coil is held together with plastic wire ties. I have a special way of protecting all my coax connectors that are exposed to the elements. I take a 4 inch piece of ¾ inch inside diameter by 1 inch outside diameter clear flexible PVC tubing and slide it over the PL-259 connector onto the coax before attachment. After tightening the coax fitting, I slide the tubing up over the electrical joint, and then

Table 1
Measured SWR vs Frequency

Frequency (MHz)	*SWR*
3.60	4.9
3.65	3.8
3.70	2.6
3.75	1.8
3.80	1.4
3.83	1.2
3.85	1.3
3.90	1.4
3.95	1.8
4.0	2.5

shoot bathtub silicon calking into the end of the clear tubing. I watch it extrude up to and past the fittings as shown in Figure 4. This makes a completely waterproof flexible fitting when hardened. It is easily removed by cutting the clear tubing with a knife. I have kept soldered wires protected in saltwater boat bilges this way for years.

Firing it Up

The dipole ends were attached to insulators and raised about 30 feet above the ground, while the center height was about 50 feet. With this construction I ran to the shack to get the SWR charted over frequency. My initial results had the center frequency almost off the bottom of the band. What to do? Well, I began cutting off 18 inch end sections of the dipole. Slowly the acceptable band began to move up to the phone band. After cutting off 10 feet of the antenna, 5 feet on each end, I ended up with the SWR values in Table 1. Unfortunately with all my trimming there may be a few inches of unequal length to the two sides, giving me an SWR that is not as low as possible. I suggest that you mark the antenna element ends with length marks so that you can trim the lengths equally.

The SWR measurements in Table 1 were obtained with an MFJ-259B SWR analyzer at the shack end of about 80 feet of Belden RG-213 coax. The radiation pattern is typical of a fairly low dipole with *EZNEC*-predicted elevation and azimuth patterns shown in Figures 5 and 6.[3]

Finally, here's why I like the antenna. I have put legal limit power to the antenna with good results. The lower broadband SWR gives any tuner more efficiency and ease when tuning to unity, but all the phone band becomes usable without a tuner and with little line loss.[2] The 10 foot shorter antenna may be important to some; nevertheless I would advise constructing the antenna longer than needed; trimming is easier than adding on to the ends. My on-the-air reports are very good. The antenna was fun and easy to make and I truly like the results. I'm on the 80 meter band now!

Acknowledgment

I wish to express my appreciation to Will Anderson, AA6DD, for his design work and ongoing advice during the construction of this antenna.

Notes

[1]J. Hallas, W1ZR, "Getting to Know Your Radio — Internal Antenna Tuners" *QST*, Apr 2006, pp 66-67.

[2]See Note 1.

[3]**www.eznec.com**.

George Prince, N6DNA, was born in Santa Monica, California in 1932 and raised on a ranch. He is a retired physician-anesthesiologist with three grown children. He first became a ham in 1961 and has held a number of calls, receiving the present one after he upgraded to an Advanced class license. His Amateur Radio pursuits have included maritime mobile operation from his 40 foot homebuilt sailboat, building vacuum tube power amplifiers, and making his own welded towers and monoband Yagi antennas. George continues to enjoy wonderful friendships he has established with fellow hams over the years. You can reach George at 600 Terry Lynn Pl, Long Beach, CA 90807 or at **gmpsr@earthlink.net**.

A No Compromise Off-Center Fed Dipole for Four Bands

An easy to build single wire antenna for 40, 20, 10 and 6 meters.

Rick Littlefield, K1BQT

Figure 1 — The 2.8:1 transformer.

Many believe the off-center fed dipole (OFCD) is a compromise antenna, but that appraisal may not be deserved. When done right, these antennas can really perform!

Understanding the OCFD

Half wave ($\lambda/2$) dipoles are generally fed in the center, a point at which the driving resistance is low enough to provide a convenient match for coaxial feed line. Dipoles will, however, efficiently accept RF power at any point along their length as long as the source is matched to the load. The key to a successful OCFD design is finding that magic point where similar driving resistances appear for multiple bands. Opinions may vary about where that best point is, but most designers locate it roughly ⅓ of the way down the wire and transform it down to 50 Ω using a broadband transformer.

The trouble begins when builders try to cover multiple bands with the antenna too close to ground, or use matching transformers with incorrect ratios. After modeling various designs on *EZNEC* and evaluating a prototype, I found driving resistances tend to converge in the 120 to 140 Ω range at the 33% feedpoint location.[1] These values suggest a transformation ratio of under 3:1, which is significantly lower than the 4:1 or 6:1 transformers often encountered.

Building a 2.8:1 RF Transformer

The simplest way to achieve a suitable match to the OCFD may be with a conventional 2.8:1 transformer as shown in Figure 1. This device has a 3:5 turns ratio and provides a match at the secondary to 138 Ω. Mutually coupled transformers require more careful design than their transmission line counterparts and generally exhibit slightly higher insertion loss. Once the right combination of inductance and core permeability is found, however, construction becomes easy because you don't need to link multiple windings together through a labyrinth of phasing connections.

I made the transformer using a binocular core consisting of two 1⅛ × ¼ inch ID 43-mix EMI sleeves (Fair Rite 2643540002 or equivalent).[2]

The relatively high core permeability of 850 yields good performance over a wide frequency range with a minimal number of turns. The cores are relatively inexpensive and widely available since they are often used as feed line chokes for RG-8X and LMR-240 coax cable.

I used 16 gauge stranded wire covered with a Teflon jacket for the 3-turn primary because it provides a high dc breakdown voltage across the device. The secondary is wound with 5 turns of 18 gauge double coated enameled wire. I found it easier to install the solid wire secondary first, saving the slippery jacketed Teflon wire for when space becomes tight inside the cores. Note that EMI sleeves may have sharp mold seams that can scrape off enamel coating, so use caution when winding.

To test the transformer for SWR response, I attached two 68 Ω resistors in series across the secondary to make up a 136 Ω load. I then connected an analyzer to the primary winding and swept it from 1.8 to 50 MHz. The transformer delivered virtually flat SWR from 2.2 to 24 MHz. The SWR began to slowly creep up beyond that point.

In order to test for insertion loss and power handling, I wound a second identical transformer and connected it back-to-back to the first. Using a signal generator and spectrum analyzer, I measured approximately 0.2 dB of insertion loss per device through 14 MHz, with losses slowly increasing beyond that point. The plot shown in Figure 2 tracks the combined loss for the binocular transformer plus a tandem 1:1 current balun (described below). This small amount of attenuation should have negligible impact on real world signal strength or antenna performance.

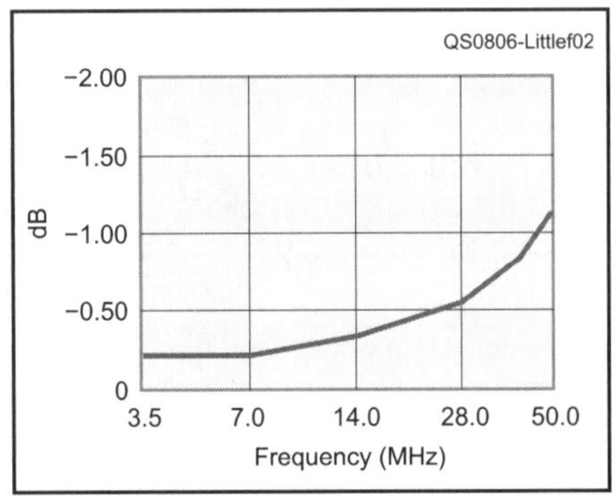

Figure 2 — Total loss of transformer and balun versus frequency.

Finally, to test power handling, I connected a dummy load to the transformers and applied a 14 MHz, 1000 W test carrier for a 10 second interval. The cores became quite warm to touch but never too hot to handle. More importantly, there were no telltale changes in SWR to signal core saturation. At 0.2 dB insertion loss, a 1000 W carrier will result in about 47 W of heat, or roughly three times what the transformer can safely handle over time allowing 7 W dissipation per core. Based on this finding, I use my AL-80A linear amplifier when I need to, but limit high power operation to casual SSB or CW contacts. I also avoid prolonged amplifier tune-ups.

The 1:1 Current Balun

Because OCFDs are fed asymmetrically, they are especially prone to radiate RF energy from the feed line. To prevent this undesired condition, I installed a 1:1 current balun in tandem with the balanced matching transformer. While the ferrite matching transformer may provide some limited blocking of the undesired common-mode path, it lacks sufficient cross sectional area to provide really good isolation. To enhance isolation, I added a

[1]Notes appear on page 140.

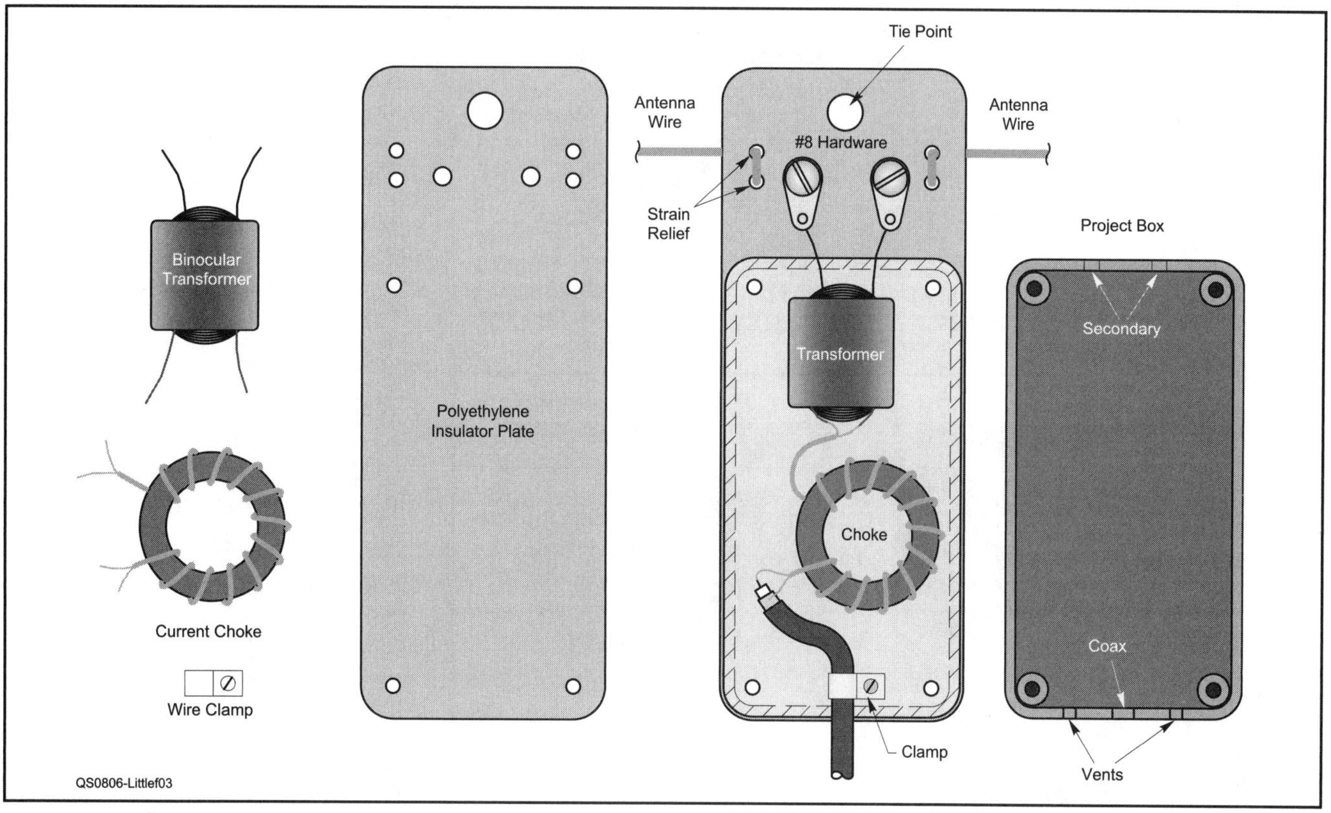

Figure 3 — Construction details of center insulator assembly.

lightweight 1:1 transmission line style current balun in tandem.

The balun core is made from two 1¼ inch outside diameter 43-mix toroids (Fair Rite 5943001601 or FT120-43) stacked together and secured with high-temperature Kapton tape. A light coating of 5 minute epoxy could be used to secure the cores if you don't have tape. The transmission line consists of 18 gauge high-temperature armature wire wound together at 4 to 6 turns per inch with an electric drill. I wound 12 turns of this twisted pair onto the form to complete the balun. Later checks with an RF current probe confirmed good common mode rejection along the feed line on all four bands. Construction details are shown in Figure 3.

Center Block and Weather Enclosure

The center insulator was made from a ⅛ inch thick piece of black marine polyethylene. Other materials may be used, but this particular plastic is very strong and provides good UV protection. I mounted the transformer, balun, and feed line attachment directly onto the polyethylene base and covered it for weather protection with an inexpensive styrene project box. The box is attached via mounting holes normally used to secure its cover. The cover isn't used, but does provide a useful drilling template. I added two ¼ inch vent holes on the bottom side of the project box to permit air circulation and used a round file to create a mouse hole to admit the coax. I also added two small notches at the top to pass the secondary transformer lugs. A couple of dabs of sealant around the secondary leads at the top will prevent water from running in around the lugs.

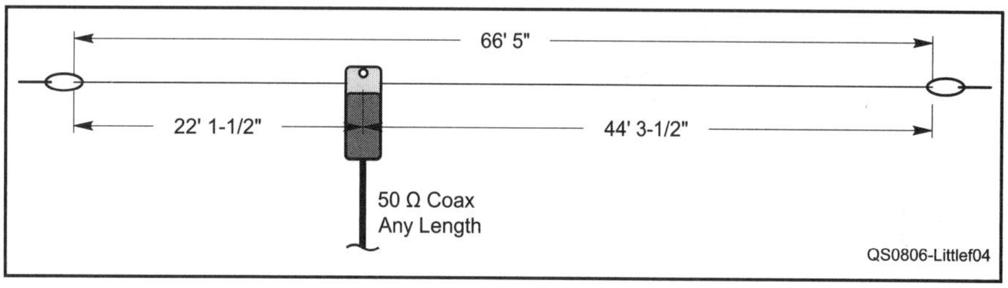

Figure 4 — Cutting dimensions for OCFD.

Making the Antenna Flat Top

I used jacketed wire and, from mid-center block, cut the legs to 22 feet 1½ inches and 44 feet 3½ inches for a total span of 66 feet 5 inches (see Figure 4). If you use bare copper with a higher velocity of propagation, increase these measurements by roughly 2.5% for a total span length of 68 feet.

Note that the antenna wire is wrapped through strain relief holes and attached on the back side of the insulator block with solder lugs. The support tether at the feed point may be used to reduce stress across the span of the flat top. By shopping around, you may be able to locate some inexpensive Teflon jacketed wire that does a very nice job of shedding water and ice. For end insulators, I used two 6 inch strips cut from black polyethylene and ⅛ inch parachute cord for support.

OCFDs and Mounting Height — the Elephant in the Room

If the OCFD has been touted as a compromise antenna, it may be because builders fail to consider the profound impact of ground proximity on the lower frequency bands. With that caveat in mind, please resist the temptation to double the wire lengths for this project to add 80 meters! It's true that the OCFD is an even-harmonic radiator and that 160, 80, 40, 20 and 10 meter bands are all harmonically related. At normal backyard mounting heights, however, Mother Earth perturbs the fundamental response more than the harmonic responses (see OCFD mounting height data in Table 1). As a result, unless you have very tall trees, you can count on 80 meters resonating

Table 1

Antenna Height versus Resonant Frequency (MHz) and Load Impedance (Ω)

Height (feet)	*40 Meters*		*20 Meters*		*10 Meters*		*6 Meters*	
70	7.11	87	14.24	150	28.68	128	50.33	139
60	7.12	108	14.18	147	28.64	127	50.34	139
50	7.06	122	14.20	127	28.70	132	50.3	137
40	6.95	114	14.29	135	28.64	125	50.33	137
30	6.87	84	14.26	175	28.67	129	50.33	140
20	6.88	47	14.06	156	28.71	137	50.38	138
10	6.99	15	14.05	63	28.42	124	50.26	151

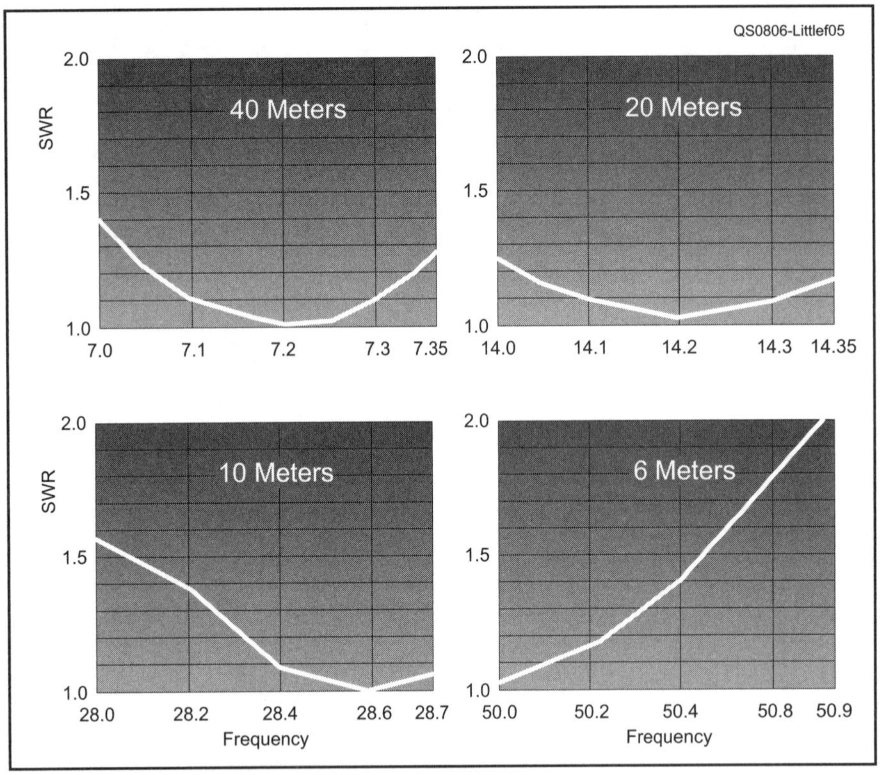

Figure 5 — OCFD SWR versus frequency plots.

below the edge of the band at well under the anticipated 130 Ω driving resistance.

What about Other Bands?

This antenna is not particularly usable on the 30, 17, 15 or 12 meter bands. The lowest modeled SWR I found was 23:1 on 17 meters. If the antenna is fed through 100 feet of RG-8X coax, the cable loss of almost 6 dB would result in an SWR at the radio of about 6:1 — likely usable with a wide range antenna tuner. There would likely be additional loss in the balun and transformer. Thus something less than 25% of the transmitter power would reach the antenna. On the other bands the mismatch is considerably worse.

Going on the Air

Figure 5 shows the measured SWR readings for my 45 foot high installation as seen through 100 feet of RG-8X. These favorable plots confirm *EZNEC*'s prediction that the 2.8:1 transformation ratio and 50 foot mounting height are a winning combination.

When it comes to on-air performance, the OCFD has the advantage of being electrically large, efficient and broadbanded. These qualities translate into having the ability to work almost any station you can hear — plus the freedom to hop from mode to mode or band to band without suffering power reduction from radio's final amplifier reducing power due to high SWR.

On 40 meters, the OCFD functions like any dipole with peak current occurring at mid element. As such, it models with 5.7 dBi gain at 42° elevation and works well for both domestic and DX contacts. On the harmonic bands, the radiation patterns develop progressively more peaks and nulls at higher octaves — much like a G5RV or a center fed dipole. As a result, the antenna will favor some directions with upward of 7.8 dBi gain on 20 meters and 9.2 dBi gain on 10 meters. Note that it is not omnidirectional and will exhibit weaker performance in directions at which nulls occur. The radiation angle on 6 meters is very low and SWR favors the bottom end of the band where horizontally polarized SSB, CW and AM signals prevail.

Finally, because this antenna has low visibility when tucked away among the trees, it might work well for hams living with covenants or apartment dweller restrictions. If you don't mind slitting turf and burying low loss cable in the dark of night, you could install the OCFD up to several hundred feet from your building. You'll lose a couple of dB to feed line loss on the higher bands, but you should suffer no additional transmission losses from high SWR. Best of all, the electrical racket from your complex as well as any consumer gadgets your signal might disable, will be several wavelengths away. Food for thought for the brave of heart!

Conclusion

This article presents a practical approach for achieving excellent multi-band performance and low SWR on its bands between 40 and 6 meters using a simple OCFD design. It doing so, it describes an alternative OCFD matching solution and raises awareness of the potentially negative impact of ground proximity on lower-frequency OCFD performance.

There have been many OCFD configurations described in the amateur literature. Serge Stroobandt, ON4AA, provides an excellent compendium of them on his Web site at **www.stroobandt.com** I also recommend reading the recent paper by L.B. Cebik, W4RNL, *The Isolated Off-Center-Fed Antenna: Some Less-Explored Facets*, available on his Web site at **www.cebik.com**. This comprehensive discussion of OCF behavior offers a wealth of new and useful information to OCF modelers and designers.

Notes

[1]Several versions of *EZNEC* antenna modeling software are available from developer Roy Lewallen, W7EL, at **www.eznec.com**.

[2]**www.fair-rite.com**.

Rick Littlefield, K1BQT, has an Amateur Extra class license. He was first licensed at age 13 in 1957. An avid builder and writer with over 100 technical articles in print, Rick especially enjoys designing antenna and low power projects and was an early inductee into the ARCI QRP Hall of Fame. His professional resume includes extensive work for familiar Amateur Radio manufacturers such as MFJ Enterprises, Ten-Tec and Cushcraft Corporation. He holds a master's degree from the University of New Hampshire and is currently employed as a product design engineer at Laird Technologies in Manchester, New Hampshire. You can contact Rick via e-mail at **k1bqt@arrl.net**.

A Pneumatically Switched Multiband Antenna

This creative design uses pneumatically activated switches to increase or decrease the electrical length of an antenna.

Gary S. Kath, N2OT, and Craig Bishop, WB2EPQ

For a number of years we have been experimenting with wire antennas trying to come up with that ideal multiband antenna that is fed with a single length of coaxial cable. Some two band 80 and 40 meter dipole antennas we tried and rejected were:

▪ *Multiband antenna using paralleled dipoles.* One approach is to cut a dipole for each band and connect the feed points of the two dipoles in parallel and to a single feed line as shown in Figure 1.[1,2] The free ends of the dipoles can be tied to different supports to keep the wires separated or insulators can be used between the wires. When the antenna is operated on a given band, one dipole is resonant having a low feed-point impedance while the other dipole is out of resonance, presenting a high feed-point impedance.

Our experience has shown that interaction between the two dipoles tends to pull the frequency away from the theoretical resonant length described by the usual dipole equation. Tuning the antenna is required by trial and error adjustment of the length of each dipole to get the best standing wave ratio (SWR). The antenna also requires making a number of insulating spacers to ensure the parallel dipoles do not touch one another on windy days.

▪ *Multiband trap antennas.* A second method to design a multiband antenna is to use *traps*. A trap is simply a parallel resonant circuit consisting of an inductor and capacitor.[3,4] In a two band parallel resonant trap dipole, one end of the traps is connected to the ends of the 40 meter dipole and the opposite ends of the traps are connected to extension wires forming the 80 meter dipole as shown in Figure 2. The trap is designed to resonate on 40 meters. The trap impedance at resonance is very high, electrically disconnecting the 80 meter section of wire while on 40 meters. When the antenna is operated on 80 meters the traps are not resonant. The trap then acts as an inductive reactance electrically extending the length of the extension wire forming the balance of the 80 meter dipole.

Traps are somewhat lossy, more involved to fabricate and can be affected by moisture. The traps also affect the ideal resonant length of the antenna making trial and error adjustment of the wire lengths necessary.

Multiband Antenna Using Switched Wire Segments

The ideal multiband antenna would use remotely controlled low loss switches to extend the length of the antenna for multiband use. On 40 meters the switches are open and on 80 meters the switches are closed.

Wired Relays

Our first thought was to simply put relays at the end of the 40 meter dipole legs to allow remote connection to the 80 meter wire sections. The problem is the control wires oper-

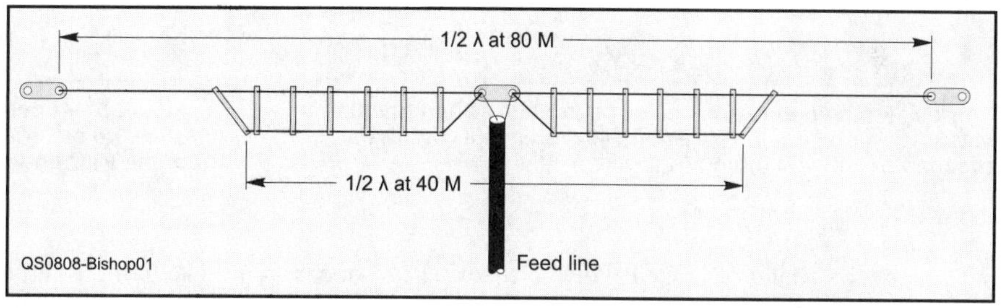

Figure 1 — Configuration of a multiband antenna made from two parallel dipoles.

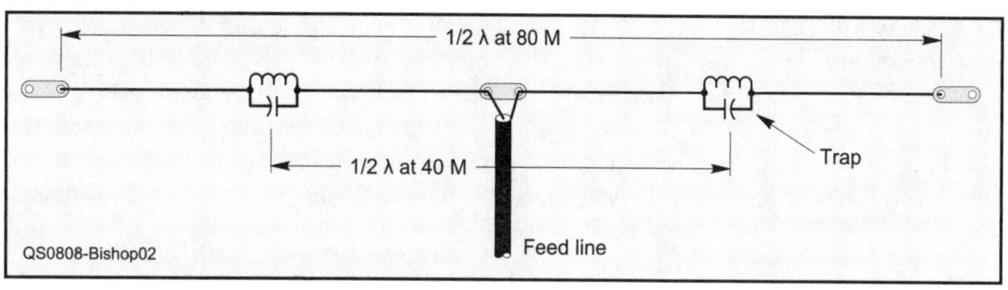

Figure 2 — Configuration of a multiband dipole using traps to separate the segments.

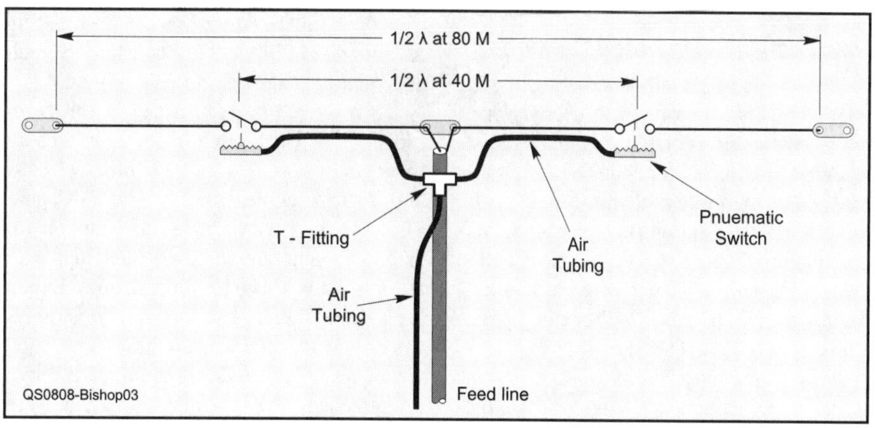

Figure 3 — Diagram of the design of a multiband antenna using pneumatically controlled switches.

[1]Notes appear on page 143.

Figure 4 — Surplus Micro Switch snap action switch used for band changing.

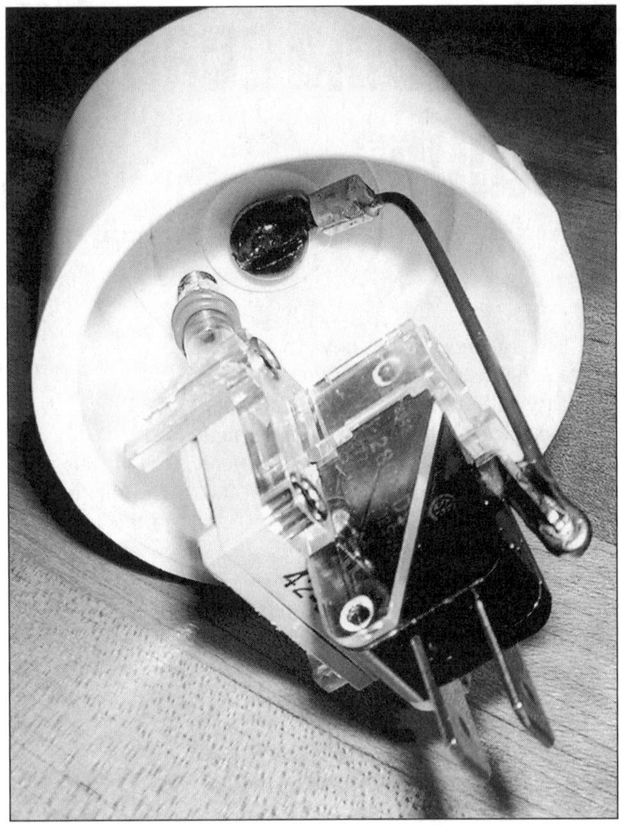

Figure 5 — Interior view of protective switch housing showing electrical and pneumatic connections through one of the PVC end caps.

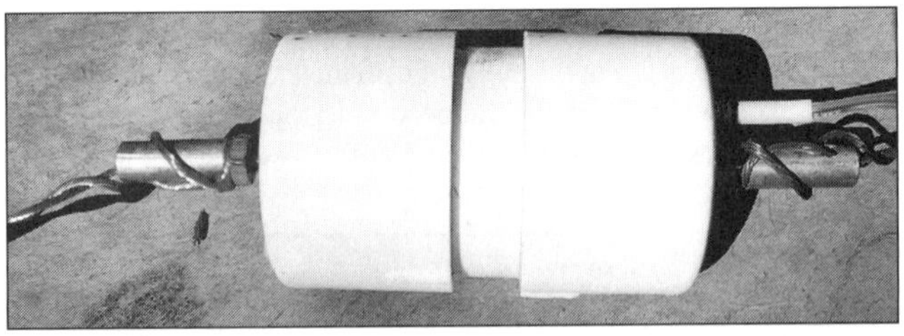

Figure 6 — Completed and sealed pneumatic switch housing with 80 meter extension wire (left), 40 meter wire (right) and attached pneumatic tubing (right).

ating the relays would be in the RF field of the antenna and interfere with the operation of the antenna.

RF Activated Switch

The next idea was to develop wireless RF activated switches. The idea was to design a circuit that could sense the RF frequency and automatically close the switches when operating on 80 meters and open the switches when on 40 meters. This would require battery powered RF sensing switches mounted at the ends of the 40 meter dipole. Although we did develop a rough prototype of this concept, it is not ready for prime time.

Pneumatically Operated Switches

Finally, we decided to experiment with pneumatically controlled switches. A pneumatic switch has a diaphragm that expands when air pressure is applied. The diaphragm movement then pushes against the actuator button of a snap action switch. A wide variety of pneumatically actuated snap action switches is available with current handling ranges from 3 to 25 A and operating pressures from 0.05 to 45 PSI.[5]

This configuration offers a cleaner looking installation, fast band switching and requires no coaxial switches to flip between antennas.

The advantage of pneumatically operated switches is there are no interfering control wires but only plastic tubing running up along the antenna. The idea was to place a pneumatic switch at the end of each leg of the 40 meter dipole and run pneumatic tubing from the switches into the ham shack as shown in Figure 3. To switch bands from 40 to 80 meters, simply apply air pressure to the tubing causing the remote switch to actuate, thereby extending the length of the dipole.

Construction of the 80 and 40 Meter Pneumatically Switched Dipole

For this application we selected surplus low pressure pneumatic switches having an integrated 20 A, 277 V ac Micro Switch snap action switch as shown in Figure 4.

To protect the pneumatic switch from moisture, it was housed inside an enclosure made up of two 1.5 inch PVC pipe end caps and a short segment of 1.5 inch PVC pipe. The end caps were drilled and tapped for size 12-24 machine screws. Round lugs with extension wires were slipped onto 1 inch long brass size 12-24 screws. The screws were then threaded through the end caps. Caulk was placed around the threads and a nut attached. The wires from the lugs were soldered to the normally open (NO) switch contacts on the pneumatic snap action switch as shown in Figure 5.

To route the air control line into the sealed pneumatic switch, a pneumatic bulkhead feed-through fitting was fabricated from a size 12-24 brass machine screw by cutting off the head, drilling a 1/16 inch diameter hole down its length and machining off the threads on the ends. One of the PVC end caps was drilled and tapped for a size 12-24 thread and the fitting was installed, sealed with caulk and locked with nuts. A short length of tubing was connected from the bulkhead fitting to the port on the pneumatic switch. The opposite side of the bulkhead fitting attaches to pneumatic tubing running back to the radio shack. If necessary, the design can be simplified by passing the tubing through a hole drilled in the end-cap eliminating the bulkhead fitting.

To complete the design, the PVC pieces were cemented together with PVC cement, forming a watertight seal as shown in Figure 6. The extending brass screws are attached between the 40 meter and 80 meter wires of the dipole using brass nuts. The dipole segment lengths were determined using the usual 468/f relationship.

The air tube feeding the pneumatic switches was 1/8 inch OD, 1/16 inch ID clear PVC tubing (McMaster Carr part number 5233 K514). This low cost tubing is trans-

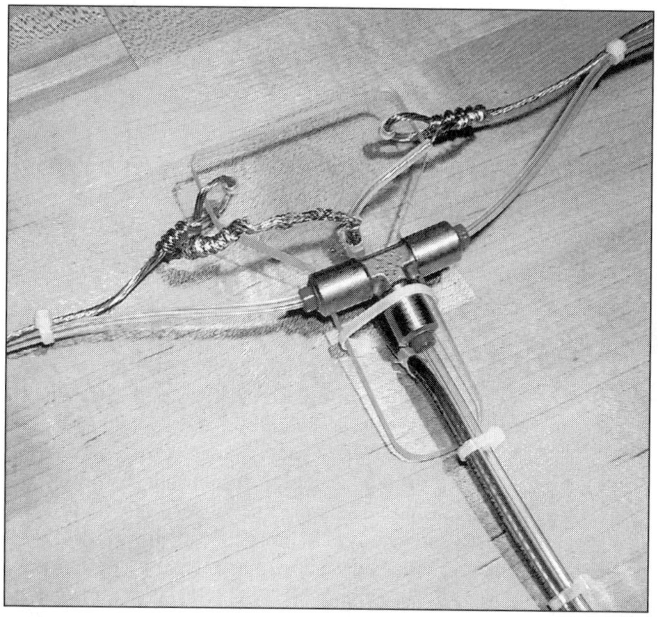

Figure 7 — Coaxial feed line attached to center of dipole. Left and right pneumatic tubing lines attached to T fitting and a single pneumatic line running down to the shack.

parent and has good ultraviolet resistance. A single length of tubing was taped along the coaxial feed line up to the dipole center insulator. A pneumatic T fitting was then used to split off the air line to the left and right segments of the dipole as shown in Figure 7. The tubing was attached along the dipole wires using electrical tape and then attached to the air bulkhead fitting on the pneumatic switch housing.

Low pressure air is required to activate the pneumatic switch from the shack. Low pressure pneumatic switches can even be activated by blowing into the air line. (This is not recommended, especially for SSB, since you will have to hold your breath while operating on 80 meters!) We simply used a small balloon hand pump to activate the switch for 80 meter operation and vented the air line for 40 meter operation. A balloon, bicycle pump, small fish pump and other air pressure providers could also be used as the air source.

Performance

Our dipole was erected between two large oak trees. One hundred feet of coax and pneumatic tubing ran down to the shack. We pulled out our old trusty MFJ Antenna Analyzer and to our surprise found the 40 meter antenna resonated with 1:1 SWR exactly at our design frequency.

We switched over to 80 meters by applying air to our pneumatic line. Checking the SWR again showed a 1:1 SWR on our 80 meter design frequency. There was no interaction between the 80 and 40 meter wires and no further adjustment was required.

On-air testing of the antenna was our next concern. Was the high voltage at the ends of the 40 meter dipole going to flash across our little snap-action switch? We fired up our 100 W transceiver and called CQ on 40 meters. Jack, K9GZK, responded to our call and we talked with no problems and no switch arcing.

Our final concerns were how long the pneumatic connections will last and whether moisture will condense in the tubing? After six months of operation, the only problem was that water entered one of the PVC enclosures via the capillary action along the brass screw threads, due to a poor caulking job. The switch was dried out and silicon caulk was reapplied over the threads. Since then, the antenna has worked fine through a variety of hot and cold weather conditions.

Other Ideas

The successful operation of the pneumatic switches generated other possible uses to explore in the future:

- *Phone/CW broadband dipole* — Use pneumatic switches to slightly extend the length of a dipole allowing broadband operation in both the phone and CW portions of the band. This is particularly suitable for 80 and 75 meter operation.
- *Multiband vertical* — Use pneumatic switches and the PVC housing to make a multiband vertical. Run the control air up the center of the antenna tubing to create a sleeker design.
- *Switching directions of a wire beam* — Use pneumatic switches to change the length of the reflector and driven elements on a three element Yagi or quad to rapidly change antenna direction by 180°.

We will leave it up to our fellow amateurs to explore other uses for low cost pneumatic switches. Meanwhile, work continues on our second generation multiband antenna design using RF activated switches.

Notes

[1]R. D. Straw, Editor, *The ARRL Antenna Book*, 21st Edition, p 7-9. Available from your ARRL dealer or the ARRL Bookstore, ARRL order no. 9876. Telephone 860-594-0355, or toll-free in the US 888-277-5289; **www.arrl.org/shop/**; **pubsales@arrl.org**.

[2]Dipole length in feet = $468/(f_{MHz})$.

[3]J. Carr, *Practical Antenna Handbook*, 2nd Edition, TAB Books, 1994, p 141.

[4]See Note 1, p 7-11.

[5]If surplus switches are not available, another source is Air-Logic (**www.air-logic.com**). Their switch number F-4100-50-B80-15A should be suitable. Others may be available from Micro Technologies, SA, Pompano Beach, FL; **www.pressureswitch.com**.

Photos by the authors.

Gary S. Kath, N2OT, was first licensed in 1968 as WN2EPP and currently holds an Amateur Extra class license. He earned a BSEE from Widener University and an MSEE from Carnegie Mellon University. Gary has 28 years of experience as an instrumentation and automation engineering manager. He enjoys DXing, contesting, low power (QRP) operation, circuit design, software development and experimenting with HF wire antennas. Gary can be reached at 2671 Sky Top Dr, Scotch Plains, NJ 07076 or at **njklepper@att.net**.

Craig Bishop, WB2EPQ took his first license exam with Gary in 1968 and was assigned consecutive call WN2EPQ. After earning his Advanced class license, Craig focused primarily on antenna experimentation, DX and contesting. He was also very involved in his high school and college radio clubs. Craig holds a BSEE from Lafayette College and is an information technology executive at a large corporation. You can reach Craig at 28 Black Birch Rd, Scotch Plains, NJ 07076 or at **cbishop@att.net**.

The W6TC DX Loop

An easy way to add 30, 40, 60, 80 and 160 to your SteppIR — or any Yagi.

George Badger, W6TC

The sunspots are at their minimum so the low bands are where the DX action is. Soon the high bands will become increasingly important. Now is the time to have an antenna system with all-band capability.

I have been building quad antennas for some years. The 5BDXCC and 5BWAZ awards provided incentive to add low band capability. I found that large kite-shaped quad loops worked very well on 30 and 40 meters, and as explained later, on 80 and 160 meters as well. I mounted these loops on extended quad spreaders in the diamond configuration.

These low-frequency loops were in the plane of conventional diamond quad loops and extended to a tuning box at the base of my 60 foot tower. Thus the wire loops were in the shape of a kite, wide at the top and coming together at the base of the tower. I matched these large loops with remote controlled networks to cover 30 and 40, and with changes to loop connections, 80 and 160 meters.

I vividly remember looking up at my high-band quad elements, wishing I could tweak their dimensions to optimize performance on each band. However, there was no way to do that from down on the ground.

W6TC

Figure 1 — Three element SteppIR with W6TC loop. The loop wire is barely visible. The loop extends from ground level to the boom extension on the left, to the mast extension at the top and back to ground level via the boom extension on the right. See Figure 2.

The SteppIR Appears

And then along came the SteppIR antenna. It was the answer to my dreams. Imagine motor driven Yagi element lengths, adjustable from my operating position! Imagine an antenna that tracks transceiver frequency, and automatically optimizes performance every 50 kHz across all the bands from 20 to 6 meters! The ads sounded too good to be true. The SteppIR seemed to be electrically and mechanically complicated, so I was concerned about reliability. However, I have two good friends, W6OD and N6ATD, who have SteppIRs, and have had no problems. They convinced me to install the three element version. The SteppIR has lived up to my fondest expectations. It is a great antenna for the high bands.

Low Bands

But what about the low bands? I have only one tower, and I live on a relatively small suburban lot, with neighbors sensitive to the appearance of our neighborhood.

Would my new SteppIR preclude operation on 30, 40, 60, 80 and 160 meters? I knew from my experience with the quads that the large low-band loops interacted with the high-band loops (they were in the same plane). They coupled and therefore required compensating adjustment.

I wanted to put a large loop on the same tower but I did not want the loop to spoil the superb performance of the SteppIR. For this reason I decided to mount the loop at 90° to the SteppIR elements, in the plane of the boom. This turned out to be a good decision, as discussed later.

W6TC DX Loop

The W6TC loop is shown in Figure 1 and the schematic diagram of Figure 2. The loop is supported on two 5 foot × ½ inch fiberglass boom extensions, and an 8 foot × 1 inch fiberglass mast extension. The wires from the ends of the boom extensions extend downward and terminate at the base of the tower. Figure 2 is a schematic diagram of the W6TC DX loop.

End Fed Dipole, aka Delta Loop

Figure 3A shows a basic dipole, the grandfather of most antennas. The current in a resonant dipole is sinusoidal. Most of the current (and therefore most of the radiation) is from the central portion of the dipole. Current at the ends of a conventional dipole at any frequency is always zero, so there is little contribution to the radiation from the ends. The major useful contribution made by the ends is to bring the antenna to resonance. The ends can be placed in any shape or position (Figure 3B) with little effect on the radiation efficiency or pattern. Figure 3C depicts loading or traps typically used in shortened dipoles and triband Yagis. Figure 3D shows the configuration of the SteppIR 30/40 meter add-on, which is proven to be an excellent radiator.[1] Note that the dipole ends are folded back almost to the center. Figure 3E depicts the SteppIR element end-loaded dipole by KL7CW.[2]

The end loaded dipole of Figure 3F shows the dipole end wires extending downward. The end loading wires can be extended downward farther so that the antenna evolves into a symmetrically end-fed dipole W6TC loop (as found in Figure 3G and Figure 3H). The dipole is fed at both ends instead of in the middle. The slanted vertical wires serve as end-loading near the top of the loop, and evolve into a transmission line as the wires come together near the bottom.

Note from Figure 3G that the feed point is now at the bottom. This is convenient because all tuning and matching for all bands can be accomplished from the ground (no more tower climbing). Further, when it is used in the Figure 3G configuration, supported at the ends, there is no requirement for a heavy transmission line. Most of the radiation comes from the center, so you do not want to sacrifice height by weighing down the center.

Wideband Pattern

One of the very important characteristics of the W6TC loop is that there is always a maximum radiation point exactly opposite the feed point regardless of loop size and frequency. Because the loop is fed in the center at the bottom, there is always a point of maximum radiation at the top center, the highest point on the antenna, at any frequency. I published an article explaining this in some detail in 1979.[3] Bill Orr also describes my loop antenna in his *Antenna Handbook*.[4]

The importance of the centered symmetrical current distribution of the W6TC

[1]Notes appear on page 146.

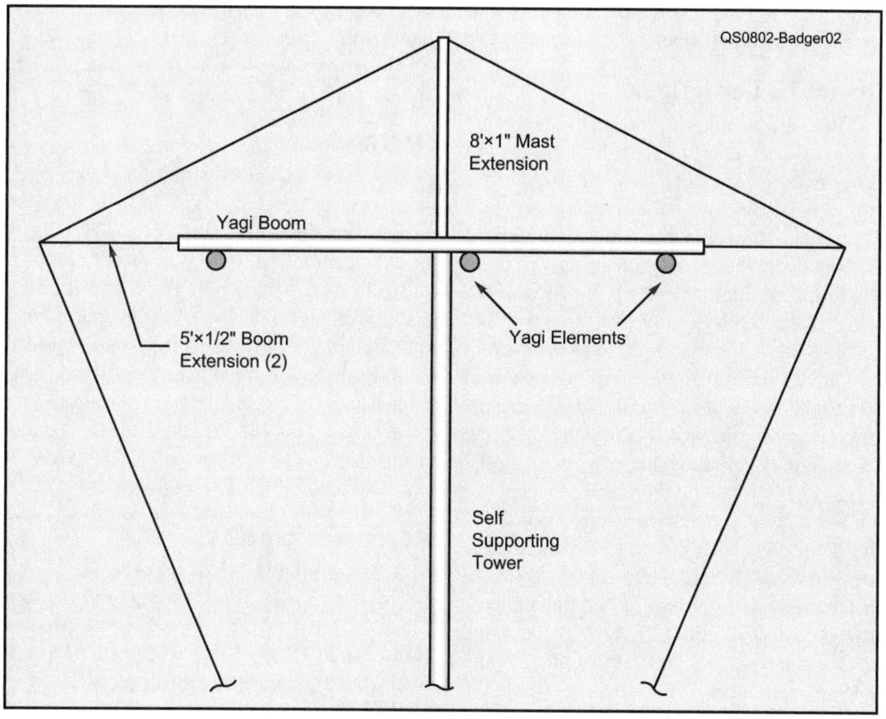

Figure 2 — Schematic diagram of the W6TC loop over SteppIR.

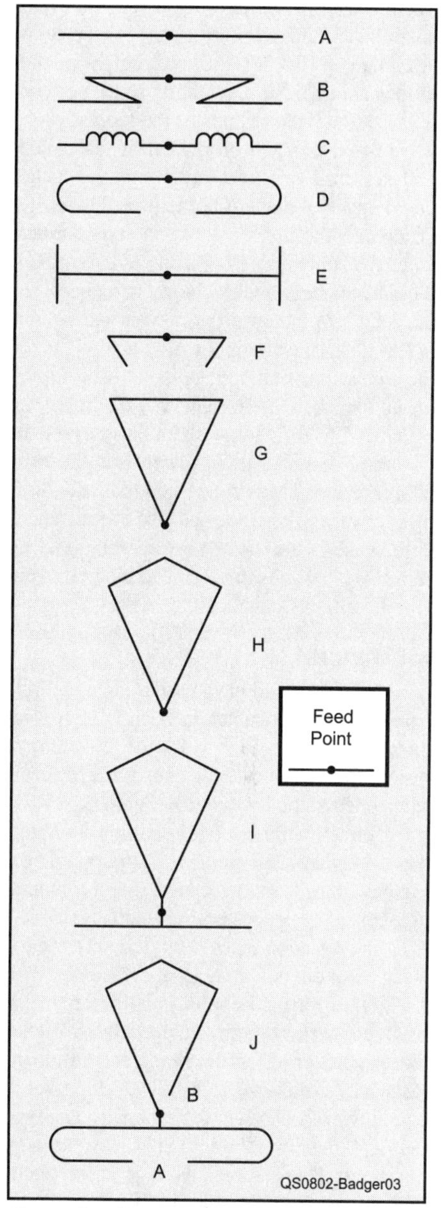

Figure 3 — How the W6TC DX loop evolved from the basic dipole.

loop antenna cannot be overemphasized. The radiation pattern over a wide frequency range is clean and stable. By comparison, the pattern of a conventional center fed dipole is far from clean as the frequency is increased over a wide range. Not only does the W6TC loop always have a maximum radiation point at the top center, the vertical components of the maximum current points on the vertical wires always cancel, regardless of frequency. The horizontal component of the radiation from the vertical wires nearest the dipole aids the radiation from the top horizontal wire, as explained in my 1979 paper.[5]

Loop Pattern Compared to that of a Wideband Dipole

A conventional multiband dipole simply does not work this way. Current at the ends is always zero, so the maximum radiation points are forced to move across the dipole, spoiling the radiation pattern at high frequency.

Dipoles or inverted Vs on the same tower with a Yagi should be avoided.[6] They should be low on the tower to prevent adversely affecting Yagi performance. Low band antennas should be as high as possible, not below the Yagi.

For DXing, antenna height is more important than length. If you can get a short wire higher, end feed it W6TC style. It will be a better DX antenna.

Rotatabilty Makes All the Difference

Having a rotatable dipole is a vast improvement over a fixed dipole, because you can move the major lobe to favor your target and use the side nulls to reduce interference and noise. *EZNEC* shows that the W6TC loop has a superb bidirectional broadside radiation pattern. On 30 meters, the gain is 8 dBi, the front to side ratio is 17 dB and the takeoff angle is 21°. On the air it feels like a beam because of the high-gain, deep-side nulls and low takeoff angle. Performance is also good on 40 meters.

Of course the W6TC loop does not have to be over a SteppIR. It can be mounted over any Yagi, regardless of boom length and height, because the loop dimensions are not critical. As a matter of fact, the loop does not have to be over a Yagi. It can be supported between trees as in Figure 3G or centered on a metal tower as in Figure 3H.

Feeding the Beast

I mounted a metal box that contains separate matching networks for each band, near the ground on my rotating tower. It is remotely controlled from the shack. Remote tuning is required in my location because open wire line cannot be installed through my house, so I went with coax feed.

A simpler way to feed the antenna is with open wire line, or ladder line, using a balanced tuner or conventional tuner with balun. As an experiment, to simulate a typical installation I temporarily installed a 75 foot open wire line to the tuner at my radio location. The W6TC loop is easily matched on all bands, 10 through 160 meters, using a basic L network tuner. Arrangements must be made to keep the line at least one line width away from the tower as it rotates.

Non-rotating Version

If your tower is fixed and is self-supporting, you can still install the W6TC loop. To prevent the vertical wires from wrapping around the tower, short lengths of PVC pipe fastened low on the tower and free to flex and rotate with the loop may be used.

Isolation, Loop to SteppIR

To show that isolation between the loop and SteppIR is adequate, I tuned both to resonance at the same frequency at which the coupling is the strongest. I matched and energized the loop on 20 meters. I then swept the SteppIR elements through the same frequency. I tried this on several frequencies and never could detect the slightest change in SWR on the loop or the SteppIR. I am satisfied that the performance of the SteppIR is not compromised by the loop.

80 Meters

Vertical polarization is preferred for

transmitting on 80 and 160 meters so connections at the feed point of the W6TC loop are changed. On 80, the total height of the loop is about 3λ/4 wavelength, so for vertical polarization the two ends of the loop are connected together and brought near resonance with a raised two wire radial system in the crawl space under my home (see Figure 3I). Of course, the radial system should be more extensive, but the two opposing radials utilize the only space available. In my case, because the total loop height is somewhat more than λ/4, and because it is, in effect, top-loaded, the maximum radiation point is well above ground level, away from nearby obstacles.

With *EZNEC* I found that almost as much current is flowing in the tower as in the two vertical wires. I was concerned that radiation efficiency would suffer because of resistance at the base of the tower. However, I modeled various resistances there and found that overall efficiency is satisfactory.

160 Meters

For 160, I could have used a large coil for baseloading the 80 meter connection. However, baseloading has low radiation efficiency. Instead I elected to leave one of the vertical wires open at the bottom (see Figure 3J). This is, in effect, a form of top-loading. Folding the monopole back on itself is an example of Figures 3D. The high voltage end of a monopole folded back does not substantially detract from the radiation pattern, but it does serve to bring the monopole to resonance.

The antenna height is substantially reduced with no significant effect on the pattern. There is a decrease in radiation resistance compared to a full λ/4 high vertical. However, because the total wire is more than λ/4, maximum radiation is above nearby obstacles. There is high voltage at the open end, so use a switch with high voltage holdoff. I use vacuum relays.

A popular form of toploaded 160 meter antenna is the inverted L. For comparison, I modeled the W6TC loop connected as in Figure 3J and modeled a classic inverted L at the same height. I modeled both over perfect ground and over two wire radials 2 feet above ground. The W6TC loop radiation pattern is better. *EZNEC* predicts that the W6TC loop is omnidirectional within 0.1 dB, whereas the inverted L is not symmetrical by more than half a dB.

EZNEC has a very useful feature. It calculates efficiency comparing total radiated power to total input power. The W6TC configuration of Figure 3J over a two wire resonant raised radial system over ground is 43% compared to 36% for the inverted L, probably because the horizontal section of the inverted L is more exposed to ground. Current in the tower is small because the tower is far from resonant. Thus the W6TC 160 meter vertical has better performance than the inverted L in most respects and requires only one support.

The W6TC Loop Radial

There is not enough space in my basement for two full λ/4 160 meter raised radials. To achieve a resonant radial system, I folded the two radials back on themselves as shown in Figure 3J. The radials are not symmetrical, so I was concerned that the current in them would not be equal. Because the ends of the two radials are physically close together and because the voltage and phase at the ends of the radials should be the same, I connected them together at A in Figure 3J forcing the voltage at the ends of the radials to be the same. The two radials are now more electrically symmetrical and therefore cancel better. I have modeled the loop radial simulating the asymmetry by placing one radial one foot closer to ground. With 1 A at the base of the monopole, current in two radials were predicted at 0.73 and 0.27 A. With the radial ends connected, the currents were 0.51 and 0.49 A, indicating almost perfect balance resulting in good cancellation of any radiation from the radials.

I have two raised resonant radial systems in my basement, one for 80 and the other for 160 meters. The 80 meter system has two opposing radials open at the ends. The 160 meter system has two opposing radials folded back and connected at the ends. With my MFJ clamp on RF ammeter, I measured the current in the radials. The 80 meter currents differ by 40%, the 160 meter by 11%.

The Loop as a Receiving Antenna

The loop receives well on 30 and 40 meters but the vertical polarization is too noisy on 80 and 160. I found that the 30 and 40 meter horizontally polarized connection is a quieter receiving antenna on 80 and 160 meters.

Thinking about the K9AY Loop, I discovered, working with *EZNEC*, that the 160 meter connection with a 10 kΩ resistor connected at B in Figure 3J has a 17 dB null.[7] I found that adding reactance increased that to more than 30 dB. I modeled both 80 and 160 meter antennas, using different ground models and received those results consistently. The same results were obtained in free space as well as over perfect ground by adjusting the impedance parameters at both ends of the loop.

The *EZNEC* results seemed too good to be true! Imagine having that kind of results in the real world. It was clearly time for real world validation. My friend Bob Alper, W6KT, at a distance of 10 miles, put a 160 meter signal on the air. After trying different resistance and reactance values, I obtained the results *EZNEC* predicted. The actual values for a particular location can not be determined by *EZNEC* because of ground variability and the effect of the tower. In my case, the load at Figure 3 J with the tower as ground was 161 pF in series with 1.24 kΩ. The front-to-back bandwidth covered the entire CW sub-band.

Having a rotary receiving beam with superb performance is very satisfying.

Results

The SteppIR works better than I expected, and so does the loop. The loop is easy to make, conservatively handles 1500 W, is easy to match on all bands and does not compromise the performance of the SteppIR. It is also excellent on the air. To show that it gets out well, I reviewed my log since August 2006 when we installed the SteppIR and loop. I have worked world-wide DX stations on each of the 160, 80, 40 and 30 meter bands using this antenna, most with just 100 W. I am very pleased with the results.

Acknowledgments

I wish to thank Michael Bach, WB6FFC, for his major role in my antenna installation along with Bob Alper, W6KT; Larry Moore, W6OD, and Richard Baldwinson, N6ATD. I also wish to thank Roy Lewallen, W7EL, for his technically superb and exceptionally user-friendly *EZNEC* antenna modeling software and L. B. Cebik, W4RNL, for his ARRL modeling course. Special thanks to my wife, Nancy, for her patience with my passion for antennas, building things and DXing, and her encouragement on the subject of this paper.

Notes

[1]SteppIR Antenna Systems, **www.steppir.com**, 2112 116th Ave NE, Bellevue, WA 98004.

[2]R. Dwight, KL7CW, "Add a 30 and 40 Meter Dipole to Your SteppIR Yagi," *QST*, Jun 2007, pp 37-39.

[3]G. Badger, W6TC, "Compact Loop Antennas for 80 and 40 meter DX," *Ham Radio*, October 1979, p 24.

[4]W. Orr, W6SAI, *The Amateur Antenna Handbook*, 1993, p 120.

[5]See Note 3.

[6]"The Doctor is IN,"*QST*, Apr 2006, p 61.

[7]G. Breed, K9AY, "The K9AY Terminated Loop — A Compact, Directional Receiving Antenna," *QST*, Sep 1997, pp 43-48.

George Badger, W6TC, was originally licensed as W6RXW in 1939 and has been a member of the ARRL continuously for 68 years. After WWII service with the 89th Infantry Signal Company in Europe, George graduated from the University of California at Berkeley with a degree in electrical engineering. He holds seven patents on microwave tube design and was Marketing Director for EIMAC, President of Svetlana, and now consults for CPI Econco. He has published many technical articles professionally and in the amateur press. He is a fellow in the Radio Club of America. His Amateur Radio interests include building equipment, antennas and working DX. He holds DXCC Honor Roll #1, 5BDXCC and 5BWAZ.

George can be reached at 341 La Mesa Dr, Portola Valley, CA 94028.

The *No Excuses* 160 Meter Vertical

Stop procrastinating.
Build this fun antenna and get on the top band train.

John Miller, K6MM

"My lot size is too small." "I can't compete with the Big Guns." "I don't have the time." "Too complicated." "Too expensive." Sound familiar? It's easy to become a top band curmudgeon — avoiding putting up a 160 meter antenna because it may be more work than fun. Well, if you're having a difficult time putting a decent signal on 160 meters, here's a possible solution to get you up and running on the "gentleman's band," while leaving all those excuses behind.

Background

My first exposure to a helically wound vertical (HWV) was Gary Ellingson's 1972 *QST* article describing a 75 meter antenna.[1] His unique *no loading coil* approach eliminated the need for guying. I first tried this antenna design while living in Pennsylvania years ago and obtained reasonable results for local contacts using low power.

For many years, I dropped the HWV approach in favor of dipoles or inverted Vs for 80/75 meters, but became interested again when reading about Jack Swinden's, W5JCK, clever *broomstick with a top hat portable antenna* design.[2] He used ½ wavelength of wire for a targeted resonance frequency of 3.800 MHz and emphasized the importance of carefully calculating and measuring the number of turns around the antenna. Jack's meticulous attention to construction detail was inspiring, and this became my second homebrew HWV project. My Yaesu FT-847 HF/VHF/UHF transceiver at 100 W with this little portable antenna was a fun combination for ARRL Field Day and short trips. There's always something satisfying about a homebrew antenna that generates memorable contacts.

Overcoming My Excuses

I'm an avid contester, but had no antenna for 160 meters. In fact, I was a bit cynical about ever being able to put up an effective 160 meter antenna on my rather small California city lot. My Northern California Contest Club (NCCC) contesting buddies, however, convinced me that I was missing out on some big time fun with great conditions during our sunspot lull, not to mention the ARRL 160 Meter, CQ 160 Meter and Stew Perry top band Challenge contests. No more excuses. It was time for me to get on the top band train too.

A review of the literature on 160 meter antenna designs leads to the usual discussion of dipoles, inverted Ls, Ts, Vs, loops, deltas and verticals. After thinking about my own location's constraints I found myself revisiting the HWV option and settled upon a design often discussed but not often deployed in the US — a helically wound vertical antenna using PVC tubing.

It appears that most HWV antennas for 160 meters have been homebrewed in the UK, where the limitations of their small gardens are common. Phil Sidwell, MØVEY, describes his top band homebrew helical, including a fairly elaborate earth ground system.[3] This type of antenna appears to have gained widespread popularity throughout the UK and Europe.

Wire Wisdom

There is no hard and fast formula for determining the amount of wire needed to establish resonance in a helical antenna. The relationship between the length of wire needed for resonance and a full quarter wave at the desired frequency depends on several factors. Some of these are wire size, diameter of the turns and the dielectric properties of the form material. Experience has indicated that a section of wire approximately one half wavelength long, wound on an insulating form with a linear pitch (equal spacing between turns) will come close to yielding a resonant quarter wavelength. Therefore, an antenna for use on 160 meters would require approximately 260 feet of wire, when spirally wound on a support.[4]

Add other possible challenges such as narrow SWR bandwidth, low feed-point impedance and radiation resistance, efficient top hat capacitance, mechanical constraints, sufficient ground radial system — and you could easily become a top band curmudgeon. But then you'd miss out on building this fun antenna — one that really works!

To try to get a first approximation on a final HWV design, I used modeling software developed by Reg Edwards, G4FGQ.[5] His program models and predicts the performance of a helically wound vertical antenna, mounted immediately above a ground plane, top-capacitance-loaded with a vertical rod or whip. Enter these variables — height and diameter of the helical coil, number of turns and diameter of wire, length and diameter of the end loading rod — and you get back theoretically useful data. Outputs include ¼ wave resonant frequency, length of wire needed, helix wire pitch, capacitance and inductance data, feed-point impedance and expected bandwidth.

Version 2.0 — Looking Better

This article describes the construction and

[1]Notes appear on page 151.

performance of a very simple but effective HWV antenna for 160 meters. In a nutshell: The antenna is made by telescoping three 10 foot PVC sections together, helically winding it with ½ wavelength of antenna wire, attaching a capacitance hat to the top and feeding it with a 50 Ω feed line against eight ground radials. The entire construction can be easily completed in just one day using very simple tools.

Construction

Step 1 — PVC Painting

The antenna is made from three 10 foot sections of readily available PVC tubing in three diameter sizes: The top section is 1 inch, the middle section 1½ inches and the bottom section 2 inches. To make this antenna stealthy and environmentally friendly, the three PVC sections were spray painted green by suspending each 10 foot section from two pieces of nylon rope between two branches of a convenient backyard tree. Brown paint might work just as well, depending on the hues of your foliage. All three 10 foot sections were allowed to dry thoroughly before I proceeded.

Step 2 — Bottom Section — Coax, Antenna and Ground Connections

The bottom 2 inch PVC section is prepared for both ground and coax connections by drilling the necessary mounting holes. A PVC cap is placed on the bottom of the 2 inch diameter PVC tube and then, using a felt tip marker, a circle is drawn around the PVC just above the border between the bottom cap and PVC section. This marker ensures that subsequent drilled holes will clear the bottom PVC cap.

Coax Connection: The PVC cap is then removed, and holes are drilled for the SO-239 UHF type bulkhead connector and four attachment screws. The SO-239 hole is centered about 2½ inches above the marker (see Figure 1).

Bottom Antenna Binding Post: One ⅛ inch hole is drilled for the antenna binding post, placed 2 inches above the marker. A red binding post was used for the antenna connection.

Ground Binding Posts: Two ⅛ inch holes are drilled for the ground posts. Each is placed 1½ inches above the marker. Black binding posts were used for ground connections.

Summary: The three binding post holes (one for the antenna wire, two for the radial connections) are placed equidistant from each other around the PVC section. The antenna post and ground posts are staggered vertically by about ½ inch to avoid any possibility of shorting (see Figure 2).

Step 3 — Wiring the Coax Connector and Antenna Post

One end of a 4 inch piece of 14 gauge wire is soldered to the center connector of the SO-239. The other end is then soldered to either a spade or ring lug. The wire is then pushed through the prepared SO-239 hole in the 2 inch PVC tube, and the SO-239 connector secured to the PVC tube using only three of the four mounting holes. The free end of the insulated wire is connected to the inner section of the red antenna post using the spade or ring lug. After you have secured the antenna post with a binding nut, the connection can be soldered.

Step 4 — Wiring the Coax Connector and Ground Post

A 6 inch section of 14 gauge insulated wire is soldered (or crimped) to spade lugs on both ends. One end is connected on the outside of the PVC to the remaining SO-239 screw and secured to the PVC. The other end of the #14 wire is connected to the closest black ground binding post on the outside of the PVC.

Figure 2 — Binding posts and internal wiring arrangement.

Inside the PVC, another piece of 14 gauge wire is attached between the two ground binding posts. This essentially connects both ground binding posts and the coax base together. Check to be sure the antenna and ground connections inside the PVC are clean and not touching each other. Braided coax, such as RG-58, can also be used instead of the #14 wire for ground post connections. At this point, the SO-239 and all binding posts should be tightened and secured. For extra strength and protection, the binding posts can also be glued to the PVC, both inside and outside.

Step 5 — PVC Mast Preparation and Assembly

The top, middle and bottom sections are assembled using a high tech solution — duct tape. I actually used Gorilla Tape 10 for wrapping because it uses two layers of adhesive and two layers of fabric backing to make it much stronger than standard duct tape.[6]

First, the 1 inch diameter PVC tubing is shortened from 10 feet to 7 feet 6 inches by cutting off 2 feet 6 inches from one end. Duct tape is then wrapped around the tubes as follows:

- For the top section apply two wrappings. The first wrap should be 2 inches from bottom of tube. The second wrap from 9½ to 11½ inches from the bottom of the tube (see Figure 3).
- For the middle section also apply two wrappings. The first wrap at 2 inches from the bottom of the tube. The second wrap from 22 to 24 inches from the bottom of the tube.

The three PVC sections are then telescoped together. When assembled, the middle section will extend 24 inches into the bottom section, and the top section will extend 11½ inches into the middle section. It's important to use enough duct tape to ensure a good fit between the PVC sections (see Figures 4 and 5).

Figure 1 — Details of SO-239 UHF coax connector attachment.

Figure 3 — PVC sections wrapped with duct tape to make a tight fit.

Figure 4 — PVC sections telescoped together and ready for final assembly.

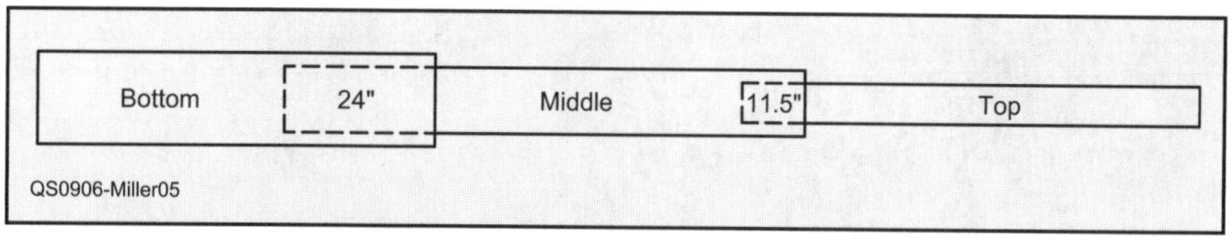

Figure 5 —Dimensions of PVC section telescoping overlap.

Figure 6 — Wire wrapped around pole and taped to maintain position.

Figure 7 — Detailed view of inside of top of PVC cap showing soldered connections.

The next step in PVC assembly is to further secure the joints with a bolt and nut. The lower joint, between the middle and bottom sections, is secured by drilling a ¼ inch hole through both PVC sections about 12 inches from the top of the 2 inch diameter top PVC section, and using a 3¼ inch bolt, washer and self locking (acorn) nut to fasten the sections together. Avoid over-tightening.

The middle joint, between the middle and top sections, is secured by drilling a similar hole, about 6 inches down from the top of the 1½ inch diameter middle PVC section. Use a similar attachment arrangement, this time with a 2¾ inch bolt.

Top antenna binding post: This is similar to the bottom antenna post previously described. The top antenna post is prepared by drilling a ⅛ inch hole 1 inch from the top of the top PVC section. A red capped binding post is attached to it using a nut and glued to secure it. The helically wound antenna wire will be connected to this post, which will also be the antenna to capacitance hat attachment point.

Step 6 — Helically Winding the PVC

With the sections assembled and fortified, the antenna is ready to be helically wrapped with wire. As previously mentioned, experimentation with HWVs has shown that a ½ wavelength of wire is often needed for ¼ wave resonance, assuming the turns are evenly spaced. At a desired resonant frequency of 1.825 MHz, 256 feet 5 inches of wire is required for a 160 meter vertical, using the formula 468/f. For this first version of the antenna, I chose 22 gauge insulated wire — I had a good supply sitting in the garage.

Using our kitchen table, which measures 5 feet long and a coffee can with 2 large screws protruding from sides at the top and bottom 180° apart to keep the wire from falling off the can as it was being wound, my wife unwound the wire from the supply spool while I wound it onto the coffee can. We went 50 times across the kitchen table for 250 feet plus an additional 6 feet 5 inches. The wire was cut, adding a few extra inches for experimentation, but keeping the 256 feet 5 inch point marked.

Wrapping begins by first attaching the antenna wire to the bottom antenna binding post of the 2 inch PVC section using a spade or ring solder lug. The wire is then wound from bottom to top, being careful to keep the winding pitch as consistent as possible, and avoiding the bolts near the two PVC joints. A spacing of about ½ inch seemed to work well. Wire wrapping is not a difficult step, but does require a bit of patience. It's best not to rush this part of the project. Duct tape is helpful here every few feet to keep the windings secure (see Figure 6). In a later version of this antenna, I hot-glued the antenna wire to the PVC for even better protection.

The end of the wire at the top of the antenna is then soldered to a spade or ring terminal and attached to the top antenna binding post with the red cap at the top of the 1 inch PVC section.

Option for Portability: It's possible to make the antenna portable by cutting the antenna wire at the two PVC joints. Then, after removing wire insulation, alligator or quick disconnect clips are attached to the antenna wire ends. The antenna can then be pulled apart, moved, re-telescoped together, and the full antenna wire length restored by

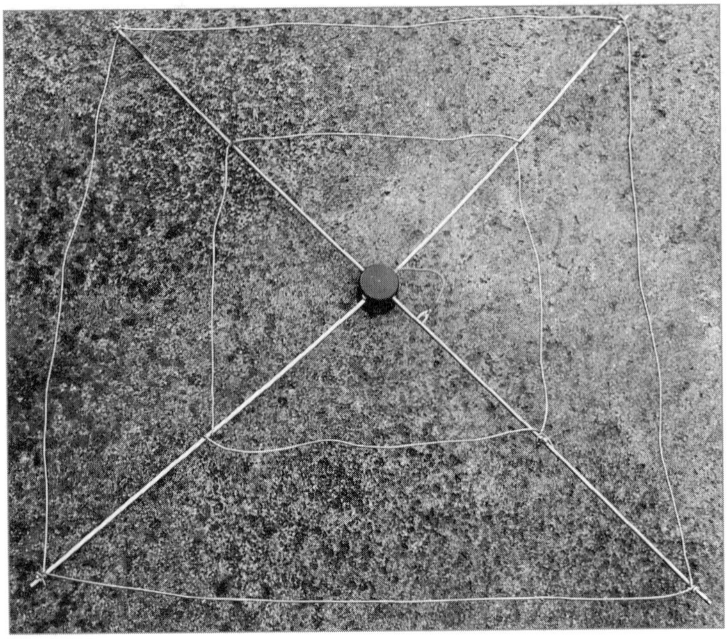

Figure 8 — Completed square capacitance hat ready for installation.

simply connecting the antenna clips together.

Step 7 — Top Cap Preparation, Capacitance Hat

There are several designs for a suitable HWV capacitance hat to provide capacity termination and reduce noise. At first, I chose the circular hat design described by Jack Swinden, W5JCK, in which six 12 inch brass rods are spaced equally around a PVC cap, and soldered together.[7] I eventually settled on a simpler square hat design using two 36 inch brass rods spaced 90° apart and connected together with 14 gauge copper wire.[8] Either of these methods work well.

The square hat construction begins by drilling four 1/8 inch holes 90° apart in the 1 inch PVC cap, about 1 inch from the bottom. An additional 1/8 inch hole is drilled next to one of these holes. The brass rods are inserted into the cap, forming an ×. A pair of pliers is helpful here, as it will be a snug fit, which is what you want.

Next, a 6 inch piece of 14 gauge insulated wire is stripped on one end, and soldered to a spade or ring lug on the other end. The stripped end is slipped through the remaining 1/8 inch hole and wrapped securely around the × junction of the two brass rods inside the PVC cap, where everything is securely soldered (see Figure 7). The brass rods are tied together externally by connecting them together with 14 gauge bare copper wire in two places — the tips of the rods and also midway between the rod ends and the PVC cap. The bare copper wire is soldered to the brass rod at all eight intersections to complete the square hat (see Figure 8).

Finally, the PVC cap is attached to the top section of the antenna, and the capacitance wire secured to the top antenna binding post using the spade or ring lug. For best results, it's important to have a good electrical connection between the antenna wire and the capacitance hat.

Figure 9 — Bottom cap on plywood lower support member.

Step 8 — Bottom Cap Preparation

The bottom cap is used to support and protect the antenna. A 1/4 inch hole was drilled in the center of a 1 foot square piece of scrap plywood. Another 1/4 inch hole was drilled in the bottom of the 2 inch PVC cap. The threaded aluminum rod was trimmed to 12 inches and run through the bottom PVC cap, and then through the plywood (see Figure 9). Nuts and washers were attached on the threaded rod inside the cap and also on the other side of the plywood. When it was tightened, only 2 inches of rod was left inside the cap, to ensure that the antenna and ground wiring in the bottom section of the mast would not be disturbed (see Step 2). About 10 inches of threaded rod was left sticking out from the bottom of the plywood (see Figure 10). The plywood base serves as a stabilizing platform to ease final installation of the vertical. By gently standing on it and pushing, you can easily drive the 10 inches of threaded rod into most ground.

Figure 10 — View of threaded rod that will secure the antenna to the ground.

Step 9 — Erecting the Antenna

After the PVC sections were bolted together, and completely wire wrapped, the capacitance hat was attached to the top of the antenna, including the capacitance wire to antenna binding post connection. The antenna is now ready for final installation (see Figure 11). The bottom 2 inch PVC cap with the plywood base was set in the ground at its mounting location. Bracing the bottom against the ground, the 27 foot antenna was carried to the PVC cap base and carefully set into the PVC cap. One person can carry and mount the antenna but it's a bit easier with two folks.

My location required bracing the mast to my back fence and securing it at the 6 foot point with nylon rope. To keep the antenna vertical, a section of nylon rope was also attached at 12 feet using a convenient tree limb and the rope secured at ground level. Final guying and bracing details will depend upon your antenna placement.

Radial Wires

This antenna does require some ground radials. Of course, use as many as you can fit in your location. I started with four 1/4 wavelength ground radials cut for 160 meters and have expanded that number now to eight, using 16 gauge stranded insulated wire. Spade lugs are soldered to ground radials that are then attached to one of the two ground posts. Because of the geometry of my property, my radials cover only a 180° arc but they work pretty well.

Initial Readings

After attaching a 6 foot piece of 50 Ω coax, an MFJ 249B antenna analyzer showed resonance close to 1.790 MHz. The antenna wire was adjusted at the bottom to bring the resonance closer to 1.830 MHz. Running 500 W through this antenna without a tuner showed a 50 kHz bandwidth, with less than

Figure 11 — Antenna fully assembled and ready to erect.

Figure 12 — Latest version of 160 meter vertical. Installed and neighbor friendly.

a 2:1 SWR. With a tuner, the antenna can achieve an SWR under 2:1 anywhere from 1.800 to 1.900 MHz.

Version 3.0 — Update

After a few months of use, I took down the antenna and decided to fortify the antenna wire by hot gluing it to the PVC. While on the ground again, the entire antenna was re-wrapped using four conductor 18 gauge stranded wire — I had a good supply sitting in the garage. At each end of the antenna the ends of the four conductor wire were twisted and soldered together before wrapping and attaching to the top and bottom antenna binding posts. I also added two elevated radials around the fence line (see Figure 12). While it seems to work better with four conductors, the single conductor is just fine.

On the Air

So how does this helically wound vertical perform? From the West Coast, it's a solid performer throughout North America. I have worked all 50 states, Canada and Mexico during the last year with it, almost all confirmed via Logbook of The World (LoTW). I was awarded First Place, Single Operator, Low Power for the Santa Clara Valley section in the 2007 ARRL 160 Meter Contest. In the 2009 CQWW 160 Meter Contest, I worked 46 states and 7 countries using 600 W in just a few hours of operating. For DX, with limited operating time, I have worked 30 countries. Overall, this antenna plays well to the Far East, South Pacific, Eastern Russia, Caribbean and Central/South America. Europe is the most difficult region to reach from my location, but that's generally true for most West Coast stations.

Am I the loudest signal on the band? No. Can I compete in pileups with folks having better antennas or higher power? No. But am I having fun on top band using a homebrew antenna that generates memorable contacts. You bet!

Summary

A helically wound vertical is not the perfect antenna for 160 meters, but for a small lot or where antenna restrictions are enforced, this easy to build vertical is a good alternative to an inverted L or dipole. During the last year, I have helped other hams around the country get on the air with this HWV design for 160 meters.[9]

This unsolicited comment from Armand Sun, K6IP, is typical of the feedback I've received:

"I finally put up the HWV antenna and I'm happy to report that it works FB. Mine has two feed options: ladder line or coax. I'm currently feeding with ladder-line and one elevated radial from leftover wire on the spool and the results are excellent! It takes a kW from 1.8-1.9 MHz. I painted mine olive drab with black 14 gauge wire so it's pretty stealthy. I would imagine brown would be good too. Sometimes the traditional designs just don't blend well with the existing antenna farm. A helically wound vertical is a good option for small lots or for those with antenna restrictions. Thanks for the design. It was fun to build and just what I needed for a top band solution."[10] So, no more excuses for Armand — nor me. Now how about you?

Notes

[1]G. Ellingson, WAØWHE, "A Helically Wound Vertical Antenna for the 75 Meter Band," *QST*, Jan 1972, p 32.

[2]J. Swinden, W5JCK, **w5jck.com/broomstick_antenna/80m_helical_antenna.html** and **www.dxzone.com/cgi-bin/dir/jump2.cgi?ID=19786**.

[3]P. Sidwell, MØVEY, **uk.groups.yahoo.com/group/topband-helical/**.

[4]R. D. Straw, N6BV, Editor, *The ARRL Antenna Book,* 21st Edition, p 6-38. Available from your ARRL dealer or the ARRL Bookstore, ARRL order no. 9876. Telephone 860-594-0355, or toll-free in the US 888-277-5289; **www.arrl.org/shop**; **pubsales@arrl.org**.

[5]R. J. Edwards, G4FGQ, "Model and Predict Helically Wound Vertical Antennas," **www.smeter.net/antennas/helical-modeling.php**, Aug 2, 1997, and "Very Short, Helically Wound, Monopole Antennas," **www.smeter.net/antennas/short-helical.php**, May 19, 2006.

[6]Gorilla Tape. **www.gorillaglue.com/tapes.aspx**.

[7]See Note 2

[8]I am especially grateful to Jon Sims, N7ON, for sharing his ideas and experiences with me. J. Sims, N7ON — Personal communications, Jan 2006.

[9]A downloadable construction manual is posted on the author's Web site at **k6mm.com/antennas/160M.pdf**.

[10]A. Sun, K6IP, e-mail communication, Jan 29, 2009. Quoted with permission.

John Miller, K6MM, is an ARRL member and the President of the Northern California Contest Club, which sponsors the California QSO Party every October. He was first licensed in 1958 as WV2BQJ in Syracuse, New York, Ohio (WB8CHZ), Pennsylvania (WA3VTM) and California (WA6OMA and KE6MI), and now holds an Amateur Extra class license.

John recently retired after careers in research, marketing and sales in the pharmaceutical, medical diagnostics and human resources industries. He holds BS and MS degrees from Syracuse University, an MBA from Pepperdine University and a PhD in chemistry from Case Western Reserve University. His hobbies today, in addition to his four grandchildren, include the radiosport of contesting, experimenting with antennas, and exploring the back roads of the Bay Area on his Harley. His personal Web site is at **www.k6mm.com**. *You can reach John at 6349 Slida Dr, San Jose, CA 95129 or at* **k6mm@arrl.net**.

A Backpacker's Delight—The Folding J-Pole

A new twist on the old J-pole.

Michael Heiler, KAØZLG

As the Emergency Coordinator (EC) for Clay County, Minnesota, I wanted to make up a jump kit that I could just grab and head out the door with. I needed a good VHF/UHF antenna that would give me some gain and portability and my inclination was to use a J-pole, but these antennas are somewhat long and cumbersome.[1] I had made one out of 450 Ω ladder line and I also made one from 300 Ω twinlead, but I now needed a version for outdoor use that was physically strong and durable, yet easily transportable.

Building the Antenna

I like the copper J-pole, so I started a design using that antenna as a model. The idea of a folding version of the copper antenna appealed to me. I went to the local lumberyard and hardware store (a home building supply will work, as well) and picked up what I needed to make the portable antenna project a reality. The necessary parts for this project can be found in Table 1, and Figure 1 can be used as a basic assembly guide for building the antenna.

[1]Notes appear on page 154.

Table 1
Materials List: Folding Copper J-Pole Antenna

- 1 each, 10 feet of $^{1}/_{2}$ inch copper pipe.
- 3 each, $^{1}/_{2}$ inch copper end caps.
- 1 each, $^{1}/_{2}$ inch right angle (elbow).
- 1 each, $^{1}/_{2}$ inch T angle.
- 2 each, $^{1}/_{2}$ inch copper pipe clamps (wall clamps).
- 2 each, $^{1}/_{2}$ inch copper pipe couplings.
- 4 each, 6-32 bolts × $^{3}/_{4}$ inch, depending on Plexiglas thickness.
- 4 each, external star washers, #6 and #8.
- 10 flat washers, #6 and #8.
- 2 each, $^{3}/_{16}$ inch eyebolts.
- 4 each, 10-24 brass nuts.
- 1 each, 2$^{3}/_{4}$ × 2$^{3}/_{4}$ inch Plexiglas, $^{1}/_{8}$ to $^{1}/_{4}$ inch thick.
- 3 ring terminals for #14-16 wire for #6-#8 bolts.
- 5 feet of $^{3}/_{16}$ inch OD bungee cord.

The antenna components are shown in Figure 2. You will first need to cut five 19$^{1}/_{4}$ inch sections and one 2$^{3}/_{4}$ inch section from a 10 foot length of copper pipe. Soldering the pipe joints is the next step. Be aware that, when soldering to copper pipe, you will need to clean the area of the pipe sufficiently by sanding. 3M scouring pads also work well for this operation. Soldering paste/flux will have to be applied to the joint ends to ensure a good joint. Solder a coupling to one end of two of the longer pieces of copper pipe. Next, attach two brass nuts to each of the eyebolts and tighten them against each other. Place the eyebolt assemblies inside two of the end caps and solder them in place using a propane torch. Fill the end caps with solder to cover the brass nuts. Figure 3 shows the eyebolt assembly, anchoring the bungee cord. Make sure you use brass nuts because the solder will not adhere well to steel.

When all the pieces are cut, lay them out, unassembled, as if they were joined. That technique ensures that you don't make any mistakes when assembling everything. You can then start to assemble the antenna and solder the rest of the copper J-pole—two 19$^{1}/_{4}$ inch pieces, one elbow (the elbow must be attached to the $^{1}/_{4}$ wave length side), a 2$^{3}/_{4}$ inch piece for spacing, the T section and one other $^{1}/_{4}$ wave section. Refer to Figure 1 for guidance—that drawing shows you how the antenna components fit together.

Radiator and Mast Assembly

After basic soldering, what you have will look much like a squared off U. After all the soldering is completed, you will need to tie a large enough knot in one end of the bungee cord so it won't pull through an eyebolt. Run the bungee cord through an eyebolt, through the other pieces of copper pipe, and place the eyebolt cap in the copper pipe, tapping the copper cap into place. Make sure that the couplings are facing down, so that when it rains the antenna doesn't fill with water.

A view of the bungee cord running through two of the elements can be seen in Figure 4. On the lower support section (below the actual J-pole radiator), you will want to add the last 19$^{1}/_{4}$ inch length so you have plenty of room for mounting the antenna to a mast. Run the bungee cord through this section also, and pull it tight...so that it holds all the sections together. With the bungee cord tight, run it through the other eyebolt and tie a knot in it, stuff it back into the pipe, place the cap on the pipe, and tap it into place. *Do not* solder the caps when putting them on the copper pipe because the eyebolts will come loose. These should only be force fitted to the pipe ends.

Feed Point

For the feed point support, use a piece of Plexiglas, cut to 2$^{3}/_{4}$ × 2$^{3}/_{4}$ inches and drill a 1/2 inch hole in the center to mount the SO-239 coaxial socket. Refer to Figure 5, which shows the feed point assembly. Before you add the washers and nuts, you may want to clean them off with sandpaper to remove any oxidation. You'll need to place the SO-239 connector on the Plexiglas and mark it for the mounting bolts. Drill the mounting holes, and mount the SO-239 socket, using 6-32 bolts, washers and nuts.

Add the copper pipe clamps to the J-pole radiator (clean the copper pipe and the pipe clamps with sandpaper on the outside and inside, as well as the radiator and stub match), as shown in Figures 1 and 2. You will have to bend them around the pipe so that you can attach them to the Plexiglas. Mark the location of the holes and drill them out. Replace the pipe clamps and attach them with 6-32 bolts, star lock washers, flat washers and nuts. Cut two short pieces of wire (no. 16 or 14)—I used no. 16 stranded wire. Attach ring terminals on one end of one wire and on both ends of the other wire. Make these wires long enough so that they can go from the center terminal of the SO-239 to a clamp (I used the clamp on the 3/4 wavelength section). Also connect one of the mounting bolts on the SO-239 socket to the other bolt on the second clamp (the matching stub). Keep the wires as short as possible. An

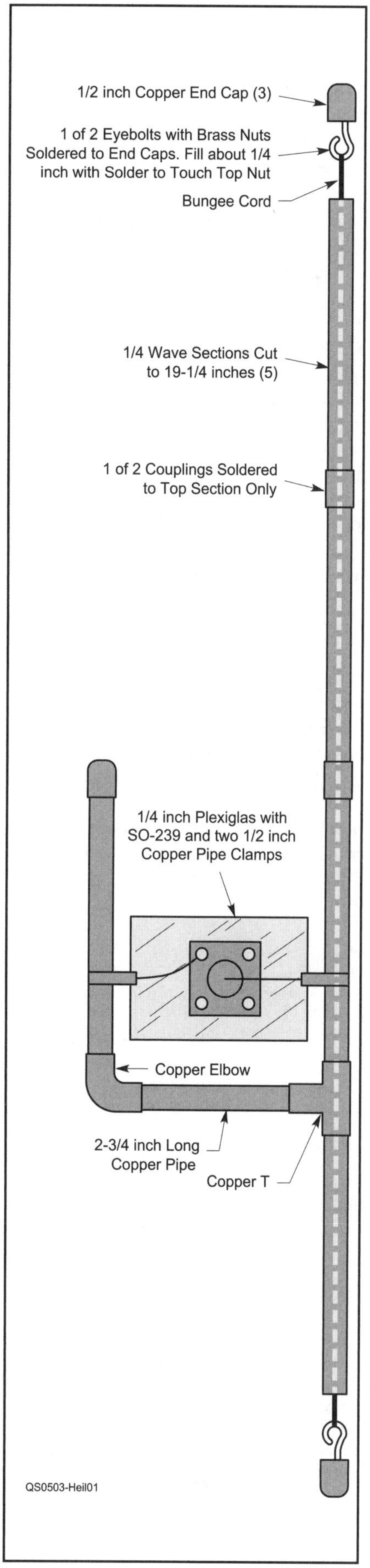

Figure 1—The assembly drawing for the VHF/UHF folding J-pole antenna.

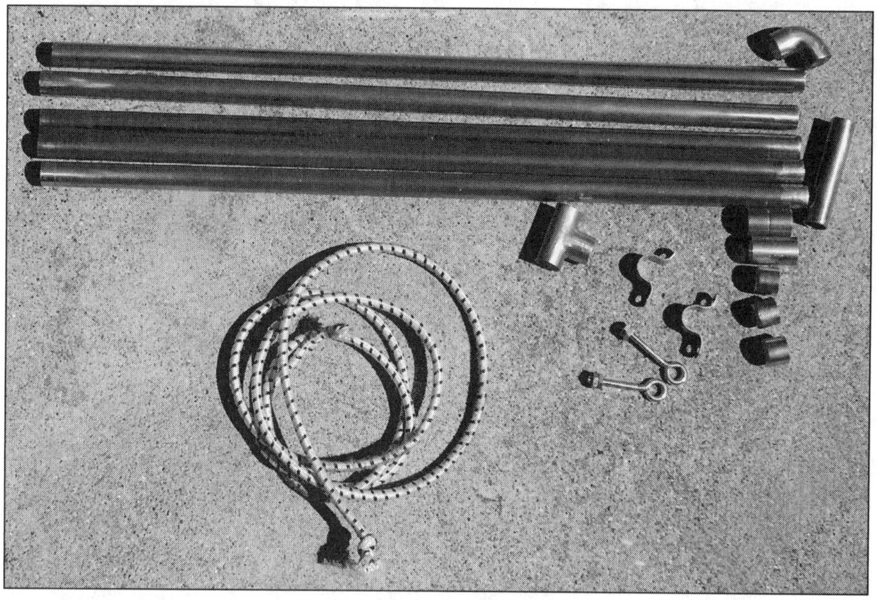

Figure 2—The antenna components—ready for construction.

Figure 3—A completed eyebolt assembly. These serve as an anchor point for the bungee cord.

Figure 4—The bungee cord runs through each of the antenna elements.

alternate method of attachment would be to solder the wire to the pipe clamps. Figure 6 shows the antenna in the process of being folded.

Tuning

Once the antenna is assembled, you can tune for minimum SWR. Start with both pipe clamps about 3 inches up from the bottom of the radiator and the matching stub and work your way down. You will want to tighten the clamps reasonably well so you get a reliable RF analyzer reading when testing. I was able to get the SWR to 1:1 at 146.6 MHz using an MFJ-269 antenna analyzer. At 446.90 MHz the SWR was 1.1:1. Once the SWR is where you want it, tighten the clamp bolts. You should be able to duplicate these results without too much trouble.

After the antenna is away from surrounding objects, recheck the SWR to see if it has changed. If it has, you will have to adjust again, as surrounding objects and your own proximity to the antenna will affect the SWR. I found it best to loosely tighten the pipe clamps so they are loose enough to move downward by a slight tapping on the Plexiglas support with a screwdriver. After you make the SWR adjustments, tighten the clamps and recheck the SWR again. You'll want to seal the exposed parts of the SO-239 connector rear with liquid electrical tape to keep out any moisture.[2] When folded, the antenna measures less than 21 inches long. The completed antenna, ready for packing, can be seen in Figure 7.

The cost of the parts for the folding J-pole is less than $15. All of the parts were obtained from a local lumberyard and hardware store. I used stainless bolts, nuts and washers because they were readily available. Stainless hardware was about $6 a pound so I grabbed a small handful of each hardware variety for future J-pole projects. You'll enjoy the portability and ease of assembly of this transportable antenna that's ideal for Field Day, casual operating or emergency use.

Figure 5—The antenna feed point assembly. The feed point can be slid up and down on the radiator and the matching section to adjust the SWR. The SO-239 socket rear needs to be weatherproofed.

Figure 6—The J-pole in the process of being "folded."

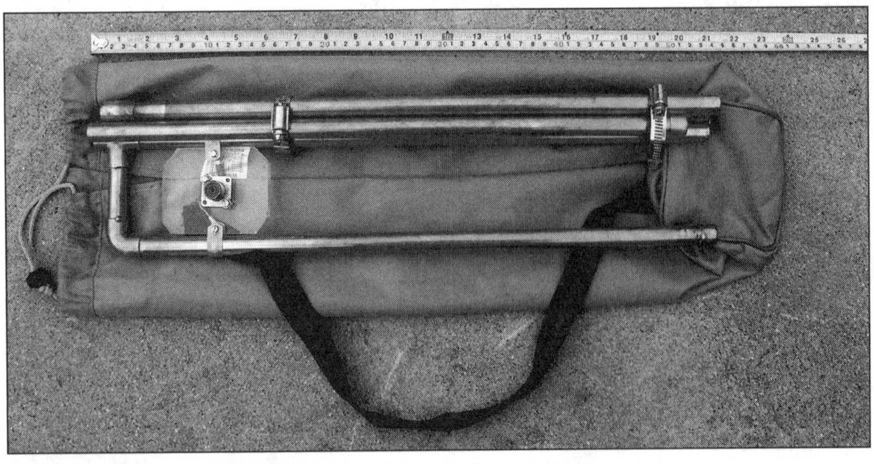

Figure 7—The completed folding J-pole, ready for packing.

Notes

[1]The 2 meter J-pole is a third harmonic antenna on 70 cm (432 MHz). As such, it will have a higher radiation angle component on 70 cm than on 2 meters, so 70 cm repeater access will not be optimum for stations far from the repeater.—*Ed.*

[2]A cautionary note—the UHF series of connectors (the PL-259 and SO-239, among others) is not waterproof. Make sure you adequately insulate both the plug and socket assemblies of these connectors, especially at VHF and UHF. Scotch 33 electrical tape, liberally stretched over the connector and cable, works well for the PL-259 plug and epoxy or liquid electrical tape (as pointed out here) works well for the SO-239 socket rear.—*Ed.*

Photos by the author.

Michael Heiler, KAØZLG, has been a ham since 1987. He is an ARRL member and is the Emergency Coordinator (EC) for Clay County, Minnesota. Mike has an electronics degree from Northwest Technical College in Moorehead, Minnesota, is a Volunteer Examiner and teaches Amateur Radio classes several times a year. Mike works on fire alarm systems for SimplexGrinnell in Fargo, North Dakota. He can be reached at **ka0zlg@arrl.net**.

The DBJ-2: A Portable VHF-UHF Roll-Up J-pole Antenna for Public Service

WB6IQN reviews the theory of the dual band 2 meter / 70 cm J-pole antenna and then makes detailed measurements of a practical, easy to replicate, "roll-up" portable antenna.

Edison Fong, WB6IQN

It has now been more than three years since my article on the dual band J-pole (DBJ-1) appeared in the February 2003 issue of *QST*.[1] I have had over 500 inquires regarding that antenna. Users have reported good results, and a few individuals even built the antenna and confirmed the reported measurements. Several major cities are using this antenna for their schools, churches and emergency operations center. When asked why they choose the DBJ-1, the most common answer was value. When budgets are tight and you want a good performance-to-price ratio, the DBJ-1 (*D*ual *B*and *J*-pole–*1*) is an excellent choice.

In quantity, the materials cost about $5 per antenna and what you get is a VHF/UHF base station antenna with λ/2 vertical performance on both VHF and UHF bands. If a small city builds a dozen of these antennas for schools, public buildings, etc it would cost about $60. Not for one, but the entire dozen!

Since it is constructed using PVC pipe, it is UV protected and it is waterproof. To date I have personally constructed over 400 of these antennas for various groups and individuals and have had excellent results. One has withstood harsh winter conditions in the mountains of McCall, Idaho for four years.

The most common request from users is for a portable "roll-up" version of this antenna for backpacking or emergency use. To address this request, I will describe how the principles of the DBJ-1 can be extended to a portable roll-up antenna. Since it is the second version of this antenna, I call it the DBJ-2.

Principles of the DBJ-1

The earlier DBJ-1 is based on the J-pole,[2] shown in Figure 1. Unlike the popular ground plane antenna, it doesn't need ground radials. The DBJ-1 is easy to construct using inexpensive materials from your local hardware store. For its simplicity and small size, the DBJ-1 offers excellent performance and consistently outperforms a ground plane antenna.

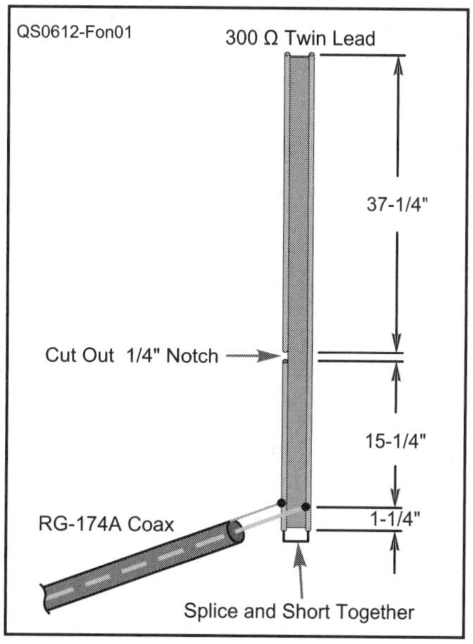

Figure 1 — The original 2 meter ribbon J-pole antenna.

Its radiation pattern is close to that of an ideal vertical dipole because it is end-fed, with virtually no distortion of the radiation pattern due to the feed line. A vertically polarized, center-fed dipole will always have some distortion of its pattern because the feed line comes out at its center, even when a balun is used. A vertically polarized, center-fed antenna is also physically more difficult to construct because of that feed line coming out horizontally from the center.

The basic J-pole antenna is a half-wave vertical configuration. Unlike a vertical dipole, which because of its center feed is usually mounted alongside a tower or some kind of metal supporting structure, the radiation pattern of an end-fed J-pole mounted at the top of a tower is not distorted.

The J-pole works by matching a low impedance (50 Ω) feed line to the high impedance at the end of a λ/2 vertical dipole. This is accomplished with a λ/4 matching stub shorted at one end and open at the other. The impedance repeats every λ/2, or every 360° around the Smith Chart. Between the shorted end and the high impedance end of the λ/4 shorted stub, there is a point that is close to 50 Ω and this is where the 50 Ω coax is connected.

By experimenting, this point is found to be about 1¼ inches from the shorted end on 2 meters. This makes intuitive sense since 50 Ω is closer to a short than to an open circuit. Although the Smith Chart shows that this point is slightly inductive, it is still an excellent match to 50 Ω coax. At resonance the SWR is below 1.2:1. Figure 1 shows the dimensions for a 2-meter J-pole. The 15¼ inch λ/4 section serves as the quarter wave matching transformer.

A commonly asked question is, "Why 15¼ inches?" Isn't a λ/4 at 2 meters about 18½ inches? Yes, but twinlead has a reduced velocity factor (about 0.8) compared to air and must thus be shortened by about 20%.

A conventional J-pole configuration works well because there is decoupling of the feed line from the λ/2 radiator element since the feed line is in line with the radiating λ/2 element. Thus, pattern distortion is minimized. But this only describes a single band VHF J-pole. How do we make this into a dual band J-pole?

Adding a Second Band to the J-pole

To incorporate UHF coverage into a VHF J-pole requires some explanation. (A more detailed explanation is given in my February 2003 *QST* article.) First, a 2 meter antenna does resonate at UHF. The key word here is

[1]Notes appear on page 157.

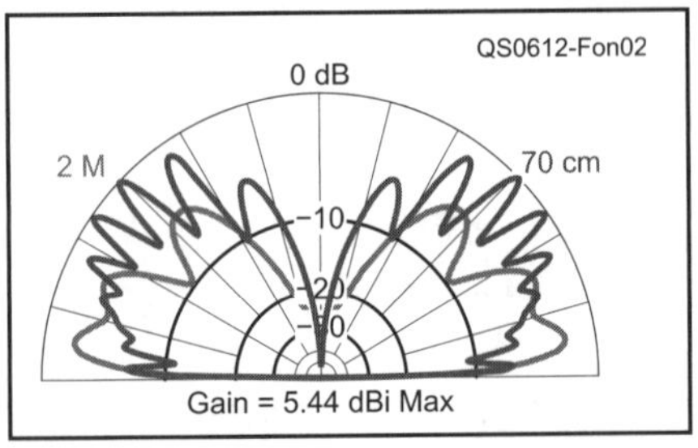

Figure 2 — Elevation plane pattern comparing 2 meter J-pole on fundamental and on third harmonic frequency (70 cm), with the antenna mounted 8 feet above ground. Most of the energy at the third harmonic is launched at 44°.

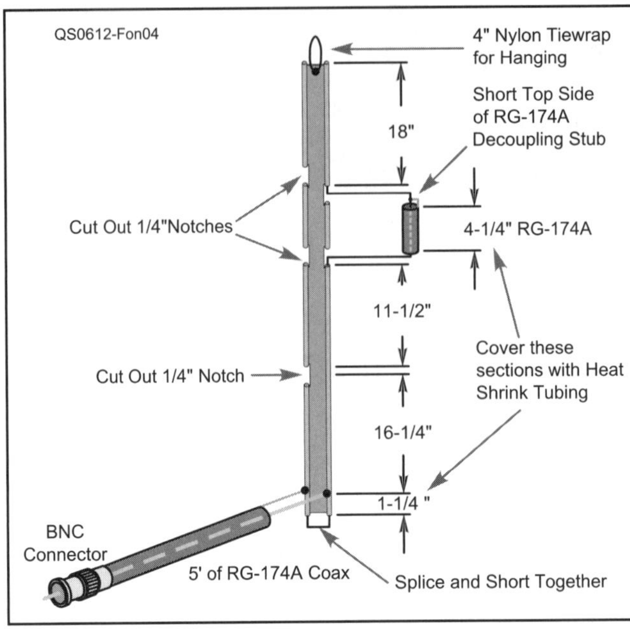

Figure 4 — The dual-band J-pole modified for portable operation — thus becoming the DBJ-2. Note that the dimensions are slightly longer than those in Figure 3 because it is not enclosed in a PVC dielectric tube. Please remember that the exact dimensions vary with the manufacturer of the 300 Ω line, especially the exact tap point where the RG-174A feed coax for the radio is connected.

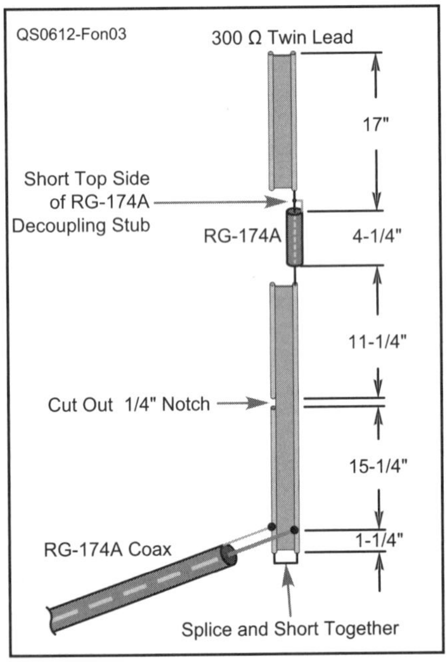

Figure 3 — The original DBJ-1 dual-band J-pole. The dimensions given assume that the antenna is inserted into a ¾ inch Class 200 PVC pipe.

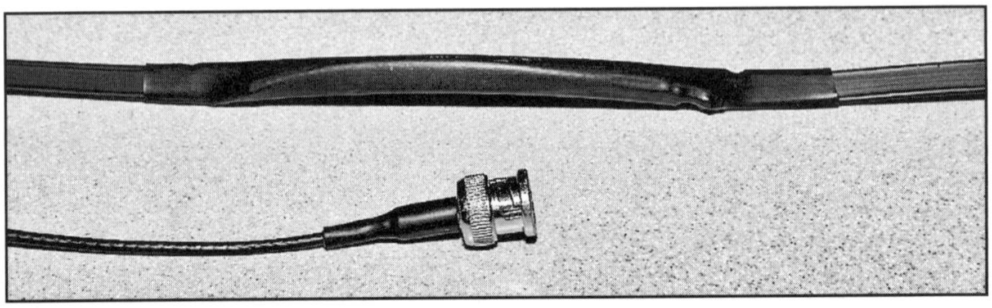

Figure 5 — The λ/4 UHF decoupling stub made of RG-174A, covered with heat shrink tubing. This is shown next to the BNC connector that goes to the transceiver.

resonate. For example, any LC circuit can be resonant, but that does not imply that it works well as an antenna. Resonating is one thing; working well as an antenna is another. You should understand that a λ/4 146 MHz matching stub works as a 3λ/4 matching stub at 450 MHz, except for the small amount of extra transmission line losses of the extra λ/2 at UHF. The UHF signal is simply taking one more revolution around the Smith Chart.

The uniqueness of the DBJ-1 concept is that it not only resonates on both bands but also actually performs as a λ/2 radiator on both bands. An interesting fact to note is that almost all antennas will resonate at their third harmonic (it will resonate on any odd harmonic 3, 5, 7, etc). This is why a 40 meter dipole can be used on 15 meters. The difference is that the performance at the third harmonic is poor when the antenna is used in a vertical configuration, as in the J pole shown in Figure 1. This can be best explained by a 19 inch 2 meter vertical over an ideal ground plane. At 2 meters, it is a λ/4 length vertical (approximately 18 inches). At UHF (450 MHz) it is a 3λ/4 vertical. Unfortunately, the additional λ/2 at UHF is out of phase with the bottom λ/4. This means cancellation occurs in the radiation pattern and the majority of the energy is launched at a takeoff angle of 45°. This results in about a 4 to 6 dB loss in the horizontal plane compared to a conventional λ/4 vertical placed over a ground plane. A horizontal radiation pattern obtained from *EZNEC* is shown in Figure 2. Notice that the 3λ/4 radiator has most of its energy at 45°.

Thus, although an antenna can be made to work at its third harmonic, its performance is poor. What we need is a simple, reliable method to decouple the remaining λ/2 at UHF of a 2 meter radiator, but have it remain electrically unaffected at VHF. We want independent λ/2 radiators at both VHF and UHF frequencies. The original DBJ-1 used a combination of coaxial stubs and 300 Ω twinlead cable, as shown in Figure 3.

Refer to Figure 3, and start from the left hand bottom. Proceed vertically to the RG-174A lead in cable. To connect to the antenna, about 5 feet of RG-174A was used with a BNC connector on the other end. The λ/4 VHF impedance transformer is made from 300 Ω twin lead. Its approximate length is 15 inches due to the velocity factor of the 300 Ω material. The λ/4 piece is shorted at the bottom and thus is an open circuit (high impedance) at the end of the λ/4 section. This matches well to the λ/2 radiator for VHF. The 50 Ω tap is about 1¼ inches from the short, as mentioned before.

For UHF operation, the λ/4 matching stub at VHF is now a 3λ/4 matching stub. This is electrically a λ/4 stub with an additional λ/2 in series. Since the purpose of the matching stub is for impedance matching and not for radiation, it does not directly affect the radiation efficiency of the antenna. It does, however, suffer some transmission loss from the additional λ/2, which would not be needed if it were not for the dual band operation. I estimate this loss at about 0.1 dB. Next comes the λ/2 radiating element for UHF, which is about 12 inches. To

Table 1
Measured Relative Performance of the Dual-band Antenna at 146 MHz

VHF λ/4 GP 4 radials	*VHF Flexible Antenna*	*Standard VHF J-Pole*	*Dual-Band J-Pole*
0 dB reference	–5.9 dB	+1.2 dB	+1.2 dB

Table 2
Measured Relative Performance of the Dual-band Antenna at 445 MHz

UHF λ/4 GP 4 radials	*UHF Fexible Antenna*	*Standard VHF J-Pole*	*Dual-Band J-Pole*
0 dB reference	–2.0 dB	–5.5 dB	0.5 dB

make it electrically terminate at 12 inches, a λ/4 shorted stub at UHF is constructed using RG-174A. The open end is then connected to the end of the 12 inches of 300 Ω twinlead. The open circuit of this λ/4 coax is only valid at UHF. Also, notice that it is 4½ inches and not 6 inches due to the velocity factor of RG-174A, which is about 0.6.

At the shorted end of the 4½ inch RG-174A is the final 18 inches of 300 Ω twinlead. Thus the 12 inches for the UHF λ/2, the 4½ inches of RG-174A for the decoupling stub at UHF, and the 18 inches of twinlead provide for the λ/2 at 2 meters. The total does not add up to a full 36 inches that you might think. This is because the λ/4 UHF RG-174A shorted stub is inductive at 2 meters, thus slightly shortening the antenna.

Making it Portable

The single most common question that people asked regarding the DBJ-1 is how it could be made portable. The original DBJ-1 had the antenna inserted into Class 200 PVC pipe that was 6 feet long. This was fine for fixed operation but would hardly be suitable for portable use. Basically the new antenna had to have the ability to be rolled up when not in use and had to be durable enough for use in emergency communications.

The challenge was to transfer the concepts developed for the DBJ-1 and apply them to a durable roll-up portable antenna. After much thought and experimenting, I adopted the configuration shown in Figure 4.

The major challenge was keeping the electrical characteristics the same as the original DBJ-1 but physically constructing it from a continuous piece of 300 Ω twinlead. Any full splices on the twinlead would compromise the durability, so to electrically disconnect sections of the twinlead, I cut small ¼ inch notches to achieve the proper resonances. I left the insulating backbone of the 300 Ω twinlead fully intact. I determined the two notches close to the λ/4 UHF decoupling stub by experiment to give the best SWR and bandwidth.

Because this antenna does not sit inside a dielectric PVC tube, the dimensions are about 5% longer than the original DBJ-1. I used heat shrink tubing to cover and protect the UHF λ/4 decoupling stub and the four ¼ inch notches. Similarly, I protected with heat shrink tubing the RG-174A coax interface to the 300 Ω twinlead. I also attached a small Teflon tie strap to the top of the antenna so that it may be conveniently attached to a nonconductive support string.

Figure 5 shows a picture of the λ/4 UHF matching stub inside the heat shrink tubing. The DBJ-2 can easily fit inside a pouch or a large pocket. It is far less complex than what would be needed for a single band ground plane, yet this antenna will consistently outperform a ground plane using 3 or 4 radials. Setup time is less than a minute.

I've constructed more than a hundred of these antennas. The top of the DBJ-2 is a high impedance point, so objects (even if they are nonmetallic) must be as far away as possible for best performance. The other sensitive points are the open end of the λ/4 VHF matching section and the open end of the λ/4 UHF decoupling stub.

As with any antenna, it works best as high as possible and in the clear. To hoist the antenna, use non-conducting string. Fishing line also works well.

Measured Results

I measured the DBJ-2 in an open field using an Advantest R3361 Spectrum Analyzer. The results are shown in Table 1. The antenna gives a 7 dB improvement over a flexible antenna at VHF. In actual practice, since the antenna can be mounted higher than the flexible antenna at the end of your handheld, results of +10 dB are not uncommon. This is the electrical equivalent of giving a 4 W handheld a boost to 40 W.

The DBJ-2 performs as predicted on 2 meters. It basically has the same performance as a single band J-pole, which gives about a 1 dB improvement over a λ/4 ground plane antenna. There is no measurable degradation in performance by incorporating the UHF capability into a conventional J-pole.

The DBJ-2's improved performance is apparent at UHF, where it outperforms the single band 2 meter J-pole operating at UHF by about 6 dB. See Table 2. This is *significant*. I have confidence in these measurements since the flexible antenna is about –6 dB from that of the λ/4 ground plane antenna, which agrees well with the literature.

Also notice that at UHF, the loss for the flex antenna is only 2.0 dB, compared to the ground plane. This is because the flexible antenna at UHF is already 6 inches long, which is a quarter wave. So the major difference for the flexible antenna at UHF is the lack of ground radials.

Summary

I presented how to construct a portable, roll-up dual-band J-pole. I've discussed its basic theory of operation, and have presented experimental results comparing the DBJ-2 to a standard ground plane, a traditional 2 meter J-pole and a flexible antenna. The DBJ-2 antenna is easy to construct, is low cost and is very compact. It should be an asset for ARES applications. It offers significant improvement in both the VHF and UHF bands compared to the stock flexible antenna antenna included with a handheld transceiver.

If you do not have the equipment to construct or tune this antenna at both VHF and UHF, the antenna is available from the author tuned to your desired frequency. Cost is $20. E-mail him for details.

Notes

[1]E. Fong, "The DBJ-1: A VHF-UHF Dual-Band J-Pole," *QST*, Feb 2003, pp 38-40.

[2]J. Reynante, "An Easy Dual-Band VHF/UHF Antenna," *QST*, Sep 1994, pp 61-62.

Ed Fong was first licensed in 1968 as WN6IQN. He later upgraded to Amateur Extra class with his present call of WB6IQN. He obtained BSEE and MSEE degrees from the University of California at Berkeley and his PhD from the University of San Francisco. A Senior Member of the IEEE, he has 8 patents, 24 published papers and a book in the area of communications and integrated circuit design. Presently, he is employed by the University of California at Berkeley teaching graduate classes in RF design and is a Principal Engineer at National Semiconductor, Santa Clara, California working with CMOS analog circuits. You can reach the author at **edison_fong@hotmail.com**.

The Handy Yagi Antenna

A merger of transceiver and antenna eliminates many design issues.

Tom Hart, AD1B

After 38 years in Amateur Radio, I finally decided that it was time to try the 440 MHz band. Some decisions should not be rushed.

After using my new Yaesu FT-60 dual band handheld transceiver for a few months, I decided to investigate antenna options on 70 cm. My goal was to build an antenna that travels easily and requires minimal setup. A little daydreaming resulted in a 38 inch long design that supports the handheld transceiver and four directors (see Figure 1). A bicycle handle bar grip permits easy aiming. Finally, a speaker-mic simplifies aiming the antenna while talking or listening.

Simplest is Best

The key to simple feed line and mechanical connections was to mount the transceiver on the antenna in place of a driven element. Director size and spacing data was taken from *The ARRL Antenna Book*.[1] The FT-60 mounts easily using the belt clamp on the rear of the radio (see Figure 2).

Making it Happen

Construction is straightforward. A 38 inch long ¾ inch diameter dowel serves as the boom. The directors were cut from ⅛ inch steel rod stock. The radio holder is made from two pieces of ⅛ inch fiberboard and ¼ inch rubber spacers. Two bolts hold the bracket in place.

Checking it Out

In the absence of proper analytical instruments, I resorted to on-air reports and *EZNEC 4.0* analysis.[2] My initial test involved calling the K1BFD 440 MHz repeater. I am unable to raise Barry's machine from my home location with the FT-60 alone. By adding the Handy Yagi, I had no trouble accessing the machine. The receiver meter was nearly full scale with the antenna pointing at the repeater. By turning the antenna in an arc toward and away from K1BFD/R, I could see the received signal strength rise and fall.

The second field test involved climbing to the top of the Blue Hill Ski Area, near Boston. I called Dick, K1HC, on 446.0 MHz simplex. According to my Magellan *Topo 3D* software, K1HC is 3.05 miles from the top of the ski lift platform at a heading of 282°. I ran ½ W for the tests.

A QSO with K1HC verified my earlier observation that the Handy Yagi shows directivity. While making a long call, I swung the antenna back and forth slowly. Dick reported that the signal strength increased and decreased with the changes in direction. When K1HC transmitted, the same effect was apparent as I moved the antenna back and forth while receiving.

Model Analysis

The *EZNEC* computations involved entering a driven element and four directors. In a compromise, my model uses a ½ wavelength ($\lambda/2$) center fed dipole to represent the FT-60. The *EZNEC* predictions provide reasonable agreement with the field observations. Azimuth and elevation plots show good front to back gain.

EZNEC computes a front to back ratio (F/B) of 4.65 dB. For calculations involving power, the relationship is:[3]

$$dB = 10 \log (P2 / P1) \quad \text{[Eq 1]}$$

Then, the power ratio may be computed from:

$$(P2 / P1) = \text{antilog} (dB / 10) \quad \text{[Eq 2]}$$

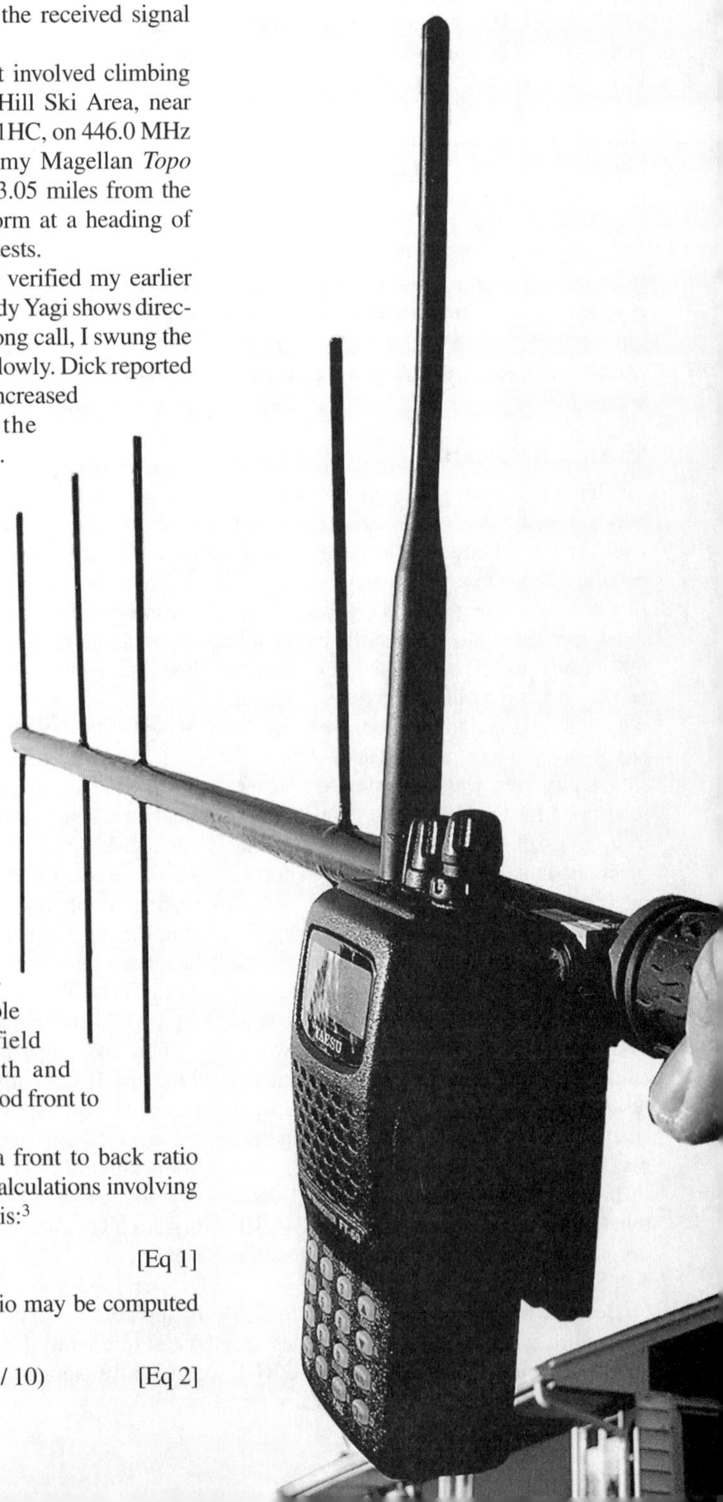

[1]Notes appear on page 155.

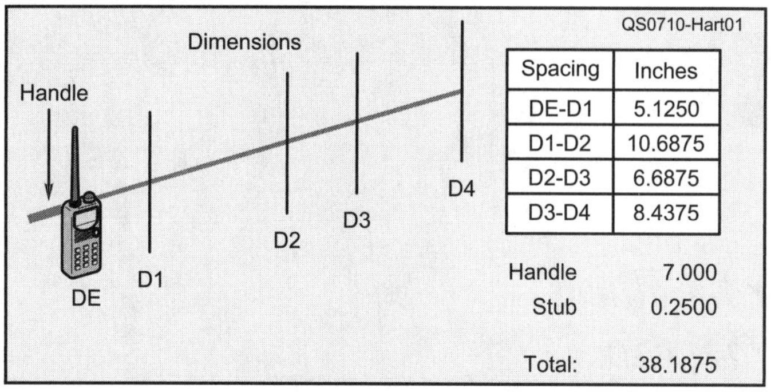

Figure 1 — Orientation and dimensions of the Handy Yagi.

Figure 2 — Details of attachment point for handheld transceiver (see title photo).

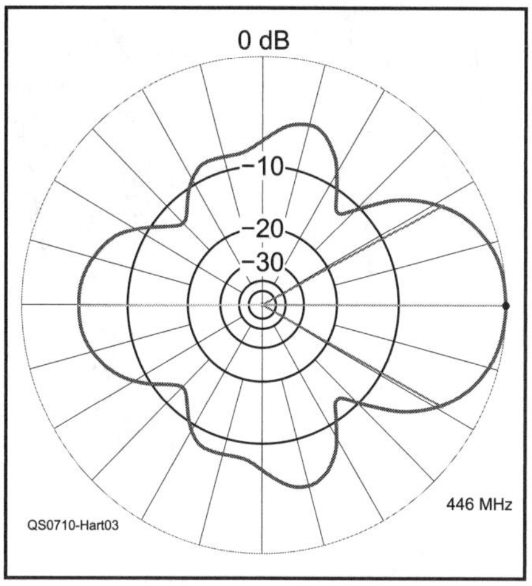

Figure 3 — *EZNEC*-predicted azimuth plot of antenna using dipole driven element.

The *EZNEC* F/B yields a power ratio of:

$$4.65 = 10 \log (P2 / P1) \qquad \text{[Eq 3]}$$

$$\text{antilog } (4.65/10) = (P2 / P1) = 2.9$$

Therefore, the calculated front to back power ratio is nearly 3:1.

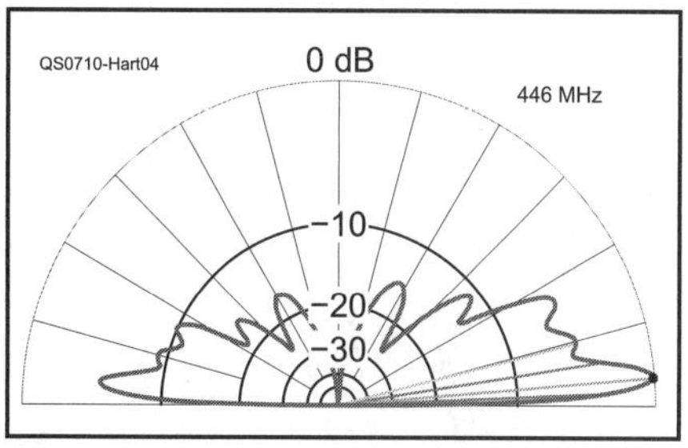

Figure 4 — *EZNEC*-predicted elevation plot of antenna using dipole driven element.

In Conclusion

For anyone interested in temporary or portable operations on the 440 MHz band, the Handy Yagi might be a simple antenna solution. The design may also be useful in foxhunting activities. Larger implementations of the basic design could be mounted on a tripod or suspended from an overhead support. The basic design is very simple and adaptable. The key is the substitution of a transceiver for the driven element and feed line.

Notes

[1]R. D. Straw, Editor, *The ARRL Antenna Book,* 21st Edition, p 18-45. Available from your ARRL dealer or the ARRL Bookstore, ARRL order no. 9876. Telephone 860-594-0355, or toll-free in the US 888-277-5289; **www.arrl.org/shop/**; **pubsales@arrl.org**.

[2]**www.eznec.com**.

[3]J. Hallas, W1ZR, "Making Sense of Decibels" *QST*, Apr 2007, p 61.

Tom Hart, AD1B, began listening to short wave broadcasts in 1961. He received his Novice class license, WN1JGG, in 1968 and has been active on CW, SSB, RTTY, FM and packet ever since. He still monitors the short wave bands, listening to Radio Australia or Radio China International while driving to work every morning.

Tom has a BS from Tufts and an MS from Northeastern. He is an accountant who would rather be chasing or giving out counties on 20 meters. You can reach Tom at 54 Hermaine Ave, Dedham, MA 02026 or via **tom.hart@verizon.net**. QST

Dual Band Handy Yagi

Thomas M. Hart, AD1B

The popular handheld Handy Yagi can now work on both VHF and UHF.

Recently, I started operating through the SO-50 and AO-51 satellites with my dual band Yaesu FT-60 handheld transceiver. The standard flexible antenna worked, but I decided to investigate handheld Yagi antennas to improve conditions. My goal was to build a simple dual band 2 meter and 70 cm Yagi without driven elements, a matching network or a feed line. In short, my plan was to mount the FT-60 on the antenna as I had with the original 70 cm Handy Yagi.[1]

After testing several configurations on Roy Lewallen's *EZNEC* program (see **www.eznec.com**), I settled on a design with seven directors on 70 cm and three on 2 meters. My Yaesu FT-60 dual band handheld serves as the driven element. Figures 1 and 2 show the configuration.

Construction

The final design balances performance and size. The elements for both bands are interlaced and mounted in parallel. There is no driven element or reflector. Instead, the FT-60 and a bicycle handlebar grip occupy the usual reflector and driven element end of the boom. The 2 meter elements can be rotated parallel to the boom to simplify storage. A screw eye at one end can be used to hang the antenna when not in use.

Computer modeling indicated that $\lambda/4$ element spacing works reasonably well on both bands. This allows the use of a 55 inch boom. The handheld is attached firmly in place by its belt clip. A speaker microphone makes transmitting and receiving very simple. See the illustrations for additional details.

All elements were cut from ⅛ inch diameter steel rod. An article by Ron Hege, K3PF, provided the dimensions for the 2 meter elements shown in Table 1. The 70 cm element dimensions are found in *The ARRL Antenna*

[1]Notes appear on page 161.

Figure 1 — Dual band Handy Yagi with 2 meter elements folded for storage.

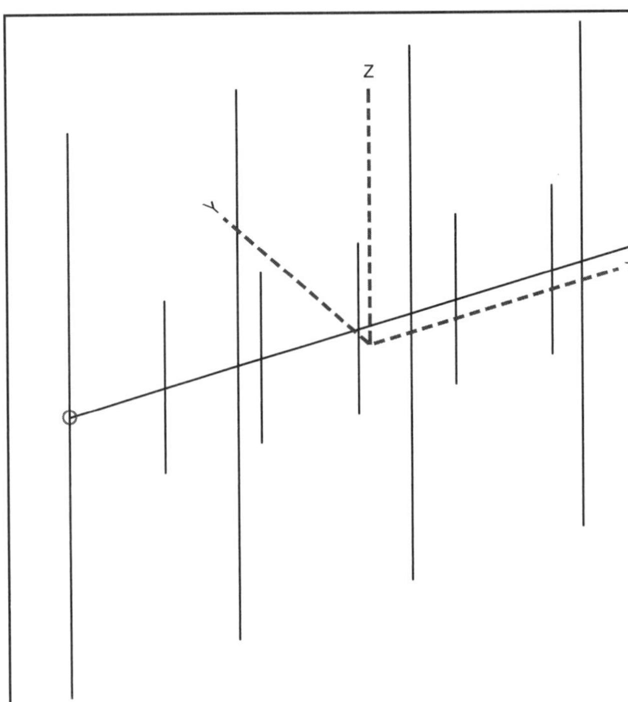

Figure 2 — An *EZNEC*-generated diagram of the antenna shows the element placement. The Yaesu FT-60 is represented by the half-wave loaded element on the left.

Table 1
Length and Spacing of 2 Meter Elements
All dimensions are in inches.

Director:	*D1*	*D2*	*D3*
Length	37.5	36.375	36.0
Element Spacing	DE to D1	D1 to D2	D2 to D3
Spacing	12	12	12
Cumulative	12	24	36

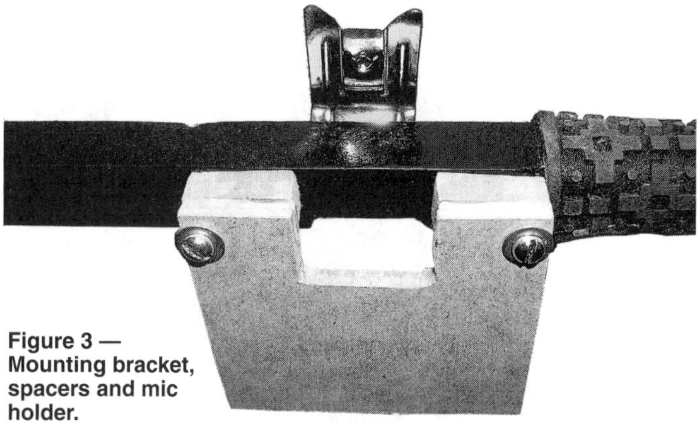

Figure 3 — Mounting bracket, spacers and mic holder.

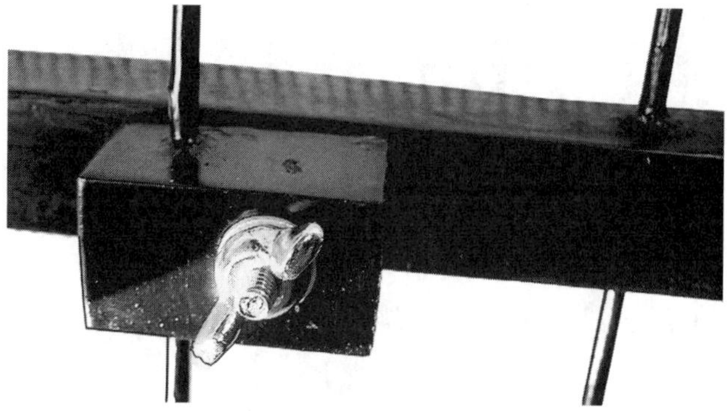

Figure 4 — The wing nut and bolt allow the 2 meter element to rotate.

Table 2

Length and Spacing of 70 cm Elements

All dimensions are in inches.

Director:	*D1*	*D2*	*D3*	*D4*	*D5*	*D6*	*D7*
Length	11.750	11.688	11.625	11.563	11.500	11.438	11.375
Elements	DE to D1	D1 to D2	D2 to D3	D3 to D4	D4 to D5	D5 to D6	D6 to D7
Spacing	6.78	6.78	6.78	6.78	6.78	6.78	6.78
Cumulative	6.78	13.56	20.34	27.12	33.9	40.67	47.45

Table 3

Dimensions of Other Antenna Assemblies

All dimensions are in inches.

Boom	0.75 × 0.75 × 55
Handle grip	5
Handheld bracket	3.5 × 3.0 × 0.25 (WHD)
Notch for handheld	1.125 W × 0.75 H
Bracket spacers	0.375
2M element mounts	0.75 × 0.75 × 1.5

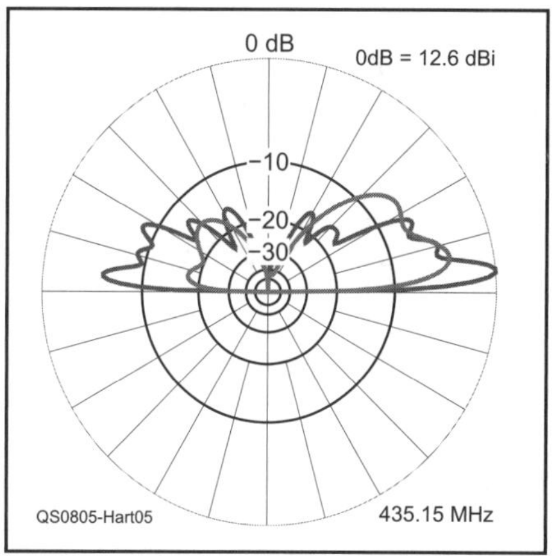

Figure 5 — Predicted elevation patterns of the 2 meter (red) and 70 cm (blue) Yagis.

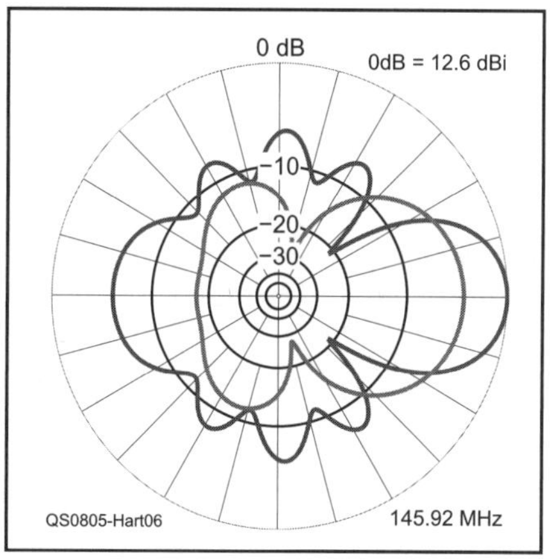

Figure 6 — Predicted azimuth patterns at the peak of the elevation response for the 2 meter (red) and 70 cm (blue) Yagis.

Book and reproduced as Table 2. Other dimensions are shown in Table 3.

Two glued together pieces of ⅛ inch pressed fiberboard form the handheld bracket shown in Figure 3. The elements are held in place with epoxy. After cutting all 10 to length, drill seven holes through the boom and three more through the 2 meter element mounts. Slide the elements into place and apply a bead of epoxy on both sides. Several light applications of epoxy will hold the rods in place.

All 2 meter element mounts require two holes. One secures the element while the second is used for the bolt and wing nut that allow the element to rotate during storage or transport as shown in Figure 4. Finally, two coats of black satin paint give the antenna a more professional appearance.

Testing

Faced with an absence of analytical instruments, the testing process involved *EZNEC* computations and operational observations. *EZNEC* computed the front-to-back ratios as 14 dB for 2 meters and 5.5 dB for 70 cm. *EZNEC* azimuth and elevations patterns are shown in Figures 5 and 6.

Field testing on 2 meters involved contacting the 146.97 MHz repeater in Paxton, Massachusetts. The repeater, according to my *Magellan Topo* 3D GPS software, is 38.0 miles away at a bearing of 251°. The FT-60 helical antenna cannot reach the repeater from my house. With the dual band Handy Yagi, I had no trouble reaching the repeater. Moving the antenna in an arc toward and away from the repeater produced corresponding signal strength changes on the meter corresponding to the predicted directivity.

On 70 centimeters, field tests involved reception of signals from the AO-51 and SO-50 satellites. Signal strength increased dramatically when the antenna approached and centered on the target. My conclusions are that operations on both bands are in agreement with *EZNEC* predictions.

The antenna that I built is designed for satellite contacts. However, the basic concept can be used for portable operations, fox hunting and emergency use as well.

Notes

[1]T. Hart, AD1B, "The Handy Yagi," *QST*, Nov 2007, pp 37-38.

[2]R. Hege, K3PF, "A Five-Element, 2-Meter Yagi for $20," *QST*, Jul 1990, pp 34-36.

[3]R. D. Straw, Editor, *The ARRL Antenna Book*, 21st Edition, p 18-45. Available from your ARRL dealer or the ARRL Bookstore, ARRL order no. 9876. Telephone 860-594-0355, or toll-free in the US 888-277-5289; **www.arrl.org/shop/**; **pubsales@arrl.org**.

All photos by the author.

Tom Hart, AD1B, began listening to short wave broadcasts in 1961. He received his Novice class license, WN1JGG, in 1968 and has been an active on CW, SSB, RTTY, FM and packet ever since. Tom has a BS from Tufts and an MS from Northeastern. He is an accountant who would rather be chasing or giving out counties on 20 meters. You can reach Tom at 54 Hermaine Ave, Dedham, MA 02026 or via **tom.hart@verizon.net**. QST

A Better Way to Work Low Earth Orbit FM Satellites

You can try a handheld, but if you want to get through, a bit more gain will help.

Juan Antonio Fernández Montaña, EA4CYQ, and Pedro Pérez Álvarez-Cienfuegos, EB4DKA

We are pleased to present an easy and inexpensive way to work FM VHF/UHF low earth orbit (LEO) satellites. While many amateurs try to communicate through LEO satellites using just their handheld transceivers and flex antennas, the results are often frustrating. You probably have almost all that's needed already, and this article will help fill in the blanks.

In addition to hardware, you will need to gain some knowledge before pushing the PTT switch in order to work through the satellites. This information is well documented in the new *ARRL Satellite Handbook* or on the AMSAT Web site.[1,2] In particular, you will need to know the uplink and downlink frequencies, how these frequencies shift as a consequence of the Doppler effect and when you will be within the satellite's footprint. There is free software available on the Internet in different programming languages to help with this. The typical software requires us to input our locator, actual time and the latest satellite Keplerian elements, also available at the AMSAT Web site. If you do not want to load software, you can also visit the AMSAT Web site and just input your Maidenhead grid locator.[3] You will get a listing of all the passes over your location.

The Hardware Suite

Once we know how to tell when and where a satellite will be within range, we can start thinking about the necessary radio equipment and antennas. Three broad categories of stations will be found on these satellites, as follows:

■ The *basic* user has a dual band handheld transceiver and a collapsible high gain whip antenna. These users will make some contacts, but are most successful using overhead passes typically for less than five minutes. These contacts will tend to be limited by weak signals and noise. Still, the basic user will have an opportunity to get to know the satellites without spending a lot of money.

■ The *intermediate* user is a former basic user who has improved station capabilities so it can maintain communication throughout each pass. While enhanced radio equipment is a possibility, antenna improvements for the handheld at this stage provide a more cost effective improvement. Commercial handheld gain antennas are available such as those from Arrow, however, there are a number of easy to build antennas that work well.[4] One that we like is the IOio 2 meter/70 cm antenna described in an AMSAT publication.[5] The IOio is named after the shape of the elements. The elements include a 2 meter linear reflector (I), then a full wave loop (O), followed by a repeat for 70 cm (io).

■ The *advanced* satellite operator is able to work through the birds comfortably from an indoor shack. This user generally employs two crossed Yagis, one for VHF and another for UHF. Cross or remotely controllable polarization is required to deal with tumbling satellites and the UHF receive system generally includes a low noise mast mounted preamplifier.

At the basic level you have to struggle to extract the modulation from the noise. While contacts are possible, they are not always easy. At the intermediate level you generally need to spend perhaps a quarter of an hour setting the antenna up before the satellite pass and another quarter hour taking it apart after the satellite is gone. Our favorite IOio is a bit quicker but it does take up space in the trunk of your car. In addition, although the reception is quite good, the uplink with only 5 W sometimes is a bit poor, especially with heavy traffic.

How Can We Make it Easier?

Considering the alternatives, your authors were thinking how to work in portable conditions keeping the good reception of the Arrow and IOio, but improving the

EB4DKA

EA4CYQ operates his portable satellite station. Mobile 2 meter uplink via a $\lambda/4$ whip and 50 W radio, downlink with the CJU antenna and a handheld 70 cm radio.

[1]Notes appear on page 164.

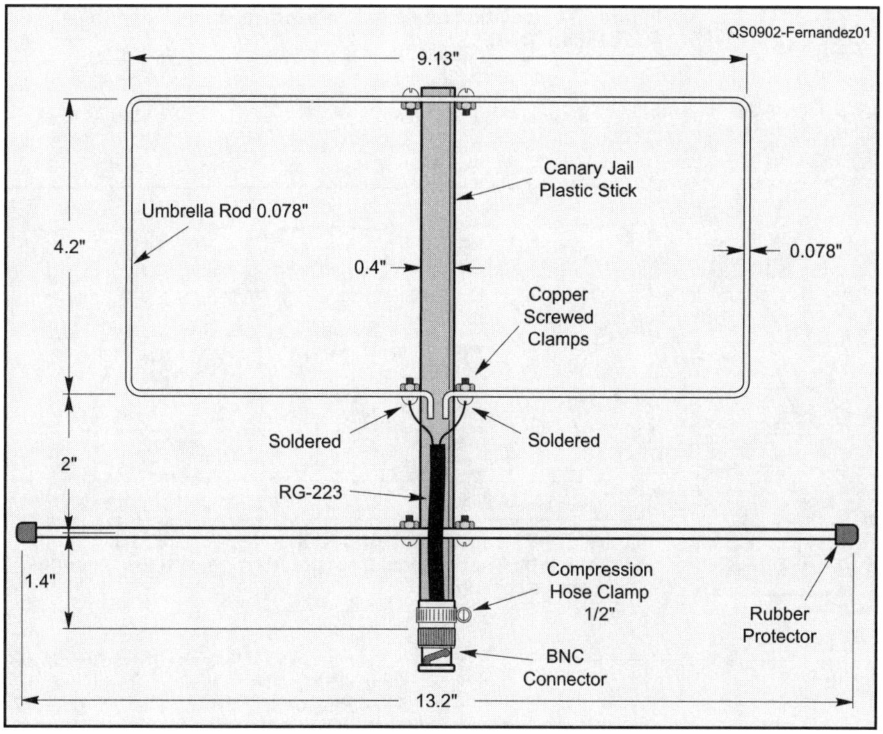

Figure 1 — Construction details of the CJU antenna.

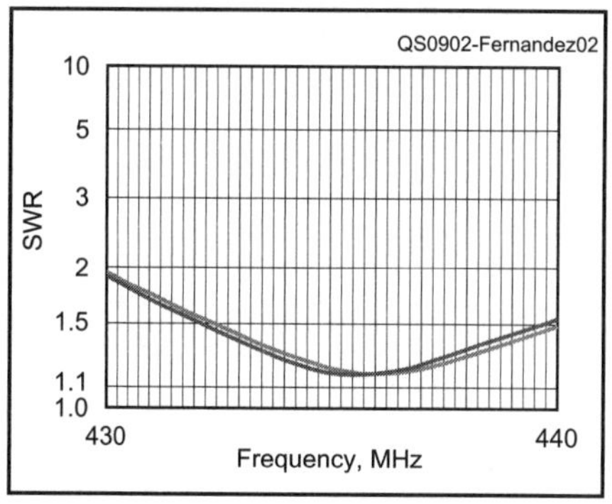

Figure 2 — SWR plots of the CJU antenna with horizontal (blue) and vertical (red) polarization at a height of 5 feet.

three drawbacks by increasing our uplink level, avoiding antenna assembly time, and leaving our trunk clear.

ARRL Education Coordinator Mark Spencer, WA8SME, authored a helpful *AMSAT Journal* article in which he analyzed the minimal conditions required to uplink and downlink successfully through the LEO satellites.[6] Mark came to the conclusion that the uplink is relatively easy, but receiving the downlink comfortably is often more difficult, especially in the UHF band. He suggests a receive antenna with at least 10 dB of gain and a low loss feed line in order to support operation through the whole satellite pass.

With this as a starting point, we considered various options to improve our satellite operation in a straightforward way. We often operate in portable locations using our parked car as a base. We use a 50 W VHF mobile transceiver with a quarter wave ($\lambda/4$) whip antenna on the car roof. We were testing its uplink performance for weeks and determined that it worked better than the 5 W into a directive handheld antenna. This makes sense because we compensate for the changing polarization with an additional 10 dB of transmitter power. This seemed like a good solution to the uplink side of the path.

The extra transmit power, however, doesn't help on the tougher downlink path. We will need a mobile or portable radio capable of receiving the FM UHF downlink. We will also need an antenna with sufficient gain to bring in enough signal from the satellite to provide a sufficient signal to noise ratio for solid reception. We were hoping to find a high gain antenna that could be attached directly to a handheld, but we didn't find anything with more gain than a telescoping whip. One advantage of a handheld antenna is that it is not necessary to have circular or cross polarization because we can rotate a linear antenna with our wrist to get the best signal.

There are some available units — commercial choices from Arrow and the IOio. The Arrow is about 4 feet long, although we could reduce the number of elements from the original seven and still have sufficient gain. The UHF band part of the IOio has a square radiator and a linear reflector set over a boom only 8 inches long. It has enough gain according to our earlier experience. But we need to prove that the gain is at least 10 dB and the radiation pattern is suitable.

Performance of the "CJU" Antenna

We got down to work with *EZNEC* taking advantage of our experience modeling other kinds of antennas.[7] We must model it considering both horizontal and vertical polarization, because in real life it needs to match the best polarization from the rotating satellite. The plots are obtained in free space and at 5 feet over real ground, the height we usually use with a handheld antenna. We will discuss why we named our "half an IOio" a "CJU."

Figure 1 is a detailed construction drawing of our compact antenna. Figure 2 shows the SWR plots of the antenna in horizontal and vertical polarization. Note that the impedance is not significantly affected by the real ground as we change polarization. Figure 3 shows the gain plots in free space, azimuth and elevation at 0°. Figure 4 shows the gain lobes over real ground. Here we can clearly see how it changes when we choose horizontal or vertical polarization because the horizontal elements are more influenced by ground reflection. The predicted gain is between 12.3 and 9.6 dBi. This gain is suitable for working satellites based on WA8SME's calculations.

With a handheld antenna, feed line losses are negligible. At the antenna resonant frequency, the impedance is over 60 Ω resulting in an SWR of about 1.2:1. Across the satellite frequency range, the SWR is 1.4:1 or less, so we didn't invest effort in making it any closer to perfection.

Building the CJU

We started our project by searching for the materials. For the boom, we needed a light round section of plastic or PVC tubing with an inside diameter to just fit over the BNC connector on the handheld. We needed a tube with an ID of about 0.4 inches for the task. Juan was talking about it with his father, EA4ABV, who has another hobby in addition to ham radio. He raises canaries. Juan saw something that gave him an idea. His father's canary cage ("jail") had a plastic perch inside that was a perfect fit. You may find something else that will do.

Now Juan just had to find some thick stiff wire for the antenna elements. On the way home from his father's house he found a broken umbrella tossed in a trash can. The rods were perfect for the elements — a very lucky morning walk. To connect the driven element to the BNC connector he used a short piece of RG-223 coax. Any 50 Ω coax will do since the length is only about 4 inches.

Two evenings later he had built the prototype. When he showed the antenna to

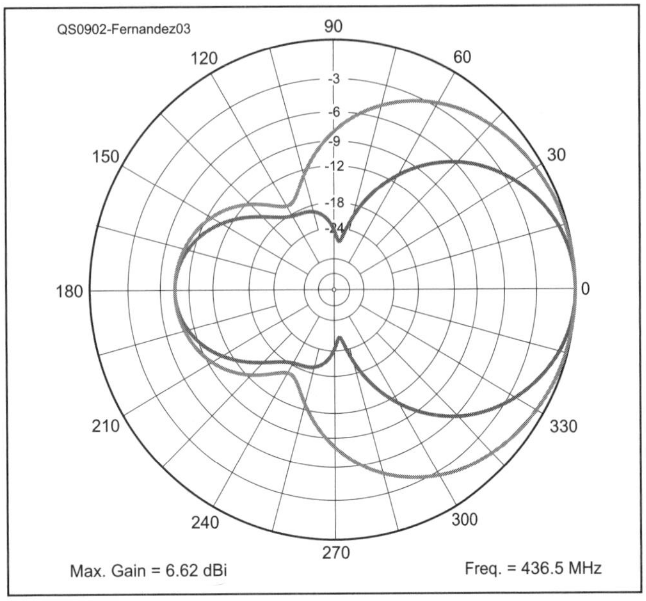

Figure 3 — Radiation lobes in free space at 0° elevation (blue) and azimuth (red).

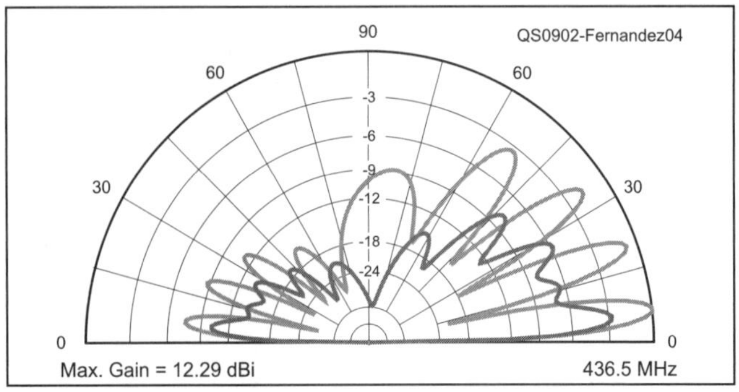

Figure 4 — Elevation plots of CJU antennas oriented in horizontal (blue) or vertical (red) polarization with the antenna 5 feet above real ground.

Pedro and told him that it was made with a "canary jail" stick and some rods of a broken umbrella, he laughed for a while and quickly named it the CJU for *canary jail umbrella*. Both of us had the same question in mind: Will it work? It was Pedro's turn to test it because Pedro is our portable conditions "expert." If you do not believe it you can see amazing video recordings at his Web site.[7]

Making it All Play

Pedro tested the CJU for several weeks now using both FM and SSB V/UHF satellites and he obtained amazing results. Only the larger Arrow performs as well. This antenna has two factors in its favor: a high dB/size relationship and the negligible feed line losses. The antenna seems to work so well, it acts as if it has a preamplifier. The signals received from FO-29 using a TH-F6 were up to S-5 and the antenna was able to receive the SO-50 signal across the whole pass with no problem at all.

Many hams already have a mobile transceiver in their cars and a handheld transceiver. With the addition of the CJU they can work satellites. The mobile radio can provide an effective uplink, while the handheld and CJU provide the needed downlink gain. Those with all-mode mobile transceivers such as the ICOM IC-706 or Yaesu FT-857/897 series and an all mode handheld can even work SSB through the linear LEO satellites.

You can see the whole station in the photo. For the uplink subsystem, he uses a 50 W mobile transceiver in the car and a λ/4 whip antenna on the trunk. To receive, the CJU is attached to a TH-D7 in my right hand, with a pair of headphones. In my left hand is the microphone of my mobile transceiver. I oriented the car so the driver's side faced the maximum elevation of the satellite pass.

EA4CYQ

EB4DKA's complete satellite station, uplink using a λ/4 whip and 50 W, 2 meter mobile, downlink using the CJU antenna and a 70 cm handheld transceiver.

We usually store the handheld in the glove compartment and the CJU in some corner, perhaps the magazine rack.

The CJU has now been widely tested by hundreds of hams who have sent us their photos. Pedro has posted some of them on his Web site. They have even renamed it the *magic antenna*.

We hope to have stimulated your curiosity for this branch of Amateur Radio. With this tool in your hand we are sure we will listen to you on the birds very soon. Thanks to the whole AMSAT community for their support of our project.

Notes

[1,2]The AMSAT Web site is at **www.amsat.org/amsat/AmsatHome.htm**. S. Ford, WB8IMY, *The ARRL Satellite Handbook*, available from your ARRL dealer in the ARRL Bookstore, ARRL order no. 9857. Telephone 860-594-0355, or toll free in the US 888-277-5289, **www.arrl.org/shop; pubsales@arrl.org**

[3]See **www.arrl.org/locate/gridinfo.html**.

[4]Arrow antennas can be found at **www.arrowantennas.com**.

[5]J. Montaña, EA4CYQ, "The IOio Antenna," *The AMSAT Journal,* Nov/Dec 2005, p 6. Information is also available at **eb4dka.laserenadigital.com**.

[6]M. Spencer, WA8SME, "Minimal Conditions Required to Successfully Uplink and Downlink Through the LEO Satellites," *The AMSAT Journal,* Jan/Feb 2007, p 6.

[7]Several versions of *EZNEC* antenna modeling software are available from developer Roy Lewallen, W7EL, at **www.eznec.com**.

[8]**eb4dka.laserenadigital.com**.

Juan Antonio Fernández Montaña, EA4CYQ, has been licensed since 1981 when he was 16 years old. He has held an "A" class license since 1984. Juan Antonio holds an Electrical and Industrial Electronics Engineer degree although his real interest is telecommunications. He belongs to Unión de Radioaficionados Españoles, ARRL, AMSAT and local radio clubs. He has published several articles about electronic accessories, antennas and satellite communications at URE, The AMSAT Journal, CQ-DL, Megahertz Magazine *and on several Web sites. Juan Antonio can be reached at P.Box: 138, 06400-Don Benito (Badajoz), Spain;* **ea4cyq@amsat.org**.

Pedro Pérez Álvarez-Cienfuegos, EB4DKA, has been licensed since 1988 when he was 17 years old. Pedro holds an ATP pilot degree. He is a charter member of Radio Club Francisco de Orellana and joined Unión de Radioaficionados Españoles and AMSAT-EA in 1998. He also has published several articles about satellites in URE, AMSAT-NA, CQ-DL *and on Internet Web sites.*

A Weather Vane Antenna for 2 Meters

VHF incognito! Here's a 2 meter antenna your neighbors won't recognize.

John Portune, W6NBC, and Fred Adams, WD6ACJ

Figure 1—The completed weather vane antenna. Can you find the loop?

Keeping a homeowners' association happy isn't easy. Many hams burdened with CC&Rs often live with badly compromised antennas. Why not combine a compact magnetic loop with a functional weather vane and put high-performance right out in full view? A neighbor even asked me where I got mine, so she could get one, too.

Compact loops, often called magnetic loops, have been around for years. They're tiny open-ended ring radiators, always less than roughly $^1/_{10}$ wavelength in diameter. Recently we've seen them mostly on HF where their small size is of great appeal. But they work very well on other bands too; here on 2 meters. Remarkably, this VHF version achieves an efficiency of 93%, yet radiates as if it were a full-sized dipole or a J-pole, but from a space only 7 inches square. What's best, it is disguised as a nice-looking weather vane. Your neighbors won't guess that your rooster has a call sign.

What makes a magnetic loop so easy to camouflage besides its size is its radiation pattern. It may surprise you, but a horizontal compact loop radiates as if it were a full-sized half-wave dipole on the vertical axis of the loop. Yet, here it looks like nothing more than the support for the weather vane's direction letters (Figure 1).

There is a restriction, however. Compact loops require low conductor resistance as compared to full-sized dipoles. (See the technical discussion later in this article.) This requires that we use copper water pipe for construction. Also, compact loops have narrow bandwidth (BW)—in this case, a theoretical BW of 600-700 kHz. Practically, however, the BW is wider. Figure 2 shows an actual SWR plot of my loop.

On the positive side, a magnetic loop's narrow bandwidth increases the received signal-to-noise ratio (SNR). In a noisy environment a compact loop will better discriminate than a dipole or J-pole. [While horizontal loops do better in noisy situations because that local noise tends to be mainly E-field oriented, the perceived SNR improvement of a narrowband antenna is principally due to the reduction in out-of-band signals that cause IMD, rather than any reduction of the receiver noise floor.—*Ed.*] They also tend to work better close to the ground or near other objects.

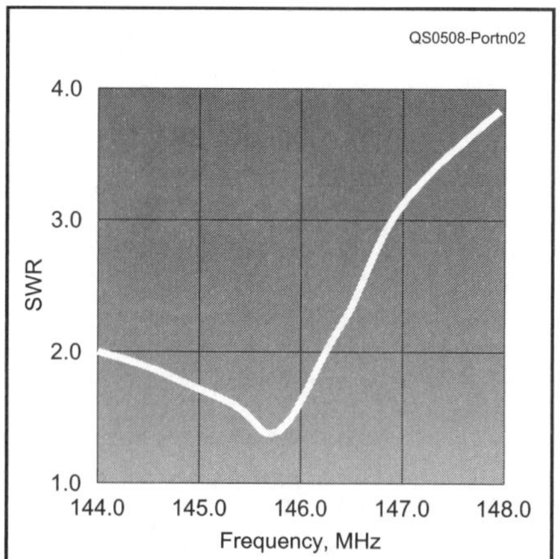

Figure 2—An SWR plot of the completed loop. A 1:1 match at the resonant frequency is possible by compressing or expanding the feed loop.

Table 1
Relative Room Temperature Conductivity of Common Metals

Metal	Conductivity
Silver	100
Copper	95.2
Gold	68.3
Aluminum	60.3
Steel	9.5
Stainless Steel	2.4

Construction

All materials (Figures 3A and 3B) are common hardware store items. I recommend a tubing cutter for cutting the pipe, copper and PVC. Also, a different brand of copper fittings may require slight adjustments to the cutting dimensions shown. But, don't worry; high precision isn't required. The loop will tune up easily if you are roughly within $^1/_8$ inch of these dimensions.

The only part of this design that requires special attention is the $^3/_4$-$^1/_2$ inch reducer that acts as a tuning capacitor across the open ends of the loop. Drill or file the inside so that it will slide freely on the pipe ($^{21}/_{32}$ inch, roughly). Otherwise you will not be able to tune the antenna easily. As purchased, stock pipe fittings are too tight. Also, cut the slot shown for a securing screw.

Important Note: Do not cut the two end pieces at first. Instead, cut just one piece the same size as the opposite side. Later, after soldering, you'll cut a gap in center. This assures good alignment of the tuning capacitor.

For soldering, clean all joints thorough-ly and apply a little flux. Use a propane torch and common solder, not silver solder (Figure 4). Also, solder sparingly. It isn't necessary to make watertight

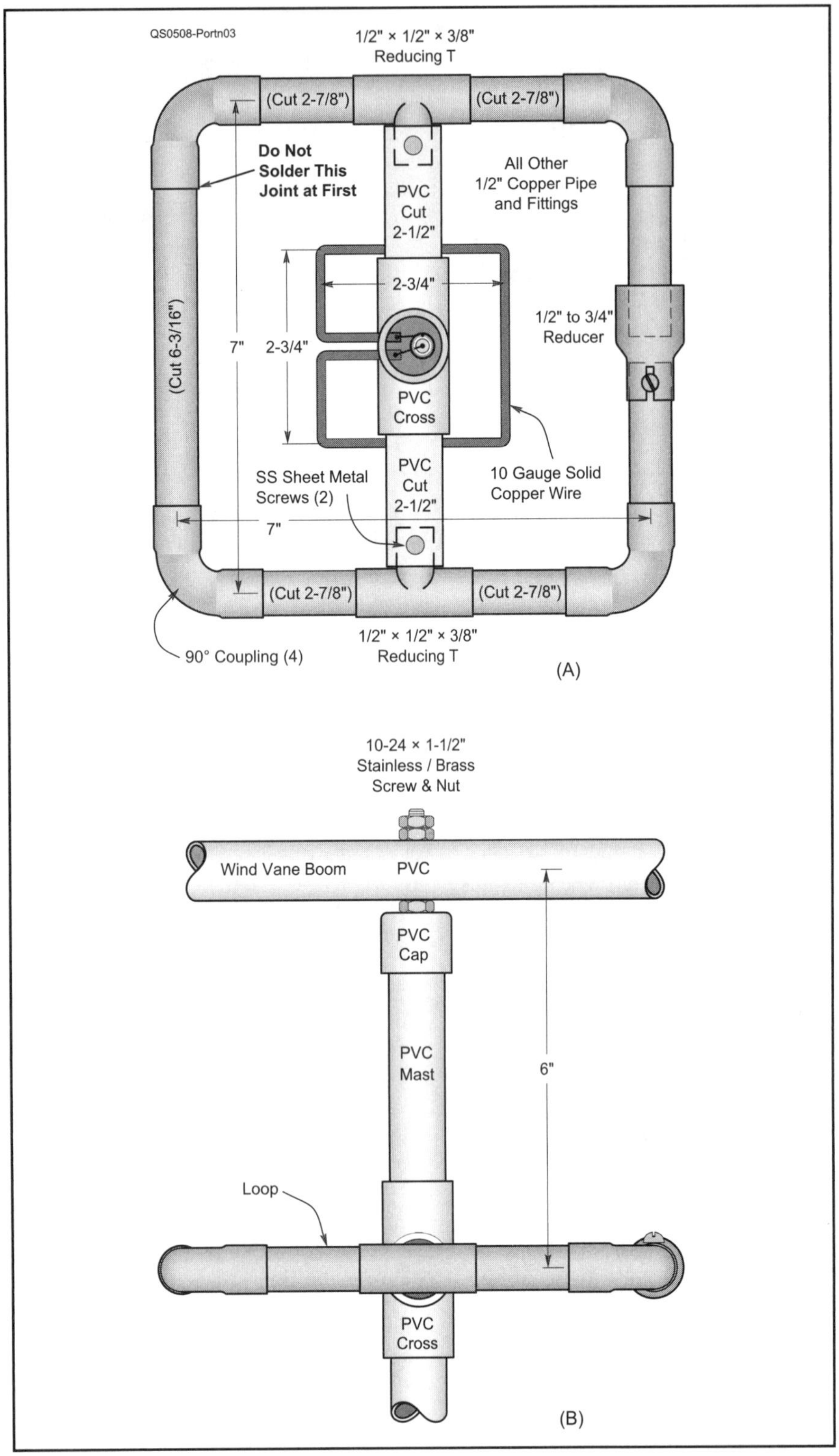

Figure 3—Parts and assembly details for the 2 meter loop are shown in (A). Note that most material is 1/2 inch copper pipe and fittings, except as noted. The inductive feed loop is made of 10 gauge solid copper wire. Basic details of the PVC weather vane support are shown in (B).

connections—tack soldering is fine. At RF frequencies, skin effect makes low dc resistance at the joints completely unnecessary. Keep the loop flat on a heat-resistant surface while soldering.

Important Note: Do not have the tuning capacitor installed while soldering. You will put it on later. Also, be very careful not to solder the one joint indicated (Figure 3A). Otherwise you will not be able to assemble the antenna.

After soldering, cut a 1/2 inch gap in the center of the copper pipe on the tuning capacitor side of the loop. Also, make a hole for a sheet metal screw to secure the tuning capacitor. Then cut the two short PVC center support arms. Note that the copper T-fittings are of the reducer type. The arms of both are 1/2 inch, but the necks are 3/8 inch. Also, add two more sheet metal screws (Figure 3A) to keep the loop from rotating.

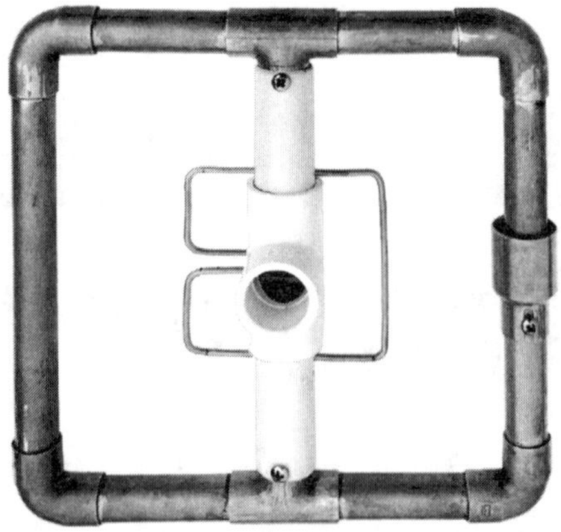

Figure 4—The copper pipe 2 meter loop. Note the inductive feed loop at the center.

Next, install the tuning capacitor, glue the PVC support arms to the center PVC cross, put all the parts together, and then finally solder the one remaining copper joint, again keeping the loop flat. For the final solder joint, wrap wet rags around the T-fittings and PVC pipe to prevent damage to the PVC.

The Feed Loop

For matching to the coax, I prefer an inductive loop, as shown. To me, this is the easiest technique for feeding a compact loop. Other methods also work, such as a gamma match, but are generally more difficult to fabricate. I only had to make a couple of trial loops during the prototype stage to find the correct size. Loops seem to normally need a coupling loop that is roughly 1/3 the size of the main loop. The dimensions will yield a good match without any further experimenting.

You can, however, after final tuning, make minor adjustment to SWR by squeezing or stretching the feed loop. Fabricate the feed loop from bare 10 gauge or 12 gauge solid copper wire. Drill holes in the PVC pipe, then thread the coupling loop through, progressively bending the corners as you go.

For weather protection, solder the coax to the coupling loop inside the PVC cross fitting. Ideally, the loop should have a balun. In practice, however, I have experienced no discernable difference with or without it. But if you prefer, heat-shrink several small VHF RF toroidal

Figure 5—(A) The weather vane "head" and "tail" cutouts and compass points. (B) The rooster cutout. These examples are not to scale—use your imagination!

cores over the coax, inside the PVC cross, to form a feed line choke.

The Weather Vane

I cut the decorative parts of the vane (Figure 4) with tin snips from an inexpensive heavy-duty plastic storage tub. The Web contains a wealth of wonderful weather vane designs, or you may use the one shown. You are obviously free to be innovative here. Do *not*, however, use metal for any of the pieces. This will detune the antenna.

For the rotating arm, keep two things in mind. First, the tail end must have more wind surface than the head end so that the vane will point into the wind. Therefore, the head end must be weighted. I used six-ounce fishing sinkers, silicone glued inside the boom. To hold the head and tailpieces on the arm, cut slots in the ends and use a little more silicone glue. Second, the pivot hole must be drilled at the balance point. Otherwise the arm will not rotate easily. The length of the vane is not critical. Mine is roughly 16 inches long.

Attach the rooster (Figure 5), or whatever "critter" you choose, with plastic cable clamps and some additional glue. Punch two small appropriately located holes in the direction letters and attached them with wire ties. Finally, give the entire structure a coat of black outdoor spray paint. This will give the impression that your weather vane is made of wrought iron, a final touch for the disguise.

Tuning

For tuning, you'll need no special tools, only a VHF SWR bridge and a 2 meter transceiver. In the design phase, I did use a FET dip oscillator to get the prototype in the ballpark. Subsequently I've only needed an SWR bridge and a handheld transceiver to tune up a new vane.

The basic tuning technique is simple. Begin with the tuning capacitor as far away from the 1/2 inch end gap as the securing screw will permit. Apply low power with your transceiver and measure the SWR at the top and again at the bottom of the 2 meter band. Write down the difference. With the capacitor all the way out, the SWR should be worse at the low end. That's because the loop is tuned too high in frequency. Remember, always keep the securing screw tight when making the SWR measurements. Also, at first, both SWR readings will be poor. Only the difference between high and low matters.

Progressively move the tuning capacitor inward, in small increments, again taking high/low readings. As you reach the desired operating frequency, high and low readings should become equal. Still, they won't necessarily be low. If you go too far, the SWR will become better at the low end, the reverse of above. Your objective is to find the position where both readings are equal and lowest. To obtain a final 1:1 match you may need to squeeze or stretch the feed loop a little. Generally, I have not had to bother with the dimensions given. Adjusting the tuning capacitor alone has yielded an adequate SWR.

A Technical Discussion

Theoretically, a compact/magnetic loop antenna is a parallel-tuned LC "tank" circuit. The variable capacitor across the ends (1/2-3/4 inch reducer) resonates the inductance of the loop to the frequency of operation. A multi-turn inductor can't radiate because of its small size. But, with a large single-turn inductor, such as a compact loop, good radiation characteristics can be achieved.

A compact loop, however, *must* be smaller than about 0.1 wavelength in diameter and 0.35 wavelength in circumference. This causes it to radiate in a unique way. The classical RF bible, *Electronic and Radio Engineering* by F. Terman, states that: "The directional pattern of a small loop is identical with that of an elementary doublet (dipole). The only difference is that *the electric and magnetic fields are interchanged*." A small horizontal loop, therefore, radiates as if it were a vertical dipole.

If we make the loop larger than 0.1 wavelength in diameter, the radiation pattern will slowly take on the more familiar form. A full half wavelength dipole, for example, folded into a loop, radiates horizontally. This is the popular "halo" antenna, once common during the early days of 2 meter mobile.

On the other hand, within limits, a compact loop may be made much smaller. A tiny loop compared to wavelength is easily implemented by merely increasing the capacitor across the loop's ends. In this way, lower frequencies or even multiple bands can be achieved. This technique is often used on HF, for example, and several good small multi-band HF compact loops are currently on the market for hams with limited space.

Radiation Resistance

There is, however, a practical limit to how small a loop can be made, due to another important characteristic of all small antennas—radiation resistance. Radiation resistance is not resistance in the usual sense. It is a kind of virtual resistance created by the actual loading of space on an antenna. Said another way, radiation resistance is a measure of how well an antenna couples to space.

But, as far as your transmitter is concerned, radiation resistance looks just like an ordinary resistor at the end of the coax. But instead of converting transmitter power into heat, which an actual resistor would do, radiation resistance does what we want, and converts RF power into a radio signal. Therefore, radiation resistance is the kind of resistance we want in an antenna. And, here's why.

In free space, a full-sized dipole has moderately high radiation resistance, roughly 72 Ω. Smaller antennas like compact loops have a radiation resistance that is much lower. The bigger the antenna, the larger a piece of space it captures, and the higher the radiation resistance. Specifically, as the size of an antenna decreases, the radiation resistance decreases —roughly as the square of size. Our weather vane is less than 1/10 wavelength in diameter. It has an actual radiation resistance of only 1.7 Ω.

But, as long as we correctly match the transmitter's output impedance (normally 50 Ω) to the radiation resistance of the antenna, efficient radiation will take place. Hence, a compact loop with a very low radiation resistance can theoretically radiate just as well as dipole or a J antenna with a much higher radiation resistance.

There is a fly in the ointment, however—conductor resistance. That's because conductor resistance is in series with radiation resistance. And both forms of resistance must share the power from the transmitter. The power the conductor resistance (a real resistance) receives is wasted as heat. The power the radiation resistance (a virtual resistance) receives makes useful radio waves.

Therefore, in a dipole, with a relatively high radiation resistance of 72 Ω, conductor resistance isn't of much concern. Even for a thin wire of poor conductance, conductor resistance is only a small fraction of the radiation resistance. Thus it shares very little of the transmitter's power. But for a compact loop, where the radiation resistance may be very low, conductor resistance can easily waste significant power. That's why we should construct small antennas with large diameter conductors, not thin wire. This is also one of the reasons why mobile HF whip antennas are characteristically low in efficiency.

Metal Conductivity

Also for a small antenna, the kind of metal can be important. Table 1 illustrates how poorly conductive some common metals are, and why copper is really the only good choice for a compact loop. Two interesting points to note are that gold is a poorer conductor than either silver or copper. Gold is used on electronic connectors not because it is the best conductor, but because it does not corrode. Also, notice that stainless steel is 40 times poorer than copper, and that aluminum is about 60% as good.

Bandwidth

The final theoretical limitation of a small antenna, as mentioned earlier, is its low bandwidth. Bandwidth is a function of an antenna's natural LC ratio. In a dipole, this ratio is low. That is, the native self-inductance of a straight wire is relatively small compared to its self-capacitance. Q, therefore, is low, and bandwidth high. For a compact loop the opposite is true. A one-turn coil has high L compared to C. Hence, Q is high and bandwidth low. This is equally true for a loaded mobile HF whip. The large loading coil is very inductive, giving the antenna high Q and low bandwidth.

So, as an overall guideline, if we use large copper conductors and can live with a small bandwidth, a compact loop is a high efficiency choice for a small space. It is also easier to camouflage. My 40 meter rose trellis loop is also rarely noticed.

Designing a Loop

And, in case you would like to try your hand at a compact loop on another band, I have placed on **www.w6nbc.com/loop.html** the loop design equations and an excellent little Basic program that quickly calculates all parameters, including efficiency, of compact loops of almost any size and frequency. It is one of the many freeware programs for compact loop design available on the Web. It is a simple DOS program, but emulates well on a modern computer running *Windows XP*.

My Operating Experience

I have had two of the weather vane loops in service for several months now, one on my mobile home and the other at a friend's house. At my location I also have a comparison J-pole antenna (basically a full-sized dipole) at the same height, and I can detect little difference. Also, the weather vanes have experienced significant wind and weather exposure during that time and show no tendency to detune. Give one a try. They've been an "undercover" winner for me.

Photos by the authors.

John Portune, W6NBC, first received his Advanced class amateur license in 1965, but has held an FCC First Class General Radiotelephone commercial license since 1961. He spent the first five years in Amateur Radio in England, as G5AJH. John upgraded to Amateur Extra in 1985. He is active on all bands through 440 MHz and has written articles for various ham and electronics journals. John is a retired TV broadcast engineer (KNBC, Burbank, California) and technical instructor (Sony Electronics, San Jose, California). He is currently an active VE team leader and participates in Central California ARES/RACES activities. You can reach him at 1095 W McCoy Ln, #99, Santa Maria, CA 93455 or **jportune@aol.com**.

Fred Adams, WD6ACJ, was first licensed as WN6AIH in 1961. Fred says that he was interested in CB radio prior to that time and that his career has mainly been in the mechanical engineering area. He advised and collaborated with W6NBC on this antenna—principally in the area of construction. Fred says he's built many antennas, but this design was a new concept to him. You can reach him at **fredeadams@verizon.net**.

Lowe's Loop for 432 MHz

Your brand-new rig covers 432 MHz, but you don't have an outside antenna for that band? Here's an easily constructed antenna that will get you on 70 cm.

Ben Lowe, K4QF

Figure 1 — Completed 432 MHz "Lowe's Loop."

Introduction

Many of the small modern transceivers on the market today are supplied with 432 MHz capabilities, but often operation on this band is not utilized due to the lack of a suitable antenna. This can be a great band for local contacts, and when propagation is good almost any simple antenna with the right characteristics — horizontal polarization and a low SWR — will permit some long-range contacts.

The purpose of this article is to present an antenna that is easy to construct in a couple of hours and is easy to match. It uses inexpensive material readily available at Lowe's Home Improvement, Home Depot or your local hardware store.[1] After all, putting up a 432 MHz antenna and generating some activity on this band would have to be considered a "home improvement" by any true-blue ham! The antenna described in this article is a *halo* loop with a matching scheme that is easy to implement.

Construction

I fabricated this loop with an emphasis on ease of construction and low cost. To that end, I used standard ¾ inch diameter copper water pipe, 5 or 10 feet long for the mast and ¼ inch copper tubing, normally used for ice makers on refrigerators, as the radiating element. Icemaker tubing normally comes in a 10 foot roll at Lowe's, but I only use about 1 foot unless I construct a multi-element array. One roll of tubing will be plenty for you and several of your friends to construct antennas.

This project also requires some #6 brass wood screws, some #8 brass wood screws both ½ inches long, and a ¼ inch nylon spacer 1 inch long. These parts are also available at Lowe's. Table 1 lists the parts needed.

Fabricating the Pipe

The basic approach to construction is straightforward. Use a tubing cutter to cut the radiating element from the ¼ inch tubing for a length of 13 inches. Center punch all holes before drilling so that the drill bit does not "walk off" the desired location. Exactly at the center of the element, 6.5 inches from each end, drill a small tap hole using a drill bit of 3/32 inches. The precise size of this hole isn't critical, since a screw will hold the element in place until it is soldered. Once you've drilled the center hole, the element is bent around a 4 inch form to create the *halo* loop. Keep the tap hole on the inside portion of the circle. That is, place the hole directly against the circular form facing in toward the surface of the form instead of facing out.

The precise diameter of the halo isn't too critical, since the element diameter can later be adjusted by slightly bending it. I shaped the element into a circular form using a piece of 4 inch diameter PVC pipe obtained from Lowe's and wrapped the 13 inch long, ¼ inch diameter, ice maker tubing around the pipe. However, any 4 inch diameter circular form can be used. Figure 1 shows the loop element.

You will be drilling a total of six holes in the ¾ inch mast. Using a sharp felt tip pen or a pencil, at a distance of 2 inches from one end of the ¾ inch pipe, mark the position of the first hole to be drilled. At this point along the pipe you will gauge the precise circumference using a piece of masking tape. See Figure 2C. Wrap the masking tape around the pipe, offsetting the edge as you wrap just enough so you can see the starting point. Mark the beginning and end points on the tape.

Inserting the Elements

Remove the masking tape and temporarily stick it to a flat surface. Use a ruler to measure

Table 1
Parts List

Description	*Lowe's Part Number*
Round slotted brass wood screw, #6, ½″ long.	490120
Round slotted brass wood screw, #8, ½″ long.	491257
Nylon spacer, ¼″ × 0.140″ × 1″.	491257
Copper pipe, ¾″ × 10′ long.	23791
Copper ice maker pipe, ¼″ × 5′ long.	22100004

[1]The author is not related to, nor does he represent, Lowe's Building Supply.

the distance along the masking tape. Calculate ¼ of the total circumference distance. Mark points along the tape, each spaced ¼ of the circumference around the pipe. Mark the center point in between two marks. This mark will be used to line up with the tap hole you've already marked on the pipe. (The tap hole is for the #6 element screw that will be used to temporarily hold the radiating element in place while it is soldered.) Rewrap the masking tape on the pipe just below the 2 inch circle previously marked on the pipe using the tap center hole as a reference.

On each side of the center hole, at 90° to each other, you will drill two other holes for the radiator to pass through the mast. Mark the positions for these holes on the pipe, along with the holes 1 inch farther down the pipe used for the nylon insulating spacer, and ¼ inch farther from that down the pipe for the #6 ground tap screw. See Figure 2 again.

Drill the two mounting holes for the radiator, starting with a small starter bit, following with larger diameter bits. You will need to use a rattail file to open up the holes to the shape of ovals to allow the radiating element to pass through the pipe.

Insert the radiating element through the oblong holes and fasten it temporarily in place with a #6 brass set screw. Then, solder the element into place using a high wattage iron or small propane torch. After the element cools to the touch, the element can be bent to make it exactly horizontal when the mast is vertical.

Insert a ¼ inch diameter, 1 inch long, nylon spacer, available from Lowe's, into the spacer hole. This spacer extends approximately ⅛ inch out each side of the pipe mast, forming a standoff insulator for the balun center conductors that make up the antenna feed system. The matching scheme itself is taken from a well known technique for the delta match using the "plumber's delight" approach. It got this name because the radiating element does not require insulation from the mast and thus can be fabricated from plumbing fixtures.

Wiring it Up

Next, construct the 4:1 balun from RG-58 coax cable that is ½λ long as shown in Figure 3. While coax with foam insulation exhibits slightly lower loss, the foam can also act like a sponge for water, changing the characteristics of the dielectric and corroding the shield. Therefore, be sure to use RG-58 coax with solid polyethylene insulation, which isn't nearly so susceptible to water. Even though the loss is slightly higher, the length is so short that the difference is negligible.

The RG-58 coax is rated at 70 W at 432 MHz. With SSB or CW operation, the duty cycle is well below 50%, so I wouldn't

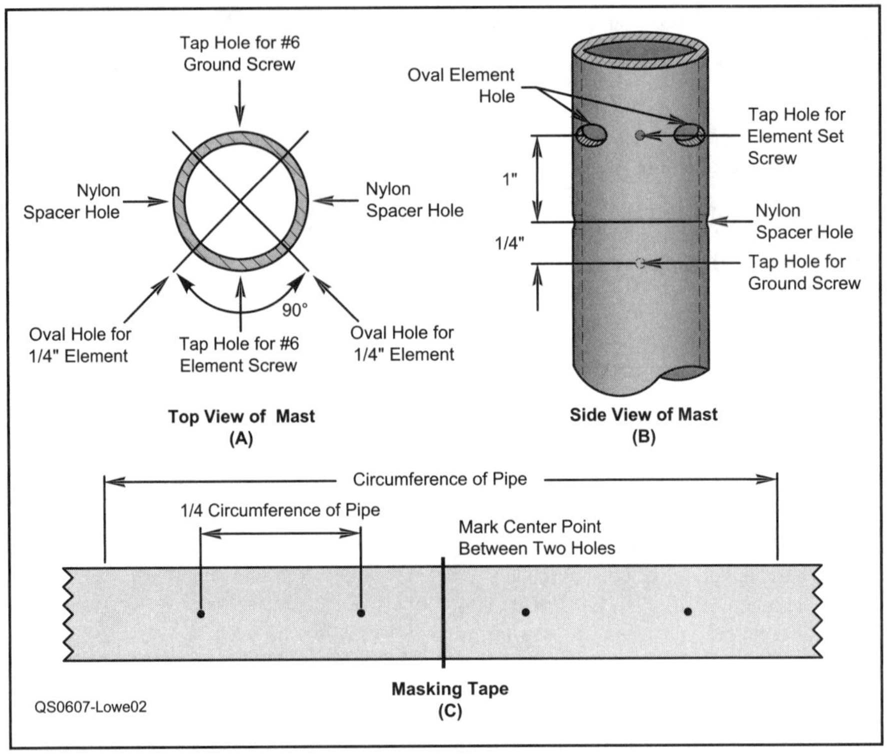

Figure 2 — Hole placement for Lowe's Loop mast. At A, top view of mast. At B, side view of mast. At C, masking tape used to measure exact circumference of mast, and then to mark off four equally spaced hole positions.

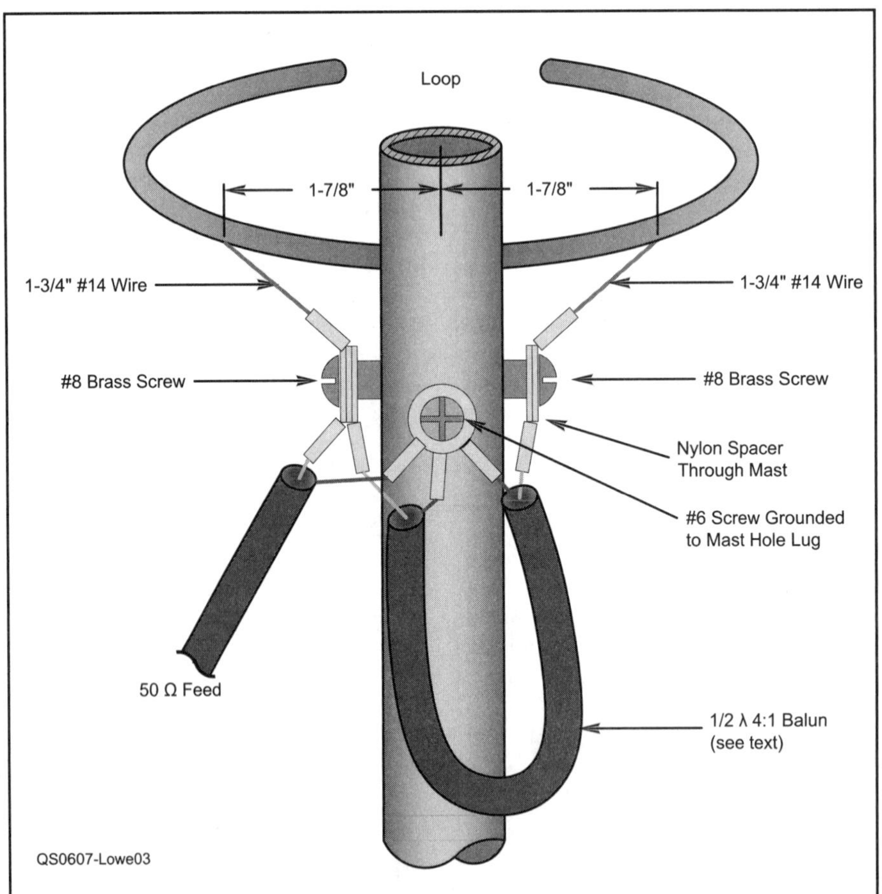

Figure 3 — Feed system, including coaxial 200 to 50 Ω coax balun.

hesitate to run as much as 100 W of RF into this antenna. If you plan to run more than 100 W to the antenna, a larger, higher power coax should be used. Also, with a typical station installation using a nondirectional antenna such as this, most setups will probably have something in the order of 3 dB feed line loss from the shack to the antenna. Therefore, even a transmitter with a couple of hundred watts of output can supply only 100 W to the antenna. As described later, a four-bay array of loops, each at 70 W capability, could take 280 W of average power, or 560 W peak power supplied to the antenna array.

The delta match feed is constructed from the 4:1 balun, two 1¾ inch lengths of 14 gauge copper house wiring and four solder lug terminals. The purpose of the 4:1 balun is to transform the 200 Ω balanced line of the delta match feed point to a 50 Ω unbalanced coax line.

You should mark the elements at the proper distance with a felt tip pen before soldering the delta wires to the element. Attach the 50 Ω feed line's center conductor to one of the lugs at the #8 brass screw that is inserted into the nylon insulator and its shield to the common ground point with the two shield connections from the balun.

The balun uses a piece of RG–58 coax that is 11⅞ inches long. Carefully remove ¾ inches of the black outside covering from each end of the coax, leaving the black covering on the coax 9⅝ inches long. Using a scribe or small sharp object, spread the braided shield on each end of the coax, tightly twisting the shield to form a solid conductor, and then carefully remove one half of the polyethylene insulation remaining on each end of the coax. Solder a small terminal lug suitable for a #6 brass wood screw on the twisted shields on each end of the coax. On the center conductor of each end of the coax, solder terminal lugs suitable for a #8 brass wood screw, along with the two 1¾ inch lengths of #14 copper house wiring as shown. Soldering is done before later attaching the lugs to the nylon spacer in order to not melt the nylon spacer. The final coax balun is shown in Figure 3.

After you've completed the coax balun and bent it into a loop, pass #8 brass wood screws through the solder lugs on the center conductors into each end of the nylon spacer on each side of the mast. If the holes for the nylon spacer barely have clearance, the #8 screws spread the nylon spacer just enough to hold the spacer securely in place. This may require a screwdriver on both ends of the nylon spacer at the same time. Then attach the #6 ground lugs on the shields to the tap hole located ¼ inch below the nylon spacer with a screw. The final step is to solder the #14 buss wires to the bottom side of the element. These wires are attached to the element at a distance of 1⅞ inch from the brass set screw used to hold the element in place.

Final Test

If the loop and balun dimensions have been followed precisely, very little tuning should be necessary to obtain a low SWR for the antenna. One factor that could require tuning, however, is a variation in the velocity factor of the particular run of coax used to construct the balun. If this is the case, one way to compensate is to make slight adjustments in the length of the balun. This is easier said than done since it requires cutting and soldering the balun each time an adjustment is made. An easier way to make slight adjustments is to vary the spacing between the ends of the loop. You can make slight tweaks in the resonant frequency of the antenna by slightly changing the diameter of the loop. If the SWR is no more than 1.6:1 or 1.7:1, change this spacing by simply holding both sides of the loop and spreading or squeezing the spacing. This should allow the SWR to be brought down to an acceptable level.

For the test antenna and subsequent models, I easily adjusted the SWR to 1.2:1. Most of the models fabricated did not require any adjustments to achieve this SWR. One model required that the spacing be adjusted to 2 inches to obtain an acceptable SWR.

Figure 4 — Four Lowe's Loops mounted in a stacked array for additional horizontally polarized gain.

Quick Fix Four-Bay Array

One useful feature of the Lowe's Loop is the ability to stack these loops vertically to narrow the elevation beamwidth and achieve omnidirectional gain in the azimuth plane. Stacking four of the loops separated by ⅝ λ, or 16 inches between loops at 432 MHz, yields a total height of 48 inches for the four loops shown in Figure 4 and should yield about 6 dB gain, minus any losses in the power divider. While you can fabricate one loop on a ¾ inch copper mast, it isn't much more effort to prepare the mast and cut the elements for three more of the loops. If you pay careful attention during the construction, you can achieve a low SWR for all of the loops. Then, it is necessary to provide four equal-amplitude and in-phase drive signals to the four elements.

If all four loops in the array are fed with equal length feed lines, then each element receives the same amplitude and in-phase signal provided by the four way divider.

Tying four identical 50 Ω feed points in parallel results in a feed impedance of 12.5 Ω. I fed the four loops in my array with equal lengths of 50 Ω, RG-58 coax, each with a BNC connector on its end. This provided an easy mechanism to check each antenna individually before combining them into the four bay array. I drilled four holes in the side walls of a ¾ inch cooper end pipe cap to permit insertion of BNC female chassis mount connectors into the pipe cap, as the divider in Figure 5 shows. I sweat soldered the BNC connectors to the copper end cap.

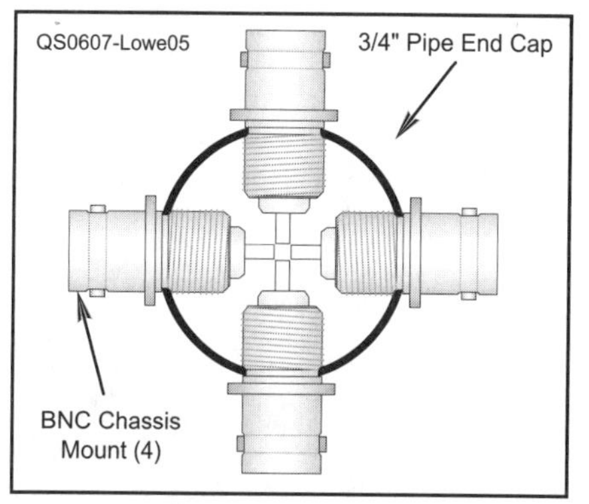

Figure 5 — Four way divider/transformer to feed stacked array of four Lowe's Loops.

For the particular version of the BNC connectors used, the center conductor pins had to be cut off a slight amount to allow all four connectors to join together at the center point and still have the ground rings of the connectors seat against the sides of the pipe cap. I did this easily with a pair of wire cutters on the soft metal center pins.

I needed to transform the 12.5 Ω impedance back up to 50 Ω. I decided to use a $\lambda/4$ Q section, which requires a 25 Ω characteristic impedance line. You can make a line with this impedance simply by paralleling the center conductors and the shields of two $\lambda/4$ lengths of 50 Ω coax. The photo in Figure 6 shows four BNCs mounted through the bottom pipe end cap, together with the two paralleled Q section coaxes before inserting them into the ¾ inch copper pipe and top end cap to finish the final assembly.

I started out by cutting two 5¼ inch lengths of RG-58. I removed ¾ inches of the outside jacket from each end of each coax. I then removed half of the center insulation at each end to expose the bare center conductors and soldered these center conductors together. After some experimentation, the final length of the paralleled cables was 3¾ inches, as shown Figure 7. The difference in length from the calculated value was likely due to the end sections, which act like additional short transmission lines with a different Z_0.

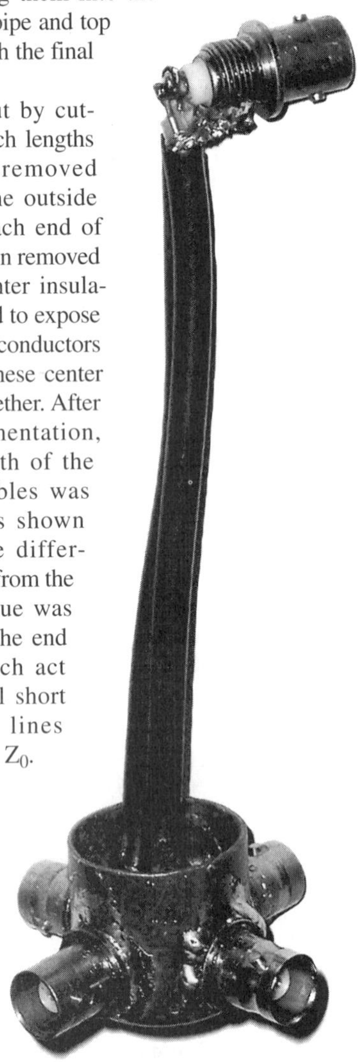

Figure 6 — Photo of four way divider/ transformer before mounting in ¾ inch copper pipe with end cap.

Incidentally, the same approach should be applicable for building two way dividers using paralleled pieces of 72 Ω coax. The paralleled coax characteristic impedance should be 35 Ω to transform the 25 Ω impedance of two 50 Ω antennas in parallel to 50 Ω. The required 35 Ω is very close to two paralleled pieces of 72 Ω coax cut to $\lambda/4$ in length.

Once I cut the Q section coax lines to length, I soldered their center conductors to the junction of the BNC center pins, and then soldered the four short ground lead pigtails onto the ground of each BNC connector. I soldered another BNC connector to the top end of the transformer at 90° to the coax to minimize lead lengths.

Figure 8 — Completed four way combiner/ transformer assembly.

It was now time to test the divider/transformer assembly. I connected a 50 Ω termination to each of the four BNC connectors and measured the SWR, which was 1.2:1. After testing, I placed the entire Q section inside a 3 inch long, ¾ inch diameter copper water pipe, together with an end cap with a hole in its side for the top BNC connector. Figure 8 shows a photo of the completed divider/transformer assembly.

Conclusion

This antenna is relatively easy to construct and to put into service. The materials are available at almost any building supply or hardware store. While one antenna provides omnidirectional coverage, the four bay array should approach 6 dB gain over a single element in all azimuth directions. This gain comes from compressing the elevation pattern, the same way that TV transmitting stations obtain gain in their omnidirectional antennas.

I've taken the same approach for a 2 meter halo simply by multiplying all dimensions by three. I've also constructed a 6 meter version, but that required capacitive loading to keep the dimensions of the antenna from getting too large. So far, my best DX on 70 cm has been Texas from Alabama. I've used the 2 meter Squalo to log Toronto to the northeast and Houston to the southwest, among the 42 grids that I've worked since putting it into operation a year ago.

I want to express my appreciation to Dieter Schliemann, KX4Y, for the use of his antenna analyzer and SWR meter, and to George Hall, W4BUW, for review of this manuscript.

Ben Lowe, K4QF, was licensed as KN4VOW in 1958 at the age of 14. He holds an MS degree from Southern Methodist University in Dallas and a BS degree from The University of Tennessee at Knoxville, both in Electrical Engineering. He is a registered Professional Engineer in the states of Alabama and Texas. Ben designs and implements microwave systems, assists the military with GPS applications and wireless architecture development, and conducts RF environmental analysis and measurements for Engineering Systems & Planning of Huntsville, Alabama.

In addition to Amateur Radio, he is a licensed pilot, has an advanced SCUBA certification, and is an avid bicyclist. His primary ham radio interests are in weak signal VHF/ UHF operation. He has written numerous articles for QST, CQ, Ham Radio *and* The ARRL Handbook. *Previous call signs include K4VOW, WA5UVM, ZL1AQC, G5EPZ, K4QF/HB/DL/ HBØ and ZF2BL. He has two adult children, a daughter who is a research scientist in genetics and a son who is a computer engineer. You can reach Ben at 848 County Rd #138, Scottsboro, AL 35768, or at* **K4QF@arrl.net**.

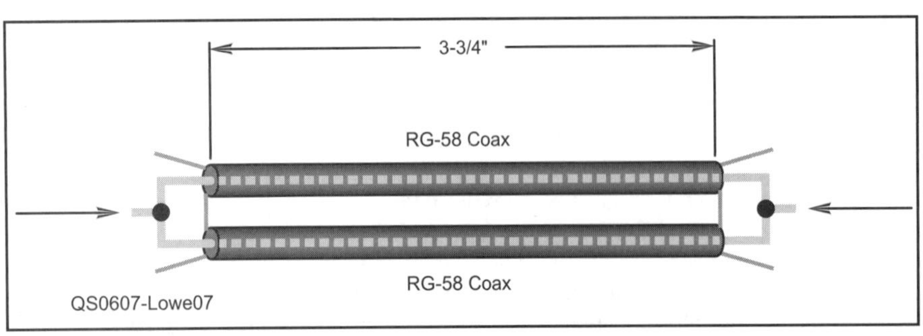

Figure 7 — Making a 25 Ω, $\lambda/4$ transmission line by paralleling two RG-58 lines.

An EZ-Lindenblad Antenna for 2 Meters

This easy to build antenna works well for satellite or terrestrial communication, horizontal or vertically polarized.

Anthony Monteiro, AA2TX

Lindenblad is the name of a type of antenna that is circularly polarized yet has an omnidirectional radiation pattern. With most of its gain at low elevation angles, it is ideal for accessing low earth orbit (LEO) Amateur Radio satellites. Because it is omnidirectional, it does not need to be pointed at a satellite so it eliminates the complexity of an azimuth/elevation rotator system. This makes the Lindenblad especially useful for portable or temporary satellite operations. It is also a good general purpose antenna for a home station because its circular polarization is compatible with the linearly polarized antennas used for FM/repeater and SSB or CW operation.

This type of antenna was devised by Nils Lindenblad of the Radio Corporation of America (RCA) around 1940.[1] At that time, he was working on antennas for the then nascent television broadcasting (TV) industry. His idea was to employ four dipoles spaced equally around a λ/3 diameter circle with each dipole canted 30° from the horizontal. The dipoles are all fed in phase and are fed equal power. The spacing and tilt angles of the dipoles create the desired antenna pattern when the signals are all combined. Unfortunately, the start of World War II halted Lindenblad's TV antenna work.

After the war, George Brown and Oakley Woodward, also of RCA, were tasked with finding ways to reduce fading on ground-to-air radio links at airports.[2] These links used linearly polarized antennas. The maneuverings of the airplanes often caused large signal dropouts if the antennas became cross-polarized. Brown and Woodward realized that using a circularly polarized antenna at the airport could reduce or eliminate this fading so they decided to try Lindenblad's TV antenna concept.

Brown and Woodward designed their antenna using metal tubing for the dipole elements. Each dipole element is attached to a section of shorted open-wire-line, also made from tubing, which serves as a balun transformer. A coaxial cable runs through one side of each open-wire-line to feed each dipole. The four coaxial feed cables meet at a center hub section where they are connected in parallel to provide a four-way, in-phase power-splitting function. This cable junction is connected to another section of coaxial cable that serves as an impedance matching section to get a good match to 50 Ω. While the Brown and Woodward design is clever and worked well, it would be quite difficult for the average ham (including this author!) to duplicate.

The major cause of the difficulty in designing and constructing Lindenblad antennas is the need for the four-way, in-phase, power splitting function. Since we generally want to use 50 Ω coaxial cable to feed the antenna, we have to somehow provide an impedance match from the 50 Ω unbalanced coax to the four 75 Ω balanced dipole loads.

Previous designs have used combinations of folded dipoles, open-wire lines, twin-lead feeds, balun transformers and special impedance matching cables in order to try to get a good match to 50 Ω. These in turn increase the complexity and difficulty of the construction.

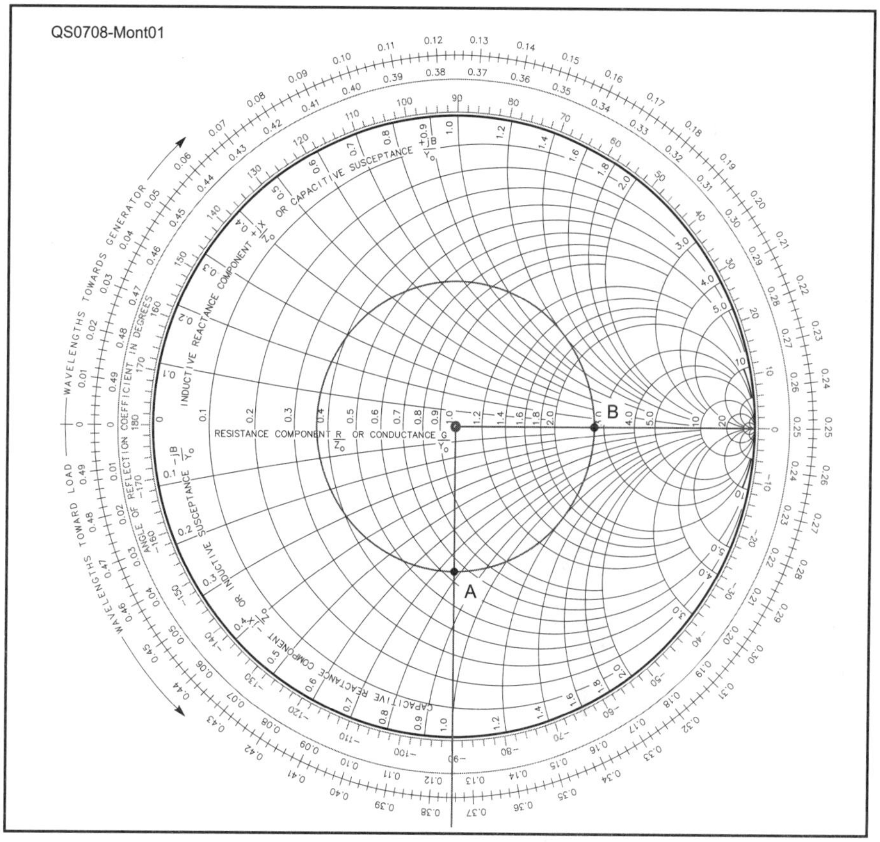

Figure 1 — Smith Chart showing transformation along a 75 Ω transmission line to yield 200 Ω.

[1]Notes appear on page 176.

Table 1

Required Materials

Quantity one, unless noted.

Aluminum tubing, 17 gauge, 6 foot length of ¾" outer diameter, quantity 3. Available from Texas Towers, **www.texastowers.com**.
Aluminum angle stock, 8" length of 2" × 2" × 1⁄16".
Aluminum angle stock, 2" length of 2" × 2" × 1⁄16" for mounting connector.
Screws, #8 × ½" aluminum sheet metal, quantity 12.
Screws, #8 × ½" aluminum sheet metal or 3⁄16" aluminum rivets, quantity 12.
PVC insert T-connector, ½" × ½" × ½" grey for irrigation polyethylene tubing. LASCO Fittings, Inc. Part# 1401-005 or equivalent. Available from most plumbing supply and major hardware stores, quantity 4.
Plastic end caps (optional), black ¾", quantity 8.
N-connector for RG-8 cable, single-hole, chassis-mount, female.
Cable ferrite, Fair-Rite part # 2643540002, quantity 4 (Mouser Electronics #623-2643540002).
RG-59 polyethylene foam coax with stranded center conductor, 10' length.
Copper braid, 4" long piece.
Ring terminal, uninsulated 22-18 gauge for 8-10 stud, quantity 4.
Ring terminal, uninsulated 12-10 gauge for 8-10 stud, quantity 4.
Heat shrink tubing for ¼" cable, wire ties, electrical tape, as needed.
Ox-Gard OX-100 grease for aluminum electrical connections.

The EZ-Lindenblad

An antidote to these difficulties is the EZ-Lindenblad. The key concept of the EZ approach is to eliminate anything that is electrically or mechanically difficult, leaving only things that are *easy*. This leads to the idea of just feeding the four dipoles with coax cable and soldering the cables to a connector with no impedance matching devices at all. This would certainly be *easy* but we also want the antenna to work! Without the extra impedance matching devices, how is it possible to get a good match to 50 Ω?

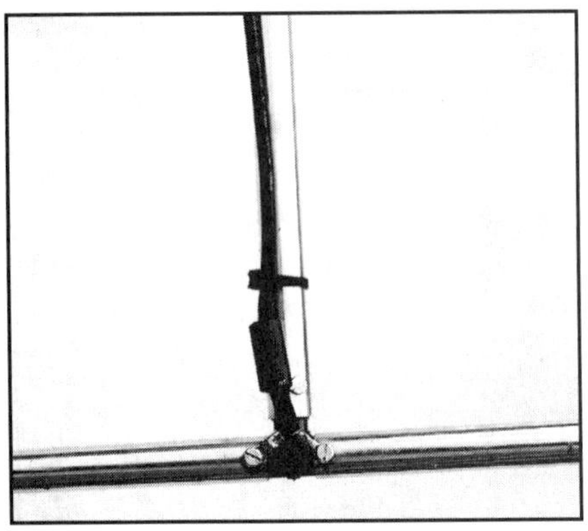

Figure 2 — Close-up of dipole electrical connections.

If we could get each of the four coax feed cables to look like 200 Ω at the connector, then the four in parallel would provide a perfect match to 50 Ω. We could do this if we used quarter-wavelength sections of 122 Ω coax to convert each 75 Ω dipole load to 200 Ω. Unfortunately, there is no such coax that is readily available.

But we can accomplish the same thing with ordinary 75 Ω, RG-59 TV type coax if we run the cable with an intentional impedance mismatch. By forcing the standing wave ratio (SWR) on the cable to be equal to 200/75, or about 2.7:1, we can make each cable look exactly like 200 Ω at the connector as long as we make them the right length. It is easy to make the SWR equal 2.7:1 by just making the dipoles a little too short for resonance. An *EZNEC* antenna model can be used to determine the exact dipole dimensions.[3]

Figure 3 — View of cross booms mounted to mast.

The conversion from the balanced dipole load to the unbalanced coax cable can be painlessly accomplished by threading each cable through an inexpensive and readily available ferrite sleeve making essentially a choke balun. The only remaining issue is the required length of the feed cables, and this can be easily determined using a *Smith Chart.*

Smith Chart

The Smith Chart was invented by Phillip Smith of The Bell Telephone Laboratories in 1939.[4] As a high school student, Phillip had been a ham radio operator and used the call sign 1ANB. After graduating from Tufts College, he went to work for Bell Labs in the radio research department. As part of his job, he needed to make many impedance calculations that in those days required doing many complex computations by hand. Phillip realized he could create a chart that would allow the solution to be plotted on a graph making his job a lot easier. Phillip Smith's chart was soon adopted by other radio engineers and quickly became a standard engineering tool that is still in use today. Technical information and free downloadable Smith Charts are available via links on the ARRL Web site.[5] Please see the Smith Chart in Figure 1.

With a Smith Chart, we can easily determine the required length of 75 Ω coax to provide a 200 Ω load. The Smith Chart shown is normalized to 1 Ω in the center so we must multiply all impedance values by our coax impedance of 75 Ω. An *EZNEC* antenna model was used to simulate cutting the dipole lengths until the SWR on the line reached 2.7:1. The model showed that the dipole load impedance would then be 49 –*j*55 Ω and this is plotted at point A on the chart. The desired 200 Ω impedance at the connector is plotted at point B on the chart. A constant 2.7:1 SWR curve is drawn between the two impedance points. The length of cable needed is read clockwise along the scale labeled, WAVELENGTHS TOWARD GENERATOR from the lines drawn through points A to B. From the chart, the length of the line needed is 0.374 λ.

The EZ-Lindenblad was designed for a center frequency of 145.9 MHz to optimize its performance in the satellite sub band. At 145.9 MHz, a wavelength is about 81 inches and since the coax used has a velocity factor of 0.78, we need to make the feed cables 81 × 0.374 × 0.78 = 23.6 inches long.

There are several computer programs available today that can also be used to do the Smith Chart calculations. These include *TLW*, provided with *The ARRL Antenna Book* and *MicroSmith* formerly offered by the ARRL, which was used as a cross check.[6]

Construction

This antenna was designed to be rugged and reliable yet easy to build using only hand

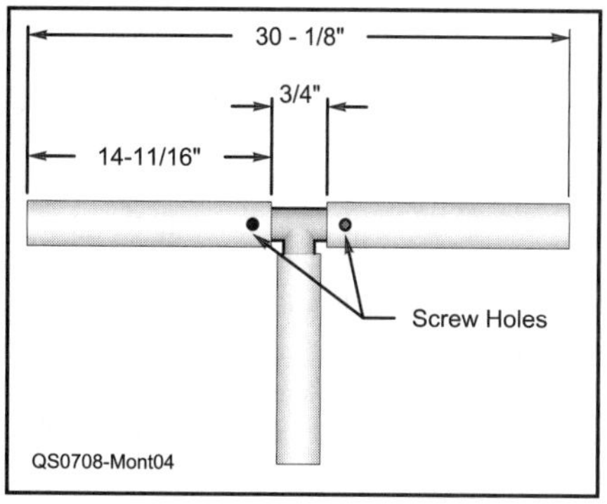

Figure 4 — Dipole dimensions.

Table 2
Aluminum Tubing Lengths

Description	*Quantity*	*Length*
Dipole rods	8	14$^{11}/_{16}$"
Cross booms	2	23"

tools with all of the parts readily available as well. Please see the parts list in Table 1. Although not critical, the construction will be easier if the specified 17 gauge aluminum tubing is used since the inner wall of the tubes will be just slightly smaller than the outer wall of the PVC insert Ts used to connect them. If heavier gauge tubing is used, it will be necessary to file down the PVC insert Ts to make them fit inside the aluminum tubes.

Start by making a mounting bracket to mount the N-connector and the cross booms. Cut a ⅝ inch hole in one side of the short piece of angle stock and rivet or screw it to the bottom of the long piece of angle stock. The completed bracket with the connector and cables attached can be seen in Figure 3.

Next, cut the aluminum tubing to make the cross booms and dipole rods as shown in Table 2. Drill holes for the machine screws at each end of the cross booms but do not insert the screws yet. Attach the cross booms to the long section of angle stock with rivets or screws. One cross boom will mount just above the other as can be seen in Figure 3. The cross booms should be perpendicular to the mounting bracket so that they will be horizontal when the antenna is mounted to its mast. Make sure that the centers of the cross booms are aligned with each other so that the ends of the cross booms are all 11½ inches from the center cross.

Make the dipoles by inserting a PVC insert T into two dipole rods. It should be possible to gently tap in the rods with a hammer but it may be necessary to file down the insert T a little if the fit is too tight. Applying a little PVC cement to the insert T will soften the plastic and make it easier to insert into the aluminum tubing if the fit is too tight. The overall dipole length dimension is critical so take care to get this correct as shown in Figure 4.

Drill holes for machine screws in each dipole rod but do not insert the screws yet.

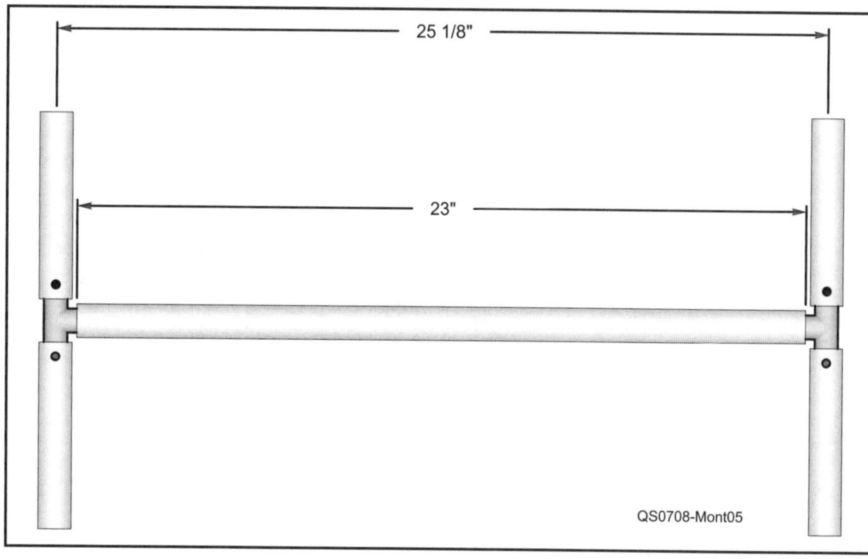

Figure 5 — Dimension of dipole spacing on cross-boom.

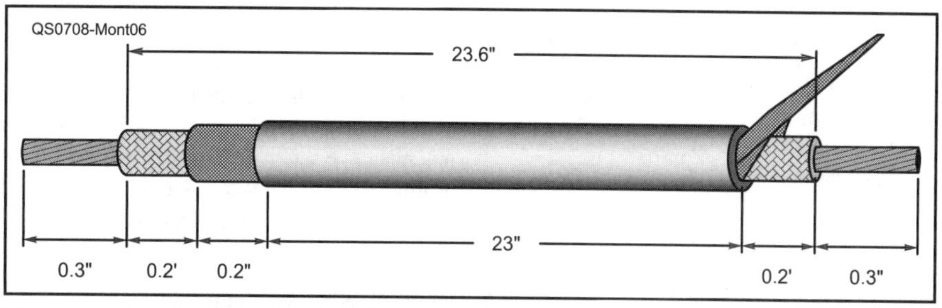

Figure 6 — Feed cables stripping dimensions.

The screws will be used to make the electrical connections to the dipoles at the center. The screw holes should be about ⅜ inch from the end of the tubing.

The dipole assemblies are attached by gently tapping the PVC insert-T into the end of each cross-boom with a hammer. The dimensions are shown in Figure 5.

Next, temporarily attach the mounting bracket to a support so that each of the cross booms is perfectly horizontal. Measure this with a protractor. Now, using the protractor, rotate the dipole assemblies to a 30° angle with the right-hand side of the nearest dipole tilting up when you are looking toward the center of the antenna. Drill a small hole through the existing cross-boom holes into the PVC insert-Ts and then use the machine screws to fasten the dipole assemblies into place. For a nice finishing touch, the dipole ends can be fitted with ¾ inch black plastic end caps.

Next, make the four feed cables by cutting and stripping the RG-59 coax as shown in Figure 6. On the dipole connection side, unwrap the braid and form a wire lead. Apply the smaller ring terminal to the center conductor and use the larger ring terminal for the braid. At the other end of the cable, do not unwrap the braid but strip off the outer insulation. Slip a 1 inch piece of shrink wrap over the coax and apply to the dipole side. Next,

Figure 7 — The EZ-Lindenblad as a portable or Field Day antenna. A 70 cm antenna is on the top.

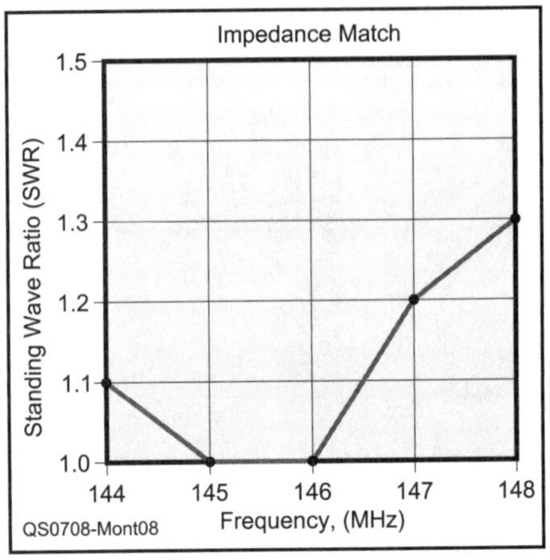

Figure 8 — Lindenblad antenna SWR at 50 Ω.

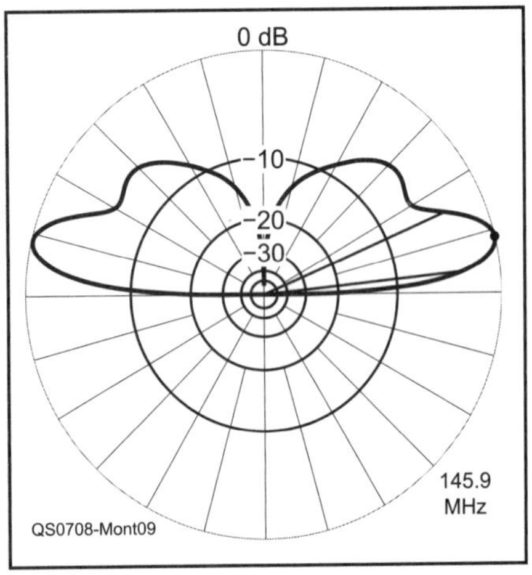

Figure 9 — *EZNEC* radiation pattern of Lindenblad antenna.

slip a cable ferrite over the cable and push all the way to the dipole end as far as it will go (ie, up to the heat-shrink tubing.) The fit will be snug and you may need to put a little grease on the cable jacket to get it started.

Prepare each dipole for its feed cable by first cleaning the area around the screw holes with steel wool and then applying Ox-Gard grease. This is to ensure a good electrical connection. The coax center conductor goes to the up side of the dipole and the braid goes to the down side. To make a connection, put a machine screw through the ring terminal and gently screw into the dipole tubing. Do not overtighten the screws or you will strip the tubing.

Apply Ox-Gard or other corrosion resistant electrical grease around the hole for the N connector. Take the 4 inch piece of braid and put the end of it through the hole for the N connector. This provides the ground connection. Secure the N connector in the mounting hole to clamp the braid. Use a wire tie or tape to hold the four feed cables together at the connector ends. Make sure to align the cables so that all the ground braids are together and the center conductors all extend out the same amount. Do not twist the center conductors together. Carefully push the four cable center conductors into the center terminal of the N connector and solder them in place. Wrap the exposed center conductors of the cables and the connector with electrical tape.

Take the piece of braid that is clamped to the N connector and wrap it around the four exposed ground braids of the coax cables. Solder them all together. This will take a fair amount of heat but be careful not to melt the insulation. After this cools, apply electrical tape over all the exposed braid and fix with wire ties. The cables should be secured to the cross booms with wire ties.

The mounting bracket provides a way to attach the antenna to a mast using whatever clamping mechanism is convenient. For a permanent mount, drill holes in the bracket to accept a pair of U bolts. The author's antenna was intended for portable operation and the bracket was drilled to accept two #8 stainless steel screws. These screws pass through a portable mast and the antenna is secured with stainless steel thumbscrews. This allows the antenna to be set up or taken down in less than a minute. The completed portable antenna as used for Field Day 2006 is shown in Figure 7. The little antenna at the top is for 70 cm.

Standing Wave Ratio

The antenna impedance match to 50 Ω was tested using an MFJ-259B SWR meter, which has a digital readout of standing wave ratio (SWR) and frequency. The frequency accuracy was verified using an external frequency counter and the 1.0:1 calibration was checked with a Narda precision 50 Ω load.

The antenna was connected to the SWR meter with a 6 foot coax jumper made of Belden 9913F7, which has very low loss. The SWR was measured at 1 MHz intervals over the 144-148 MHz range. As can be seen in the chart of Figure 8, the antenna provides an excellent match over the entire 2 meter band.

Power Handling Capability

This antenna was designed to safely handle any of the currently available VHF transceivers. The power handling capability was tested by applying a 200 W CW, signal, key down for 9½ minutes. Immediately after the test, the ferrites and cables were checked and there was no noticeable temperature rise.

Radiation Pattern

The antenna radiation pattern predicted by the *EZNEC* model is shown in Figure 9. This is the elevation plot with the antenna mounted at six feet above ground although it can be mounted higher if desired for better coverage to the horizon. As shown in the plot, the pattern favors the lower elevation angles. The –3 dB points are at 5° and 25° with the maximum gain of 4.8 dBic (with respect to an isotropic circularly polarized antenna) at around 13°. This is an excellent pattern for accessing LEO satellites. Most of the satellite pass elevations will be in this range and it is also the elevation at which the satellite provides the best chance for DX contacts.

The antenna radiation is right-hand circularly polarized, which will work with virtually any LEO satellite that uses the 2 meter band. The circularity was checked by measuring the difference between the horizontal and vertical radiation components. This was done using a linearly polarized sense antenna mounted 100 feet away feeding into an FT-817 radio with the AGC switched off. The radio audio was connected to an ac voltmeter with a dB scale. The test showed a difference of less than 3 dB, which is very good for an omnidirectional antenna. A reference horizontally polarized antenna measured nearly a 30 dB difference.

On the Air

The EZ-Lindenblad antenna has been used for SSB, FM and packet operation on the AO-07, FO-29, SO-50, AO-51, VO-52, NO-44, NO-60 and NO-61 satellites. The portable setup, as seen in Figure 5, was used for the satellite station at the North Shore Radio Association, NS1RA, 2006 Field Day effort. Field Day is an excellent test of any antenna as it is probably the busiest 48 hours of the year on the satellites and the EZ-Lindenblad performed well.

An earlier version of this antenna was published in the *AMSAT Space Symposium Proceedings* of October 2006.

Notes

[1]G. Brown and O. Woodward Jr. *Circularly Polarized Omnidirectional Antenna*, RCA Review, Vol 8, no. 2, Jun 1947, pp 259-269.

[2]See Note 1.

[3]*EZNEC+ V4* antenna analysis software by Roy W. Lewallen, W7EL, available from **www.eznec.com**.

[4]F. Polkinghorn, "Phillip H. Smith, Electrical Engineer, an Oral History," IEEE History Center, Rutgers University, New Brunswick, NJ. Available at **www.ieee.org**.

[5]Smith Chart information is available at **www.arrl.org/tis/info/chart.html**.

[6]R. D. Straw, Editor, *The ARRL Antenna Book*, 21st Edition. Available from your ARRL dealer or the ARRL Bookstore, ARRL order no. 9876. Telephone 860-594-0355, or toll-free in the US 888-277-5289; **www.arrl.org/shop/**; **pubsales@arrl.org**.

Tony Monteiro, AA2TX, was first licensed in 1973 as WN2RBM. He has worked in the electronics industry for over 25 years starting as a member of the technical staff at Bell Laboratories and later as an engineering director at several telecommunications companies. You can reach Tony at 25 Carriage Chase Rd, North Andover, MA 01845 or at **aa2tx@amsat.org**.

A New Spin on the Big Wheel

A popular 2 meter antenna returns in an improved, easier to reproduce form.

L. B. Cebik, W4RNL, and Bob Cerreto, WA1FXT

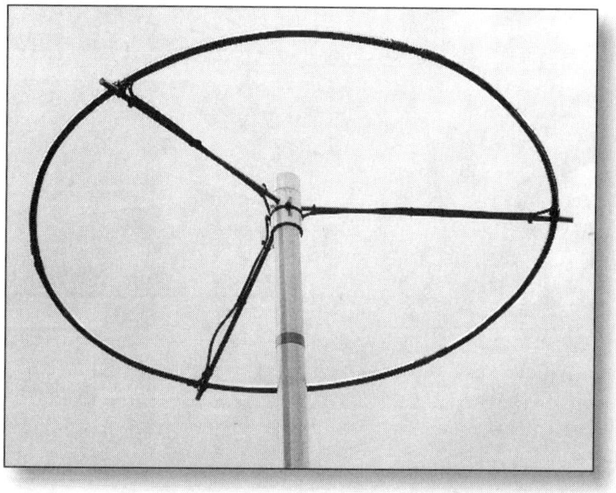

In his "Antenna Options" column in the Jan/Feb 2008 issue of *QEX*, L.B. Cebik discusses some different options for omnidirectional horizontally polarized antennas. Here he and Bob Cerreto provide the details on how to build and use two versions of an update to the Big Wheel antenna from the '60s.

Most attempts to develop a horizontally polarized omnidirectional (HPOD) 2 meter antenna have sought to minimize the antenna's size. Shapes such as circles (halos), squares and rectangles usually result in the need for either hypercritical dimensions or difficult matching conditions — or both. By turning to more conventional full size structures using three dipoles, we can reduce the number of critical parameters and ease the process of replicating the antennas in a home workshop. In fact, we shall describe two versions of the same basic antenna. One is a triangle of three dipoles that folds into a flat package, suitable for easy transport to a hilltop. The other is a circle of three dipoles that requires somewhat less space but needs greater precision in construction. Both antennas share a common feed system and display broadband characteristics that ease the builder's task.

The Basic Three Dipole Design

A 1961 *QST* article described a horizontally polarized 2 meter antenna called the big wheel.[1] The authors described it as three full wavelength (λ) loops with a parallel connection at the central hub and feed point. Unfortunately, their description proved to be off target. In fact, the antenna is a continuous loop with three high-impedance feed points, as shown by the current curves on the left in Figure 1. (All wire models of the same antenna show the same results, but are less clear to read when converted to graphics showing the current distribution.) The legs constituted transmission lines (with variable spacing in the original) that transformed the high impedance at the rim to a low impedance at the hub. By judicious sizing of the circle and the legs, the authors managed a very good omnidirectional antenna.

Unfortunately, many amateurs had difficulty replicating the design because the antenna's dimensions are critical at every point. Small changes in the leg (transmission line) spacing or even differences in the tubing curvature at the rim could throw off the impedance values at the hub.

The big wheel is difficult to model because numerical electromagnetic code (NEC) based antenna modeling tools implementation of transmission line models are not fully accurate when applied to a low current position along the antenna's geometry. The antenna proved equally difficult to build due to the sensitivity of the structure to small dimensional changes. Therefore, we decided to re-explore a territory that the big-wheel authors had set aside: the use of three dipoles to form the same HPOD patterns. The center and right outlines in Figure 1 show the triangular and circular forms that emerged. Note that the current magnitude curves place the feed points of the dipoles at high current, relatively low impedance positions, removing the big wheel's matching challenge.

Both forms are very broadband in virtually every operating parameter once the builder gets the dimensions correct. The triangle, with a wider separation between the dipole end tips, is less critical with respect to dimensions, but requires more space. The circular version, with tighter coupling between dipole tips, requires more careful construction, but results in a more compact structure. In fact, for the same performance, the circular three dipole antenna is smaller than the original big wheel.

The far-field performance of the three dipole HPODs and the big wheel are virtually identical. Therefore, the data in Figure 2 applies equally to all three designs. At a height of 20 feet above average ground, the three elements in all of the designs provide

[1]Notes appear on page 181.

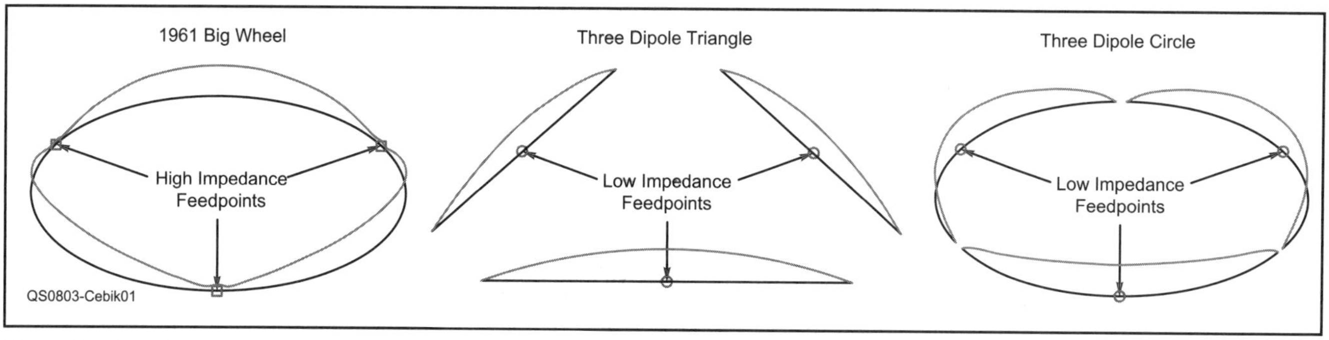

Figure 1 — Relative current magnitudes on three different three element HPOD antennas.

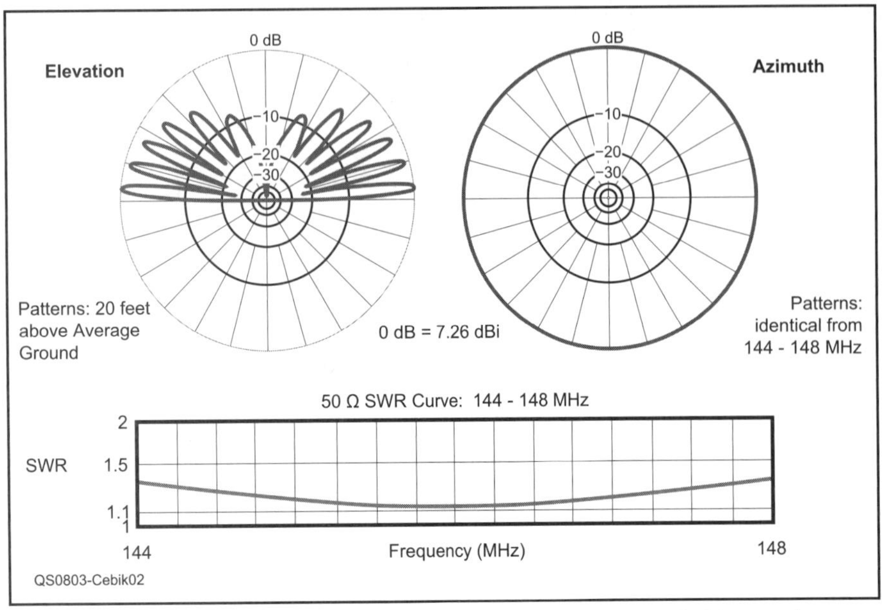

Figure 2 — Representative elevation and azimuth patterns and 50 Ω SWR curve for a three dipole HPOD antenna using either a triangular or a circular shape at 20 feet above average ground. The patterns of the conventional big wheel are virtually identical in shape and strength.

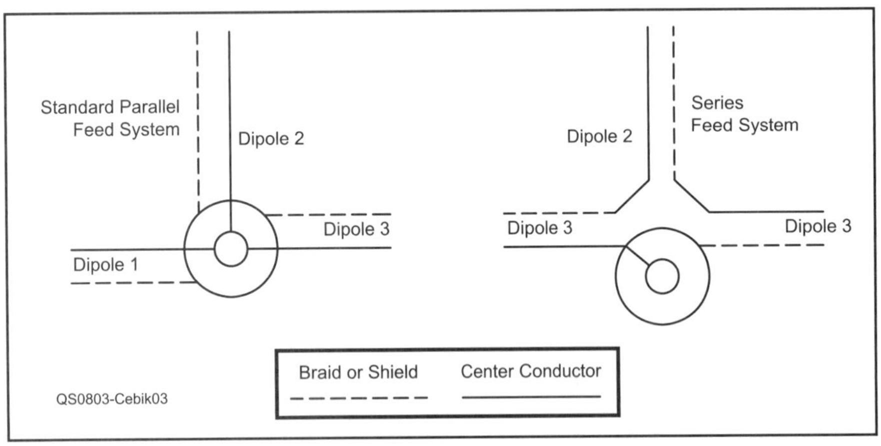

Figure 3 — Alternative methods of feeding the three dipole arrays: the prototypes employ the series feeding system for ease of matching to a 50 Ω main feed line.

an average gain in the lowest lobe of about 7.2 dBi. The azimuth pattern is as close to circular as is possible with fewer than four elements. The gain variation for the worst case was less than 0.3 dB. Although a physical antenna is unlikely to obtain the perfection of a model, the differences in gain around the horizontal will still fall below the level that an operator can detect.

The modeled SWR curve applies to both of the three dipole models.[2] Because the dipoles of the final designs present feed point impedance close to 50 Ω, we may use standard coaxial cable of virtually any length to reach the hub without changing the impedance significantly. Matched to a 50 Ω main feed point at the hub junction, the SWR curve is very flat and in the model shown in the graph, the SWR is acceptable (well under 2:1) for at least 8 MHz in the 2 meter range. Moreover, the circularity of the pattern and the gain are virtually constant across the entire 2 meter band. Even though the antenna is likely to see service only in the first MHz of the band, the broadband characteristics ease the difficulty of successfully building a version at home.

To obtain a 50 Ω main feed point impedance, the three dipole arrays use a somewhat nonstandard arrangement at the hub. As shown on the left in Figure 3, most builders would try to connect the three connecting cables in a parallel arrangement. The resulting impedance would be less than 17 Ω, a somewhat difficult value to handle. As well, any remnant reactance at the hub would constitute a significant portion of the impedance magnitude, creating a matching challenge. However, both of our three dipole designs use the scheme on the right, a series connection of the lines with the source. The resulting hub impedance is close to 150 Ω, and any stray reactances become very small portions of the impedance magnitude. Therefore, a simple λ/4 matching section can handle the impedance transformation to the 50 Ω region.

150 Ω is a borderline value. If the actual impedance of a physical antenna is below this value, then a 75 Ω line (such as RG-59) provides the best match. If the value is a bit higher than 150 Ω, then a 93 Ω cable (such as RG-62) is the better choice. For the lower half of the 2 meter band, λ/4 is about 20 inches. Multiply this length times the velocity factor of the line actually used to obtain the physical length of the matching line. If you are not satisfied with the frequency of the SWR minimum value, you can lower it by lengthening or shortening the matching line.

Based on the basic design information, W4RNL built a triangular version of the array, suitable for both home and hilltop service. WA1FXT took on the task of bending aluminum to form a circular prototype that is both smaller and perhaps more aesthetically pleasing. The differences give the builder options that may fit one or another set of shop skills and tools. Therefore, let's examine both antennas.

A Three Dipole Triangle

A very capable horizontally polarized omnidirectional antenna consists of three dipoles fed in phase. Each dipole is broadside to a direction 120° from the adjacent dipoles. The goal is to find dimensions that will achieve this goal plus provide a workable feed point impedance at each dipole. The prototype constructed to test the basic model of this arrangement used ½ inch diameter aluminum tubing as a light but sturdy material. Each dipole used a 2 inch length of 0.375 inch diameter fiberglass rod as a center insulator. The dipole halves are held in place with #6 stainless steel sheet metal screws. The gap should be as small as is feasible, ⅛ to ¼ inch. These same screws will fasten the ends of the coax cable to the element, with a stainless steel washer to prevent electrolysis between the aluminum element and the copper wires. For ease of disassembly in portable operation, the prototype used lugs under the screws.

The key is to find the correct dimensions so that each dipole presents a 50 Ω impedance at its feed point to match the connecting cable impedance. Table 1 lists some dimensions for both 0.5 and 0.375 inch aluminum tubing, perhaps the two most likely materials for this project. For the triangle, we used

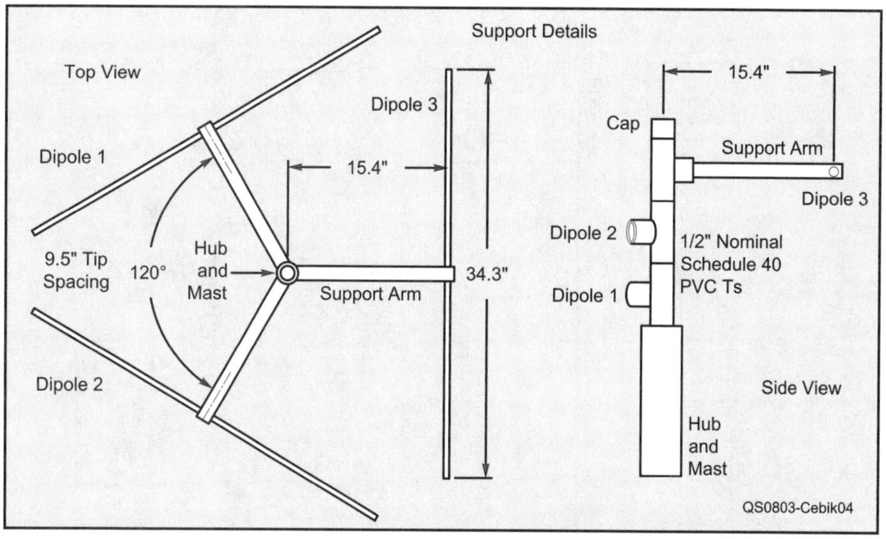

Figure 4 — Some details of the support structure used for the three dipole 2 meter triangle.

Table 1

Dimensions for a Three Dipole 2 Meter Triangle (inches)

Design Frequency (MHz)	*Element Diameter*	*Radius to Feed Point*	*Dipole Length*	*Tip-to-tip Spacing*
146	0.5	15.4	34.3	9.5
146	0.375	15.3	34.7	9.15
144.5	0.5	15.6	34.7	9.6
144.5	0.375	15.5	35.1	9.25

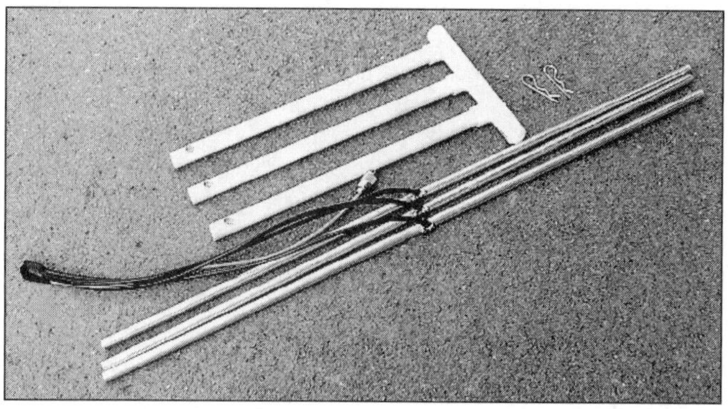

Figure 5 — The triangle HPOD disassembled for transport.

146 MHz as the design frequency because the performance and the SWR do not significantly change across the band. This center-design frequency also provided a good view of the antenna's broadband properties. However, the table also lists dimensions that are usable if the builder wishes to place the performance center of the antenna at 144.5 MHz. The prototype used the half-inch-diameter material and the 146 MHz dimensions for that material.

Note the length of the dipole. It is about 3.3 inches shorter than an independent dipole composed of the same material. The resonant impedance (50 Ω) is lower than the usual value for a standard dipole of about 70 Ω. The three dipoles in the triangle do interact by virtue of both the proximity of their feed points and the closeness of their tips. The dimensions of the triangle are therefore quite critical to successful operation of the array as designed. However, in the triangular form, they are not finicky, and cutting errors of ⅛ to ¼ inch will not materially affect performance.

In fact, the relatively relaxed conditions for the triangle prompted the particular design that emerged. The prototype may be useful for field or hilltop service, since the support structure and the elements and their cable come apart and store in a flat package for transport. Figure 4 provides a few of the support structure details. The arms are ½ inch nominal Schedule 40 PVC, with T fittings at the hub. Each fitting has a cemented stub pointing upward. For field service, do not cement the upper part of the stub to the next T. Instead, let it twist. Once you have aligned the elements as specified below, drill holes through the T fitting and the stub and install 2.5 inch hitch-pin clips to hold the antenna in alignment during use. A convenient way to accomplish this without a protractor is to predrill only the T fitting of each movable arm. Then, assemble the hub and install the elements and align the arms so that each of the three spaces between dipole ends is as equal as possible. Finally, drill through the outer T fitting holes through the stub. Install the hitch-pin clips.

To fold the structure for transport, first remove the clips, and then twist the arms into parallel alignment. Remove the elements and cables as a unit. Figure 5 shows the two piece transport package for the antenna. Assembly is simply the reverse of disassembly. In field service, each assembly will require a new set of cable ties, while disassembly will require a small knife or cutter. (Please carry away cut cable ties from the operating site.)

For field use, the elements pass through tight ½ inch diameter holes in the outer ends of the arms (carefully measured to establish the feed-point radius). Each element slides into the armhole until the cable connections touch the edge of the hole. Under most conditions, the element will not need further pinioning to remain in place. However, a small cable tie on the other side of the arm will increase holding power. For a more permanent installation, the builder can cement the hubs in final alignment and install any desired kind of brace at the armhole to position the elements more permanently. Caution — each element should be marked to ensure that the connecting cable center conductor is in the same position on each arm. Either all three center conductor screws should be touching the armhole or all three should be away from the armhole. Reversing one of the dipoles relative to the other two will produce a highly directional pattern and a feed-point impedance that is well off the target value.

The construction style vertically spaces the element at 2.5 inch intervals. Modeling tests showed that the displacement increased the distortion of the circular azimuth pattern from about 0.1 dB up to about 0.3 dB, a variation that we have not been able to detect in operation.

Figure 6 shows the cabling details of the triangle, beginning with the dipole connection at the end of the support arm. For field use, small cable ties hold the connecting cables neatly on their way to the junction. Since the connecting cable impedance matches the dipole impedance, the cable length is not at all critical. As well, the cable

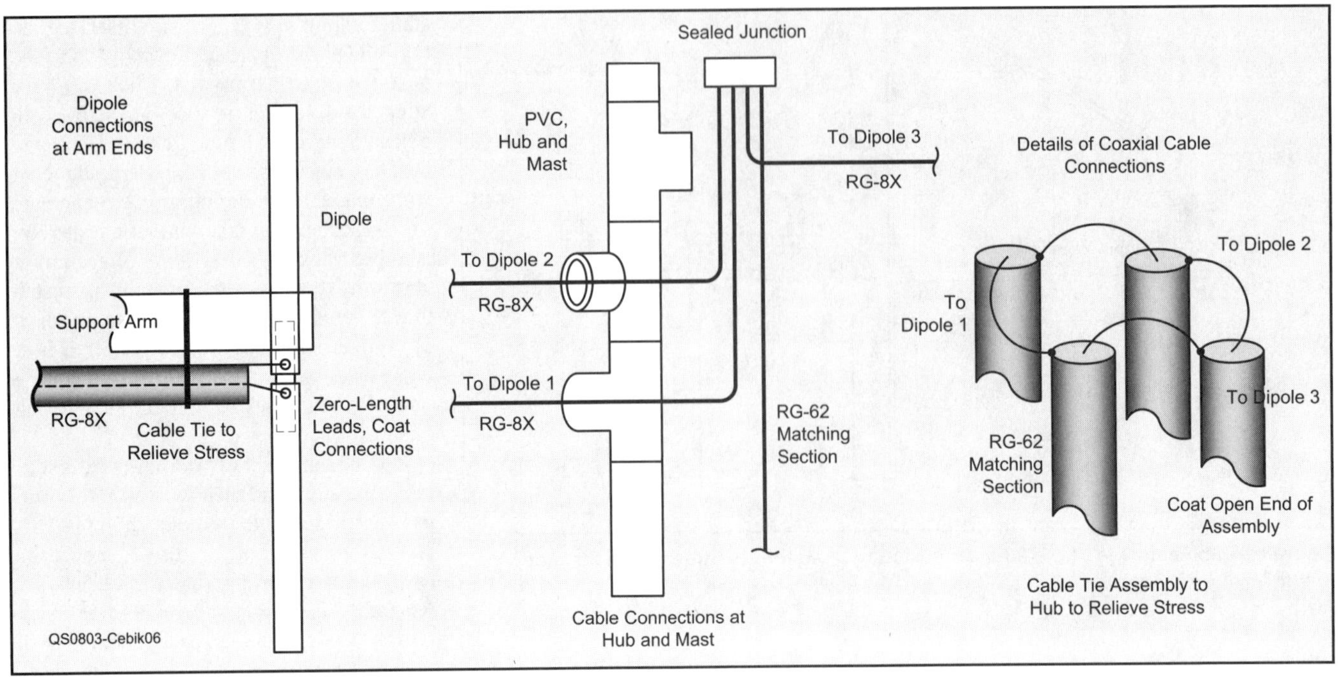

Figure 6 — Some details of the cable runs and connections used on the three dipole triangle; a similar system is used with the three dipole circle.

type also makes no significant difference to performance. Calculations show that cables ranging from RG-58 on the lossy end of the scale to lossless hypothetical cables change the array gain by about 0.1 dB for reasonable cable runs. The prototype used 20 inch connecting cable lengths to ensure that there are no stresses at the main junction of the cables. Figure 7 shows the entire assembled antenna on its test mast, including the cable ties that keep the coaxial cables neatly arranged and free of stresses. Figure 8 shows the simple connections at the dipole centers, uncoated for portable use and easy disassembly.

Since the antenna uses a series connection to arrive at a junction impedance above 50 Ω, as shown on the right in Figure 6, the antenna requires a λ/4 cable section to transform the impedance to a value that matches the main feed line. The design data showed a net junction impedance of about 150 Ω. Although a section of 75 Ω cable can be pressed into service, it will transform the impedance to a value in the 30 Ω range. A better choice is RG-62 (A or B) with an impedance of 93 Ω and a measured velocity factor of 0.84. Ideally, a 50 Ω impedance at the source end of the matching section calls for close to 170 Ω at the load end. At the design frequency, the impedance will be about 57 Ω, a very usable figure. With about 17 inches of RG-62 93 Ω cable having a velocity factor of about 0.82, the antenna provided a 50 Ω SWR of 1.3:1 or better from 144 to 145 MHz.

The SWR check is one of two tests for the assembled antenna. The other check requires a second person stationed about 10 λ (about 65 to 70 feet) or more away. One person provides a signal source and the other provides reception with a signal strength reading. It does not matter which person is at the antenna. For the test, simply hand rotate the antenna and test mast, checking for any changes in the signal strength. If the assembly has gone well, the S-meter should not budge.

Although the triangle assembly leans toward potential hilltop uses for the array, permanent installation requires only a few extra steps. Instead of using hitch-pin clips at the hub, cement the rotatable hubs. As well, cement or otherwise fasten the dipole elements into place at the armholes. For both types of use, seal the cable junction. Several thin coats of *Plasti-Dip* or liquid electrical tape work well and durably. Finally, use a brush to coat the arm ends of the connecting coaxial cables with the same material.

Figure 7 — An overall view of the HPOD triangle in use, showing the general construction and cable routing.

Figure 8 — A close-up of the junction of the support arm, the dipole element, and the connecting cable termination for portable operation. For a permanent installation, insert a stainless steel washer between the hook connector and the element and seal the junction.

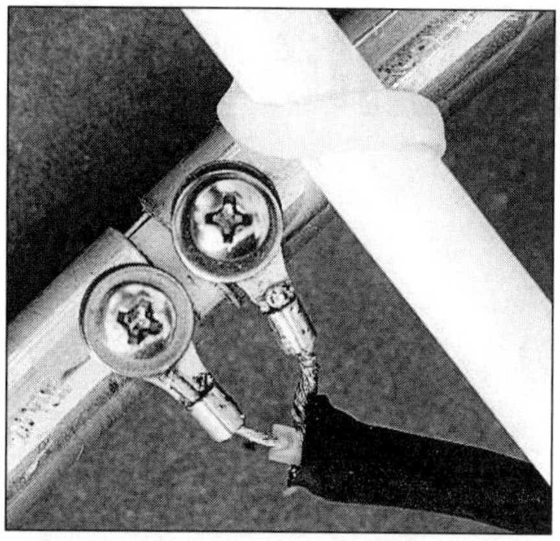

Figure 9 — A close-up of the method used to connect the element to the support arm with cable ties in the circular antenna.

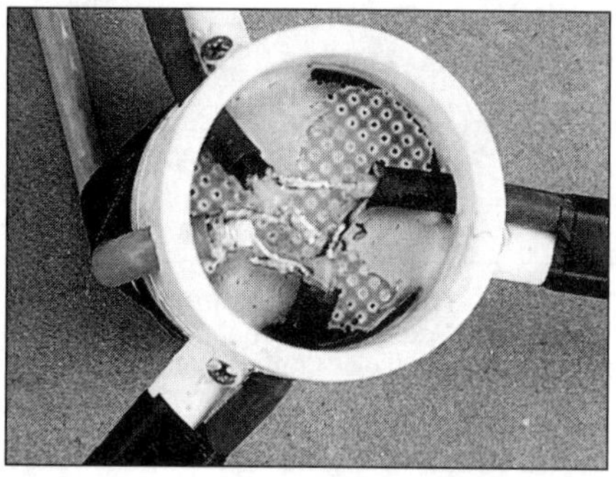

Figure 10 — The hub, support rods and the main junction of connecting cables and the λ/4 matching section before final sealing.

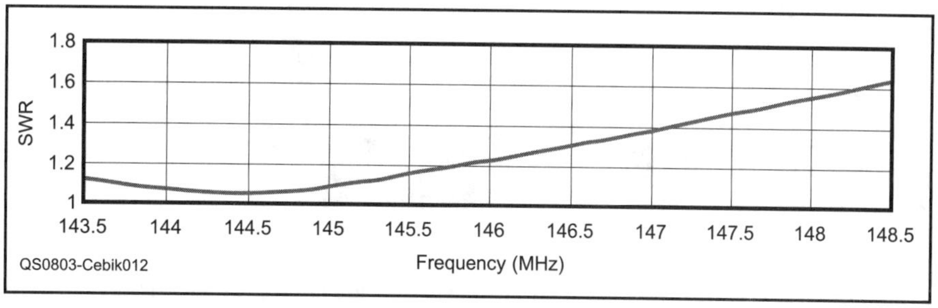

Figure 11 — Measured SWR of the circular three dipole horizontally polarized omnidirectional antenna.

Table 2

Modeled and Final Dimensions for a Three Dipole 2 Meter Circle (inches)

Design Frequency (MHz) Modeled	*Element Diameter*	*Radius to Feed Point*	*Circumference*	*Dipole Length*	*Tip-to-Tip Gap*
146	0.5	15.5	97.4	31.4	1.0625 (1 1/16)
146	0.375	16.0	100.5	32.4	1.125 (1 1/8)
144.5	0.5	15.7	98.4	31.75	1.0625 (1 1/16)
144.5	0.375	16.2	101.6	32.75	1.125 (1 1/8)
Final					
144.5	0.5	15.8	99.375	32.0	1.0625 (1 1/16)

Table 3

List of Materials and Local Sources for Circular Wheel

Material	*Suggested Source*
Two pieces 36 × 36 × 3/4 inch plywood	Old pallets or Home Center
144 inches 3/8 inch aluminum tubing	Texas Towers or **online metals.com**
12 inches 3/8 inch aluminum tubing	Texas Towers or **online metals.com**
6 feet 3/8 inch fiberglass rod	Home Center, farm and garden supply, Max Gain Systems
1 1/4 inch PVC coupler, pipe, end cap	Home Center, hardware stores
No. 6 stainless steel hardware	Home Center, hardware stores
1 3/4 inch 1/4-20 bolts	Home Center, hardware stores

A Three Dipole Circle

For a permanent installation or for mobile use, you may prefer a circle of three dipoles. The circle has no loose dipole ends and is more compact than the triangle. Indeed, it is aesthetically more pleasing. However, such pleasure comes at a cost. The construction and adjustment of the elements are somewhat more critical, although completely manageable.

The element connection arrangements for the circular version are shown in Figure 9 with the required materials in Table 3. The hub connections are shown in Figure 10. Figure 11 shows the resulting SWR — well below 2:1 across the entire band. Detailed fabrication instructions, including a fabrication jig are described in a Web version of this article found at **www.arrl.org/files/qst-binaries/.** The Web version also describes the methodology of stacking either version for additional omnidirectional gain towards the horizon.

Conclusion

Over the years, the use of three dipoles to form an omnidirectional horizontally polarized radiation pattern for 2 meters has experienced unwarranted neglect. The designs and prototypes described in these notes should restore these arrays to their rightful places as viable alternatives to many of the other possible HPOD designs. Except for using reasonable care both in cutting and adjusting the elements, the three dipole array — in either triangular or circular form — provides both a high performance antenna of its type and the potential for a satisfying antenna building experience.

Notes

[1]R. Mellen, W1IJD, and C. Milner, W1FVY, "The Big Wheel on Two," *QST*, Sep 1961, pp 42-45.

[2]*EZNEC* models of a big wheel simulation and of various versions of the three dipole triangle and circle are available at **www.arrl.org/files/qst-binaries/**.

Bob Cerreto, WA1FXT, is a Senior Engineer with a major dc power systems supplier. Back in Connecticut, Bob was first licensed in 1965 as WN1FXT. Since then, the ham radio world for Bob has been filled with many homebrew projects. He feels most comfortable in his radio lab and woodworking shop with fellow hams and grandchildren. Since he moved to Ohio in 1984, his backyard has become a three-acre antenna breadboard. No single antenna seems to stay in the air for too long before it is replaced by something new. Visitors to WA1FXT often find themselves part of a new homebrew project. Bob's wife, KC8BOM, serves as a catalyst for his ham radio ventures and is an accomplished builder too. Bob is looking forward to retirement when he has to make that difficult choice every morning: "Which project do I work on today?"

Bob can be reached at 49015 Middle Ridge Rd, Amherst, OH 44001 or **tonybob@adelphia.net**.

Since his retirement as a professor at the University of Tennessee, L. B. Cebik, W4RNL, has written extensively for radio amateurs about antennas and related topics in many venues, including his own Web site, **www.cebik.com**. *Licensed since 1954, LB has served ARRL in many capacities, including as Educational Advisor and as Technical Advisor. He is a contributing editor to* QEX *and is technical editor for* antenneX. *LB can be reached at 1434 High Mesa Dr, Knoxville, TN 37938-4443 or* **cebik@cebik.com**. QST

Double Cross — A NOAA Satellite Downlink Antenna

An easy to build antenna for ground reception of NOAA weather or amateur satellite signals.

Gerald Martes, KD6JDJ

US National Oceanographic and Atmospheric Administration (NOAA) polar orbiting weather satellites (POES) transmit data for production of gray scale images of the ground below them.[1] These automatic picture transmissions (APT) signals are sent at 137 MHz and are available twice a day to anyone on the earth. There are many free programs available that can decode the satellite signals and then produce color images on a personal computer. I prefer *APTDecoder* by Patrik Tast, available on the Internet.[2]

Antennas for POES Satellite Reception

The NOAA weather satellites are polar orbiting, so they can and do appear at all azimuth and elevation directions from any ground location. An ideal ground-based antenna for reception of NOAA satellite signals would be right-hand circularly polarized (RHCP) and have no deep pattern minimums within the hemisphere. Figure 1 shows the ideal pattern shape of a ground-based antenna for APT reception. It would have a 12 dB minimum toward zenith since the satellites will be approximately 12 dB stronger when overhead as compared to their strength when at the horizon due to path loss.

This ideal pattern cannot quite be obtained for a ground-based receiving antenna for VHF. But it is a useful guide to remind the antenna designer that the ground-based antenna pattern should have a minimum of nulls within the hemisphere, with a maximum toward the horizon.

Four dipoles can be mounted as shown in Figure 2 to produce a radiation pattern with excellent RHCP at 0° elevation in the free space radiation pattern as shown in Figure 3. The design concept for this four dipole array with hemispheric coverage with RHCP is derived from the fundamental concept that two dipoles, crossed, spaced a quarter wave apart, and fed in phase as shown in Figure 4 can be polarized to produce a pattern as shown in Figure 5.

The pattern null along the X axis can be filled in by including a second pair of crossed dipoles as shown in Figure 6. If the second pair is fed 90° later than the first pair, the four dipole array has the excellent radiation pattern shown in Figure 7.

Double Cross as an Amateur Satellite Antenna

The Double Cross can also be built as

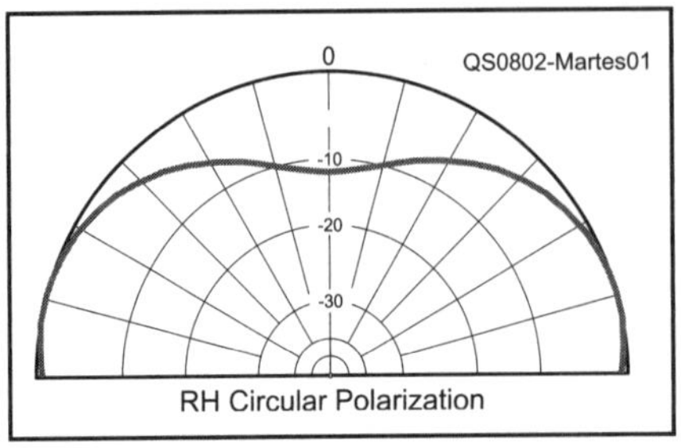

Figure 1 — Ideal radiation pattern for a ground-based antenna for reception of NOAA weather satellite images.

KD6JDJ

Figure 2 — The four dipoles that make up the Double Cross antenna prototype.

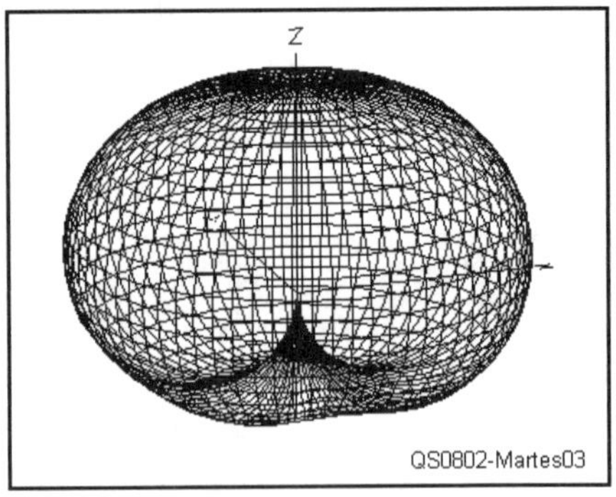

Figure 3 — The free space three dimensional pattern of the Double Cross closely matches our design goal.

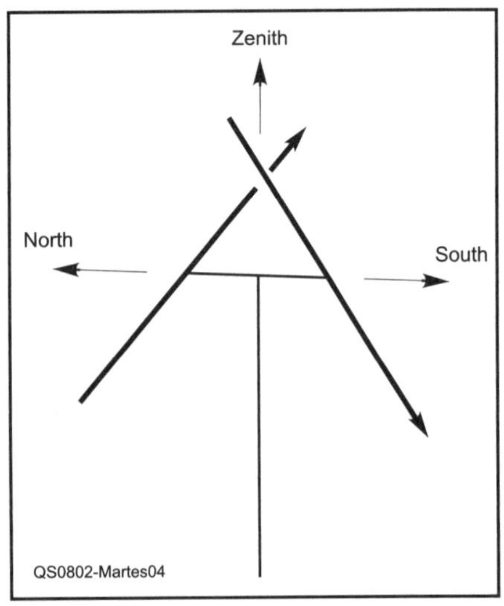

Figure 4 — One pair of crossed dipoles makes half a Double Cross.

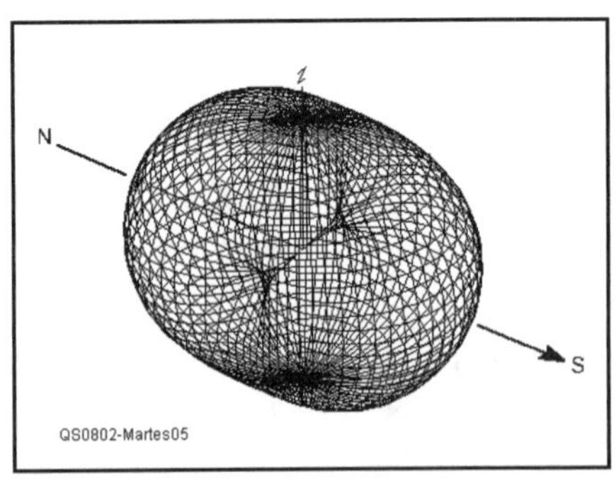

Figure 5 — Free space radiation pattern of crossed dipoles.

[1]www.oso.noaa.gov/poesstatus/index.asp.
[2]www.poes-weather.com/.

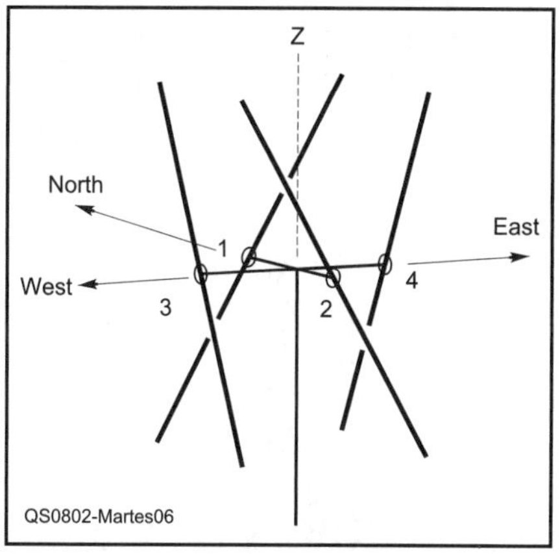

Figure 6 — Two pairs of crossed dipoles become the Double Cross.

an antenna for amateur frequencies. Table 1 shows the dimensions for both NOAA and amateur satellite bands. Note that circular polarization toward the horizon is also useful for terrestrial communication, since it responds equally well to both vertical (for FM) and horizontal (for SSB and CW) polarization.

Making it Happen

One beneficial aspect of this Double Cross antenna is that it is quite tolerant of construction variations. That is, an antenna in this configuration will almost always work well even when the dimensions are only close to the optimum design. The only thing critical is the proper connection of the harness to the dipoles.

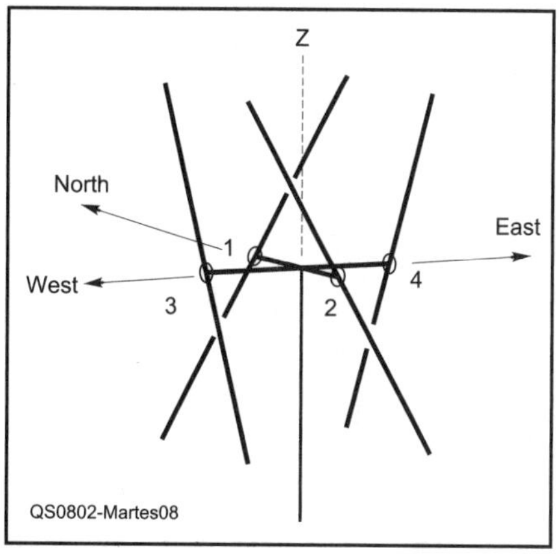

Figure 8 — Dielectric support for the dipoles and harness.

Field testing with this antenna indicates that it produces very little radiation pattern nulling within the hemisphere. Pattern nulls would be expected to produce dark horizontal lines across the image. The images, recorded with the Double Cross thus far constructed and tested, have indicated that very little image quality degradation is seen as a result from elevation plane nulls from ground reflection.

A Double Cross APT antenna can be built by constructing a dielectric support as shown in Figure 8 and attaching the dipoles and harness, as described later, to the dielectric support.

Table 1

Dimensions of Double Cross Array (Inches)

Operating Frequency (MHz)	*137*	*145*	*435*
Dipole length	38.25	37.125	12.125
Dipole diameter	0.375	0.375	0.25
Dipole spacing	21.5	20.5	6.75
Poly coax λ/4 phasing section	14.25	13.5	4.5

Build four dipoles from a convenient conductor, each about 38 inches long, and attach them to the supports. The dipole supports numbers 1 and 2 are separated by about 20 inches. The dipoles number 3 and 4 are also separated by 20 inches. Each of the four dipole supports is tilted 30° from vertical.

Dipoles 1 and 2 are fed in phase and with the proper polarity, so the upward pointing end of dipole #1 has the same polarity as the downward pointing end of dipole 2. The upward pointing end of dipole 3 has the same polarity as the downward pointing end of dipole 4.

The input impedance of each of the λ/2 dipoles when configured as shown will be very close to 50 Ω. Each pair is wired in series to have an impedance of 100 Ω. After connecting the two pairs in parallel, as shown in the harness diagram of Figure 9, we end up with the desired 50 Ω for the run to the radio.

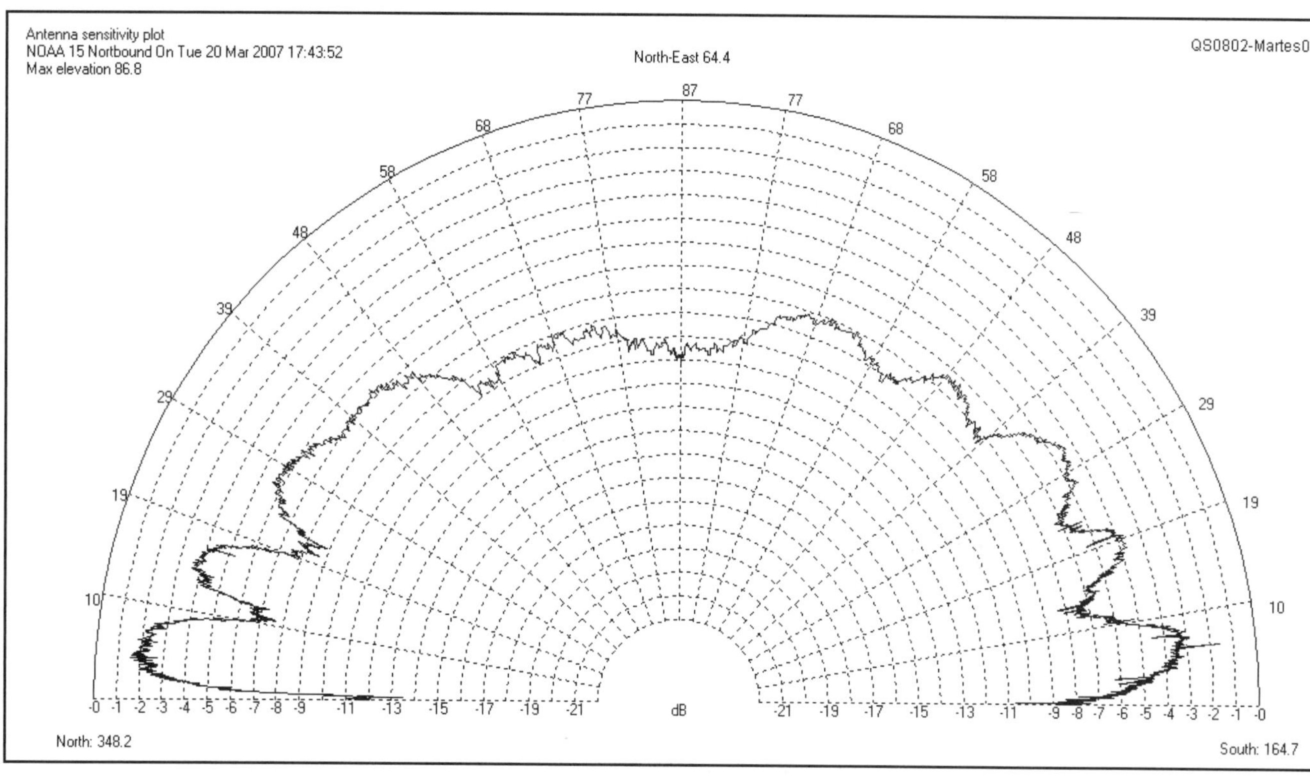

Figure 7 — The measured pattern of the Double Cross antenna closely matches the ideal of Figure 1.

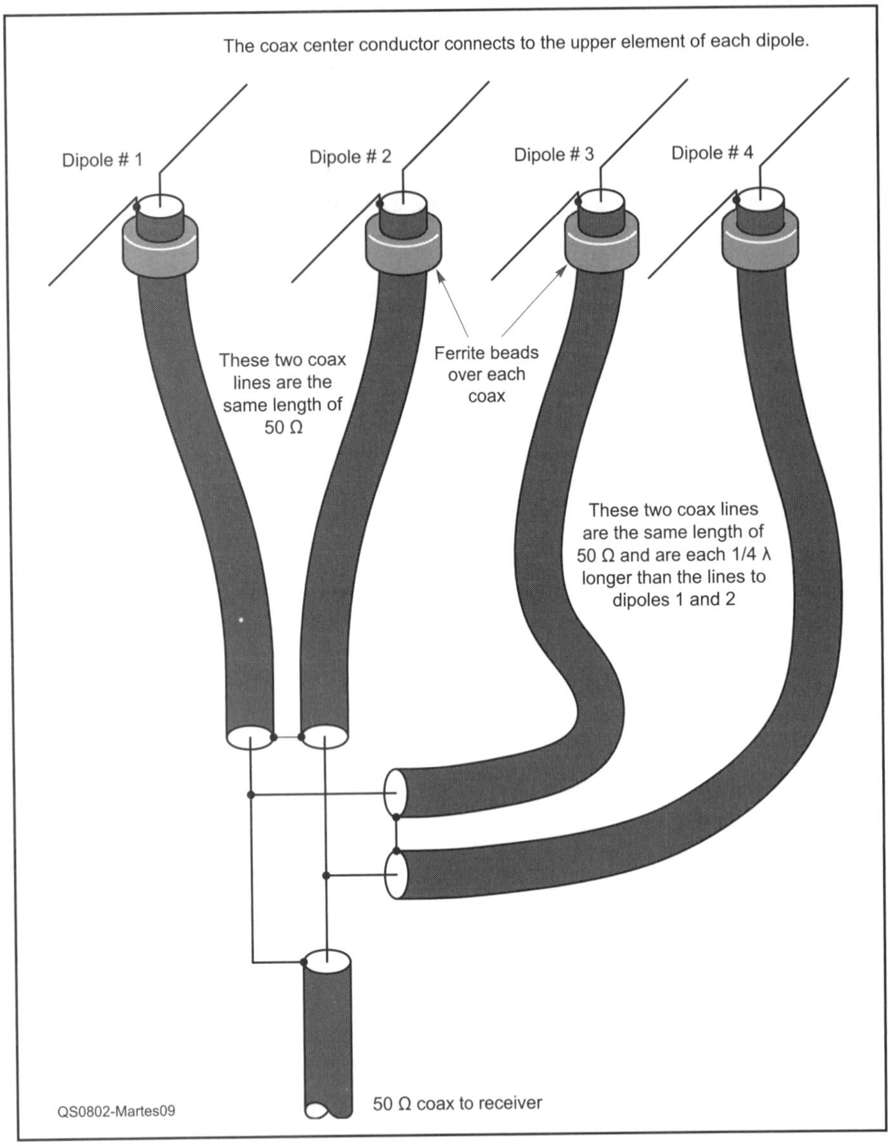

Figure 9 — Harness for Double Cross using 50 Ω coax.

Figure 11 — NOAA satellite image from Double Cross at 2 λ height.

KD6JDJ

Figure 10 — Photograph of the Double Cross antenna mounted 2 λ above ground, at the beach in Southern California.

It Works!

The antenna works very well, either from my home station or on the road, as shown in Figure 10. A sample of the received signal is shown in Figure 11.

Gerald Martes, KD6JDJ, has been licensed, on and off since 1949. He received his first license and call KL7LL while in the USAF in Alaska. He took the test again in the 1950s and had the call K6LZC. Several years ago, he took the General class exam and barely passed it but is now KD6JDJ and promises not to let this one lapse. He also held commercial First Class Radiotelephone and Radiotelegraph licenses, including CW at 20 WPM.

Gerald earned a BSEE from Pacific States University and was employed as an antenna design engineer from 1955 through 1970. He then started his own electrical business and never went back to engineering.

Gerald can be reached at 5061 Tripoli Ave, Los Alamitos, CA 90720 or at **j.jmartes@verizon.net**.

Diamonds in the Sky

Try this high performance omnidirectional antenna for 2 meters.

Skip Teller, KH6TY

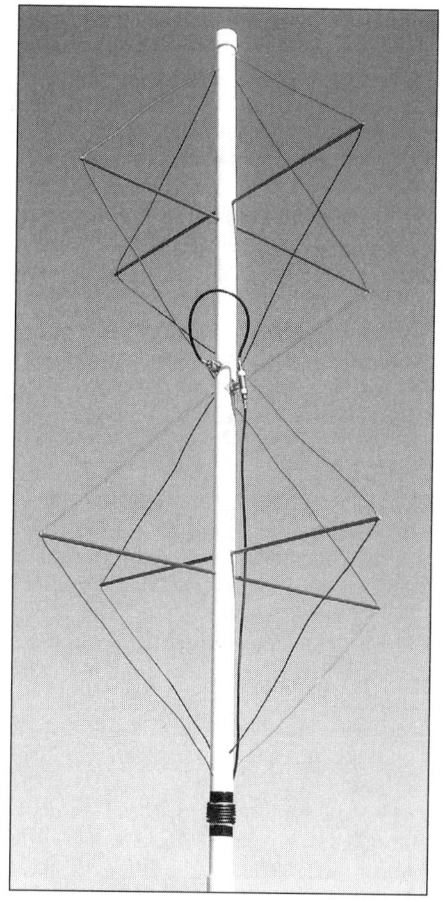

Down here in the "low country" (meaning flat as a pancake!) of South Carolina we've been encouraging digital ragchewing on 2 meter FM using the DominoEx 8 mode. We do it by running an informal net that covers almost the entire state on simplex using horizontally polarized antennas. (See WB8IMY's "Eclectic Technology" column in the March 2009 *QST*.)

As the Net Control Station (NCS) I needed a relatively high-gain, omnidirectional antenna that could cover a wide area. In our DominoEX net, all stations point their beams at the NCS and the NCS relays every transmission so that no one is left out of the conversation. The farther away a station is, the more antenna gain he will need to hear (and be heard) by the NCS. This means that while the net participants can use directional antennas, the NCS must always use an omnidirectional antenna in order to hear from any direction.

Thanks to the fellows at Google, several solutions finally turned up on the Internet and among those were the *Big Wheel*, various *turnstiles* and *stretched rectangle* antennas, and an old design called the *skeleton-slot*.

After extensive testing I ultimately settled on a stretched-rectangle derivative that was inexpensive, simple to build, had moderately high gain with horizontal polarization and a wide beamwidth. I call this design the *DDT*, or *Double Diamond Turnstile*.

The DDT antenna has a noble pedigree. It is based on designs variously described and published by Peter Dodd, G3LDO; B. Sykes, G2HCG; David Jefferies, G6GPR; Dan Handelsman, N2DT; Paul Carr, N4PC, and Brian Beezley, K6STI. They found that by stretching a square quad loop in height they could increase the gain by as much as 2 dB, while at the same time lowering the input impedance from 125 Ω for a square down to 50 Ω for an individual loop — a perfect transceiver match.

The square quad loop, fed at the bottom, may be thought of as a pair of ¼ wavelength horizontal dipoles connected by vertical ¼ wavelength phasing lines. G3LDO found that stretched quad loops could be stacked and fed in parallel without using separate phasing lines and L.B. Cebik noted in his *QST* article on the 6 meter turnstile antenna that a diamond shaped loop, fed at the bottom, is equivalent to the square quad loop (also fed at the bottom) and results in horizontal polarization. In contrast, feeding a quad loop on the side results in vertical polarization.

It was a simple matter to experiment with stacked diamond-shaped loops for 2 meters and the result is a horizontally polarized, bi-directional, antenna with good gain over a wide beamwidth (the azimuth pattern is the same as a dipole, since the diamonds are equivalent to stacked dipoles). Using *EZNEC* antenna modeling software, I sized the two diamonds to produce as much gain as possible with a 75 Ω feed impedance at their junctions. This would let me "turnstile" two 75 Ω double-diamonds, faced 90° from each other, and achieve 360° coverage. Because the antennas are at 90° to each other they can (and should) share the same mast. The two double-diamond antennas are effectively in parallel, connected by a phasing line to combine their individual bidirectional patterns into a single omnidirectional pattern as shown in Figure 1.

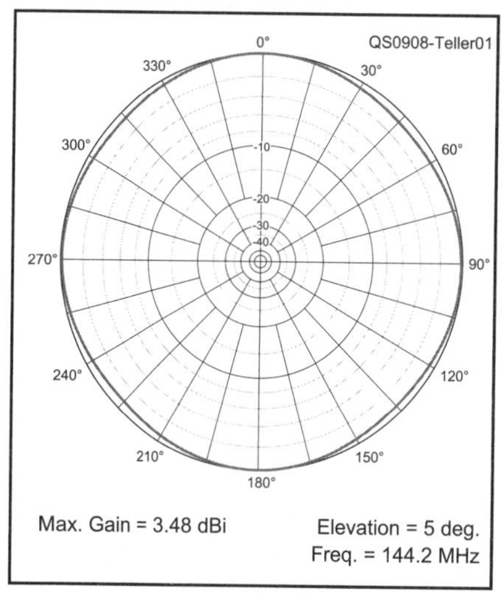

Figure 1 — The azimuth radiation pattern of the DDT antenna.

The DDT can be constructed from materials commonly available at any hardware or home improvement store. The materials needed are a 6 foot length of 1¼ inch Schedule 40 PVC pipe, 32 feet of #14 insulated, stranded, copper house wiring, four fiberglass 5/16 inch diameter rods (sold as driveway markers) for spreaders, one or two 1¼ inch PVC caps, four 6-32 × 2 inch machine screws with star washers and nuts, and two SO-239 connectors from your local RadioShack store or junk box. Fiberglass tubing of ¾ inches in diameter or greater can also be used for the mast and would withstand wind forces better when mounted to a mast or tower instead of being hung in a tree.

Building the DDT

Begin construction by drawing a reference line along the full length of the pipe, making sure the line is as much along the center of the pipe as possible. This is *very* important in order to keep the spreaders at right angles to each other. To do this, I placed the pipe, which is used as the mast for the antenna, on a flat tabletop and clamped it down at both ends so it would not move. I then cut a small block of wood, about 1 inch high, held a pencil tightly on top of the wood and, holding the center of the pipe down so it did not bow, scribed a line the length of the pipe to use as a reference for drilling the necessary holes (see Figures 2 and 3).

In order to help insure that the spreaders pass through the mast at right angles, I found it most successful *not* to drill a spreader hole all the way through the mast to the other side. Instead, I made a paper template to mark the entrance and exit holes for all the spreaders, wire, and connector mounting holes exactly 180° apart and perpendicular to the pipe. You can do the same by cutting a strip of paper 1½ inches in width. Wrap it around the mast and trim the strip until the ends *just* meet. Now fold the paper in half, and in half again, to form creases in the paper at 90°, 180° and 270° with reference to the scribed line at 0°. Use this template to mark the holes in the mast for the wires, spreaders, connectors and mounting screws.

- Starting at one end of the mast, draw short lines across the scribed line at 2, 3½, 16¼, 17¾, 30½, 32, 44¾, 46¼, 59 and 60½ inches.
- Place the template at the 2 inch mark on the line, making sure it lines up perfectly at the edges and the edge bisecting the 2 inch mark. Mark holes at 90° and 270°and label these A.
- Move the template to the 3½ inch mark, mark for holes at 0° and 180° and label these B.
- Move the template to the 16¼ inch mark, mark for holes at 90° and 270° and label these C.
- Move the template to the 17¾ inch mark, mark holes at 0° and 180° and label these D.
- Move the template to the 30½ inch mark, mark holes at 0° and 180° and label these E.
- Move the template to the 32 inch mark, mark holes at 90° and 270° and label these F.
- Move the template to the 44¾ inch mark, mark holes at 90° and 270° and label these G.
- Move the template to the 46¼ inch mark, mark holes at 0° and 180° and label these H.
- Move the template to the 59 inch mark, mark holes at 90° and 270° and label these I.
- Move the template to the 60½ inch mark, mark holes at 0° and 180° and label these J.

Before drilling the holes, re-check the locations carefully. When you're satisfied that every mark is correct, drill all the marked holes with a 5⁄32 inch bit. These will serve as wire holes and also as pilot holes for later enlargement.

Expand the holes at C, D, G and H to 5⁄16 inch diameter. If necessary, move the drill around a little to slightly enlarge the hole for the 5⁄16 inch diameter fiberglass spreaders.

Look at holes E and F. Choose two of these holes to mount your SO-239 connectors. Make sure to select holes that will position the connectors at right angles to each other on the mast. Widen your chosen holes to ⅝ inches in diameter, insert the connectors and use them as templates to drill four mounting holes (two for each connector) parallel to the scribed line. When drilling the holes the bit must go through the mast and out the other side. Each hole must be large enough to accommodate 6-32 machine screws.

For each of the four spreader rods, cut to a length of 32 inches and drill *parallel* 5⁄32 inch wire holes ¼ inch from the end of each spreader.

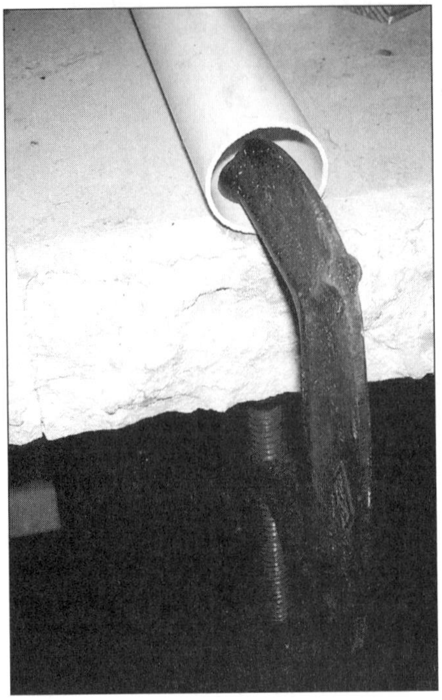

Figure 2 — Clamping the mast to a table keeps it stable as you are marking hole locations.

Figure 3 — Scribing a straight reference line along the mast is critical.

Wiring the Antenna Frame

The two antennas will be wired separately. Follow the diagram in Figure 4.

Begin wiring the first antenna by cutting off a 16 foot length of wire, stripping 2 inches of insulation off the end, tinning the end ¼ inch. Feed the tinned end through hole F opposite the hole for the SO-239 connector. Solder to the center pin of the SO-239 connector. Place the connector in the ⅝ inch hole and mount it to the PVC with two 6-32 machine screws, securing them in place on the other side of the mast with star washers and nuts.

Push a spreader through the C holes and center the spreader on the mast. Do the same for the lower spreader using the G holes. You can use epoxy glue or other strong adhesive to keep the spreaders in place, although the wire tension itself may be adequate to hold them in position.

Looking at the wire coming out the back side of the mast (from the center pin of the SO-239 connector), lead it *up and to the left,* feeding the free end of the wire through the hole in the spreader, through the A holes and out the other side of the mast. Continue feeding the wire through the right side of the top spreader and back down to top set of nuts and washers opposite the SO-239 connector. Strip off about ½ inch of insulation, loosen the top mounting screw, (wrap the wire *around* the screw between the mast, star washer and nut and tighten the nut).

Continue feeding the same wire *down and to the right*, going through the lower spreader's hole. Continue downward and through the I holes, out the other side of the mast, *up* through the left-hand spreader hole and over to the connector. Pull the wire taut, cut it 2 inches longer than needed to reach the connector and strip 3 inches. Tin the end ¼ inch to keep the wire from unraveling. Then, wrap the wire a few times around the wire coming out of hole F behind the SO-239 connector center pin and solder. This completes the wiring of the "upper" antenna.

Rotate the mast 90° about its axis and repeat the procedure, using the remaining holes for the "lower" antenna group. Be sure to lead the wire to the *left and up* to the first spreader. *Both wires attached to the connector center pins must lead to the left and up in order for the phasing to be correct.* As you thread the wire it will not cross at the connector, but instead continue down the same side for the lower diamond.

The Phasing Line

The two antennas are connected by a length of RG-59 coaxial cable that will combine the patterns of the antennas and produce an omnidirectional, horizontally polarized pattern. It is critical to use RG-59 and not some other type of coax as the impedance of the phasing line must be as close to 75 Ω as possible. I assembled a phasing line from RG-59 coax with PL-259 connectors on each end, measuring 16½ inches from the tip of one PL-259 to the tip of the other PL-259.

Viewing the antenna with the connectors toward the floor, mount a PL-259 T adapter (RadioShack 278-198) to the lower SO-239 connector. Screw one end of the phasing line into the upper SO-239 connector and the other end into one end of the coax T adapter. The coax feed line is then screwed into the other end of the T adapter as shown in Figure 5.

It is a good idea to create a choke balun to keep currents on the outside of the coax feed line from disturbing the omnidirectional pattern of the antenna. This can be done by winding five turns of the RG-58 feed line

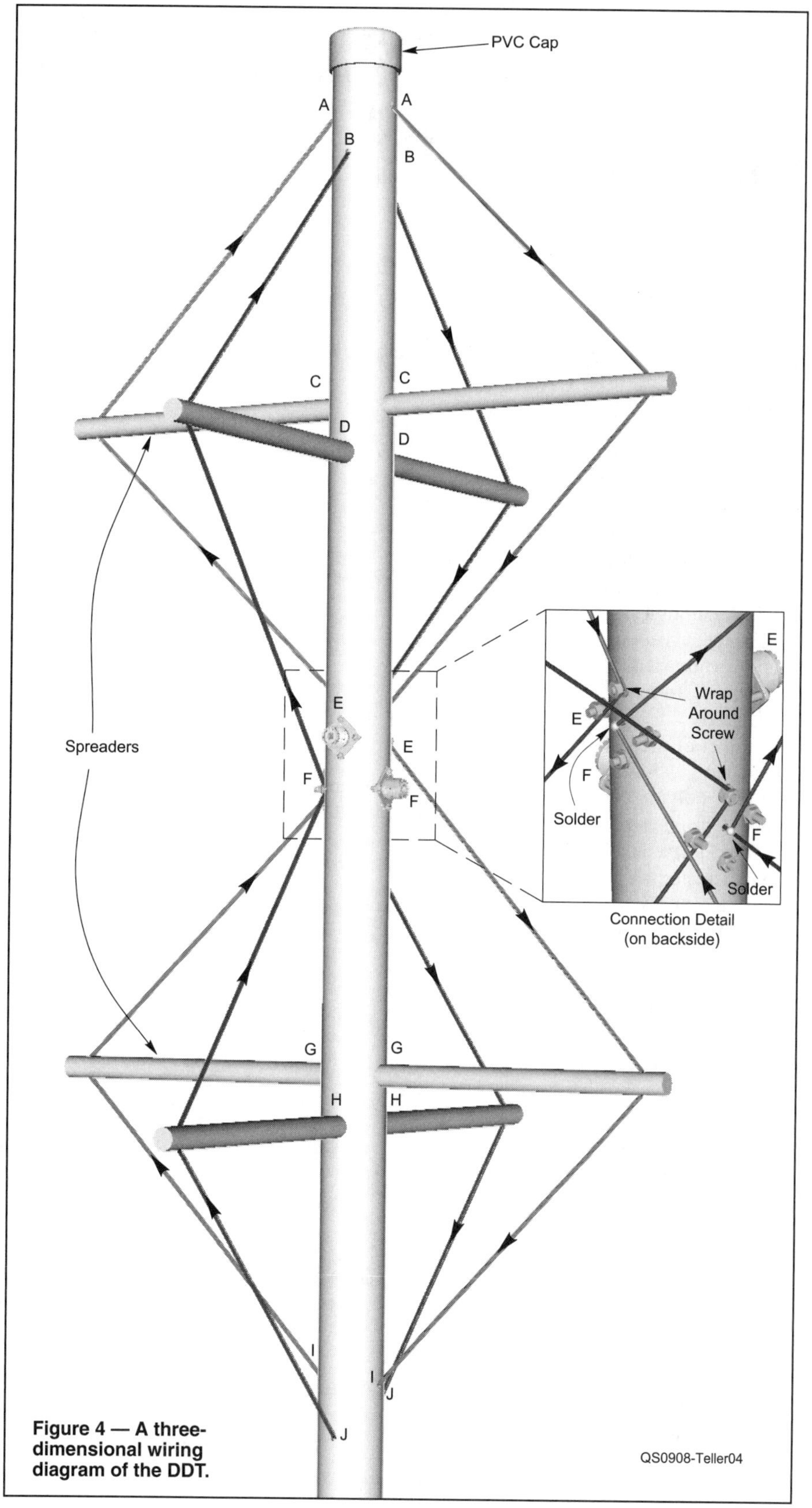

Figure 4 — A three-dimensional wiring diagram of the DDT.

Figure 5 — A close-up view of the phasing line and wire terminations with the T adapter visible.

coax around the mast. Tape the choke in place with electrical tape and weatherproof all connectors and connections.

When completed, the antenna should have an SWR of about 1.3:1 from 144 to 145 MHz since it is designed to be used in the "weak signal" portion of the 2 meter band where horizontal polarization is the convention.

The antenna can be hung from a tree or used as a "topper" for a mast or tower. No rotator is needed, of course, as the antenna is omnidirectional. If you choose to hang the DDT from a tree, drill a hole in the top cap before gluing it on and mount an "eye" bolt to hold the rope. Glue another cap to the bottom and tape the feed line in place on the lower part of the mast for a strain relief. If it is used as a "topper" for a mast or tower, use the thicker 1¼ inch PVC pipe, insert a wood dowel into the PVC as far as it can go to strengthen it against bending and breaking, glue on the top cap and spray with paint for UV protection.

References

B. Beezley, K6STI, "A Gain Antenna for 28 MHz," *QST*, Jul 1994, p 70.

P. Carr, N4PC, "The H-Double Bay Antenna," *CQ*, Sept 1995, p 28. This is an SDR for 30 meters that is horizontally polarized and fed at the lower end-wire.

L.B. Cebik, W4RNL, "A 6-meter Quad Turnstile," *QST*, May 2002, p 42.

D. Handelsman, N2DT, "The Double Rectangle," Three variations on a rectangular theme, *Communications Quarterly*, Spring 1999, p 67.

Photos by the author.

Howard ("Skip") Teller, KH6TY, is an ARRL member and was first licensed in 1954. He received his commercial First Class Radiotelephone license in 1959 and worked his way through college as chief engineer of several radio stations. He holds a BS degree in electrical engineering from the University of South Carolina and is retired from running a factory in Taiwan, where he manufactured the weather alert radio that he originated in 1974 and is still sold by RadioShack and many other companies. Skip enjoys developing digital software, such as DigiPan *and* NBEMS, *designing 2 meter transceivers and antennas. He is currently studying the potential of working 2 meter DX on FM using digital modes. You can contact Skip at 335 Plantation View Ln, Mt Pleasant, SC 29464;* **kh6ty@comcast.net**.

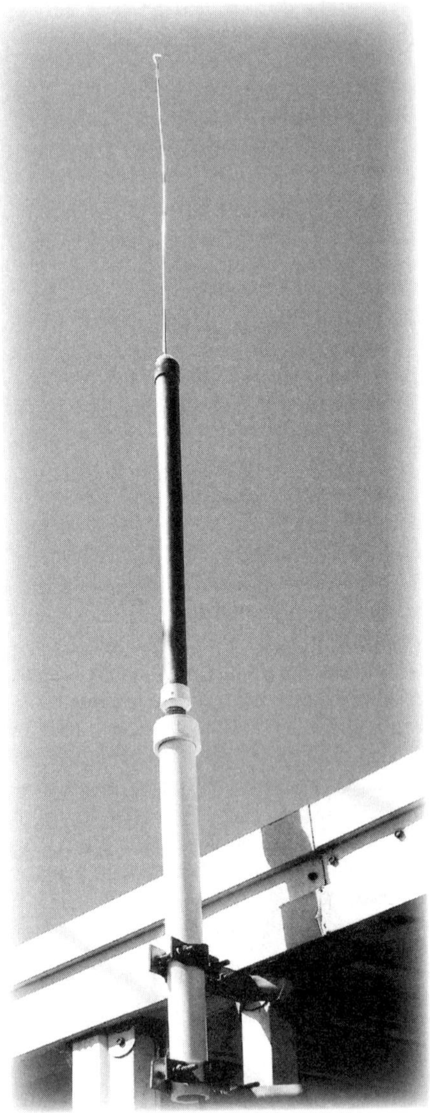

Homebrew Coaxial Dipole for VHF or UHF

Here's a base station antenna you can easily build for 146, 220 or 440 MHz. Performance is equal to a J-pole, but it's smaller, less obtrusive and more weatherproof.

John E. Portune, W6NBC

Hams like to build antennas, especially if they're made from ordinary hardware store items and can be assembled with common hand tools. Here is a homebrew coaxial dipole built from a small stainless whip, a length of threaded table-lamp tubing and some ¾ inch copper and PVC fittings. The one shown is for 440 MHz but it can readily be scaled for 146 or 220 MHz.

For homebrew vertical VHF antennas, coaxial dipoles often play second fiddle to J-poles. That's because the center connection to coax is often difficult to fabricate in the home workshop. Yet both antennas have the same performance. They're both full sized, half wave vertical dipoles, and the coaxial is shorter.

Making a Coaxial Dipole

If you start with a common half wave ($\lambda/2$) stainless whip and extend it all the way down through a $\lambda/2$ long support tubing, here made from a threaded table lamp tube, the lower part of the whip becomes the center conductor of a short length of rigid coax feeding the center of the antenna. Now connection to normal coax is easily made below the antenna. To form the rigid coax section, you'll need to insulate the center conductor (lower part of the stainless whip) from the lamp tubing with some ¼ inch inside diameter (ID) polyethylene tubing. Hardware stores normally carry it. This short length of rigid coax formed in this way isn't precisely 50 Ω characteristic impedance, but the difference is totally insignificant. The drawing in Figure 1 shows the details.

Assembly Details

The bottom half ($\lambda/4$) of the radiating dipole is a coaxial sleeve made from ¾ inch copper pipe and a pipe cap. The coax feed runs up its center to the connector at the bottom of the lamp tubing. Support and insulation of the bottom of the sleeve is provided by a ¾ inch CPVC plastic pipe cap. For those not familiar with CPVC fittings, they're made to mate with copper pipe and can handle high water temperatures. That's not true of common PVC fittings. Most hardware stores now carry CPVC. Drill a ⅜ inch hole in the center top of the copper and the CPVC caps for the lamp tubing to pass through.

The whole antenna is held together by two lamp tubing nuts and a plastic lamp finial, also readily available at hardware stores (see Figure 2). Note that a lamp tubing nut is also required inside the copper pipe cap. Drill a small hole in the middle of the lamp finial for the stainless whip. On the bottom of lamp tubing below the antenna install a 1¼ inch common PVC pipe cap, and secure it with two more lamp tubing nuts. This gives you a way to easily mount the antenna on top of any convenient length of 1¼ inch

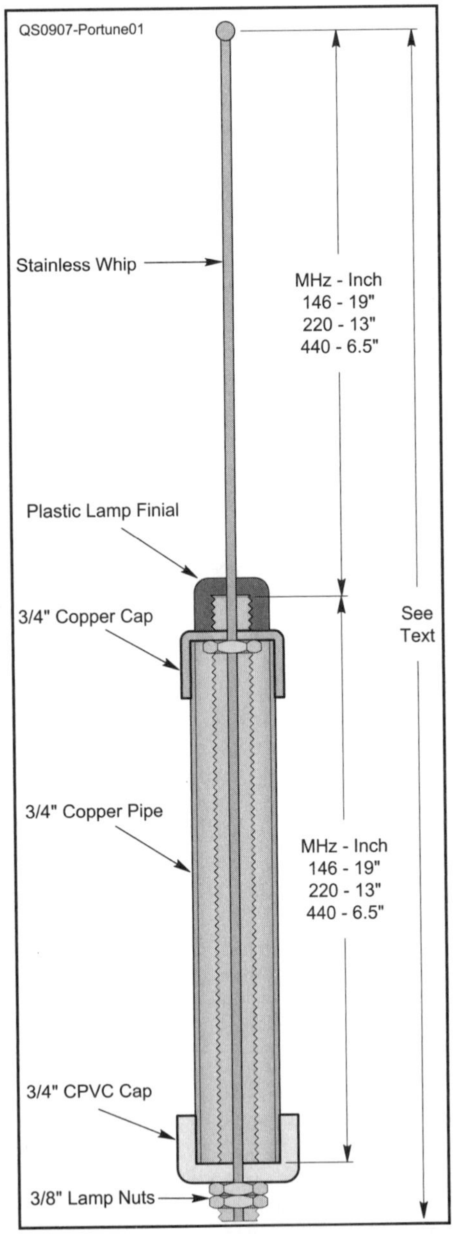

Figure 1 — Dimensioned drawing of coaxial dipole for three bands.

PVC pipe. Run the coax feed down through the PVC pipe.

Hooking it Up

A conventional PL-259 UHF type coax connector for RG-8 coax will actually screw onto the bottom of the lamp tubing. The threads are not a perfect fit, but will tighten satisfactorily. The stainless whip runs down all the way to the very tip of the PL-259 connector. Solder it in there. Before doing so, however, install all the pieces of the antenna onto the threaded lamp tubing.

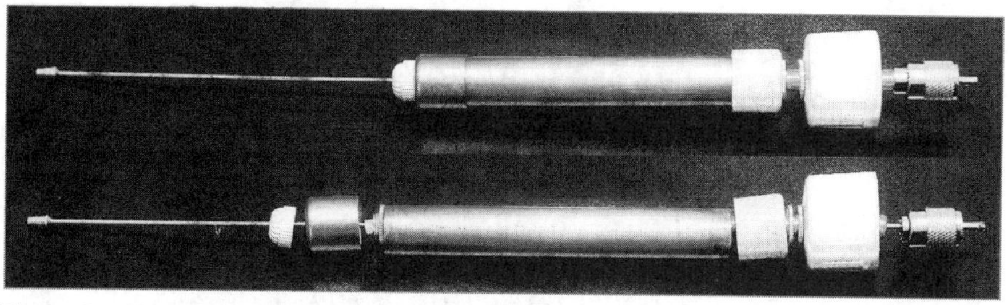

Figure 2 — Details of final assembly of coaxial dipole.

Many hams may think that stainless steel won't solder. It definitely will with a hot iron and acid flux. Scrape the end of the whip and dip it in hydrochloric swimming pool acid. With a little action from the tip of the soldering iron the whip will tin perfectly well. Before soldering, however, grind two or three small side notches in the bottom end of the whip. A Dremel tool works well for this. The notches will help the solder securely lock the whip into the tip of the PL-259 connector. Neutralize any leftover acid with baking soda solution.

Perhaps surprising to some, it really isn't necessary to solder any other parts of the antenna. There is adequate mating surface at the joints for the RF to cross over efficiently. Do, however, seal all possible water access spots with common silicone RTV glue and or plastic electrical tape.

Make it for the Band You Like

There isn't an exact length required for the lamp tubing or the stainless whip. These merely need to provide enough space for all the pieces of the antenna to go together. I had a 48 inch whip on hand that I used uncut for my 146 MHz coaxial dipole and a similar 17 inch uncut whip for 440 MHz. I merely cut the lamp tubing to an appropriate length to fit the whips. What does matter, however, is the length of the whip above the top of the lamp tubing as well as the length of the coaxial sleeve. These need to be close to a $\lambda/4$. For 440 I used 6½ inches, on 220, 13 inches, and 19 inches for 146 MHz. These antennas are quite broad band and will cover the entire band in each case with these sizes. No cutting or pruning is necessary.

For ruggedness, or perhaps for stealth, you can install the whole antenna inside of 2 inch PVC water or ABS soil pipe and close the ends with end caps. I live in a mobile home park where antennas are not permitted, but my landlord thinks my coaxial dipoles (in ABS pipe) are vent pipes.

Try out one of these homebrew coaxial dipoles. You may find you prefer its smaller size, less obtrusive appearance and superior weatherproofing as compared to a J-pole.

ARRL Member John Portune, W6NBC, received a BSc in physics from Oregon State University in 1960, his General Radiotelephone license in 1961 and his Advanced class amateur license in 1965. He spent five years in England as G5AJH and upgraded to Amateur Extra class in 1985 to become a volunteer examiner (VE). John retired as a broadcast television engineer and technical instructor at KNBC in Burbank and then from Sony Electronics in San Jose, California.

John is active on many bands and modes, predominantly from his HF RV mobile station. He has written various articles in ham radio and popular electronics magazines and remains active as a VE team leader, ham license teacher and Web site designer. You can reach John at 1095 W McCoy Ln #99, Santa Maria, CA 93455, or at **jportune@aol.com**.

The Quadrifilar Helix as a 2 Meter Base Station Antenna

Here's an easy to build VHF base-station antenna that works equally well toward the horizon or overhead.

John E. Portune, W6NBC

Designed for spacecraft use in the early days of space exploration, the quadrifilar helix (QFH) antenna has not gained much popularity on the ham bands. Yet, as a general-purpose base-station antenna, here for 2 meters, it's hard to beat. It's almost an omnidirectional antenna in both planes, like the mythical *isotropic* radiator. No matter what direction signals come from, or whether the polarization is vertical or horizontal, the QFH works them. It's good for overhead satellites, such as the International Space Station, for horizontally polarized 2 meter SSB simplex stations on the horizon and also for vertically polarized mobile and repeater stations. No, it isn't a gain antenna — no true omni can be. The joy of a QFH is it always gets out.

The QFH is often used by hams for receiving weather satellite pictures from the 137 MHz NOAA automatic picture transmitting (APT) satellites in low polar orbit. Its omnidirectional and circular polarization characteristics accommodate the constantly changing direction and polarization of the APT satellite signals. I have built several for this service. Three of these weather birds still fly by every day — NOAA 15, 17, 18 and 19. (Pictures available at **w6nbc.com**.)

My primary objective here, though, is to tame the QFH as a 2 meter base station antenna, using local hardware store materials and easy construction methods. I also will dispel some of the antenna's technical mysteries. Many existing published QFH articles feature difficult to build designs or too much unnecessary technical jargon.

Picturing the QFH

Take two vertical full wavelength rectangular loops with open feed points *at the top*. Now place them on the same vertical axis, but with one loop rotated 90° horizontally (quadrature). Also, you need to make one loop slightly larger than the other. This creates a phase shift at the feed point to compensate for the physical rotation of the loops. Next, twist both loops horizontally a quarter turn into helices. Finally connect the feed points in parallel. Voilà — you have a quadrifilar helix antenna.

The curious egg-beater-like configuration of the QFH has useful characteristics — an almost perfectly spherical radiation pattern as well as circular polarization throughout the pattern. This version is right-handed. For left, twist the loops in the opposite direction. For our general purpose 2 meter base station antenna, the twist direction does not matter. And yes, there is a small loss working linear polarized signals (vertical or horizontal) with a circularly polarized antenna, but it is quite acceptable. Commercial broadcast antennas often use this very technique to accommodate both mobile (vertical) as well as home antennas (horizontal).

Next question: Is the QFH a critical design? After building several, I've decided that it's not. The published articles made me initially think otherwise. But after experimenting ham style with square loops and tall versus thin rectangular ones, and the small size difference between the two loops as well as the amount of twist, I have concluded that the QFH is a dimensionally tolerant design. The performance changed little with all these variations. I'll say more about modification concepts later, but for now let's just build a QFH. That's the tricky part for most hams.

Building a QFH

Obtain a 4 foot length of common inexpensive 2 inch ID schedule 40 ABS pipe, the black pipe used for drains, and 16 feet of ⅜ inch soft aluminum tubing. Drill four pairs of $^{13}/_{32}$ inch holes through the ABS pipe, opposite each other horizontally, for the ⅜ inch aluminum tubing as shown in Figure 1. Carefully note that the vertical distance between the top and bottom hole pairs must be ½ inch less than the actual length measured along the helix of the loops as given in Figure 1.

Drill the top hole pairs 1 inch below the top end of the ABS pipe. This is to allow for an ABS pipe cap. Also, always keep in mind that the bottom of each loop is rotated a quarter turn compared to the top, to form the helix.

Take care to position these holes accurately. The easiest way is to wrap a sheet of paper snugly around the ABS pipe, keeping the edges square. Mark the circumference on the paper and then fold the paper circumference in quarters. You now have an accurate drilling guide. Use a center punch or scratch awl to transfer the hole positions to the ABS pipe.

Next mark the ⅜ inch aluminum tubing with the dimensions given in Figure 1. Cut the two sections to length with a tubing cutter. There is no need to compensate for the bends; this has been accounted for in the dimensions given. Next insert one of the yet-unbent lengths of aluminum tubing through the appropriate opposing bottom holes in the ABS pipe, centering the aluminum tubing in the bottom (8 inch) segment.

For the corners bends, these are most easily made by first flattening the tubing at the bend points. Pipe elbows, as specified for some QFH designs, are difficult to implement. To flatten the tubing, simply adjust the jaws of a pair of clamping pliers to gently make the flat spots. Do so directly over the marks, but only for one bend at a time, right while you are assembling the antenna. The biggest advantage of this method is that it eliminates separate loop segments, any need to solder (difficult with aluminum) and having to make adjustments in loop circumference for the bends.

Make your first bend upward at the bottom of only one loop. Now, on the same side, but this time at the top, again flatten and bend the tubing. Think carefully about the direction the tubing must bend to enter the top hole. This bend will be roughly 90° from the bottom bend. Then, after inserting the loop end into the ABS pipe, gently form that side of the loop into a smooth quarter-turn helix. It should end up with 90° bends top and bottom. The aluminum tubing will easily bend in your fingers.

Next form the other half of the loop, following the same procedure, beginning again at the bottom. Once again, before flattening and bending, consider the direction the aluminum tubing must bend. Also, keep the distance between the vertical sides of the loop roughly constant (8 inches), top to bottom. Again, form this half of the loop into a smooth helix and then repeat the procedure for the second loop.

Lastly, drill four $\frac{5}{32}$ inch holes vertically through the ends of the loops, roughly ⅜ inch from the end of the tubing, at the connection end. Remember: The feed point is at the top of the antenna. Do this with the loops installed on the ABS pipe. These holes will accept #12 × ¾ inch stainless steel sheet metal screws for connection to the feed coax. Screw them in temporarily at this time to form threads and to hold the loops in place.

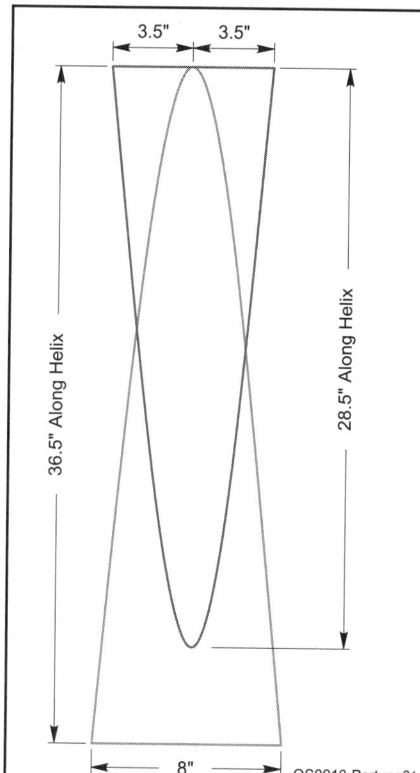

Figure 1 — Loop dimensions. Note the vertical distance between top and bottom holes is ½ inch less than the length along the helix.

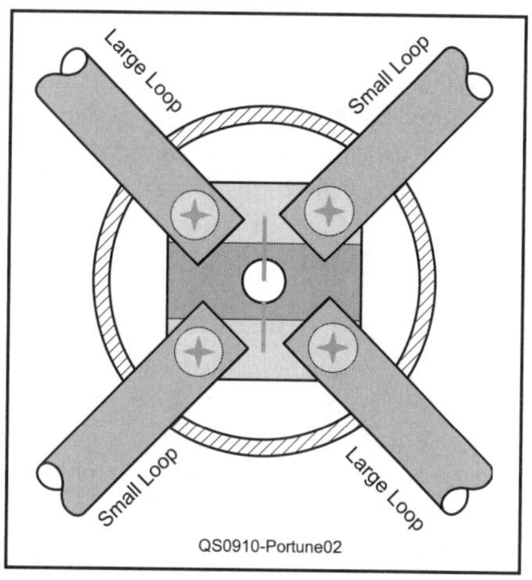

Figure 2 — Electrical connection to the loops. Observe the phase of the connection.

The Feed-Point Connection

The simplest way to connect the loops in parallel to the coax at the feed point is to fabricate the small 1⅜ × 1⅜ inch printed circuit board shown diagrammatically in Figure 2. Its exact dimensions and hole positions are not critical. But be sure to remove the copper foil from the center third. This little PCB provides mechanical rigidity as well as solder points for the feed coax. Carefully note the orientation of the copper-foil strips and the loops.

It is essential to use a balun with a QFH. Without a balun, the common mode coax current will compromise the unique radiation pattern of the QFH. The easiest type is a choke balun simply made up of several turns of the feed coax wrapped around a ¾ inch PVC pipe coupling. It is located inside the ABS pipe, just below the feed point. Simply fill the coupling with as many turns as will conveniently fit, securing the turns of coax through holes near the end of the coupling. Leave roughly 6 inches of coax free above the balun for connection to the loops. You do not need to be precise with any of this.

Assemble the balun and prepare the end of the coax for soldering. Pass the coax through the hole in the PC board from the bottom and solder it. I do not bother with weather proofing. If you wish, apply some silicon RTV sealant. With the ABS pipe cap on top, the 137 MHz QFH I've had in service for two years now shows little signs of weathering at the connections.

For final assembly, drop the balun, coax and feed-line assembly into the ABS pipe from the top. Lower the whole assembly until the balun is below the ends of the loops, but the PC board above them. Install the four #12 × ¾ inch stainless sheet metal screws into the holes in the loop ends through holes in the PC board. Pay attention to the orientation of the copper strips on the PCB compared to the loops, Figure 2. There are two possible ways, one right (as shown) and one wrong. The coax passes out through the bottom of the ABS tube. Finally place a 2 inch ABS pipe cap on top. Do not glue it on; you may wish to get back into the connection point later.

I mounted my antenna with a common antenna vent-pipe clamp and a second short piece of 1½ inch ABS pipe. Plastic is preferable to metal. The finished antenna slips down over the smaller pipe, permitting easy installation and removal.

Modifying the QFH

The existing published articles tend to make one think that the loops of a QFH must be highly precise in dimension, relative size, shape and twist. On spacecraft, where specifications must be accurately known, I suppose these are appropriate. But in my ham experience, and also in APT weather satellite service, I have found that a QFH is just as tolerant to variations as any antenna.

If one makes the loops roughly square, for example, the radiation pattern approaches spherical (in free space). If the loops are taller and thinner, but still a full wavelength in circumference, the antenna will begin to exhibit some horizontal gain, and a proportionate decrease in overhead gain. For ham and APT satellite use, this is actually desirable. One normally wants more gain toward the horizon than upward. How much off-square can QFH loops be? Frankly — almost any reasonable amount. There is no specific proportion. The radiation pattern changes only slowly as the antenna gets taller and thinner.

The only practical limit to tall and thin is feed impedance. It decreases as the loops get thinner, but again only slowly. Conveniently, QFH antennas come naturally close to 50 Ω. Additional matching is rarely required. This 2 meter version exhibits acceptably low SWR over the entire 2 meter ham band as shown in Figure 3.

Similarly, the amount of helical twist can be varied. Most designs specify 180°, a few 90°. Multiple turns are also possible. My experience is that all twist variation work much the same. The degree of circular polarization does change somewhat, but not enough to matter. I chose 90° mainly for ease of construction. Also, I find a half turn QFH more aesthetically pleasing.

The QFH on Other Bands

A QFH scales as easily to other bands as does a dipole. I was able to predict the change of size from 137 to 146 MHz merely from the frequency ratio. One only needs to make the total circumference proportional to the wavelength of operation.

In practice, the loops for a QFH antenna of this proportion and made of tubing of roughly this size, need to be roughly 5% larger than the classic metric formula —

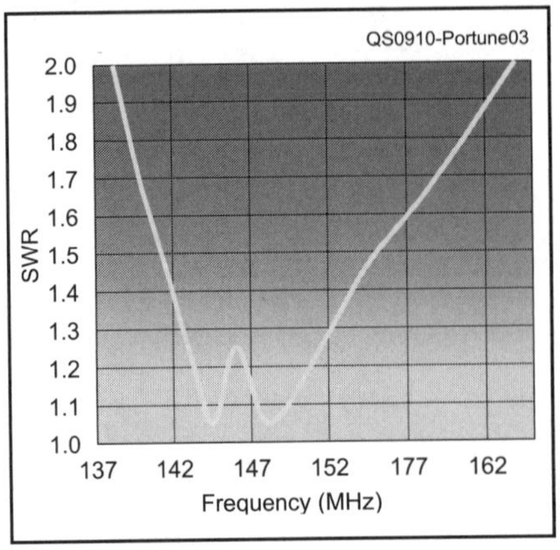

Figure 3 — Measured SWR plot of my 2 meter QFH mounted low over an aluminum carport roof.

300 divided by frequency — predicts. By cut-and-try, I found that the circumference of my smaller loop needed to be physically an actual full wavelength, as given by the formula, and the larger loop 1.1 wavelengths. Electrically, the small loop turns out to be tuned slightly high and the large loop slightly low of the operating frequency. My MFJ-259 antenna analyzer shows two nearby SWR minima, one for each loop as shown in Figure 3.

Actually having two slightly off tuned loops yields an advantage. With one loop electrically high and the other low, the bandwidth of the antenna ends up larger than one would expect from a single loop. Theoretically, the ideal frequency difference is one that causes a 90° phase shift between the loops at the feed point. But as with all other characteristics, one does not need to be highly precise here either. A rough 10% difference in loop size is fine for general ham designs.

Locating the Antenna

One of the more attractive features discovered in the original design phase for spacecraft use was the comparative insensitivity of a QFH to nearby objects. Many antenna types are worse. As one moves a QFH from free-space toward a ground plane, the pattern does change, but slowly and predictably. It becomes progressively cardioid shaped, with a shallow null downward. The upward pattern remains spherical, and the polarization does not change. This was quite convenient for spacecraft use. It is also good for ham applications. Both my QFH antennas are mounted only a short distance above a flat aluminum car port roof and neither exhibits strange characteristics.

A QFH won't make you the gain king of 2 meters, but as a general purpose base-station antenna it is hard to beat for versatility. For me it was an improvement over the vertically polarized J pole that I previously had in the same location, especially for horizontally polarized 2 meter SSB stations. It is much the same for vertically polarized FM mobiles and repeaters. But with the QFH, I am now also prepared to talk to the ISS overhead without losing the signal, as I would have with the J pole.

ARRL Member John Portune, W6NBC, received a BSc in physics from Oregon State University in 1960, his General Radiotelephone license in 1961 and his Advanced class amateur license in 1965. He spent five years in England as G5AJH and upgraded to Amateur Extra class in 1985. John retired as a broadcast television engineer and technical instructor at KNBC in Burbank and then from Sony Electronics in San Jose, California.

John is active on many bands and modes, predominantly from his HF RV mobile station. He has written a number of articles in QST, Ham Radio *and* Popular Electronics *magazines and remains active as a VE team leader, ham license teacher and Web site designer. You can reach John at 1095 W McCoy Ln #99, Santa Maria, CA 93455, or at* **jportune@aol.com**. QST

Building UHF Yagis — A Practical Approach

Jim Klitzing, W6PQL

W6PQL treats us to a detailed photo essay on how he builds his well-designed UHF Yagi antennas.

The method of construction described here can be used on most UHF Yagis, and is especially useful for the 70, 33 and 23 cm bands. I make all my antennas with hand tools, hardware store aluminum and a simple common-sense process. I'll be describing the construction of a single Yagi, the basic building block for more complex arrays of multiple Yagis, such as the one shown in Figure 1.

Figure 1 — Array of Yagi pairs for 23 and 33 cm on 6-foot booms.

Freeware Tools

I like to use the DL6WU wideband Yagi designs whenever possible. A good design tool for this is the *Yagi Calculator* written by VK5DJ.[1] Figure 2A shows a printout I used for a 23 cm DL6WU Yagi built on a boom made from ½ inch outside diameter (OD) aluminum tubing with 0.058 inch thick walls. Figure 2B shows the radiation pattern for a 24 element 1296 MHz design. I also used the calculator to design several other antennas for 70, 33 and 23 cm. After building them, I tested them on an antenna range and found that the measured performance was very close to the expected results.

You'll need to know a few things before running *Yagi Calculator*, such as:

- operating frequency
- boom diameter
- element diameter
- number of directors
- are directors in direct contact with the boom or not?
- driven element diameter
- dipole gap at the feed point
- dipole height (for folded dipoles)
- balun design

You may have to change the number of directors to get the gain and beamwidth you want, or to fill up a given boom length. Armed with the information produced from the calculator, you can proceed to the next step.

Figure 3 shows all of the parasitic elements I prepared for four 23 cm Yagis on 48 inch booms that I used in a beacon array. It does help to organize the elements like this prior to attachment to the boom!

[1]Notes appear on page 196.

For material, I usually use ⅛ inch aluminum welding rod. It is inexpensive, easily cut to size with wire or bolt cutters and holds up well in weather. Once cut to size, you should trim the ends flat with a file.

I designed these antennas for full metal-to-metal contact between the boom and the parasitic elements. Be certain to keep that in mind when designing an antenna with whichever calculator you use.[2]

Drilling Straight Holes in a Round Boom

If you ever tried drilling straight holes through a round boom with hand tools, you probably wound up with elements sticking out of the boom in many different directions. It can be very frustrating trying to keep everything in a straight plane, but with the right method it can be easily done. Prepare the boom by cutting it to the proper length.

- Using a laundry marker, mark the end of the boom on opposite sides as shown in Figure 4. The tubing on the right is the piece you are working on.
- Tape the boom securely to a marking guide, positioning one of the end marks so that it almost touches the guide. I used another piece of tubing as a guide, but you can use almost anything that is straight, even a piece of wood.
- Place the tip of your marker in the joint between the boom and the guide, and mark a line down the entire length of the

ELEMENT LENGTHS AND SPACING

Reflector
118 mm long at boom position = 30 mm

Radiator
Single dipole 106 mm tip to tip at boom posn =75 mm
Folded dipole 108 mm tip to tip at boom posn =75 mm

Dir (no.)	Length (mm)	Spaced (mm)	Boom position (mm)	Gain (dBd)	Gain (dBi)
1	103	17	92	6.1	8.3
2	102	41	133	7.3	9.5
3	101	49	182	8.4	10.6
4	100	57	239	9.3	11.5
5	99	64	303	10.1	12.3
6	98	68	371	10.8	13.0
7	97	72	442	11.4	13.6
8	96	75	517	11.9	14.1
9	95	78	596	12.4	14.6
10	94	82	677	12.9	15.0
11	94	85	762	13.3	15.4
12	93	87	850	13.6	15.8
13	93	89	938	14.0	16.1
14	92	90	1028	14.3	16.4
15	92	91	1119	14.6	16.7
16	91	91	1210	14.8	17.0

Director spacings are measured from the previous element
Tolerance for element lengths is +/- 1 mm

Boom position is the mounting point for each element as measured from the rear of the boom and includes the 30 mm overhang.The total boom length is 1240 mm including two overhangs

A half wave 4:1 balun uses 0.70 velocity factor RG-141 (PTFE) and is 79 mm long plus leads

Here are some construction details for a folded dipole

Measurements are taken from the inside of bends
Folded dipole length measured tip to tip = 108mm
Total rod length =236mm
Centre of rod=118mm
Distance HI=GF=32mm
Distance HA=GE=55mm
Distance HB=GD=79mm
Distance HC=GC=118mm
Gap at HG=15mm
Bend diameter BI=DF=30mm

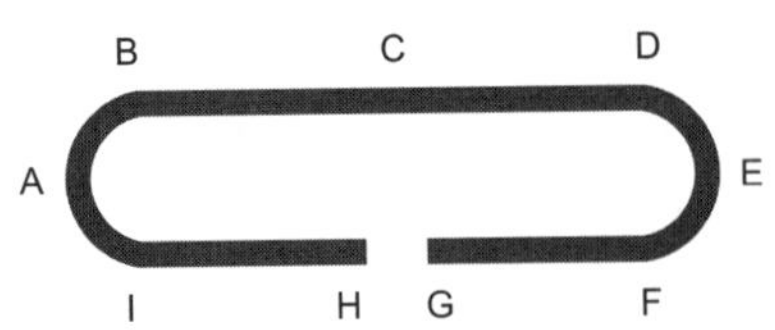

QS0608-Klitz02A

(A)

Gain estimated for E-Plane beamwidth and sidelobe levels ~ 16 dBd

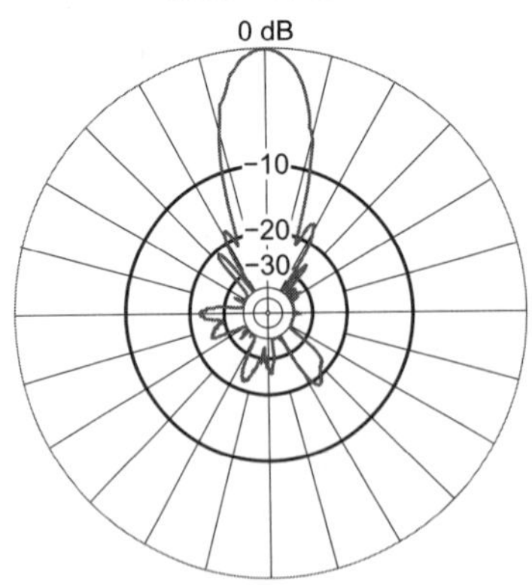

(B)

Half Power Beamwidth = 22.7°
Front to Back Ratio = 25 dB
E-Plane

QS0608-Klitz02B

Figure 2 — At A, portion of a printout from VK5DJ's *Yagi Calculator*. At B, polar pattern for 24 element DL6WU Yagi at 1296 MHz.

boom. See Figure 5. You will probably need to tape the boom to the guide in several places to prevent the marker from forcing the boom away from the guide.

- Untape the boom from the guide. Rotate the boom to the opposite side using the other locating mark you made at the end of the boom as a reference. Retape the boom to the guide and mark a line down this side also.
- Now that you have both lines drawn, place tick marks on the lines (both sides of the boom) where the holes for the parasitic elements must be drilled. Measure from the end of the boom to each mark to elim-

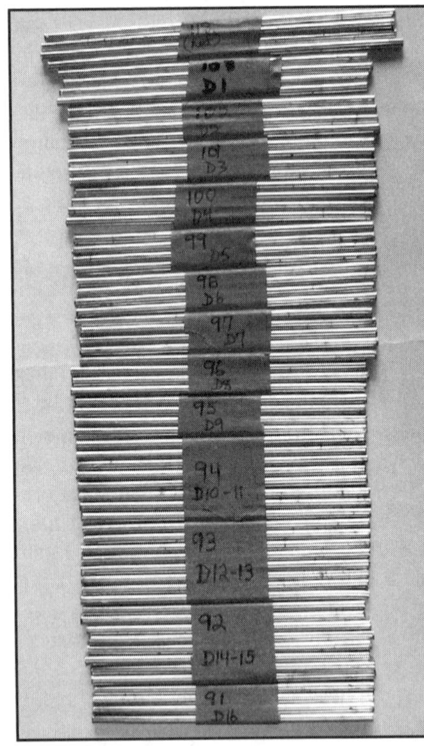

Figure 3 — Lineup of welding rod elements for four 23 cm Yagis, ready to be installed on booms.

Figure 4 — Setting up two aluminum booms for marking down the middle.

inate cumulative errors. See Figure 6.

- Finally, use a center punch to prep the locations (both sides) that will be drilled for the parasitic elements. If you are using ⅛ inch diameter elements, use a ⅛ inch drill bit.
- Drill the holes one side at a time. A common mistake is to drill all the way through the boom in one pass. This will almost always result in crooked elements unless you are using a drill press, and even then it's difficult to do properly that way.

Fastening the Elements to the Boom

- Hold the boom in a small vise as shown in Figure 7.
- Insert and center an element into the boom, beginning with the reflector. Use a small hammer and tap it through gently.
- Using a center punch, stake the boom next to the element in two places, preferably opposite one another. Stake the boom at a 45° angle toward the element as shown on the right, and at the 10:30 and 4:30 positions. See Figure 8 for a detail of what the staking dimples should look like.
- Stake the element on the other side of the boom as well.
- Attach the rest of the elements in a like manner.

If you did a good job with your humble hand tools, your antenna should look something like Figure 9.

You may find an element or two out of alignment with the others as you sight down the boom. It never goes perfectly. So, to adjust an element, grasp the end and bend it slightly to put it in alignment with the others. Looking closely, you might notice a slight "zig" in the alignment of the reflector on this antenna. I bent this into position after discovering it was just out of alignment with the first director. These small adjustments will not affect the performance of the antenna.

If you live in an environment in which corrosion is a problem (by the ocean, for example), a dab of latex paint covering the boom-to-element joints can be helpful, and will not detune the antenna.

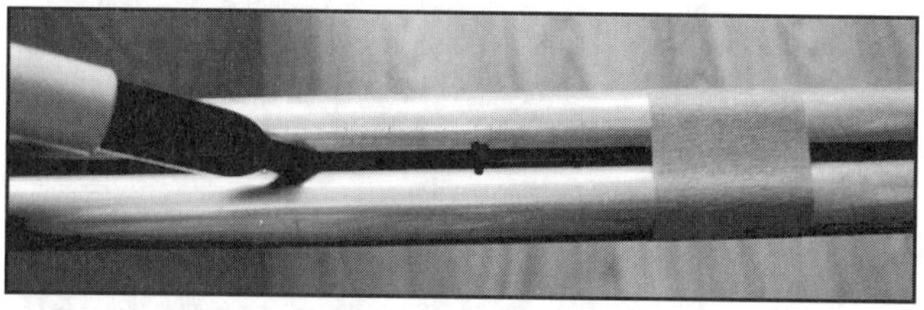

Figure 5 — Marking two aluminum booms, using each as a guide for marking the other down the middle.

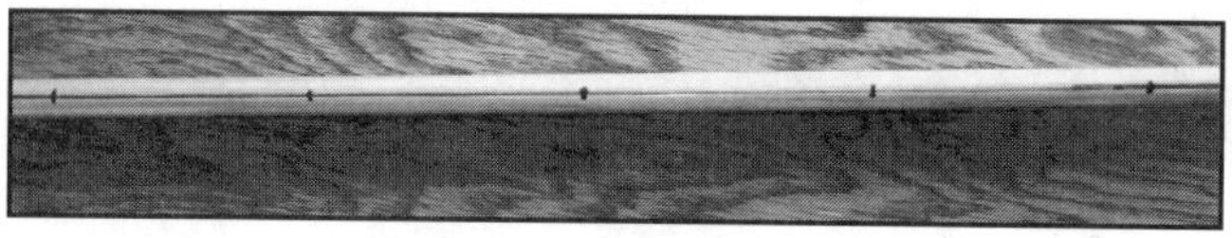

Figure 6 — One of the booms after marking.

Figure 7 — "Dimpling" the boom to capture each element.

Figure 8 — Close-up of boom dimpling.

Figure 9 — Photo of Yagi with reflector and director elements installed.

Building the Driven Element

See Figure 10 for a view of the parts used in the driven element. You will need:

- One chassis-mount style N connector (at least two opposing mounting holes are needed), with Teflon insulation (not plastic).
- Two ½ to ¾ inch aluminum or brass spacers.
- 12 gauge copper wire for the folded dipole.
- A piece of UT-141 semi-rigid coax for the ½ λ balun.

The smooth part of a screwdriver handle makes a good form for bending both the balun and the dipole element. I usually wrap the barrel and threads of the N connector with several layers of paper, and hold it in a small vise. The paper protects the threads and slows heat loss to the vise while soldering.

Solder the shield of the UT-141 coax balun to the connector as shown in the close-up photo in Figure 11. Tin the solder cup of the N connector's center pin, and solder one end of the driven element dipole into it. Making a small bend into the cup with the wire is helpful.

Last, bend the center conductors of the balun under and around the ends of the dipole and solder as shown. Be careful to leave a small gap between the outer conductor of the balun and the ends of the dipole (about 2 mm).

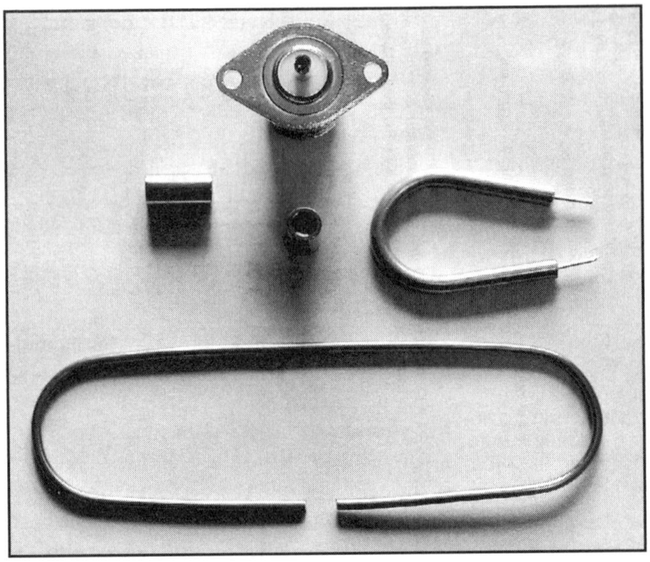

Figure 10 — Parts used to make folded dipole driven element.

Figure 14 — Driven element after painting.

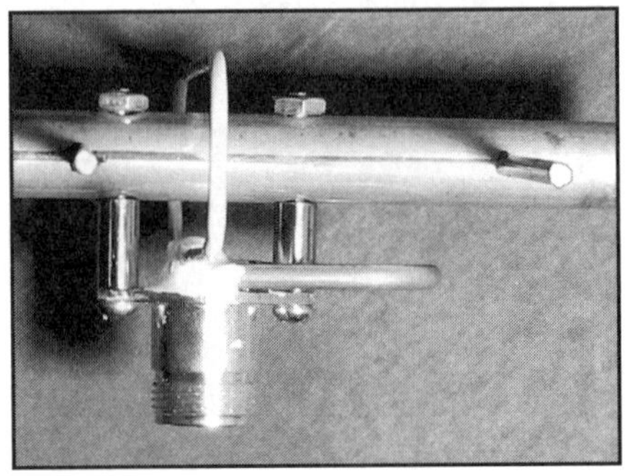

Figure 15 — Side view of the driven element mounted to the boom with spacers.

Figure 11 — Close-up of semi-rigid UT-141 coax used to make balun. It is connected to a Type N connector and the folded dipole driven element.

Figure 12 — Another view of the completed driven element.

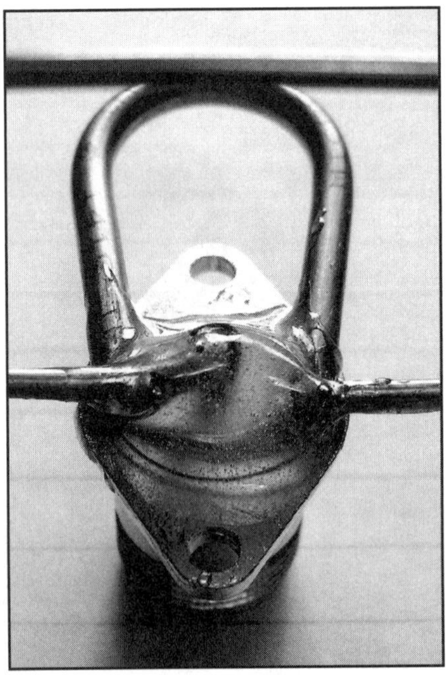

Figure 13 — Close-up of the driven element showing epoxy used as weather protection.

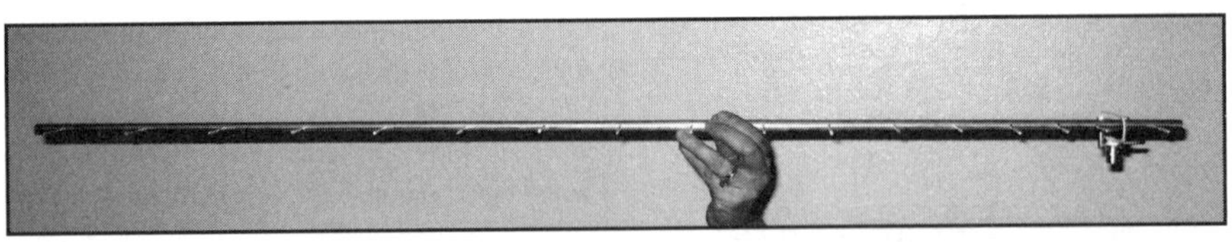

Figure 16 — Side view of the finished Yagi. Note the element alignment

Your driven element should now look like Figure 12. Apply a generous coat of epoxy to the connections for strength and weatherproofing. See Figure 13. Not all epoxies are created equal, so be certain to test a cured lump in a microwave oven before using it on your antenna. Place your epoxy sample next to a cup of water in the microwave and cook for 30 seconds. If the epoxy becomes hot, try another brand. Just warm is okay. Apply a coat of white latex paint to cover the epoxy seal and the copper dipole. See Figure 14.

All that is left to do now is mount the driven element to the boom. For this I used ½ inch aluminum spacers and 6-32 machine screws. See Figure 15. Make certain you have the dipole placed at the correct distances from the reflector and first director before drilling the mounting holes in the boom.

Figure 16 shows the completed 4 foot boom antenna, held by my wife (this is as much of her as she would allow me to photograph). You may not need to tweak anything to minimize the SWR, but if you do need to, the SWR is affected by the shape or height of the folded dipole. You can also bend the ends of the folded dipole toward or away from the reflector as an additional adjustment. You should be able to get the SWR down below 1.2:1 at mid band and below 1.5:1 at the band edges.

Notes

[1]VK5DJ's Yagi calculator is available on his Web page at **vk5dj.mountgambier.org/Amateur_radio.html**.

[2]This is also discussed in detail in *The ARRL Antenna Book*. See p 18-30 and 18-31 in the 20th Edition. Available from the ARRL Bookstore for $39.95 plus shipping. Order number 9485. Telephone 860-594-0355 or toll-free in the US at 888-277-5289; **www.arrl.org/shop/**, **pubsales@arrl.org**.

Jim Klitzing, W6PQL, was first licensed in 1964 as WB6MYC. He has been a metrologist for both the US Air Force and Hewlett-Packard Company. He also designed, manufactured and sold a line of solid state VHF/UHF linear amplifiers. He is currently manager of engineering services at Agilent Laboratories in Palo Alto, California. Jim has always enjoyed building his own equipment, even going as far as making a 40 meter kW SSB/CW transceiver as well as various VHF through microwave transverters, amplifiers and test equipment. Currently antennas are his main interest. You can reach the author at 38105 Paseo Padre Ct, Fremont, CA 94536; **qrz@w6pql.com**.

A Short Boom, Wideband 3 Element Yagi for 6 Meters

L. B. shows us how to craft an easy to reproduce 6 meter Yagi that covers the whole band.

L. B. Cebik, W4RNL

Although there are many 2 and 3 element Yagi designs for 6 meters, few do justice to the needs of the FM operator who requires some gain and directivity from 51 to 54 MHz. The 3 element design described here will actually cover the entire 6 meter band (a bandwidth of about 7.7%) with reasonable performance and a 50 Ω SWR of less than 2:1. The short 3 foot boom is a further attraction. Figure 1 is a photograph of a completed Yagi mounted on a PVC test mast.

Beam Design and User Expectations

Evaluating a beam design requires an appreciation of three ingredients: your needs on a particular band, general beam antenna behavior and reasonable expectations to have of a beam once installed. Your requirements for a beam on 6 meters change depending on your operating goals. If you only use CW, SSB or digital modes, then you tend to operate in the lowest MHz of the band and can use a beam with a narrow operating bandwidth. If you wish to use FM and want some directivity, however, then you need a wideband design with good performance from 51 to 54 MHz.

Moreover, the beam should be vertically polarized, which normally translates into a beam with vertically oriented elements. FM activity generally relies on point-to-point signal paths. Cross polarization of the sending and receiving antennas may result in weak signals or no signals at all. As shown in Figure 1, the present antenna is designed for broadband 6 meter service using vertically oriented elements.

Older designs for a truly wideband Yagi depend on the spacing between the driver and the reflector elements to obtain their broad coverage. The wider spacing produces a beneficial side effect — it raises the feed-point impedance to about 50 Ω so that the user can make a direct connection between the feed line and the antenna without needing to add a matching network. However, these designs carry a penalty. The lower gain 2 element driver-reflector Yagi needs a 3.5 foot boom, while the higher gain 3 element Yagi's boom is over 6 feet long. The design here has three elements, but uses a boom less than 36 inches long. Figure 2 shows its outline, while Figure 3 is a photo of the prototype beam.

To obtain higher gain and a shorter boom, you might turn to a driver-director Yagi design. Unfortunately, these designs have quite low feed-point impedances and very narrow operating bandwidths. Broadening the operating bandwidth of a driver-director design involves redesigning the driver, using two elements with a simple phase line between them. Now you have a beam that competes with the 3 element standard Yagi, but does so on a much shorter boom.

A driver-director design (in fact, any Yagi with one or more directors) has another advantage over the 2 element driver-reflector Yagi if you consider your needs on the band. A driver-reflector Yagi shows its highest gain

Figure 1 — The prototype 3 element wideband phase-fed Yagi on the test mast. Note that this mast is nonconductive to avoid interactions with the nearby driver elements.

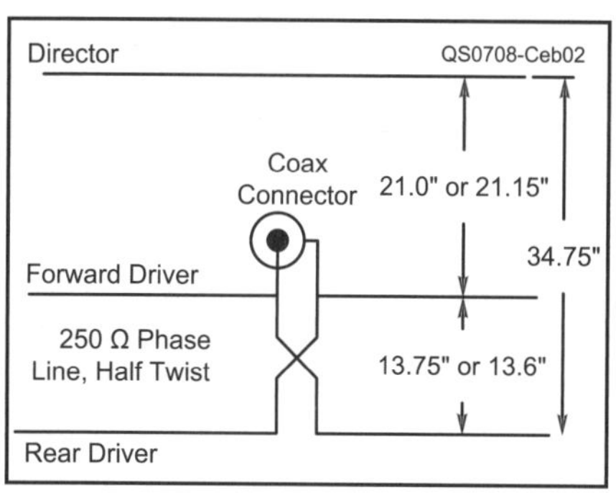

Figure 2 — General outline of the 3 element wideband short-boom Yagi for 6 meters.

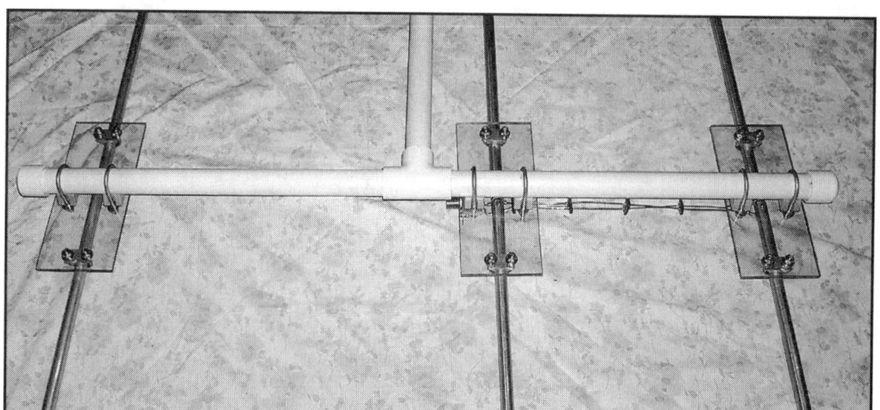

Figure 3 — Photo of the entire central beam structure, including the PVC boom and mast stub.

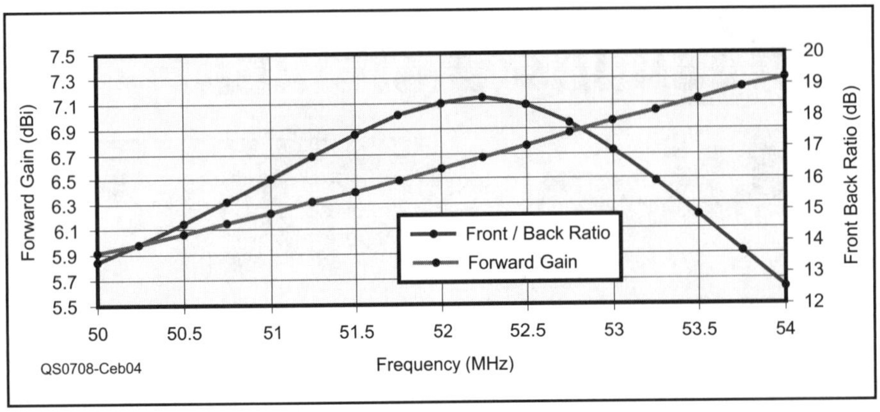

Figure 4 — Modeled free space gain and 180° front-to-back ratio of the 3 element wideband short-boom Yagi for 6 meters.

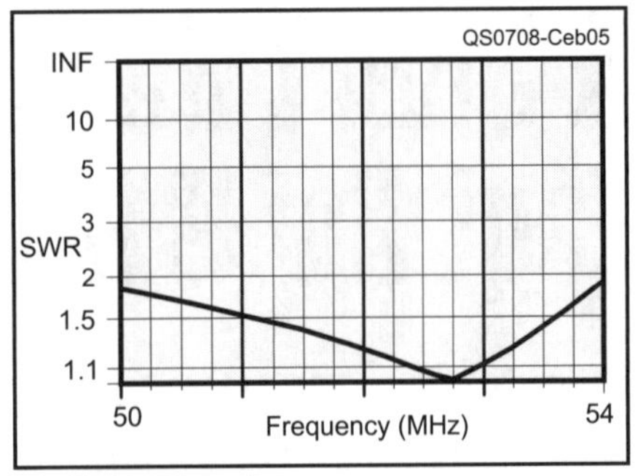

Figure 5 — Modeled 50 Ω SWR curve for the prototype 3 element wideband short-boom Yagi for 6 meters. The array requires no matching network and uses ⅝-½ inch elements with bare wire phase line. The measured SWR was identical to the modeled SWR.

Table 1
Dimensions of the Yagi for Different Construction Methods

Tapered-Element Version Using 0.625"/0.5" Elements; 250 Ω (Open Wire) Parallel Phase Line

0.625" Inner Sections = 24" Each Side of Center or 48" Total; Add 2" to Tip Lengths for Insertion

Element	*Total Length*	*Tip (0.5") Length*	*Spacing from Rear Element*
Rear Driver	110"	31"	—
Forward Driver	103.5"	27.75"	13.75"
Director	101"	26.5"	34.75"

Tapered-Element Version Using 0.625"/0.5" Elements; 250 Ω RG-63 Phase Line

0.625" Inner Sections = 24" Each Side of Center or 48" Total; Add 2" to Tip Lengths for Insertion

Element	*Total Length*	*Tip (0.5") Length*	*Spacing from Rear Element*
Rear Driver	108.5"	30.25"	—
Forward Driver	104"	28"	13.75"
Director	101"	27"	34.75"

Uniform-Element Version Using 0.5" Elements; 250 Ω (Open Wire) Parallel Phase Line

Element	*Total Length*	*Spacing from Rear Element*
Rear Driver	108.9"	—
Forward Driver	102.6"	13.6"
Director	99.8"	34.75"

Tapered-Element Version Using 0.5"/0.375" Elements; 250 Ω (Open Wire) Parallel Phase Line

0.5" Inner Sections = 24" Each Side of Center or 48" Total; Add 2" to Tip Lengths for Insertion

Element	*Total Length*	*Tip (0.375") Length*	*Spacing from Rear Element*
Rear Driver	111"	31.5"	—
Forward Driver	104.6"	28.3"	13.6"
Director	102"	27"	34.75"

Tapered-Element Version Using 0.5"/0.375" Elements; 250 Ω RG-63 Phase Line

0.5" Inner Sections = 24" Each Side of Center or 48" Total; Add 2" to Tip Lengths for Insertion

Element	*Total Length*	*Tip (0.375") Length*	*Spacing from Rear Element*
Rear Driver	109"	30.5"	—
Forward Driver	105"	28.5"	13.6"
Director	102"	27"	34.75"

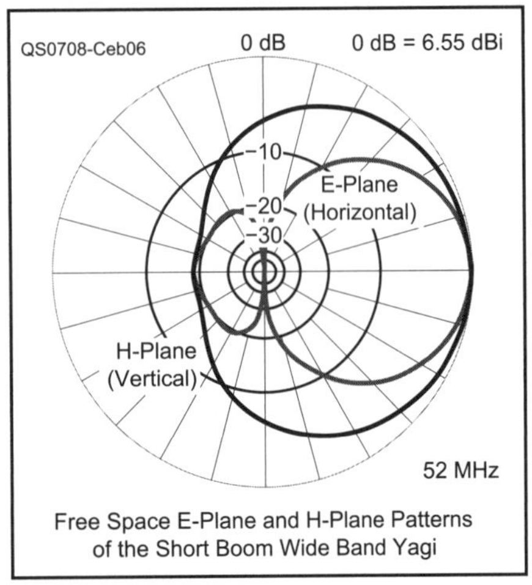

Figure 6 — Free-space E-plane (horizontal) and H-plane (vertical) patterns for the 3 element wideband short-boom Yagi at 52 MHz.

at the low end of the band, with decreasing gain as you increase the frequency. Figure 4 shows the modeled performance of the prototype antenna from 50 to 54 MHz. Like a longer standard 3 element Yagi, the gain shows a rising curve so that we have maximum gain in the FM section of the band. The graph also shows the anticipated front-to-back performance, which varies from nearly 13 dB at the band edges to a peak value of over 18 dB.

The final critical operating parameter is the 50 Ω SWR across the band. The present design requires no additional matching network. Instead, it connects directly to a 50 Ω transmission line (although I always recommend the use of a common-mode current attenuator, such as a ferrite bead choke or a 1:1 balun at the feed-point). Figure 5 shows the modeled SWR curve for the entire 6 meter band. (The test antenna curve overlaps the modeled curve too closely to justify a second line on the graph.)

Newer amateurs who read *QST* often get a wrong impression of Yagi patterns, because almost all beam patterns for HF appear with the elements horizontal (or in the E-plane). E-plane patterns differ from H-plane patterns (when the elements are vertical). Figure 6 overlays the E-plane and H-plane free-space patterns of the phased-driver design at 52 MHz to show the difference. The E-plane pattern shows a limited beamwidth and very deep nulls 90° from the axis of maximum forward gain. The H-plane pattern has the same forward and rearward gain maximums, but has a much broader beamwidth and no side nulls at all. The H-plane pattern generally represents the azimuth pattern that we should expect from a vertically oriented Yagi. Because the beamwidth is so great, the forward gain of a vertical beam over ground will be less than the forward gain of a horizontal Yagi at the same height. However, the gain of each beam over a similarly oriented dipole will be just about the same (a bit over 4 dB).

Design and Construction of the 3-Element Broad-Band Short-Boom Beam

The Boom

The 3 element short boom wideband Yagi consists of two elements that serve as a phased pair of drivers. Forward of the driver pair is a single director element. All versions of the beam have a total length of 34.75 inches from the center of the rear element to the center of the director. Therefore, you need a boom that is perhaps 38 to 40 inches long, depending upon your construction methods. An aluminum tube and a schedule-40 PVC 1 inch nominal tube are equally effective, since we shall insulate and isolate our elements from the boom with mounting plates. The prototype employed a PVC-pipe boom with a T connector for the boom-to-mast connection. The stub slips inside my main test mast, which uses 1¼ inch nominal PVC.

Element Dimensions

Table 1 lists five separate sets of dimensions for the beam, since amateurs generally have access to different material stocks. The first two sets use ⅝ inch diameter inner sections with ½ inch outer or tip sections. The middle set uses uniform-diameter ½ inch tubes throughout. The final two dimension sets use ½ inch inner sections with ⅜ inch diameter tips. The prototype in the photos uses the first set of dimensions. However, the only change in performance among the dimensions is that the fattest elements provide a very slight improvement in the SWR curve at the band edges.

Table 1 illustrates two important facts for newer beam builders. First, small design changes, such as a change in the phase line structure or a change in element diameter, can alter beam performance enough to

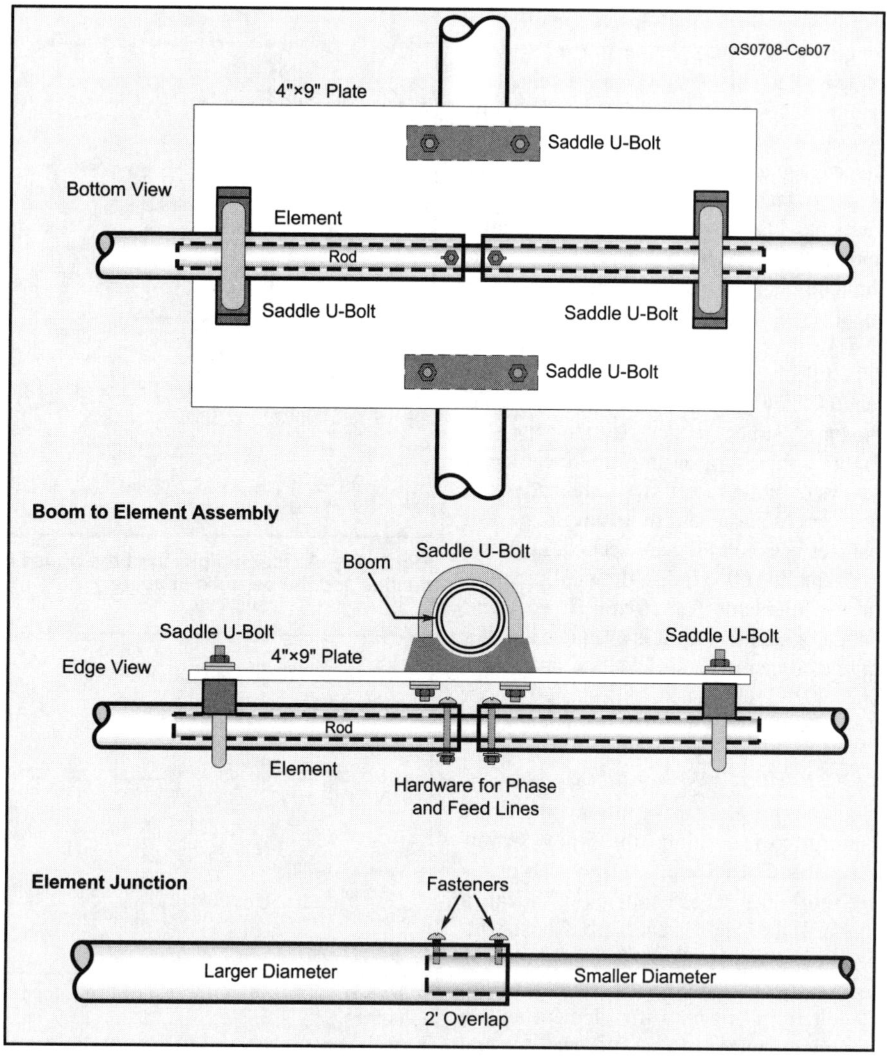

Figure 7 — Some construction details for the prototype array, including the junction of element sections and the element to boom plate assembly. The sketches apply to driver elements; omit the gap for the director.

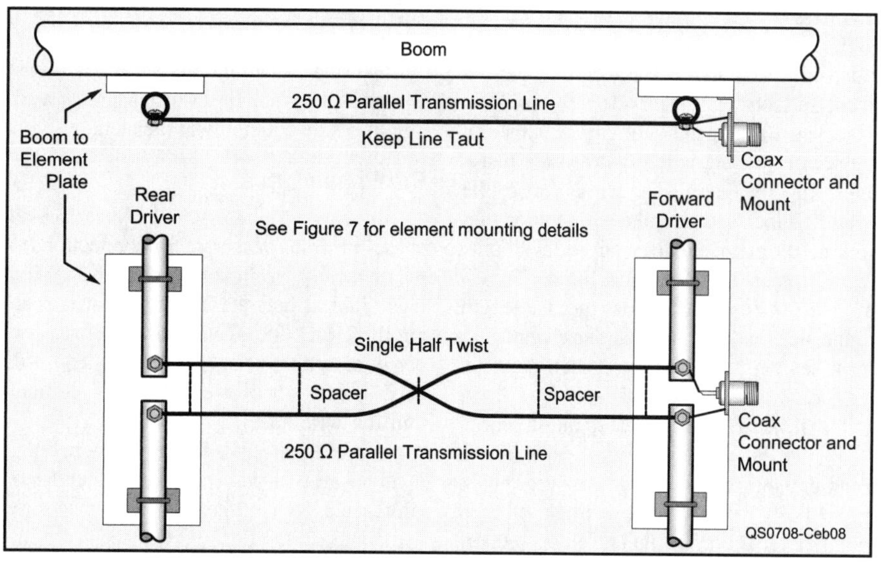

Figure 8 — The basic structure of the array feed-point and the phase lines. Except for the spacers, the same structure applies to the use of RG-63 cables for the phase lines.

require a complete redesign of the element lengths. (Element spacing is usually less critical.) Second, tapered element diameters require added length to be electrically equivalent to uniform diameter elements, even if the average diameter of the tapered element is larger than the diameter of the uniform element. Changing just the lengths of inner and tip sections can require an adjustment to the total element length for equivalent electrical performance.

If you use one of the tapered-diameter schedules, add about 2 inches to the tip lengths shown in Table 1 for insertion into the inner element section. See the bottom of Figure 7 for a typical junction. Sheet metal screws or aircraft rated rivets are among the most useful fasteners to ensure long term contact between element sections. I prefer high-quality 6063-T832 drawn aluminum tubing, available from Texas Towers and similar sources, since it is available in outer diameter increments of ⅛ inch in the US. The 0.058 inch wall thickness (just under $\frac{1}{16}$ inch) allows a smooth fit but a tight joint.

Element-to-Boom Mounting

For the element-to-boom junctions, I prefer to use ¼ inch thick polycarbonate plates. For 6 meters, 9 by 4 inch plates are more than large enough. As shown in Figure 7, the chief fasteners are U-bolts with saddles, available from DX Engineering, as one source. I use saddles to prevent the U-bolt from crushing the element and to provide a space between the element and the plate. For the driver elements, you need additional hardware to handle the phase lines. The element U-bolts have a spacing of ¾ inches, while the boom U-bolts are around 1 inch to 1¼ inch tubes. Like all of the hardware (including lock washers) in the beam, the U-bolts are stainless steel.

The sketch shows a gap between element halves for each driver. The feed line and the phase line connect to each side of the gap. The exact size of the gap is not critical, although very close spacing will help keep the phase lines properly spaced. The gap is part of the overall element length and not in addition to it. To keep the element halves aligned with only two U-bolts at the plate ends, I used a length of ½ inch diameter fiberglass rod, the length of the plate, inserted into the ⅝ inch inner element sections. Almost any ½ inch diameter non-conductive rod or stiff tubing will do, although wood is susceptible to moisture.

For the director, you may use either of two mounting methods. If you do not over tighten the U-bolts, you may use a single 48 inch length of tubing for the inner element section. If you wish to guard against crushing, you can use two 24 inch lengths with an insert of the next smaller aluminum tubing size that is as long as the plate. Sheet metal screws work well to ensure continuity in the element.

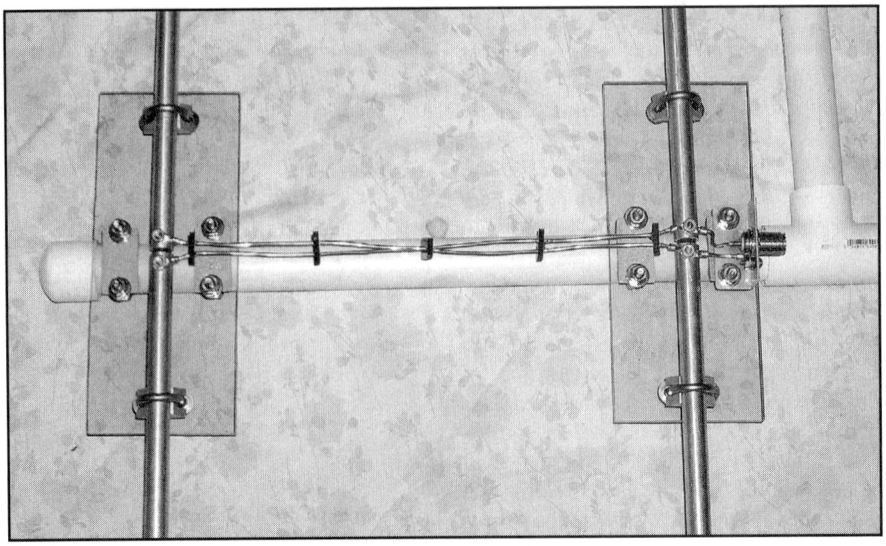

Figure 9 — A close-up photo of the phase line and driver assembly. Note the half twist in the line and the periodic spacers.

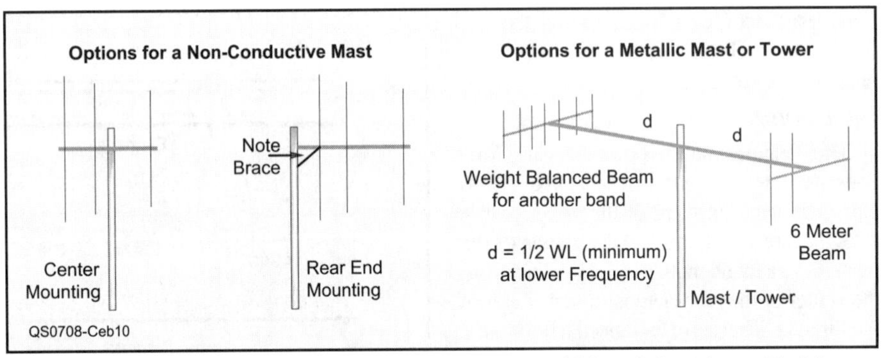

Figure 10 — Some mounting options for the 3 element wideband short-boom Yagi for 6 meters.

Table 2
250 Ω Open Wire Transmission Line Dimensions

AWG Wire Size	*Wire Diameter*	*Center-to-Center Spacing*
#14	0.0641"	0.262"
#12	0.0808"	0.330"
#10	0.1019"	0.416"
#8	0.1285"	0.525"

The Phase Lines

Since you need only 13.75 inches of phase-line (plus a bit extra for connections to the elements), you likely should make your own. Table 2 lists the center-to-center spacing for 250 Ω lines using common bare copper wire, arranged by AWG size. You will need spacers about every 3 inches to maintain the wire spacing accurately. The best way to make spacers is to drill the wire holes in a long strip of polycarbonate, Plexiglas or similar plastic. Cut the spacers to size after you complete the drilling. Do not make the holes too large; you want a tight fit. A drop of superglue at each hole will lock the spacers in place. The velocity factor of this phase line will be very close to 1.0.

I do not recommend using 300 Ω TV twinlead as the phase line. Even high quality 300 Ω line has a velocity factor of about 0.8 to go with its characteristic impedance that is already 20% higher than optimal. Using a taut line will make the TV phase line about 25% longer electrically than the value needed to create the right conditions for the drivers to operate well. Two elements with a phase line use a fairly critical combination of element dimensions and spacing — along with equally critical values of phase-line characteristic impedance and electrical length — to get the job done. This involves dividing the current at the feed point so that each driver element receives the correct current magnitude and phase angle for maximum gain from the pair (in the presence of the director element) — while yielding a 50 Ω feed-point impedance.

There is an alternative way to construct the phase-line. RG-63 coax (available from such sources as The Wireman, **thewireman.com**) has a characteristic impedance of 125 Ω. Cut two 13.75 inch sections. At each end, solder the braids together and tape them to prevent contact with the elements. You then use only the center conductors to make the phase-line connections (with the single half-twist, of course). The line now forms a series-connected dual coax cable with an impedance of 250 Ω. Since RG-63 has a velocity factor of 0.84, you must adjust the lengths of the two driver elements to compensate. In Table 1, you will find alternative dimensions for both the ⅝-½ inch and the ½-⅜ inch versions of the array. These variations on the design produce performance that is indistinguishable from the versions that use bare wire phase lines.

Figure 8 shows the connection scheme for the phase lines and the feed point. I used #8 bolts and nuts through each driver at the gap, along with ring solder lugs to fasten the phase line between doubled nuts on the bolts. Very short leads go from the forward driver connections to a coax connector mounted on a small plate that I fastened to the element mounting plate. Once I completed all final adjustments, I coated all exposed connections with one of the liquid plastics that substitutes for electrical tape. The close-up photo in Figure 9 of the phase line structure was taken before I weather-sealed the connections.

The Boom-to-Mast Connection

Although the prototype uses a PVC boom and mast, standard 1 inch or 1¼ inch aluminum tube is also suitable as a boom material. However, to minimize the metallic mass in the immediate vicinity of the forward driver, I recommend a polycarbonate plate with stainless steel U-bolts to fasten the boom to the mast. Figure 10 shows some alternative methods of mounting the beam. The two sketches at the left apply to non-conductive masts that extend at least a small distance below the tip of the lower element halves. Rear end mounting is feasible, since the beam is relatively light. However, a 45° brace is necessary to reduce boom sag.

The sketch at the right of Figure 10 shows an alternative that allows the use of a metal cross boom that is at right angles to the elements. If you have a beam for another band that is weight-balanced with the 6 meter beam, you can set up the dual mounting scheme by allowing enough room between the beams. I recommend a total starting distance of 1 λ at the lower frequency. You can then shorten the distance in small increments while maintaining proper weight distribution until you begin to notice interactions. Some pattern distortions may occur before you encounter disturbances to the feed-point impedance.

Adjustment and Performance

The beam design emerged from *NEC* software with the Leeson corrections enabled to accurately account for the tapered diameter elements on most versions of the array.[1] I tried to be as careful in the shop as at the computer. Consequently, the test version of the antenna required no post-construction adjustments. Indeed, the measured SWR curve is so close to the modeled version that I checked it several times to ensure that I was not wishfully seeing numbers that did not exist. The predicted SWR minimum occurs at about 52.75 MHz, while the prototype showed the lowest value at 52.8 MHz (give or take the tolerances of my test meters). If your minimum SWR point is lower in frequency, extend the tips of both drivers in ⅛ inch increments. If the minimum frequency is too high, shorten the driver tips in the same manner. The director, if properly constructed, should not need adjustment. Also be certain that you do not have too much metal in the boom to mast plate and that it is not too close to the forward driver. The prototype's all PVC boom and mast stub, coupled with the PVC upper section of the test mast, eliminated this concern.

The beam appears to perform just as expected, with significant forward gain (about 4 dB) over a vertical dipole (at the same height) and good rearward quieting. A gain of 4 dB can make the difference between signals that fall below or rise above the FM threshold of reception. The present design is perhaps the smallest Yagi of which I am aware that provides this level of performance.

In fact, the design is adaptable to other frequency bands that need good performance over a wide operating bandwidth. The first MHz of 10 meters and the entire 40 meter band are two such examples. Although a 40 meter version falls outside the limits of my shop and test facilities, I'll bet that I can build a horizontal version for 10 meter use in time for the new sunspot cycle.

[1] *EZNEC* models for all five versions of the array in Table 1 are available at the ARRL Web site at **www.arrl.org/files/qst-binaries/Cebik0807.zip**.

*Licensed since 1954, L. B. Cebik, W4RNL, is a prolific writer on the subject of antennas. Since retiring from teaching at the University of Tennessee, LB has hosted a Web site (***www.cebik.com***) discussing antennas — both theoretical and practical. He has written more than 15 books, including the ARRL course on antenna modeling. Serving both as a technical and an educational ARRL advisor, he's also been inducted into both the QRP and QCWA Halls of Fame. LB can be reached at 1434 High Mesa Dr, Knoxville, TN 37938 or at* **cebik@cebik.com**. QST

A Rugged VHF Beam Antenna That Resists Ice Buildup

This unique design will help you keep that Yagi up when you most need it.

Ray Abraczinskas, W8HVG

Figure 1 — A rugged 8 element 222 MHz beam antenna resists the effects of ice shorting out the gamma match and damaging the elements. The design criteria used are applicable to any beam antenna.

The Independent Repeater Association operates a 2 meter linked repeater system in Michigan (see **www.w8hvg.org**). Our 222 and 440 MHz band auxiliary link antennas have had many problems allowing successful operation during our cold Michigan winters. We operate twelve 2 meter repeaters continuously linked together 24/7 via these links. The system uses 51 antennas, 14 of which are 222 MHz Yagi beams. Due to the shortage of commercially available 222 MHz beam antennas available commercially, we've had to build our own for many of our linking applications.

The Problem

During harsh Michigan winters, we discovered a major problem with the design of those VHF beam antennas in that they have an exposed gamma or T matching arrangement. During severe storms, we found that the exposed match will get coated with ice and effectively short out the antenna feed point, rendering the antenna useless until the ice thaws. It's especially problematic on high towers where the wind-chill temperature is very low.

One such commercially available antenna that caused us difficulty had a beautifully machined aluminum box attached to the boom with the T match bars protruding through plastic electrical feed-through insulators out to the driven element. Ice would simply coat over and around the feedthrough insulators and effectively short out the match. Sometimes it would take weeks or months for the ice to melt and restore operation. This, of course, was unacceptable for a wide-area linked repeater system, hopefully with 24/7 availability. There just had to be a better way.

Beam antennas can also be damaged by falling ice from higher up on a tower installation. This can often cause catastrophic failure of the antenna elements due to the heavy ice. Large ice sheets often break loose during high wind or as part of the melting process.

Every year, after a hard winter, we would find that many of the thin elements on several of our beam antennas would be broken from falling ice, usually requiring replacing the entire antenna. In many cases they would still seem to work, albeit with reduced performance. However, it was evident that more rugged solid elements were required.

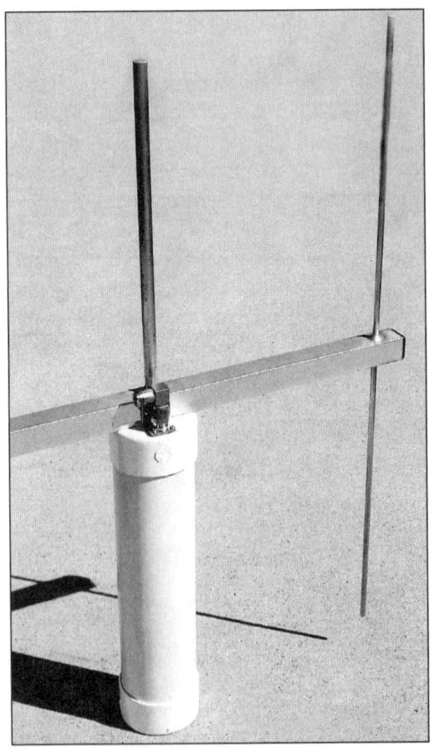

Figure 2 — A Schedule 30 PVC pipe with end caps form a shroud over the driven element and gamma match. All elements are welded or otherwise secured to the square boom. A chassis mount Type N connector and elbow are used to connect to the coax.

Some Background

There are no known commercially available 222 MHz amateur beam antennas that are specifically designed to perform while coated with ice. There are also no known commercially available beam antennas for that band that are specifically designed to reduce or resist ice buildup, especially around the feed point where it really matters.

Searching the Internet revealed that some antenna designers talk about the problem and how under certain conditions, ice buildup can alter the beam gain. In some cases, the direction of the signal can be reversed due to the ice buildup changing the director elements into reflectors.

Some designers used the phrase that an antenna must be *hydrophobic* — in other words fear or shed water — in order to reduce ice buildup. That makes sense but how does one build water aversion into an antenna?

Further investigation on the Internet revealed that certain coatings are made to shed water easier and therefore could possibly reduce the ice buildup. Coatings of epoxy, teflon or silicone were "tight skinned" and supposedly could easily shed water.

I found some information about enclosing the antenna in a protective radome as a potential solution. Commercial FM broadcast antennas and certain radar antennas use protective radomes. For a long boom amateur beam, a radome would be impractical due to increased wind loading. They would likely also add significant cost.

Quantum physics tells us that black materials radiate and absorb light energy more readily than lighter color materials. Most people know that black objects warm up faster than white objects in the sun (and they cool down faster in the dark). That's why people generally wear light colored clothes in the sunlight and also why solar heater panels are painted black inside. It seems reasonable then to assume that black antennas would absorb sunlight and melt ice faster than light aluminum colored antennas.

For many years, I have been using the do-it-yourself aluminum available in most hardware stores for making beam antennas. There are usually choices available in a rack of different sized aluminum tubes, rods, angles, channels, flat stock and sheet stock. I've used these materials to make equipment mounts, antennas and mounting brackets, including a rugged tri-band ground plane antenna described in *QST*.[1]

Building a Suitable Yagi

I wanted to use readily available materials to make our ice-resistant beam antennas. Therefore, do-it-yourself aluminum, stainless bolts and clamps and schedule 30 PVC pipe were the materials of choice.

I discovered a self-priming all-weather epoxy spray paint called VHT, available in automotive supply stores, that would aid in shedding the water as well as provide a black body for a solar heating effect. A black epoxy powder coating sprayed on the antenna might also provide a similar effect but would cost more so I opted for the epoxy spray paint. Two cans of spray paint will provide two coats.

Design Objectives

Here is a summary of the objectives I had in mind while designing the ice-resistant beam antenna. The criteria can be applied to just about any VHF or UHF beam type antenna.

- Use lightweight aluminum available in most hardware stores.
- Use square tubing for the boom, and solid rods for the elements to make it strong.
- Use welded construction if available (see sidebar for more information).
- Protect the driven element match with a shroud.
- Make it hydrophobic to shed or resist water.
- Make it black to absorb solar energy and more easily melt ice buildup.
- Use stainless bolts, screws and U-bolts in constructing and mounting the antenna.[2]
- Use a Type N coax chassis fitting and elbow for low loss and water resistance.

[1]Notes appear on page 205.

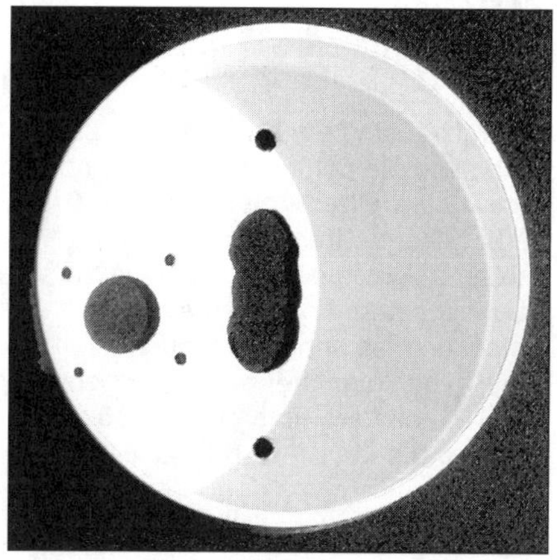

Figure 3 — End cap showing holes for the driven element weld, the Type N chassis connector and two screw holes to fasten it to the boom.

Construction

The beam is made from 1 inch square aluminum tubing stock 72 inches long. The driven element is made from ½ inch aluminum rod and all the other elements are made from ¼ inch aluminum rod, all available at most hardware stores. The solid rod elements are more durable against falling ice than tubing. The element dimensions for 222 MHz are shown in Table 1. All these techniques should also be applicable at 146 or 440 MHz with your favorite Yagi dimensions.

The gamma matching bar is made from ⅜ inch diameter aluminum tubing. It contains a short piece of polyethylene insulated center conductor removed from RG-8 coax, with one end soldered to the Type N connector.

Table 1

Element Lengths for 222 MHz Yagi (inches)

Element	Length
Reflector	26⅛
Driven element	24¹⁵⁄₁₆
Director 1	23¾
Director 2	23½
Director 3	23¼
Director 4	23
Director 5	22⅝
Director 6	22⅜

The center-to-center spacing between all the elements is 10 inches with 1 inch at each end of the boom. Square plastic caps, available at hardware stores, were used to plug each end of the boom.

The shroud covering the gamma match and driven element half is made from schedule 30 PVC pipe as shown in Figure 2. It consists of two 3 inch end caps and a short section of 3 inch pipe cut 14 inches long. Note that 3 inch PVC end caps have an actual outside diameter of 3½ inches. One end cap is drilled as shown in Figure 3 to go over the driven element weld flush against the boom. It's fastened to the boom with two #8 stainless sheet metal screws. It's also drilled to accept the chassis mount connector, which is mounted on the outside of the cap on a 1 × 1¼ inch aluminum angle 1 inch wide and drilled to accept the Type N chassis connector. The connector is fastened with four #4 stainless bolts to the 1 inch aluminum angle bracket flush against the PVC cap. The center pin of the Type N chassis connector has the RG-8 coax center conductor soldered in it and protrudes through the cap. The gamma match ⅜ inch tubing slides over the RG-8 coax center conductor and is clamped to the driven element.

The gamma match clamp can be made from two pieces of ½ inch wide ¹⁄₁₆ inch thick flat stock formed as a clamp between the ½ inch driven element and the ⅜ inch gamma tube. Two #6 stainless bolts secure the clamp after it's adjusted for the best match.

The other cap has several ¹⁄₁₆ inch moisture drain holes in it. Both caps are fastened to the PVC pipe with two #6 stainless sheet metal screws. After cutting the elements to length, chamfer the ends slightly by lightly filing.

Measure, mark and drill the boom for each element. Use a drill press to maintain alignment and straightness. Then weld each element in the boom. As an alternative method, you could fasten the elements with stainless steel panhead bolts, as described in the sidebar.

After drilling one of the PVC end caps to fit over the driven element, drill it to mount to the side of the boom with two #8 stainless sheet metal screws.

Gamma Match Construction

The gamma match tube is made from ⅜ inch diameter aluminum tubing 5 inches long. The capacitor insert is made from the center conductor of RG-8 coax and is 4 inches long. One end is soldered to the Type N chassis connector. Following assembly, the gamma tube is moved, along with the shorting clamp, for the best match — the point with the lowest SWR. Be sure and mount the beam away from metal objects

Welding Aluminum Elements

For antenna ruggedness and durability under heavy icing conditions, it's best to weld the elements to the boom. There are several ways to weld aluminum; either find someone capable, locate a friendly welding shop (in the phone book) or do it yourself using the inexpensive *Alumaloy* rods available in hardware stores or on the Internet. These allow welding with a standard propane torch. Working with them is similar to soldering copper pipe and therefore the normal propane soldering safety precautions are applicable. Make sure you wear safety glasses and heavy gloves and work in a clear safe area.

If aluminum welding is not available, an alternative construction method using stainless bolts is viable. See Figure A. After drilling the boom for the elements, prepare the elements. Put each element in a vise and lightly mark the center with a center punch, and then drill through the center of each ¼ inch element using a number 43 drill bit. Tap each element center hole using a 4-40 tap. Check to ensure a 4-40 bolt will enter the threads. Then using a ⅛ inch bit, drill a hole in the boom centered between each element hole as shown in Figure 4. On the ½ inch diameter driven element, drill the center with a number 36 drill bit and tap for a 6-32 bolt. Check to ensure a 6-32 bolt will enter the threads. Remove any burrs. Insert the elements into the boom and fasten each with a stainless 4-40 × ¾ inch panhead bolt and star washer. Use a 6-32 × ¾ inch stainless pan head bolt and star washer to fasten the driven element to the boom. Tighten each element bolt a quarter turn snug.

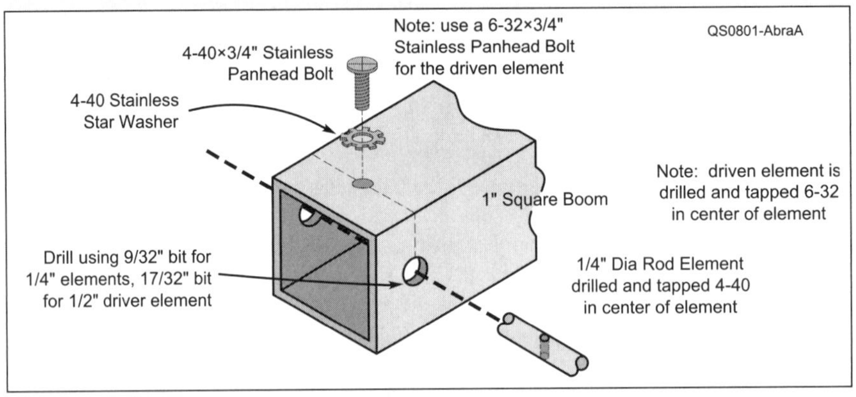

Figure A — Illustration of alternate assembly method using stainless bolts.

or hold the beam high and pointing straight up while adjusting for minimum SWR. Details of gamma construction are shown in Figure 4.

Make a 1 × 1¼ inch aluminum angle bracket 1 inch wide from $^1/_{16}$ inch thick angle stock. This will be used to mount the Type N chassis connector to the PVC cap and to the boom. Drill out the 1¼ × 1 inch side to the pattern of the Type N connector. The 1 × 1 inch side will mount to the boom with two #6 stainless sheet metal screws. Following final assembly, use some silicon caulk to seal between the PVC cap, the boom and around the Type N connector.

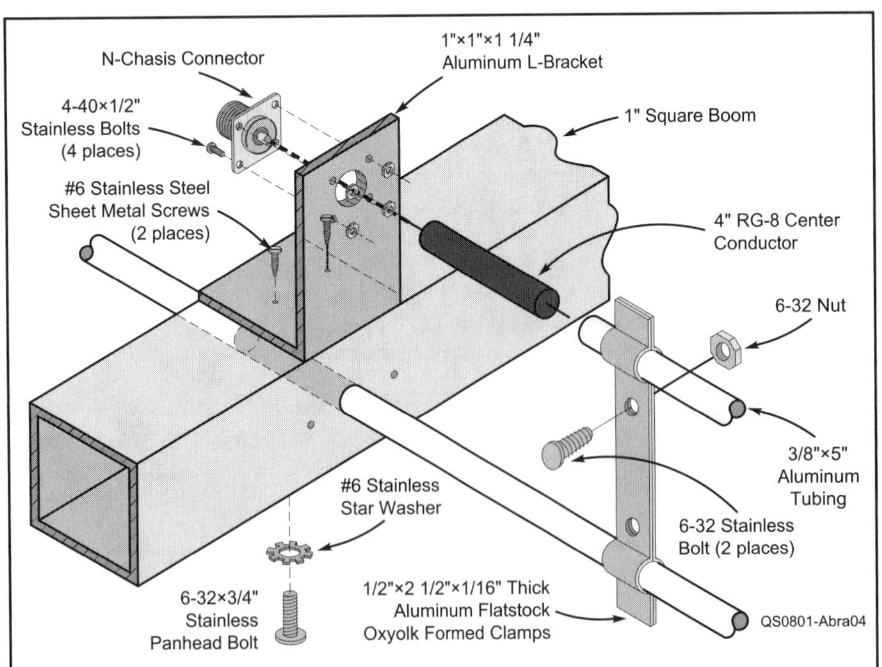

Figure 4 — Details of gamma match assembly.

Mast Attachment

The antenna balance point (center of gravity) is in between directors Dl and D2. This is the place where the mounting clamp should be installed.[2] The antenna is mounted vertically on a horizontal mast. The mounting clamp is thus drilled vertically through the square boom.

Finishing Touches

Following final assembly, check the SWR and set the gamma match. Do it with the shroud off. When it looks good enough, slip the shroud over and recheck the SWR. I expected some dielectric loading effects but found little difference. Then spray everything with the black epoxy spray paint (except the coax fittings). Spray the PVC end caps and pipe, both on the inside and outside, before final assembly. Fasten the end caps with two #6 stainless sheet metal screws in each cap. Spray the boom and elements with the black epoxy paint. Two coats should do it, then be careful handling the antenna to avoid scratching the black paint. Check the SWR again to confirm there is no, or very little, change.

Note that there will likely be a change in the SWR after the antenna is in its final position due to height and surrounding differences than when you checked it on the ground. It's the Antenna God's Law!

Finally, remember to waterproof the coax connection to the beam either by wrapping it with tape, a butyl rubber sealant, or shrink tubing. Be sure to mount the antenna with the shroud pointing down.

Performance

This antenna was installed in the fall of 2006 replacing a problematic auxiliary link beam that would continually fail to transmit during winter ice and snow storms. When we took the problem antenna down, we noticed that two of the eight elements were broken off due to falling ice.

Since installing the new shrouded beam, we've had three severe storms, one with up to an inch of ice that coated everything. The new beam antenna kept transmitting the vital link signal without failure. Although our observations are limited so far, that's all the proof we need!

Thanks to Tom Parker, KC8YOE, of YOE Antennas for helping to fabricate and weld two antennas for the link system. He welds some beautiful antennas!

Notes

[1]R. Abraczinskas, W8HVG, "A Rugged Broadband Tri-Band VHF-UHF Base Antenna," *QST*, May 2004, p 39.

[2]An excellent stainless U-bolt mast clamp is available from DX Engineering at **DXEngineering.com** or 1-800-777-0703.

Photos by the author.

Ray Abraczinskas, W8HVG, was first licensed as W3HJR in 1956, and as KA2MA in Japan in 1957 (see his listing on **www.qrz.com** *for details). He holds a BSEE and retired in 2001 as a special projects manager from Smiths Aerospace in Grand Rapids, Michigan. He enjoys ham radio, genealogy, fishing in Canada and writing. He also assists the Independent Repeater Association with building and maintaining the Michigan Link Repeater System (***www.w8hvg.org***). You can contact him at 4295 Kentridge SE, Grand Rapids, MI 49508; or* **abra@i2k.com**.

Nominal Characteristics of Commonly Used Transmission Lines

RG or Type	*Part Number*	*Nom.* Z_0 Ω	*VF %*	*Cap. pF/ft*	*Cent. Cond. AWG*	*Diel. Type*	*Shield Type*	*Jacket Matl*	*OD inches*	*Max V (RMS)*	*Matched Loss (dB/100') 1 MHz*	*10*	*100*	*1000*
RG-6	Belden 1694A	75	82	16.2	#18 Solid BC	FPE	FC	P1	0.275	300	0.3	.7	1.8	5.9
RG-6	Belden 8215	75	66	20.5	#21 Solid CCS	PE	D	PE	0.332	2700	0.4	0.8	2.7	9.8
RG-8	Belden 7810A	50	86	23.0	#10 Solid BC	FPE	FC	PE	0.405	300	0.1	0.4	1.2	4.0
RG-8	TMS LMR400	50	85	23.9	#10 Solid CCA	FPE	FC	PE	0.405	600	0.1	0.4	1.3	4.1
RG-8	Belden 9913	50	84	24.6	#10 Solid BC	ASPE	FC	P1	0.405	300	0.1	0.4	1.3	4.5
RG-8	CXP1318FX	50	84	24.0	#10 Flex BC	FPE	FC	P2N	0.405	600	0.1	0.4	1.3	4.5
RG-8	Belden 9913F	50	83	24.6	#11 Flex BC	FPE	FC	P1	0.405	300	0.2	0.6	1.5	4.8
RG-8	Belden 9914	50	82	24.8	#10 Solid BC	FPE	FC	P1	0.405	300	0.2	0.5	1.5	4.8
RG-8	TMS LMR400UF	50	85	23.9	#10 Flex BC	FPE	FC	PE	0.405	600	0.1	0.4	1.4	4.9
RG-8	DRF-BF	50	84	24.5	#9.5 Flex BC	FPE	FC	PE	0.405	600	0.1	0.5	1.6	5.2
RG-8	WM CQ106	50	84	24.5	#9.5 Flex BC	FPE	FC	P2N	0.405	600	0.2	0.6	1.8	5.3
RG-8	CXP008	50	78	26.0	#13 Flex BC	FPE	S	P1	0.405	600	0.1	0.5	1.8	7.1
RG-8	Belden 8237	52	66	29.5	#13 Flex BC	PE	S	P1	0.405	3700	0.2	0.6	1.9	7.4
RG-8X	Belden 7808A	50	86	23.5	#15 Solid BC	FPE	FC	PE	0.240	300	0.2	0.7	2.3	7.4
RG-8X	TMS LMR240	50	84	24.2	#15 Solid BC	FPE	FC	PE	0.242	300	0.2	0.8	2.5	8.0
RG-8X	WM CQ118	50	82	25.0	#16 Flex BC	FPE	FC	P2N	0.242	300	0.3	0.9	2.8	8.4
RG-8X	TMS LMR240UF	50	84	24.2	#15 Flex BC	FPE	FC	PE	0.242	300	0.2	0.8	2.8	9.6
RG-8X	Belden 9258	50	82	24.8	#16 Flex BC	FPE	S	P1	0.242	300	0.3	0.9	3.2	11.2
RG-8X	CXP08XB	50	80	25.3	#16 Flex BC	FPE	S	P1	0.242	300	0.3	1.0	3.1	14.0
RG-9	Belden 8242	51	66	30.0	#13 Flex SPC	PE	SCBC	P2N	0.420	5000	0.2	0.6	2.1	8.2
RG-11	Belden 8213	75	84	16.1	#14 Solid BC	FPE	S	PE	0.405	300	0.1	0.4	1.3	5.2
RG-11	Belden 8238	75	66	20.5	#18 Flex TC	PE	S	P1	0.405	300	0.2	0.7	2.0	7.1
RG-58	Belden 7807A	50	85	23.7	#18 Solid BC	FPE	FC	PE	0.195	300	0.3	1.0	3.0	9.7
RG-58	TMS LMR200	50	83	24.5	#17 Solid BC	FPE	FC	PE	0.195	300	0.3	1.0	3.2	10.5
RG-58	WM CQ124	52	66	28.5	#20 Solid BC	PE	S	PE	0.195	1400	0.4	1.3	4.3	14.3
RG-58	Belden 8240	52	66	29.9	#20 Solid BC	PE	S	P1	0.193	1400	0.3	1.1	3.8	14.5
RG-58A	Belden 8219	53	73	26.5	#20 Flex TC	FPE	S	P1	0.195	300	0.4	1.3	4.5	18.1
RG-58C	Belden 8262	50	66	30.8	#20 Flex TC	PE	S	P2N	0.195	1400	0.4	1.4	4.9	21.5
RG-58A	Belden 8259	50	66	30.8	#20 Flex TC	PE	S	P1	0.192	1400	0.5	1.5	5.4	22.8
RG-59	Belden 1426A	75	83	16.3	#20 Solid BC	FPE	S	P1	0.242	300	0.3	0.9	2.6	8.5
RG-59	CXP 0815	75	82	16.2	#20 Solid BC	FPE	S	P1	0.232	300	0.5	0.9	2.2	9.1
RG-59	Belden 8212	75	78	17.3	#20 Solid CCS	FPE	S	P1	0.242	300	0.2	1.0	3.0	10.9
RG-59	Belden 8241	75	66	20.4	#23 Solid CCS	PE	S	P1	0.242	1700	0.6	1.1	3.4	12.0
RG-62A	Belden 9269	93	84	13.5	#22 Solid CCS	ASPE	S	P1	0.240	750	0.3	0.9	2.7	8.7
RG-62B	Belden 8255	93	84	13.5	#24 Flex CCS	ASPE	S	P2N	0.242	750	0.3	0.9	2.9	11.0
RG-63B	Belden 9857	125	84	9.7	#22 Solid CCS	ASPE	S	P2N	0.405	750	0.2	0.5	1.5	5.8
RG-142	CXP 183242	50	69.5	29.4	#19 Solid SCCS	TFE	D	FEP	0.195	1900	0.3	1.1	3.8	12.8
RG-142B	Belden 83242	50	69.5	29.0	#19 Solid SCCS	TFE	D	TFE	0.195	1400	0.3	1.1	3.9	13.5
RG-174	Belden 7805R	50	73.5	26.2	#25 Solid BC	FPE	FC	P1	0.110	300	0.6	2.0	6.5	21.3
RG-174	Belden 8216	50	66	30.8	#26 Flex CCS	PE	S	P1	0.110	1100	0.8	2.5	8.6	33.7
RG-213	Belden 8267	50	66	30.8	#13 Flex BC	PE	S	P2N	0.405	3700	0.2	0.6	2.1	8.0
RG-213	CXP213	50	66	30.8	#13 Flex BC	PE	S	P2N	0.405	600	0.2	0.6	2.0	8.2
RG-214	Belden 8268	50	66	30.8	#13 Flex SPC	PE	D	P2N	0.425	3700	0.2	0.7	2.2	8.0
RG-216	Belden 9850	75	66	20.5	#18 Flex TC	PE	D	P2N	0.425	3700	0.2	0.7	2.0	7.1
RG-217	WM CQ217F	50	66	30.8	#10 Flex BC	PE	D	PE	0.545	7000	0.1	0.4	1.4	5.2
RG-217	M17/78-RG217	50	66	30.8	#10 Solid BC	PE	D	P2N	0.545	7000	0.1	0.4	1.4	5.2
RG-218	M17/79-RG218	50	66	29.5	#4.5 Solid BC	PE	S	P2N	0.870	11000	0.1	0.2	0.8	3.4
RG-223	Belden 9273	50	66	30.8	#19 Solid SPC	PE	D	P2N	0.212	1400	0.4	1.2	4.1	14.5
RG-303	Belden 84303	50	69.5	29.0	#18 Solid SCCS	TFE	S	TFE	0.170	1400	0.3	1.1	3.9	13.5
RG-316	CXP TJ1316	50	69.5	29.4	#26 Flex BC	TFE	S	FEP	0.098	1200	1.2	2.7	8.0	26.1
RG-316	Belden 84316	50	69.5	29.0	#26 Flex SCCS	TFE	S	FEP	0.096	900	0.8	2.5	8.3	26.0
RG-393	M17/127-RG393	50	69.5	29.4	#12 Flex SPC	TFE	D	FEP	0.390	5000	0.2	0.5	1.7	6.1
RG-400	M17/128-RG400	50	69.5	29.4	#20 Flex SPC	TFE	D	FEP	0.195	1400	0.4	1.3	4.3	15.0
LMR500	TMS LMR500UF	50	85	23.9	#7 Flex BC	FPE	FC	PE	0.500	2500	0.1	0.4	1.2	4.0
LMR500	TMS LMR500	50	85	23.9	#7 Solid CCA	FPE	FC	PE	0.500	2500	0.1	0.3	0.9	3.3
LMR600	TMS LMR600	50	86	23.4	#5.5 Solid CCA	FPE	FC	PE	0.590	4000	0.1	0.2	0.8	2.7
LMR600	TMS LMR600UF	50	86	23.4	#5.5 Flex BC	FPE	FC	PE	0.590	4000	0.1	0.2	0.8	2.7
LMR1200	TMS LMR1200	50	88	23.1	#0 Copper Tube	FPE	FC	PE	1.200	4500	0.04	0.1	0.4	1.3
Hardline														
1/2"	CATV Hardline	50	81	25.0	#5.5 BC	FPE	SM	none	0.500	2500	0.05	0.2	0.8	3.2
1/2"	CATV Hardline	75	81	16.7	#11.5 BC	FPE	SM	none	0.500	2500	0.1	0.2	0.8	3.2
7/8"	CATV Hardline	50	81	25.0	#1 BC	FPE	SM	none	0.875	4000	0.03	0.1	0.6	2.9
7/8"	CATV Hardline	75	81	16.7	#5.5 BC	FPE	SM	none	0.875	4000	0.03	0.1	0.6	2.9
LDF4-50A	Heliax – ½"	50	88	25.9	#5 Solid BC	FPE	CC	PE	0.630	1400	0.02	0.2	0.6	2.4
LDF5-50A	Heliax – ⅞"	50	88	25.9	0.355" BC	FPE	CC	PE	1.090	2100	0.03	0.10	0.4	1.3
LDF6-50A	Heliax – 1¼"	50	88	25.9	0.516" BC	FPE	CC	PE	1.550	3200	0.02	0.08	0.3	1.1
Parallel Lines														
TV Twinlead (Belden 9085)		300	80	4.5	#22 Flex CCS	PE	none	P1	0.400	**	0.1	0.3	1.4	5.9
Twinlead (Belden 8225)		300	80	4.4	#20 Flex BC	PE	none	P1	0.400	8000	0.1	0.2	1.1	4.8
Generic Window Line		450	91	2.5	#18 Solid CCS	PE	none	P1	1.000	10000	0.02	0.08	0.3	1.1
WM CQ 554		440	91	2.7	#14 Flex CCS	PE	none	P1	1.000	10000	0.04	0.01	0.6	3.0
WM CQ 552		440	91	2.5	#16 Flex CCS	PE	none	P1	1.000	10000	0.05	0.2	0.6	2.6
WM CQ 553		450	91	2.5	#18 Flex CCS	PE	none	P1	1.000	10000	0.06	0.2	0.7	2.9
WM CQ 551		450	91	2.5	#18 Solid CCS	PE	none	P1	1.000	10000	0.05	0.02	0.6	2.8
Open-Wire Line		600	0.95-99***	1.7	#12 BC	none	none	none	**	12000	0.02	0.06	0.2	—

Approximate Power Handling Capability (1:1 SWR, 40°C Ambient):

	1.8 MHz	*7*	*14*	*30*	*50*	*150*	*220*	*450*	*1 GHz*
RG-58 Style	1350	700	500	350	250	150	120	100	50
RG-59 Style	2300	1100	800	550	400	250	200	130	90
RG-8X Style	1830	840	560	360	270	145	115	80	50
RG-8/213 Style	5900	3000	2000	1500	1000	600	500	350	250
RG-217 Style	20000	9200	6100	3900	2900	1500	1200	800	500
LDF4-50A	38000	18000	13000	8200	6200	3400	2800	1900	1200
LDF5-50A	67000	32000	22000	14000	11000	5900	4800	3200	2100
LMR500	18000	9200	6500	4400	3400	1900	1600	1100	700
LMR1200	52000	26000	19000	13000	10000	5500	4500	3000	2000

Legend:

**	Not Available or varies	N	Non-Contaminating
***	Varies with spacer material and spacing	P1	PVC, Class 1
ASPE	Air Spaced Polyethylene	P2	PVC, Class 2
BC	Bare Copper	PE	Polyethylene
CC	Corrugated Copper	S	Single Braided Shield
CCA	Copper Cover Aluminum	SC	Silver Coated Braid
CCS	Copper Covered Steel	SCCS	Silver Plated Copper Coated Steel
CXP	Cable X-Perts, Inc.	SM	Smooth Aluminum
D	Double Copper Braids	SPC	Silver Plated Copper
DRF	Davis RF	TC	Tinned Copper
FC	Foil + Tinned Copper Braid	TFE	Teflon®
FEP	Teflon ® Type IX	TMS	Times Microwave Systems
Flex	Flexible Stranded Wire	UF	Ultra Flex
FPE	Foamed Polyethylene	WM	Wireman
Heliax	Andrew Corp Heliax		

BNC CONNECTORS

Standard Clamp

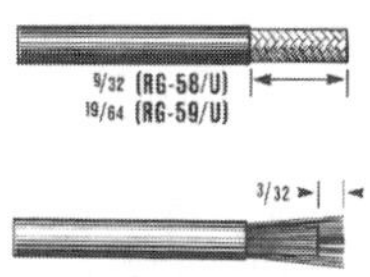

1. Cut cable even. Strip jacket. Fray braid and strip dielectric. ***Don't nick braid or center conductor.*** Tin center conductor.

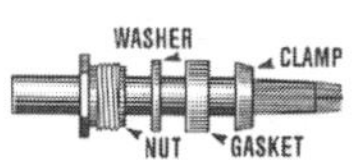

2. Taper braid. Slide nut, washer, gasket and clamp over braid. Clamp inner shoulder should fit squarely against end of jacket.

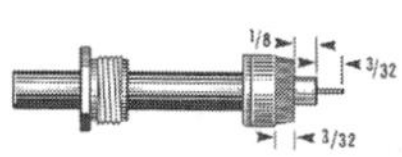

3. With clamp in place, comb out braid, fold back smooth as shown. Trim center conductor.

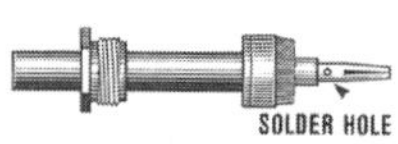

4. Solder contact on conductor through solder hole. Contact should butt against dielectric. Remove excess solder from outside of contact. Avoid excess heat to prevent swollen dielectric which would interfere with connector body.

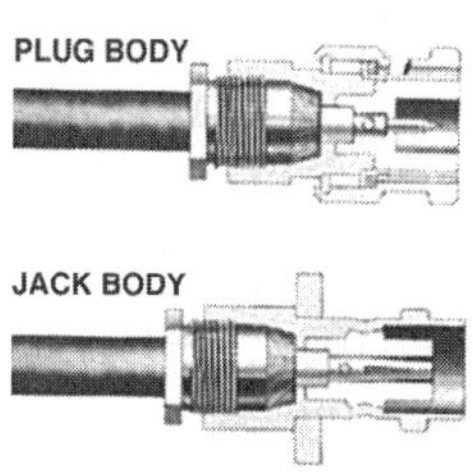

5. Push assembly into body. Screw nut into body with wrench until tight. ***Don't rotate body on cable to tighten.***

Improved Clamp

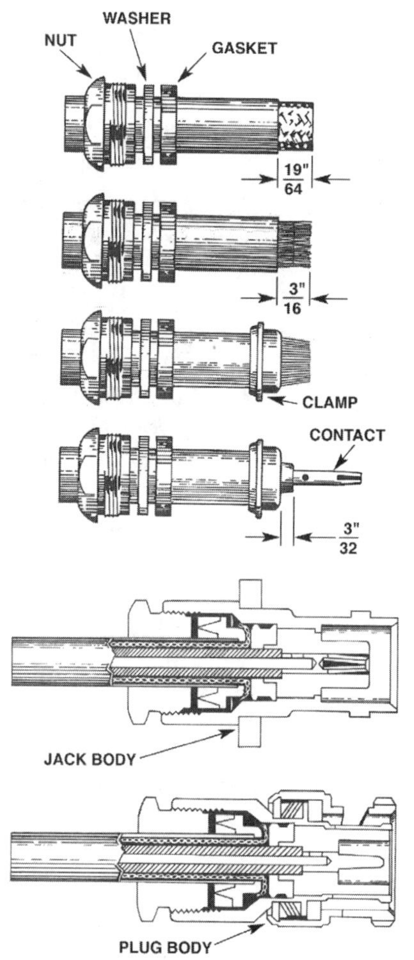

Follow 1, 2, 3 and 4 in BNC connectors (standard clamp) exceptas noted. Strip cable as shown. Slide gasket on cable *with groove facing clamp.* Slide clamp *with sharp edge facing gasket.* Clamp *should* cut gasket to seal properly.

HBK05_19-18

C. C. Clamp

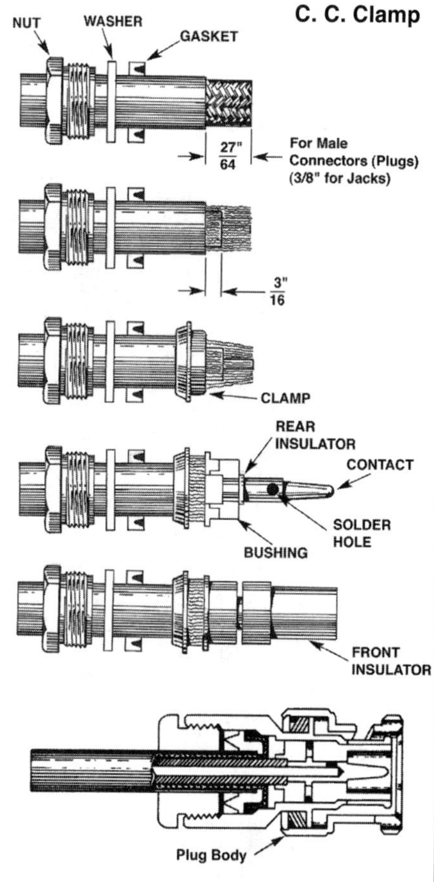

1. Follow steps 1, 2, and 3 as outlined for the standard-clamp BNC connector.

2. Slide on bushing, rear insulator and contact. The parts must butt securely against each other, as shown.

3. Solder the center conductor to the contact. Remove flux and excess solder.

4. Slide the front insulator over the contact, making sure it butts against the contact shoulder.

5. Insert the prepared cable end into the connector body and tighten the nut. Make sure the sharp edge of the clamp seats properly in the gasket.

Coaxial Cable Connectors

UHF Connectors

Military No.	Style	Cable RG- or Description
PL-259	Str (m)	8, 9, 11, 13, 63, 87, 149, 213, 214, 216, 225
UG-111	Str (m)	59, 62, 71, 140, 210
SO-239	Pnl (f)	Std, mica/phenolic insulation
UG-266	Blkhd (f)	Rear mount, pressurized, copolymer of styrene ins.

Adapters

PL-258	Str (f/f)	Polystyrene ins.
UG-224,363	Blkhd (f/f)	Polystyrene ins.
UG-646	Ang (f/m)	Polystyrene ins.
M-359A	Ang (m/f)	Polystyrene ins.
M-358	T (f/m/f)	Polystyrene ins.

Reducers

UG-175		55, 58, 141, 142 (except 55A)
UG-176		59, 62, 71, 140, 210

Family Characteristics:

All are nonweatherproof and have a nonconstant impedance. Frequency range: 0-500 MHz. Maximum voltage rating: 500 V (peak).

N Connectors

Military No.	Style	Cable RG-	Notes
UG-21	Str (m)	8, 9, 213, 214	50 Ω
UG-94A	Str (m)	11, 13, 149, 216	70 Ω
UG-536	Str (m)	58, 141, 142	50 Ω
UG-603	Str (m)	59, 62, 71, 140, 210	50 Ω
UG-23, B-E	Str (f)	8, 9, 87, 213, 214, 225	50 Ω
UG-602	Str (f)	59, 62, 71, 140, 210	—
UG-228B, D, E	Pnl (f)	8, 9, 87, 213, 214, 225	—
UG-1052	Pnl (f)	58, 141, 142	50 Ω
UG-593	Pnl (f)	59, 62, 71, 140, 210	50 Ω
UG-160A, B, D	Blkhd (f)	8, 9, 87, 213, 214, 225	50 Ω
UG-556	Blkhd (f)	58, 141, 142	50 Ω
UG-58, A	Pnl (f)		50 Ω
UG-997A	Ang (f)		50 Ω

Panel mount (f) with clearance above panel

M39012/04-	Blkhd (f)	Front mount hermetically sealed
UG-680	Blkhd (f)	Front mount pressurized

N Adapters

Military No.	Style	Notes
UG-29,A,B	Str (f/f)	50 Ω, TFE ins.
UG-57A.B	Str (m/m)	50 Ω, TFE ins.
UG-27A,B	Ang (f/m)	Mitre body
UG-212A	Ang (f/m)	Mitre body
UG-107A	T (f/m/f)	—
UG-28A	T (f/f/f)	—
UG-107B	T (f/m/f)	—

Family Characteristics:

N connectors with gaskets are weatherproof. RF leakage: –90 dB min @ 3 GHz. Temperature limits: TFE: –67° to 390°F (–55° to 199°C). Insertion loss 0.15 dB max @ 10 GHz. Copolymer of styrene: –67° to 185°F (–55° to 85°C). Frequency range: 0-11 GHz. Maximum voltage rating: 1500 V P-P. Dielectric withstanding voltage 2500 V RMS. SWR (MIL-C-39012 cable connectors) 1.3 max 0-11 GHz.

BNC Connectors

Military No.	Style	Cable RG-	Notes
UG-88C	Str (m)	55, 58, 141, 142, 223, 400	

Military No.	Style	Cable RG-	Notes
UG-959	Str (m)	8, 9	
UG-260,A	Str (m)	59, 62, 71, 140, 210	Rexolite ins.
UG-262	Pnl (f)	59, 62, 71, 140, 210	Rexolite ins.
UG-262A	Pnl (f)	59, 62, 71, 140, 210	nwx, Rexolite ins.
UG-291	Pnl (f)	55, 58, 141, 142, 223, 400	
UG-291A	Pnl (f)	55, 58, 141, 142, 223, 400	nwx
UG-624	Blkhd (f)	59, 62, 71, 140, 210	Front mount Rexolite ins.
UG-1094A	Blkhd		Standard
UG-625B	Receptacle		
UG-625			

BNC Adapters

Military No.	Style	Notes
UG-491,A	Str (m/m)	
UG-491B	Str (m/m)	Berylium, outer contact
UG-914	Str (f/f)	
UG-306	Ang (f/m)	
UG-306A,B	Ang (f/m)	Berylium outer contact
UG-414,A	Pnl (f/f)	# 3-56 tapped flange holes
UG-306	Ang (f/m)	
UG-306A,B	Ang (f/m)	Berylium outer contact
UG-274	T (f/m/f)	
UG-274A,B	T (f/m/f)	Berylium outer contact

Family Characteristics:

Z = 50 Ω. Frequency range: 0-4 GHz w/low reflection; usable to 11 GHz. Voltage rating: 500 V P-P. Dielectric withstanding voltage 500 V RMS. SWR: 1.3 max 0-4 GHz. RF leakage –55 dB min @ 3 GHz. Insertion loss: 0.2 dB max @ 3 GHz. Temperature limits: TFE: –67° to 390°F (–55° to 199°C); Rexolite insulators: –67° to 185°F (–55° to 85°C). "Nwx" = not weatherproof.

HN Connectors

Military No.	Style	Cable RG-	Notes
UG-59A	Str (m)	8, 9, 213, 214	
UG-1214	Str (f)	8, 9, 87, 213, 214, 225	Captivated contact
UG-60A	Str (f)	8, 9, 213, 214	Copolymer of styrene ins.
UG-1215	Pnl (f)	8, 9, 87, 213, 214, 225	Captivated contact
UG-560	Pnl (f)		
UG-496	Pnl (f)		
UG-212C	Ang (f/m)		Berylium outer contact

Family Characteristics:

Connector Styles: Str = straight; Pnl = panel; Ang = Angle; Blkhd = bulkhead. Z = 50 Ω. Frequency range = 0-4 GHz. Maximum voltage rating = 1500 V P-P. Dielectric withstanding voltage = 5000 V RMS SWR = 1.3. All HN series are weatherproof. Temperature limits: TFE: –67° to 390°F (–55° to 199°C); copolymer of styrene: –67° to 185°F (–55° to 85°C).

Cross-Family Adapters

Families	Description	Military No.
HN to BNC	HN-m/BNC-f	UG-309
N to BNC	N-m/BNC-f	UG-201,A
	N-f/BNC-m	UG-349,A
	N-m/BNC-m	UG-1034
N to UHF	N-m/UHF-f	UG-146
	N-f/UHF-m	UG-83,B
	N-m/UHF-m	UG-318
UHF to BNC	UHF-m/BNC-f	UG-273
	UHF-f/BNC-m	UG-255

US Customary Units and Conversion Factors

Linear Units

12 inches (in) = 1 foot (ft)
36 inches = 3 feet = 1 yard (yd)
1 rod = $5^1/_2$ yards = $16^1/_2$ feet
1 statute mile = 1760 yards = 5280 feet
1 nautical mile = 6076.11549 feet

Area

1 ft^2 = 144 in^2
1 yd^2 = 9 ft^2 = 1296 in^2
1 rod^2 = $30^1/_4$ yd^2
1 acre = 4840 yd^2 = 43,560 ft^2
1 acre = 160 rod^2
1 $mile^2$ = 640 acres

Volume

1 ft^3 = 1728 in^3
1 yd^3 = 27 ft^3

Liquid Volume Measure

1 fluid ounce (fl oz) = 8 fluid drams = 1.804 in
1 pint (pt) = 16 fl oz
1 quart (qt) = 2 pt = 32 fl oz = $57^3/_4$ in^3
1 gallon (gal) = 4 qt = 231 in^3
1 barrel = $31^1/_2$ gal

Dry Volume Measure

1 quart (qt) = 2 pints (pt) = 67.2 in^3
1 peck = 8 qt
1 bushel = 4 pecks = 2150.42 in^3

Avoirdupois Weight

1 dram (dr) = 27.343 grains (gr) or (gr a)
1 ounce (oz) = 437.5 gr
1 pound (lb) = 16 oz = 7000 gr
1 short ton = 2000 lb, 1 long ton = 2240 lb

Troy Weight

1 grain troy (gr t) = 1 grain avoirdupois
1 pennyweight (dwt) or (pwt) = 24 gr t
1 ounce troy (oz t) = 480 grains
1 lb t = 12 oz t = 5760 grains

Apothecaries' Weight

1 grain apothecaries' (gr ap)
= 1 gr t = 1 gr
1 dram ap (dr ap) = 60 gr
1 oz ap = 1 oz t = 8 dr ap = 480 gr
1 lb ap = 1 lb t = 12 oz ap = 5760 gr

Conversion

Metric Unit = Metric Unit × US Unit

(Length)		
mm	25.4	inch
cm	2.54	inch
cm	30.48	foot
m	0.3048	foot
m	0.9144	yard
km	1.609	mile
km	1.852	nautical mile
(Area)		
mm^2	645.16	$inch^2$
cm^2	6.4516	in^2
cm^2	929.03	ft^2
m^2	0.0929	ft^2
cm^2	8361.3	yd^2
m^2	0.83613	yd^2
m^2	4047	acre
km^2	2.59	mi^2
(Mass)	**(Avoirdupois Weight)**	
grams	0.0648	grains
g	28.349	oz
g	453.59	lb
kg	0.45359	lb
tonne	0.907	short ton
tonne	1.016	long ton
(Volume)		
mm^3	16387.064	in^3
cm^3	16.387	in^3
m^3	0.028316	ft^3
m^3	0.764555	yd^3
ml	16.387	in^3
ml	29.57	fl oz
ml	473	pint
ml	946.333	quart
l	28.32	ft^3
l	0.9463	quart
l	3.785	gallon
l	1.101	dry quart
l	8.809	peck
l	35.238	bushel
(Mass)	**(Troy Weight)**	
g	31.103	oz t
g	373.248	lb t
(Mass)	**(Apothecaries' Weight)**	
g	3.387	dr ap
g	31.103	oz ap
g	373.248	lb ap

Multiply →

Metric Unit = Conversion Factor × US Customary Unit

← Divide

Metric Unit ÷ Conversion Factor = US Customary Unit

Reflection Coefficient, Attenuation, SWR and Return Loss

Reflection Coefficient (%)	*Attenuation (dB)*	*Max SWR*	*Return Loss, dB*
1.000	0.000434	1.020	40.00
1.517	0.001000	1.031	36.38
2.000	0.001738	1.041	33.98
3.000	0.003910	1.062	30.46
4.000	0.006954	1.083	27.96
4.796	0.01000	1.101	26.38
5.000	0.01087	1.105	26.02
6.000	0.01566	1.128	24.44
7.000	0.02133	1.151	23.10
7.576	0.02500	1.164	22.41
8.000	0.02788	1.174	21.94
9.000	0.03532	1.198	20.92
10.000	0.04365	1.222	20.00
10.699	0.05000	1.240	19.41
11.000	0.05287	1.247	19.17
12.000	0.06299	1.273	18.42
13.085	0.07500	1.301	17.66
14.000	0.08597	1.326	17.08
15.000	0.09883	1.353	16.48
15.087	0.10000	1.355	16.43
16.000	0.1126	1.381	15.92
17.783	0.1396	1.433	15.00
18.000	0.1430	1.439	14.89
19.000	0.1597	1.469	14.42
20.000	0.1773	1.500	13.98
22.000	0.2155	1.564	13.15
23.652	0.2500	1.620	12.52
24.000	0.2577	1.632	12.40
25.000	0.2803	1.667	12.04
26.000	0.3040	1.703	11.70
27.000	0.3287	1.740	11.37
28.000	0.3546	1.778	11.06
30.000	0.4096	1.857	10.46
31.623	0.4576	1.925	10.00
32.977	0.5000	1.984	9.64
33.333	0.5115	2.000	9.54
34.000	0.5335	2.030	9.37
35.000	0.5675	2.077	9.12
36.000	0.6028	2.125	8.87
37.000	0.6394	2.175	8.64
38.000	0.6773	2.226	8.40
39.825	0.75000	2.324	8.00
40.000	0.7572	2.333	7.96
42.000	0.8428	2.448	7.54
42.857	0.8814	2.500	7.36
44.000	0.9345	2.571	7.13

Reflection Coefficient (%)	*Attenuation (dB)*	*Max SWR*	*Return Loss, dB*
45.351	1.0000	2.660	6.87
48.000	1.1374	2.846	6.38
50.000	1.2494	3.000	6.02
52.000	1.3692	3.167	5.68
54.042	1.5000	3.352	5.35
56.234	1.6509	3.570	5.00
58.000	1.7809	3.762	4.73
60.000	1.9382	4.000	4.44
60.749	2.0000	4.095	4.33
63.000	2.1961	4.405	4.01
66.156	2.5000	4.909	3.59
66.667	2.5528	5.000	3.52
70.627	3.0000	5.809	3.02
70.711	3.0103	5.829	3.01

$$\rho = \frac{SWR-1}{SWR+1}$$

where ρ = 0.01 × (reflection coefficient in %)

$$\rho = 10^{-RL/20}$$

where RL = return loss (dB)

$$\rho = \sqrt{1-(0.1^{X})}$$

where X = A/10 and A = attenuation (dB)

$$SWR = \frac{1+\rho}{1-\rho}$$

Return loss (dB) = $-8.68589 \ln(\rho)$
where ln is the natural log (log to the base e)

Attenuation (dB) = $-4.34295 \ln(1-\rho^2)$
where ln is the natural log (log to the base e)

FEEDBACK

Please use this form to give us your comments on this book and what you'd like to see in future editions, or e-mail us at **pubsfdbk@arrl.org** (publications feedback). If you use e-mail, please include your name, call, e-mail address and the book title, edition and printing in the body of your message. Also indicate whether or not you are an ARRL member.

Where did you purchase this book? ☐ From ARRL directly ☐ From an ARRL dealer

Is there a dealer who carries ARRL publications within:

☐ 5 miles ☐ 15 miles ☐ 30 miles of your location? ☐ Not sure.

License class:

☐ Novice ☐ Technician ☐ Technician with code ☐ General ☐ Advanced ☐ Amateur Extra

Name ______________________ ARRL member? ☐ Yes ☐ No

______________________ Call Sign ______________

Address ______________________

City, State/Province, ZIP/Postal Code ______________________

Daytime Phone ( ) ______________________ Age ______________

If licensed, how long? ______________________

Other hobbies ______________________ E-mail ______________

____

Occupation ______________________

For ARRL use only	ACV8
Edition	1 2 3 4 5 6 7 8 9 10 11 12
Printing	1 2 3 4 5 6 7 8 9 10 11 12

From ______________________________

____

____

Please affix postage. Post Office will not deliver without postage.

EDITOR, ANTENNA COMPENDIUM, VOL 8
ARRL—THE NATIONAL ASSOCIATION FOR AMATEUR RADIO
225 MAIN STREET
NEWINGTON CT 06111-1494

please fold and tape